EVOLUTIONARY SYSTEMS

EVOLUTIONARY SYSTEMS

Biological and Epistemological Perspectives on Selection and Self-Organization

Edited by

GERTRUDIS VAN DE VIJVER
Department of Philosophy,
University of Ghent, Belgium

STANLEY N. SALTHE
Department of Biology,
City University of New York - Binghamton University, U.S.A.

and

MANUELA DELPOS
Konrad Lorenz Institute, Vienna, Austria

KLUWER ACADEMIC PUBLISHERS
DORDRECHT / BOSTON / LONDON

A C.I.P. Catalogue record for this book is available from the Library of Congress.

ISBN 0-7923-5260-2

Published by Kluwer Academic Publishers,
P.O. Box 17, 3300 AA Dordrecht, The Netherlands.

Sold and distributed in North, Central and South America
by Kluwer Academic Publishers,
101 Philip Drive, Norwell, MA 02061, U.S.A.

In all other countries, sold and distributed
by Kluwer Academic Publishers,
P.O. Box 322, 3300 AH Dordrecht, The Netherlands.

Printed on acid-free paper

Printed in the Netherlands

TABLE OF CONTENTS

II. EVOLUTIONARY SYSTEMS: AN EPISTEMOLOGICAL POINT OF VIEW

CAUSALITY AND EXPLANATION

SYMBOLS, SIGNS AND MEANING

PREFACE

The three well known revolutions of the past centuries — the Copernican, the Darwinian and the Freudian — each in their own way had a deflating and mechanizing effect on the position of humans in nature. They opened up a richness of disillusion: earth acquired a more modest place in the universe, the human body and mind became products of a long material evolutionary history, and human reason, instead of being the central, immaterial, locus of understanding, was admitted into the theater of discourse only as a materialized and frequently out-of-control actor.

Is there something objectionable to this picture? Formulated as such, probably not. Why should we resist the idea that we are in certain ways, and to some degree, physically, biologically or psychically determined? Why refuse to acknowledge the fact that we are materially situated in an ever evolving world? Why deny that the ways of inscription (traces of past events and processes) are co-determinative of further "evolutionary pathways"? Why minimize the idea that each intervention, of each natural being, is temporally and materially situated, and has, as such, the inevitable consequence of changing the world?

The point is, however, that there are many, more or less radically different, ways to consider the "mechanization" of man and nature. There are, in particular, many ways to get the message of "material and evolutionary determination", as well as many levels at which this determination can be thought of as relevant or irrelevant.

History has illustrated how fundamentally different the viewpoints on evolution can be. It has shown above all how these differences depend on divergent and shifting philosophical, ethical and scientific backgrounds. Therefore, it is necessary to attempt to elucidate these backgrounds, because they alone allow us to understand the relevance or irrelevance of certain arguments; they alone allow us to answer the question of why certain doors keep closed, and why, at precise moments, certain doors are opened. Philosophical and scientific backgrounds determine and encapsulate the relative values of arguments.

The aim of the present book is to elucidate the scientific and philosophical backgrounds that play a role in one of the major debates now taking place in evolutionary thinking, namely that on self-organization and selection. Our focus is therefore on Evolutionary Systems, a field that attempts to address as coherently and as broadly as possible the scientific and epistemological challenges of the concepts of change and evolution. Its main aim is to understand how complex dynamical systems, living and non-living, linguistic or non-linguistic, come to be organized as systems — how, in other words, their inherent dynamical nature gives rise to organizations and forms that have found a balance between potentiality for change and evolution on the one hand, and requisite stability in a given environment on the other. Studies in Evolutionary Systems aim at making explicit the various constraints at work in the development of such systems, as well as at clarifying why and how they can co-determine the modalities of their further evolution.

Evolutionary Systems is essentially concerned with the debate on self-organization and selection in two basic ways. First, it applies insights from theories of complex self-organizing systems to the current theories of evolution, which leads to the direct confrontation of self-organization and selection as possible explanatory principles. Second, its concern is with the most radical philosophical consequences of a complex, self-organizational view on evolution. This leads, amongst other things, to a questioning of the classical scientific picture of causality, as well as to the development of a basically semiotic viewpoint on evolution.

This book is organized in two parts, that roughly follow, as just sketched, the basic lines of thought present in Evolutionary Systems. Let us therefore explain in some more detail the scope of both parts.

EVOLUTIONARY SYSTEMS FROM A BIOLOGICAL POINT OF VIEW

The possible implications of complex dynamical systems theories for the existing biological theory of evolution make up a vast subject of discussion. If we follow Depew and Weber's viewpoint (1995), these new mathematical tools are likely to lead to a basic paradigm shift in evolutionary biology. Whereas Darwin had found his main inspiration in Newtonian mechanics, and neo-Darwinism in classical thermodynamics, time has come for complexification: the basic simplifications of large parts of the Modern Synthesis have to be abandoned, and evolutionary biology has to open itself up to influences from complex systems dynamics to be able to provide a non-reductionist, and basically non-atomistic, account of evolution and development.

Self-organization and *selection* together play a key-role in these new developments. One's interpretation of their respective explanatory functions is indicative of the more, or less, "neo-Darwinian" position one occupies in the field of evolutionary thinking. Indeed, ever since the neo-Darwinian interpretation, genetic variations and evolution of populations by natural selection have been considered as the conceptual and explanatory core of the biological theory of evolution: variation functions as the necessary "shuffling of cards", whereas natural selection serves to explain the genesis of biological order and the phenomenon of adaptation. If one adopts this viewpoint, there is no need to take into account either the internal structure of the organism, or its developmental dynamics, or any causal mechanisms operative in its concrete material development.

Precisely this viewpoint has been under severe attack these last years, for the idea that the evolution of biological systems can be adequately described in terms of variation and natural selection alone, is now challenged. The issue at stake is thus whether one should complement, or even substitute, the 'classical' picture of evolution by new ones informed by mathematical and physical models of complex dynamical, self-organizing systems.

The contemporary stances on the self-organization versus selection debate are roughly the following: 1. some authors still believe that the formation of order and adaptation on all biological levels can essentially be explained by the mechanisms of variation and natural selection (Richard Dawkins, John Maynard Smith, Szathmáry, George Williams, ...); 2. others view selection and self-organization as partly complementary principles, whose respective importance must be detailed in various situations (Robert Brandon, Michael Conrad, David Depew & Bruce Weber, Stuart Kauffman, ...). This viewpoint presupposes that there are organizing principles that determine in a certain way the efficiency of natural selection. They can either enhance or weaken this efficiency. This interpretation is sometimes called the null-hypothesis of self-organization (cf. Robert Richardson & Richard Burian, 1990); 3. still others attribute a more prominent role to self-organization than to selection in the production of biological order (Dan Brooks & Ed Wiley, John Collier, Brian Goodwin, Stan Salthe, Peter Saunders, Rod Swenson, ...).

The supportive arguments given in Evolutionary Systems are generally the following.

First, even though the Modern Synthesis is in a position to elucidate an organism's local capacity to become adapted to its environment (fine-tuning), its potential to adequately account for the inception of a large-scale biological order is still disputed. In this regard, the role of ecological evolution is stressed, and major problems concern similarity — or more generally, the role of systemic, constitutive and generic processes in evolutionary change.

Secondly, it is deemed essential to take into account the concrete material principles of development in thinking about the evolution of biological systems.

Thirdly, it is necessary to address the question of the origins of the living. Evolutionary Systems does so in terms of general organizational principles related to the emergence of catalytic and metabolic functions and of reproduction. These general processes of organization are seen to have been prior to, or at least to have appeared along with the origin of genetic replication. As a consequence, it is necessary to understand how natural selection of the reproductively fit emerges from chemical and physical principles and constraints.

Therefore, mechanical and thermodynamical constraints have to be taken into account, and other types of selection are conceptualized as well — i.e. physical or chemical. Evolution is thus to be conceived of in terms of the stable, the efficient and the evolvable, instead of solely on the basis of naturally selected fittingness. A theory of evolution that starts from the existence of living beings and black-boxes their origin, their internal structures and material development, can never be an adequate theory of evolution.

EVOLUTIONARY SYSTEMS FROM AN EPISTEMOLOGICAL POINT OF VIEW

Besides using theories of complex systems dynamics as descriptive tools, Evolutionary Systems carries the seeds of a much more radical paradigm shift. The aim of Evolutionary Systems is to understand how complex dynamical systems come to be organized as systems — that is, how their inherent dynamics give rise to organizations and forms that are in-between fixity and flux. As a consequence, the focus is on the developmental stability of systems acquired through *interaction*, and evolution is conceived on the basis of a *closure* between the system and the environment. As a matter of fact, the evolution of the system can change the conditions of its further adaptation. Hence, the evolution of species is no longer describable in terms only of the selective power of the environment, and the objects of evolution are not just the sequences of species. The object of evolution is viewed as ultimately the whole network of mutually interacting species — i.e. the history of the whole Biosphere.

Because of the shift towards an interactionist viewpoint, causality, as interpreted in classical scientific terms, comes under discussion. Here it has indeed come to be important to grasp the causal powers of *systems*. It is important to link the interpretation of causality to the way in which evolutionary systems acquire a certain form of individuality (a "self") through an interactive history. Biological organizations are not passive, but highly active beings, co-organizing their own evolution. That Aristotelean and Spinozist causal views are being reintroduced in this context, can, then, hardly come as a surprise.

Moreover, it is of central importance to arrive at characterizing the types of interaction between systems. Which interactions can be considered as meaningful, which as perturbations? Is the attribution of meaning to a type of interaction inevitably externalist and global, or is there some other way of defining it, related, for instance, to the values, functions and purposes inherent to the local dynamics within systems, or to local systemic interactions?

It is important to see that describing the interactive history of systems in terms of stability and evolvability — in terms of fixity and flux — fundamentally rearranges any discussion on values, aims and purposes, and on individuality and subjectivity. Clearly, the aim cannot be control, it will rather be creativity — i.e. promotion of further dynamics. The semiotic viewpoint attempts to arrive at clear concepts and clear answers precisely on these points. It establishes an inherent link between the temporal and material situatedness of natural systems and the way in which their interactions might be meaningful to them, as well as meaningfully described. It shows that the difficulty is related to the construction of global descriptions that try to adequately account for local dynamics.

In this way, Evolutionary Systems can lead to the elucidation of frequently misunderstood expressions, such as: "We, the expected", "At home in the universe", or of claims such as "We may recover our sense of worth" (Kauffman, 1995, p. 5). These certainly need not to be understood in teleological or vitalistic terms, even if the language used sometimes seems to suggests this, but contain the promise of another way of conceiving science: one in which atomistic mechanicism, backed up by the ideals of reductionism, is replaced by a historical, i.e. dynamic and interactive view on man and nature. A genuine historical science is one that acknowledges that "history arises when the space of possibilities is too large by far for the actual to exhaust the possible" (*ibid.*, p. 181). It is one that considers the weight of the basic choices, values and purposes of observers, always local, always temporally situated, as no longer negligible.

ACKNOWLEDGEMENTS

The present volume, as well as the extensive discussions preceding it during the ISES-95 meeting (International Seminar on Evolutionary Systems, Vienna), were made possible by a grant from the European Communities (PSS*0816). We are indebted in particular to Prof. P. Fasella, Director General of DG XII for Science, Research and Development. We are also grateful to the Konrad Lorenz Institute, and in particular Prof. R. Riedl, for valuable support during and after that meeting. We also gratefully acknowledge the support of the Fund for Scientific Research Flanders.

The Editors

REFERENCES

Depew, D.J.; Weber, B.H., 1995, *Darwinism Evolving: Systems Dynamics and the Genealogy of Natural Selection*, Cambridge MA, Bradford, The MIT Press.

Kauffman, S., 1995, *At home in the universe. The Search for Laws of Self-organization and Complexity*, Viking, London.

Burian, R.M.; Richardson, R.C., 1991, Form and Order in Evolutionary Biology, *PSA 1990: Proceedings of the 1990 Biennial Meeting of the Philosophy of Science Association*, A. Fine, M. Forbes & L. Wessels (eds.), East Lansing, vol. 2, pp. 267-287.

VILMOS CSÁNYI

EVOLUTION: MODEL OR METAPHOR?

INTRODUCTION

Distinguished developmental stages of scientific thinking are clearly recognized in the history of the natural sciences. The most basic forms are unorganized sets of isolated beliefs, conjectures, experiences, observations. At the next higher level a "name" appears; it is a metaphor which can be understood in many ways, nevertheless beliefs, and conjectures are interconnected in it. The metaphor is a "black box" whose internal structure and characteristics are not yet known exactly, but it has already been recognized as "something" by being given a name, therefore the metaphor becomes thinkable and can be studied.

A metaphor could be transformed into a model during scientific investigation. A model simulates the behavior of the studied system. The relation of the model and the studied system is characterized by an isomorphism which is based upon simplification and analogy. This isomorphism is used as an explanation of the modelled system and also for predicting or calculating the operation of the complex system.

DEVELOPMENT OF THE EVOLUTION METAPHOR

The development of the evolution metaphor has been the subject of many books (a very good one was written by Mayr, 1982), and I will not review them here, not even by a sketch, but I would like to discuss a certain aspect of the development of this metaphor, which led to many important questions of biology, some of which are still unanswered, and are connected to my own theoretical research in this field (Csányi, 1978, 1982a,1982b, 1985, 1988, 1989a, 1989c, 1989d).

J.B.A. Lamarck used the notion of evolution as a metaphor for characterizing the transformation of lineages of different species. Evolutionary transformation explained the existence of fossils of extinct species because these could be regarded as earlier states of a continuous transformation. However Lamarck was unsuccessful in forming a model from the metaphor of evolution, mostly because he erroneously assumed the inheritance of acquired characteristics.

It is noticeable that the concept of evolution in this stage lacked those concrete mechanisms, logical structures which could make it a model. The metaphor of evolution meant transformation, change, development, emergence of a more complex higher form in the biological and social realms, without the ability to predict, or even to explain the concrete mechanisms of these changes.

Lamarck's metaphor was transformed into a logically correct, to some extent testable, model by C.R. Darwin (1859). This was done by filling up the black box of the metaphor with concrete biological mechanisms and their logical connections.

G. Van de Vijver et al. (eds.), Evolutionary Systems, 1–12.

For the last 130 years, the development of biology has been revolving around the Darwinian model, and until the last decades it seemed that evolutionary theory had become the unshakable fundament of biology. Every other theory in biology somehow joined to the evolutionary model, supporting it or getting its own support from it.

Then what is the problem? Why have the fierce arguments renewed about the evolutionary theory?

The answer to this question can be found in the interpretation of the evolutionary metaphor. Both Lamarck and Darwin considered evolution as the transformation of the lineages, that is they regarded sequences of species as the objects of evolution. In the classic or in the synthetic model some genetical processing is made on these objects and the result is the transformation of the species. Who can argue against the concept, that the transformation of species is accompanied by genetical changes? Nobody. Even those who interpret the metaphor of evolution in an extended way, can accept the concept of continuous genetical changes of species.

The development of biology after Darwin created not only population genetics and molecular biology, but also other branches of the biological sciences like ecology, ethology and biological system science. We come to realize that there are organizational levels in the biosphere. Living organisms mutually influence each other; moreover this is the most important effect which shapes their patterns of existence and survival.

The classic evolutionary metaphor was therefore transformed: evolution is now becoming understood as the *history of the whole Biosphere*.

This new metaphor refers not only to individual species but to the whole mutually interacting network of all species. Among its concepts there are hints of the direction of evolution, of system, ic. interaction, and of organizational levels, but where are the concrete mechanisms? How will this metaphor be transformed to a scientific model?

Many unanswered questions have been raised. Is there a direction to evolutionary changes? If there is no direction, as many scientists claim just on the basis of population genetics models, then how were the levels of organization formed?

What is the source of the creative nature of evolution? How could the environment, which always has been assumed to be a continuously changing entity, create such a complex and seemingly well organized hierarchically divided entity, the biosphere which, as assumed by many, sometimes acts as a unified whole?

What are the evolving entities, the species or the genes or the whole biosphere? The classic Darwinian metaphor, in spite of all its other success, is unsuitable to answer these questions.

I want to emphasize, that it is not the concept of evolution which is being argued, as many who are unfamiliar with the scientific side of the subject suppose. No. The debate is about a new interpretation of the evolution metaphor. It can be recognized clearly that the new extension of the metaphor will eventually lead to an entirely new paradigm in biology.

EVOLUTIONARY SYSTEMS

The reformulation of the metaphor needs concepts from the systems sciences. Let us take an easy example. For explaining the origin of life there are various models based upon genetic mechanisms provided by the current evolutionary paradigm. There are explanatory models about the appearance of the first cells, the emergence of protein synthesis and the formation of the genetic code. If some cell which has heredity mechanisms has already appeared the formation of the present biosphere seems to be easily explained by the theory of lineages and the classical evolutionary model.

That is, the subjects of evolution are the lineages of proto-cells in these models. But there is no reason to suppose that the first proto-cells have such isolating mechanisms which supply them with a satisfactory identity for forming a population of individuals, differing in heritable characteristics, which is the basis of the Darwinian theory. It is much more plausible that individuality is the result of still further evolution. Therefore, at the origin the proto-cell forming molecules could easily have shuttled among the various proto-cells without significant barriers. Therefore the following question is justified: What was the subject of the evolution of the proto-cell or "primeval soup"? The latter most probably was a chemical system which was spread over the whole surface of the primitive Earth. Was it able to develop, to transform, to evolve? If so, then what model could describe its behavior?

Here we have three scalar levels of organization, the soup in which the proto-cells are the components, the proto-cells, and the molecules which are the components of the proto-cells. Taking these levels of organization in pairs, there is an analogy between the relations of molecules with proto-cells and these of proto-cells with the soup.

If proto-cells were subjects of evolution then molecules were as well. But for the latter we have to define molecular lineages, which can be done only in a systemic framework. If the evolution of macromolecules has led to the emergence of an organized entity, the cell, then the evolution of proto-cells and proto-cell species must also have led to the emergence of an organized biosphere (proto Gaia?). What were the relationships of these species to the whole biosphere? Is the biosphere a mere set of species or are there criteria on the basis of which the biosphere appears as a unified whole, as a living system built up from the species as components (Lovelock, 1919)?

For a correct answer to these questions it is necessary to look again at the classical evolutionary theory. The motor of evolution is Darwinian natural selection, but the theory says almost nothing about the nature and pattern of the selective forces, except that they originate from the environment.

The environment is the universal outer agent which is responsible for the evolutionary changes. The whole classic metaphor being to turn "upside down" if we realize, on the basis of current biological knowledge, that the most important parts of the environment of a given species are other living species. It would be foolish to argue against this. But if we accept this, it immediately becomes clear that we put a very special feed-back into the model of evolution. The evolution of a given species is dependent on the selective influence on its environment, but this statement is true for every species, and therefore the most important power in evolution is evolution itself. Every species is changing, adapting continuously, and this process changes the environment and the conditions of selection, in the way hypothesized in the Red Queen theory (Van Valen, 1973). Can the behavior of a self-organizing system be explained by the models of population genetics, which only answer questions concerning the distribution of alleles? If the conditions of selection change then the frequencies of alleles also change and this flow of change is considered to be evolution. This is certainly not a satisfactory answer, we need new models which transcend the mechanisms of population genetics without discarding them.

For example it is reasonable to suppose for the behavior of a system which is self-organizing a direction of change. It could attain some equilibrium state, it could have organizational levels, etc. Moreover, and this is most important, the outer agent of the classic model of environment becomes an internal agent which itself is the subject of the same evolutionary process. Beyond that, the question is not only how a changed environment formed some species, but why and how the environment has changed, although this is clearly a "vicious circle". In the classical paradigm, which is accepted now in biology, there can be no answer to such questions.

It is of interest that the system-level extension of the evolutionary metaphor has already occurred outside biology. The first notable extension has been formulated by Herbert Spencer (1862) who is definitely out of favor with biologists now. Spencer's general definition of evolution is as follows: "Evolution can be defined as a change from an incoherent homogeneity into a coherent heterogeneity accompanying the dissipation of motion and integration of matter" (Spencer, 1862, p. 18).

It is clear from this definition, and from the accompanying text, that Spencer conceives evolution not as transformation of certain components of a system during time, but a change in the whole system.

It is worth citing here a definition by an outstanding Darwinian theoretician John Maynard Smith (1969, p. 5): "We shall regard as alive any population of entities which have the properties of multiplication, heredity and variation. The justification for this definition is as follows: any population of any entities with these properties will evolve by natural selection so as to become better adapted to its environment. Given time, any degree of adaptive complexity can be generated by natural selection".

In this definition Maynard Smith mentions only the components of the biosphere, because only these fit into the classical model, but for the interpretation of the new extended metaphor we need much more. A system-model is necessary, in which the selective forces acting on the components are not outer agents but inferred consequences of the internal, larger structure, the organization and self-evolution of the biosphere. In the classic model the drive of the transformation of the components originates from the need to adapt to the environment. If we substitute the concept of the environment with the concept of the whole system then this drive originates from the need to fit, to adapt to the great whole. How can evolution be modeled if the great whole itself is changing. What is the drive for changes of the "whole"?

Genetics, population genetics, molecular genetics provide excellent and testable mechanisms for the explanation of certain features of the living world, but they operate only with changes of isolated components of the system. We need a new model which reflects the systemic nature of living world. A short sketch of such a model (Csányi, 1989d) is given in the following.

NATURAL ENTITIES AS COMPONENT SYSTEMS

If we look at any of the natural entities on Earth which are organized in some way, we find that all these entities can be characterized by their *components*, their *processes* and their *organizations*.

Concerning *components* a single-celled organism is built up from various molecules as components. A higher organism, as a plant or an animal consists of various cells. Ecosystems are composed of organisms of various species. Human societies also contain components, such as humans, artefacts, and other living beings. We may call such entities component systems.

Examining the various components of entities as cells, organisms, societies for *processes*, it can be found that they exist by self-maintenance, achieved by the continuous renewal of their own components. The components are assembled and decomposed at the expense of a continuous energy flow through the system. The cell, for example, can be considered as a functionally closed network of molecules and molecule-producing processes which continually produces the same network of components and processes. The same is true for organisms or ecosystems. Production of the same components and sub-components by a functionally closed network of processes is also the basic mechanism of reproduction, the other characteristic trait of these entities. It can be found in case of cells, organisms, ecosystems and societies.

Concerning *organization*, all biological and cultural entities are characterized by the mutual interactions of their components. Various molecules, components of a cell, are interacting chemically; the cells, components of an organism, are influencing each other's life processes. People, artifacts, ideas, the components of a society, are also mutually interacting. In general the components of a natural entity, through their interactions, participate in the maintenance of the whole. The particular mechanisms of the components' interactions are different but there is a common aspect: through their interactions the components always *influence the probability of each other's genesis or existence* (in one or both directions). Each of these interactions can be rendered to particular components and named as *functions*. A digestive enzyme in a cell, for example, splits the structure of other proteins (components), i.e. it decreases the probability of their existence. A predator taking its prey also decreases the probability of the existence and genesis of progeny of the prey animals. The existence of cars increases the probability of genesis of gas stations and repair shops in society. The common aspect of these various functions can be expressed as a *general function* by which they promote both the system's and their own maintenance. By analyzing self-production and reproduction it can be shown that the underlying principle of organization in both cases is the formation of a network of the various components with special functions resulting in copying or replication of the very components of these networks, and replication of the whole system.

In the functional organization of the cells, molecular components are copied over and over again. In an organism the cell components are originated by cell divisions, that is by the copying of the cell components. Therefore we can call this kind of organization *replicative organization*, and the general function by which each component joins to the replicative organization we can call *replicative function*.

THE AUTOGENETIC REPLICATIVE MODEL

Based upon the above mentioned considerations, a common general model of these natural systems can be worked out. The basic general process in this model system is *replication*. Replication is generally considered a synonym for copying, where a constructor produces a copy of a component or a given subsystem. To do so, the constructor needs a description, the information necessary for this copying process. The essence of replication is the function of copying, regardless of the particular mechanisms of storage and retrieval of the information it uses. It does not matter whether this information is stored separately (as in the case of DNA for example) or distributed through the whole system. It is also the same whether a separate object, a component, is copied, or if the constructor itself, the fact of copying emerges from the functional organization, not from the particular mechanisms. Based upon this argument about self-production and reproduction, the concept of replication will be used in two senses.

We define *temporal replication* as the continuous renewal of the system in time. This is the uninterrupted existence of the system which is manifested via the sequential and functional renewal of components of the system. The components are assembled and decomposed, but the extent to which these take place is always sufficient to maintain the unity and identity of the system and its organization. The definition of *spatial replication* is identical to that of reproduction. The system produces its own copy, which becomes separated from it in space. From one system or unit, two units are formed. Spatial replication also proceeds by copying components of the system, as in the case of replication in time, but increases its numbers of copies, while the original organization of the system remains unchanged.

THE ORIGIN OF REPLICATIVE SYSTEMS

A system of components that has not yet developed functions is considered to be a *zero-system*. A zero-system has no organization. A precondition for starting a self-organizing process in a zero-system is the presence of a minimal set of components which is able to replicate and which fulfills the following criteria: 1. it contains at least one cycle of component-producing processes, 2. at least one of the components participating in this cycle can be excited by the energy flux flowing through the system.

Such minimal sets of components are termed *autogenetic system precursors* (AGSP). On the basis of various previous considerations and data collected from real biological systems, it has been inferred that the functional informational content of a zero-system containing an appropriate AGSP will increase in time, this process is called *autogenesis* (Csányi, 1985; Csányi & Kampis, 1985).

During autogenesis, as time advances, an increasing part of functional information becomes replicative information. This can appear only as an extension of AGSP, that is, additional replicative cycles appear which are interconnected with the AGSP. Their formation occurs in a *super-cycle*. As time passes, replicative coordination of super-cycles develop, and fidelity of replication increases. We can speak of some kind of functional differentiation and cooperation which results in the formation of communities of simultaneously replicating components, that is sub-systems, called *compartments*. Their components are separated from others by their participation in co-replication. The emergence of compartments is equivalent to the organizational, i.e. functional closure of the network of component-producing processes and components having a replicative function. This succession of events is called the *compartmentalization and convergence of replicative information*. Components and compartments embodying them may co-exist with different degrees of replicative fidelity. For a while components may replicate with high accuracy, but the compartments formed by them do so with a low level of fidelity. As time passes the fidelity of replication of both may increase and perhaps a next scalar level of organization may be created, a "compartment of compartments", which also replicates with an increasing fidelity. A system may contain several different kinds of compartments, which are all replicative units with diverse fidelities. Among these, interrelationships develop, and as a result, their replication becomes coordinated. Gradually, the whole system will start replicating as a final replicative unity.

In the autogenetic process, the organization of the system (and of its parts) changes due to the functions of new emerging components. Thus, autogenesis is possible only while the state of identical replication has not yet been achieved globally. In that state, the system has become functionally closed and its replication in time continues as long as the environment does not change. There are no further organizational changes initiated by organizational causes because new functions can no longer originate. In the state of identical replication, the system is an autonomous, self-maintaining unity, a network of components and component–producing processes which, through the functional interaction of the components, produces exactly the same network which produced them. Its organization is almost closed and cyclic. Its input and output are subordinated to its replication, but through them, its existence depends on the invariance of the environment. Such a system exists in the state of continuous self-renewal, and its definition is exactly the same as an autopoietic system described by Varela et al. (1974) and Maturana (1975). The relations and differences between the autopoietic and autogenetic systems are discussed in Csányi and Kampis (1985).

ORGANIZATIONAL LEVELS

A replicative system is thus characterized by the fact that its functionally more or less closed replicative network embodies various replicative subnetworks; its components embody various sub-components. They form *organizational levels*. An organism, for example, is a more or less autonomous network of components and component-producing processes which is replicating in time, i.e. the organism is a component system. As a unity it is also a component of higher organizational levels, like ecosystems or the whole biosphere. The components of the organism are the cells, which themselves are, in turn, autonomous replicative networks of molecules, components of the lowest level of organization, and atoms are the elementary building units of the biological system.

The uselessness of the classic model devoid of the system concept is especially obvious in applications of the evolution metaphor outside biology. The new metaphor has been used for describing the operation and interpretation of the behavior of mind (Csányi, 1989d), culture (Boyd & Richerson, 1985), technology (Sahal, 1976) whole societies (Csányi, 1989b, 1989d, Csányi & Kampis, 1987, 1991; Pantzar & Csányi, 1991; Kampis & Csányi, 1992), and even for modelling the Universe (Chaisson, 1988). All of these interpretations are based upon the *changes of systems* embodying interacting components. Kampis (1987a, 1987b, 1991) gives an outstanding review and evaluation of these problems.

The problematic questions of evolutionary research can be answered only by a paradigm shift. We need new explanatory and predictive models, like the one described above. There is no space here to review the whole area of these problems. I have done so in several theoretical publications (Csányi, 1989d, 1992a; Kampis & Csányi, 1988, 1990), but I would like to discuss, or at least to touch on some questions which I think are the most important ones.

THE QUESTION OF THE "UNITS" OF THE EVOLUTIONARY PROCESS

There is a continuous debate among biologists concerning the units of evolution. In the classical theory the organism is the unit of selection and of evolution itself. The organism is influenced by the effects of the environment, the organism propagates or dies, and so it is logical to regard it as a unit of evolution. More recently mostly on the basis of the elegant arguments of Dawkins (1976), the gene is also seriously assigned for the role of a unit, because the lifetime of genetic information surpasses that of organisms, organisms are not copied into the next generation, while genes are, etc. Others search for the units of selection at higher organizational levels. According to Stanley (1975) the species themselves are under selection. The possibility that ecosystems are subjects of selection has also been discussed (Dunbar, 1972).

The main points of argument in the classic theory for defining the units of selection were that a unit must be an individual, member of a population, be variable, and must possess heredity, that is, be able to replicate itself physically in space. By these requirements in the rigorous definition certain entities are excluded from evolution. For example the biosphere exists only in one copy, on Earth at least, therefore variants do not exist, therefore it is unable to evolve (Dawkins, 1982). If we accept this argument, evolution cannot be invoked for the history of the biosphere on Earth.

Human personalities as concrete individual entities also exist only in single copies, their developmental process could not be regarded as evolution on the basis of the above train of thoughts. In the same way social systems also can be denied from the possibility of evolution because they exist only as individual copies. It is an other question that many different social systems can exist in the same time and their differential selection can be accepted as evolution by the classical theory, but I refer to concrete individual social systems. In spite of these arguments of Darwinian biologists, the metaphor of evolution is used in a wide context for characterizing changes of personalities and societies in the social sciences.

Who is right?

The metaphor of evolution can be extended to all of the above mentioned entities without any problem, if variability is not restricted to "variability in space" that is to the simultaneous presence of many variants at the same time, but if we accept "variability in time" as I suggested in the replicative model (Csányi, 1989d). The basis of this concept is that evolution may occur in a system in which the components are interacting with each other, they are produced and decay in an interconnected network of processes and components. Evolution occurs in such system without the loss of continuity of its identity and autonomy. The system could change in time by the transformation of the network of the interactions which occurs because of autogenesis, new components may be created, old ones cease to exist, etc. This transformation is a real evolutionary process in one individual and autonomous entity, which exists only in a single copy. This inference is valid for the evolution of personality and society, and above all, the biosphere.

It is of interest that Dawkins (1976), who is conservative, accepts the replicative evolution of memory traces ("memes") which create the human personality, but he could not reach the logical conclusion that if memes are able to evolve, then capacity to evolve cannot be denied a personality which is based upon the memory traces. If a personality can evolve, then of course many entities which are single systems can as well. In the same way, if species are able to evolve, then what is the proper term for the transformation of the whole system in which species are only components? Memory traces compose a dynamic system which is the most important part of our personality and it constantly evolves from birth to death.

If a higher organization is able to evolve, then the constraints which created the higher level of organization must have some role in the models describing evolution. The classical models of evolution do not recognize such higher level constraints.

This drawback of the classic model can be traced to a wrong concept of individuality. In the arguments about evolution it seems as though individuality would be an all or nothing phenomenon. Nevertheless individuality and autonomy are evolutionary features of organizations and appear gradually during evolution. When, during the origin of life, some kind of proto-cells already had formed which consisted of certain kinds of macromolecules this does not mean that proto-cells had already acquired individuality on the same degree as modern cells did.

It is much more plausible that the different macromolecules could be exchanged by them almost freely, that is, that the place of origin and operation of these molecules were not always the same. Isolation mechanisms which constrained the exchange of macromolecules probably appeared much later, and only when these mechanisms were well developed was the transformation of proto-cells to individual autonomous cells made. Individuality therefore is not an absolute precondition of evolution as Darwin thought, but it is a highly developed product of it. The same train of thought is valid for the ecosystem (Salthe, 1985). Ecologists ceaselessly debate whether ecological entities are "systems" or mere sets of species living on the same area. This debate is also based on the wrong assumption that individuality, in this case individuality of ecosystems is an all or nothing character. If individuality is the product of evolution then it has a degree and modern ecosystems can be classified into different classes in this regard, but according to general evolutionary considerations all move to a higher degree of individuality which can be attained only over long evolutionary time periods. The same argument is valid for the biosphere. These thoughts form the basis of the assumptions about the direction, equilibrium and the end state of evolution. If the emergence of individuality is a general tendency which already reached a high degree in cells and organisms, and ecological units move in the same direction, then the assumption can be formulated that the direction of evolution always leads to the formation of highly organized autonomous individual systems and evolution proceeds until this state is attained at the highest organizational level.

Models concerning personality and society and the biosphere based upon this assumption have already been constructed (Csányi, 1989d).

CREATIVITY OF THE EVOLUTIONARY PROCESS

In the classical evolution metaphor the power of selection originates from the environment. According to the new interpretation of the metaphor, it comes from the system itself, there is a closure here.

What is, then, the origin of the creativity of evolution?

It was shown earlier (Csányi, 1991, 1992b) that creativity is the consequence of the "hidden properties" of the components of the system. The hidden properties are those which are not involved in the given organization, and their variety is infinite as was pointed out by Bunge (1963) in case of the atoms of chemical elements. Emergence of new characteristics is dependent on the given organization of the system because it is built upon a given set of properties of the components. All other characteristics of the components are hidden regarding the organization. Changes in the organizational structure, or the appearance of new components, make the activation of some of the hidden properties and their participation in the new organization possible. Emergent properties in biological evolution are finally the consequence of the hidden properties of the atoms of chemical elements. Emergence of the hidden properties push the system towards increasing complexity and ever higher levels of organization. The existing organization is always based upon the properties of the components while properties of the components depend on the existing organization.

The organization of the system and the components of the system can be conceived only in each other, and for their interpretation we do not need external agents.

Organization and complexity of an evolutionary system always show a spontaneous increasing tendency toward an equilibrium state whose conditions can be given exactly (Csányi, 1989). With these, the direction of evolution and its possible end state can also be defined.

The classical evolutionary model which was based on external causes could not provide satisfactory answers for the meaning of evolution, because the concept of external environment as the final cause denied this possibility. In the new extended evolutionary model discussed above, evolution occurs for internal reasons, and as I have already discussed earlier (Csányi, 1989d), the goal and meaning of it is unequivocally the mere "being", an internal drive to be. There is no meaning of existence from an external agent's point of view, its goal is itself, and it follows from this that it resists any kind of change.

Evolution directed by a drive to exist moves towards individuality and stability. Therefore in the final model the metaphor of evolution turns into stability and unchangeability of existence. With this the evolution metaphor turns into its own negation.

Department of Ethology, ELTE University
Budapest, Hungary

REFERENCES

Boyd, R.; Richerson, P.J., 1985, *Culture and the Evolutionary Process*, The Un. of Chicago Press, Chicago.

Bunge, M., 1963, *The Myth of Simplicity*, Prentice-Hall, Englewood Cliffs, N.J.

Chaisson, E., 1988, *Universe: An evolutionary Approach to Astronomy*, Prentice Hall, Englewood Cliff, NJ.

Csányi, V., 1978, Az evolucio altalanos elmélete, *Fizikai Szemle*, vol. 28, pp. 401-417, pp. 441-452.

Csányi, V., 1982a, *General Theory of Evolution*, Publ. House Hung. Acad. Sci., Budapest, pp. 121.

Csányi, V., 1982b, General Theory of Evolution, *Soc. Gen. Syst. Res.*, vol. 6, pp. 73-95.

Csányi, V., 1985, Autogenesis: Evolution of Selforganizing Systems, in *Dynamics of Macrosystems*, J.-P. Aubin, D. Saari & K. Sigmund (eds.), Proceedings, Laxenburg, Austria 1984, *Lecture Notes in Economics and Mathematical Systems*, vol. 257, Springer-Verlag, Berlin, pp. 253-267.

Csányi, V., 1988, Il modello replicativo dele'evoluzione biologica e culturale, in *Physis: abitare la terra*, M. Ceruti & E. Laszlo (eds.), Feltrinelli, Milano, pp. 249-260.

Csányi, V., 1989a, The replicative model of self-organization: A general theory of evolution, in *The Paradigm of Self-organization*, G.J. Dalenoort (ed.), Springer, pp. 73-76.

Csányi, V., 1989b, The Replicative Model of Cultural Evolution, *Humanbiol*, vol. 19, pp. 83-87, Budapest.

Csányi, V., 1989c, Origin of Complexity and Organizational Levels During Evolution, in *Complex Organizational Functions: Integration and Evolution in Vertebrates*, D.B. Wake & G. Roth (eds.), John Wiley & Sons LTD, pp. 349-360.

Csányi, V., 1989d, *Evolutionary Systems and Society: A General Theory*, Duke University Press, Durham.
Csányi, V., 1991, Social Creativity, *World Futures*, vol. 31, pp. 23-31.
Csányi, V., 1992a, Natural Sciences and the Evolutionary Models, *World Futures*, vol. 34, pp. 15-24.
Csányi, V., 1992, Nature and Origin of Biological and Social Information, in *Evolution of Information Processing Systems*, K. Haefner (ed.), Springer, Berlin, pp. 257-281.
Csányi, V.; Kampis, G., 1985, Autogenesis: Evolution or Replicative Systems, *J. Theor. Biol.*, vol. 114, pp. 303-321.
Csányi, V.; Kampis, G., 1987, Modelling Society: Dynamical Replicative Systems, *Cybernetics and Systems*, vol. 18, pp. 233-249.
Csányi, V.; Kampis, G., 1991, Modelling Biological and Social Change Dynamical Replicative Network theory, in *The New Evolutionary Paradigm*, E. Laszlo (ed.), Gordon and Breach Publ., New York.
Darwin, C.R., 1859, *On the Origin of Species*, Dent, London.
Dawkins, R., 1976, *The Selfish Gene*, Oxford University Press.
Dawkins, R., 1982, *The Extended Phenotype*, W.H. Freeman & Co, New York.
Dubar, M.Y., 1972, The Ecosystem as a Unit of Natural Selection, in *Growth by Intussusception: Ecological Essays in Honor of G. Evelyn Hutchinson*, E.S. Deevey (ed.), Transactions of the Academy 44, Connecticut Acad. Arts and Sci., New Haven, pp. 114-130.
Kampis, G., 1987a, Some Problems of System Descriptions I. Function, *Int. J. Gen. Syst.*, vol. 13, pp. 143-156.
Kampis, G., 1987b, Some Problems of System Descriptions II. Information, *Int. J. Gen. Syst.*, vol. 13, pp. 157-171.
Kampis, G., 1991, *Self-Modifying Systems: A New Framework for Dynamics, Information, and Complexity*, Pergamon, Oxford-New York, pp. 543+xix.
Kampis, G.; Csányi., V., 1988, A System Approach to the Creating Process, *IFSR Newsletter*, no. 20, pp. 2-4.
Kampis, G.; Csányi, V., 1990, Coevolution and the units of evolution, in *Organizational Constraints on the Dynamics of Evolution*, J. Maynard Smith & G. Vida (eds.), Manchester UP, Manchester & NY, pp. 385-399.
Kampis, G.; Csányi, V., 1992, Societies as Replicative Component-Systems, *World Futures.*, vol. 34, pp. 25-41.
Lovelock, J.E., 1979, *Gaia*, Oxford University Press.
Maturana, H.R., 1975, *Autopoietic Systems: A Characterization of the Living Organization*, Urbana, Illinois.
Maynard Smith, J., 1969, *The Theory of Evolution*, Penguin, London.
Mayr, E., 1982, *The growth of biological thought*, The Belknap Press of Harvard UP, Cambridge, Mass.
Morowitz, H.J., 1968, *Energy Flow in Biology*, Academic Press, New York.
Pantzar, M.; Csányi, V., 1991, Replicative model of the Evolution of the Business Organization, *J. Soc. Biol. Struct.*, vol. 14, no. 2, pp. 149-163.
Prigogine, I.; Nicolis, G.; Babloyantz, A., 1972, Thermodynamics of Evolution, *Physics Today*, vol. 25, pp. 23-28, pp. 38-44.
Sahal, D., 1976, System Complexity: Its Conception and Measurement in the design of Engineering Systems, *IEEE Trans. Syst. Man and Cybernet.*, pp. 440-445.
Salthe, 1985, *Evolving Hierarchical Systems*, Columbia University Press, New York.
Spencer, H., 1862, First Principles, Londres, William & Norgate.
Stanley, S.M., 1975, A theory of evolution above the species level, in *Proc. Natl. Acad.Sci. USA*, vol. 8, pp. 731-349.
Van Valen, L., 1973, A New Evolutionary Law, *Evol. Theor.*, vol. 1, pp. 1-30.
Varela, F.G.; Maturana, H.L.; Urbibe, 1974, Autopoiesis, The Organization of the Living Systems, Its Characterization and a Model, *Biosystems*, no. 5, pp. 187-196.

STANLEY N. SALTHE

THE ROLE OF NATURAL SELECTION THEORY IN UNDERSTANDING EVOLUTIONARY SYSTEMS

INTRODUCTION

The literature on evolutionary systems (e.g. Csányi, 1989; Goonatilake, 1991; Laszlo, 1987, 1994) is unclear about the role of natural selection.

Evolutionary systems is a much more complex discourse than is evolutionary biology. In evolutionary biology we consider the play of only two forces — the random generation of heritable variants, and the sifting of these variants by success. As a result, in neo-Darwinism we find represented the interaction of chance and necessity, but no real form of systematicity. Systemic or constitutive changes — reflected in biology, for example, in allometric relationships (e.g. Calder, 1984; Niklas, 1994; Peters, 1983; Schmidt-Nielsen, 1984) — signify the material embodiments of structures, and these have been quite generally ignored by neo-Darwinians. As I hope to show, selection theory is unintelligible because one could never anticipate (as one could predict a metabolic rate from facts about size using allometric relationships) what adaptations will be marshaled to eliminate particular environmental challenges.

Evolutionary systems is a comparatively materialist enterprise because there is concern for the details of the actual generation of systems and their variants, which amounts to a concern for their intelligibility. In biology such matters are assigned to developmental biology (which today unfortunately has itself been herded by neo-Darwinian influence into a narrowly constricted developmental genetics, compared with its former robustness as embryology). Selection theory is, in comparison with developmental studies, an idealistic enterprise. For example, the philosopher Elliot Sober (1984) has referred to fitness as a "non-physical property" because it is a matter of comparing the relative endurance in time (or over generations) of different forms whose material embodiments continually change. But I mean something more as well in my attribution of idealism here, as will become clear in the following.

BIOLOGY AND GENETICS

Biology in almost all of its subdisciplines has been taken up into the study of genetics (for some interesting holdouts see some of the papers in Dubbeldam et al., 1993 as well as Goodwin, 1994). Now the form of genetics discourse, from its beginnings at the turn of the last century, is the experiment aimed at explaining stable phenotypic differences. If there were no such differences between compared forms, there could be no search for "genes" that determine them. Because of this fundamental orientation, I believe biological genetics discourse can serve as a basis only for explanations of *differences* between forms, not for

G. Van de Vijver et al. (eds.), Evolutionary Systems, 13–20.

their similarities. In genetics, similarity is conventionally reduced to mere "absence of difference", and serves only as a null hypothesis. I conclude that neo-Darwinism, rooted as it is in genetics, cannot directly confront similarities (for a similar but more radical conclusion from a very different perspective, see Morss, 1993).

Now, such a bias implicitly excludes considering the results of the play of material and physical laws. If we have two unlike objects in the same locale that have been influenced only by the same laws and various material contingencies, they would not be distinguished in genetics discourse. These might include two hurricanes, two mountains or two grains of sand. Differences among these would be attributable to various material contingencies modulating the effects of lawful constraints. Such differences, being putatively non-inheritable, would not figure in genetics discourse. Even differences between two birds, if no more specified information about them than that they *were* birds was to be considered, would be unobserved in genetics. So the results of physical and material laws, seen in biology for example in allometric relationships and scaling rules, could therefore have no explicit part to play in genetics, and so the results of constitutive (i.e. material) processes would be subliminal in that discourse.

If two chains of DNA differed in sequence, but the final gene products were functionally indistinguishable, there would be reckoned no genetic difference between the two, and none would be looked for. These genetically occult differences in DNA would have been generated by physical processes, but would not be differences that make a difference to the geneticist — or even, it may be, to the organism involved.

As a result of this bias, natural selection is itself reduced to choices between immaterial differences. What is an immaterial difference? I suggest that it is a difference that could not be specified on average by differential impacts of natural and material laws acting in conjunction with local initial and boundary conditions. Material differences, then, would include differences, e.g. in the sizes of mountains or of trees, or in the *shapes* of trees to the extent that these are determined directly by the forces bearing in the environment.[1] These kinds of differences, given the subtly pejorative label "ecotypes" by Turesson (1922), were very carefully excluded from evolutionary genetics discourse. All differences not allocatable to differences in stable internal predispositions were, quite properly given its discursive intentions, rigorously excluded from genetics discourse.

Now, differences in sequence between two strands of DNA, whether "genetic" or not, would have trivial effects on their physical properties. Howard Pattee (e.g. 1972) has characterized what is sought in genetics discourse as "symbolic", as opposed to "physical", properties. As is obvious, important symbolic differences can be associated with trivial physical differences (that is, the physical differences would be so slight that it would require vast amounts of time over immense numbers of cases near equilibrium to precipitate distinctions that would make a difference between them in natural systems). Yet such trivial physical differences can make the difference, when interpreted by biological systems, between, e.g. one species, or one sex, and another (for an interesting example showing the importance of the interpretation of genes by biological systems, see Hultmark, 1994).

ADAPTATION

Now let us consider adaptation. In general there is no material necessity associated with adaptations. Consider the well-known case of sickle cell anemia in humans (e.g. Livingstone, 1971; Stamatoyannopoulos, 1972). Populations in Africa became adapted to living in areas where malaria was endemic by way of modifications of the hemoglobin carried in the red cells of the blood. As a materialist, you would never have predicted how adaptation was actually achieved here. The kind of adaptation that was constructed does not interfere with the insertion of parasites into the blood stream by mosquitoes. Nor does it interfere with the ability of the parasites to feed on hemoglobin. And it did not bolster the ability of the human organism to withstand any of the many symptoms of malaria.

What it did was to modify hemoglobin in such a way that, when the oxygen concentration within the red cell becomes lowered by the metabolic activity of increasing numbers of the aerobic parasites, it would precipitate, its crystals then deforming the red cell in such a way that it would lyse, releasing the parasites into the blood stream at a developmental stage when they are not ready to survive there. In this way individual infected cells will be sacrificed at trivial cost to the human organism. But there is more. It turns out that many populations achieved essentially the same kind of protection independently, by way of different specific mutations in hemoglobin. And so there is no one single material way to achieve even this very minute, refined adaptation.

Furthermore, some of these similar adaptations outcompete others because they will increase in the population faster (Templeton, 1982). So, hemoglobin-S protects even in the heterozygote, while hemoglobin-C, for example, requires the homozygotic combination to be effective. The latter, curiously, is much more effective in its protection of people, not being associated with serious genetic disease, while the fitter hemoglobin-S causes the eventual death of one quarter of one's offspring, therefore cutting down the growth rate of the population. In other words, the "fitter" adaptation (in the neo-Darwinian sense) is physiologically worse for the organism and its heirs than are less "fit" ones.

Now, I would maintain that adaptation too is a non-material relation, in the sense that no combination of physical or material laws with any combination of initial and boundary conditions could generally be used to predict the way that it will be achieved in any particular case. Am I then perhaps suggesting that complex systems are themselves to some degree immaterial? I think I am moving in that direction, given what materialism has historically been. To practice as a materialist one must be able to successfully interfere in an intelligible physical system, moving it in a desired direction. Adaptations may be physically embodied, to be sure, but they are also immaterial, depending as they do upon so many and such subtle habitual relationships between physical objects that they are plainly relations between historical events and occasional contingencies. So they are outside the domain of the predictive sciences, which depend upon constitutive generic, (i.e. predictable) changes in the objects of their inquiries.

It is for this reason that I claim that adaptations are semiotic. For the biologist they are signs in an observed population referring to entities existing in their mutual environment. Signs are constructed physically, yes, but the sign relation is itself immaterial, being a question of perspectives on associations of, in short, meaning.

GENERICITY

Now, it seems clear that in evolutionary systems discourse a concern with generic properties has determined its major lines of inquiry (this has led me to wonder if a better label would not have been 'developmental systems'). And to understand genericity many tools have been utilized — non-equilibrium thermodynamics, information theory, cybernetics, systems theory, non-linear dynamics, hydrodynamics, cosmogeny, computer simulations, and even systematic philosophy, history and semiotics. And these approaches are applied to many different kinds of systems — evolutionary systems is a multi-disciplinary movement, which is ultimately looking for inter- (or meta-) disciplinary patterns of change.

In my opinion, the role of natural selection theory in evolutionary systems will only be to explain or model the fine-tuning of systems to their local environments — what Darwinians call 'adaptation'. Consider another example from biology: selection theory would tell us little when comparing forms as unrelated as birds and bats, even about their differences. Comparing very different kinds of organisms is just not conducive to genetic analysis. About their similarities it could have nothing to say at all, including those related to flight. Even, despite a flourishing subdiscipline of life history theory, it would not give us a single story of why they both have notably low fecundity compared to most other kinds of organisms. This is because the adaptive meanings (I use this word deliberately instead of 'causes' since we are in semiotic discourse here) of these various similarities were very likely quite different because the environments involved would not have been the same, let alone the genes. Based in genetics, the judgment of neo-Darwinians does not extend comfortably to similarities — even to similar adaptations. Indeed, in the Darwinian world adaptations could not really be similar at all, except adventitiously, suggesting no deep meaning.

In fact, neo-Darwinian evolutionary theory is really about short term, local adaptation only, and is most useful when comparing systems closely related by descent living in different environments (when similarity need be taken only as a starting point of evolution). Evolutionary systems, on the contrary, is intensely concerned to explain similarities between widely different kinds of systems. Topics that come closest to such interests within evolutionary biology — parallel and iterative evolutions, evolutionary convergence, and ecological vicarage — have been widely ignored by recent neo-Darwinians (e.g. these topics are not even mentioned in any of the several editions of the book *The Theory of Evolution* by that dean of neo-Darwinians, John Maynard Smith. I note also that in Depew's and Weber's 1995 book on Darwinism that, of these concepts, only convergence receives a single line out of about 22,000 in the book, and is then put between quotes! They have quite correctly measured the interest of neo-Darwinism in similarity).

In general, proximate regulation by environment provides only limiting constraints, and so can determine only some aspects of what may *not* occur in the construction of those semiotic relationships called adaptations. These are influenced by local contingencies, yes, but must also be guided by systemic, constitutive, generic processes. This was for me the interesting message of Stuart Kauffman's book of 1993. Adaptations must be subject to these knowable processes as well, and without considering them, full explanations of adaptation will not be found. For example, consider that, given all the ways in which selection might have interfered with malarial debilitation, in population after population it successfully worked only on the form of hemoglobin in respect to its stability. Can Darwinians explain this? No. Do they attempt to explain this kind of thing? Not at all. In fact, since these were concurrent adaptations in different populations, we could consider this too as an essentially non-Darwinian problem of parallel evolution.

John Collier (personal communication) feels that the difference here between neo-Darwinian and evolutionary systems practice relates to scale. That is to say, Darwinian processes occur locally, while the subjects of evolutionary systems are compared globally. I think this is true, but it seems to me that systemic, constitutive processes are always having local effects, even as much as the contingencies that are alone used by Darwinians to explain adaptations. But, if we are to trust the Darwinians, these systemic forces just don't seem to matter as much as those contingencies. And yet we have had all those allometric relationships (summarized in Calder, 1984; Niklas, 1994; Peters, 1983; Schmidt-Nielsen, 1984) before us for a long time!

EVOLUTIONARY SYSTEMS

I would like to explain in detail precisely what I mean by my assertion that the results of natural selection mediate only the fine-tuning of systems to local environments. For this I require what I call the specification hierarchy (Salthe, 1993). Any organism can be viewed as a mere physical system, say, by considering only the relations of diffusion within it and/or between it and its environment. Or we could consider it to be a chemical system, by considering only various metabolic facts. Or we could take it to be a biological system per se, considering, say, its reproductive activities. Organisms are kinds of chemical systems, which in turn are kinds of physical systems. If we consider, say, birds, we will find that they are all the same kind of chemical and physical system, but that they differ in biological details. During ontogeny, physical and material laws are entrained by a nucleated dissipative structure to produce form which, at any step during its production may be modulated by genetic information so as to reflect historically acquired alterations in genetic information. And so genetic information, while it may act at any stage of development, is effective always only at the highest integrative level. It cannot change constraints imposed by the material and physical laws involved in the lower integrative levels, which if we were coarser observers, oblivious to details, might be taken as all there is involved in development. In other words, a developmental process might or might not be constrained by a given allele carrying historical information, but it must always involve, say, chemical dehydrogenation.

Now, we must realize that in neo-Darwinian discourse it is very important that there *not* be taken into account any constitutive changes or tendencies. These, for example, the "nomogenesis" of Leo Berg (1926), among many similar ideas, were definitively banished from Darwinian discourse by George Simpson (1949), under the label 'orthogenesis'. In the days when I was a neo-Darwinian myself, I spent a good deal of time construing evolutionary trends of any kind to be epiphenomenal constructs of an earlier generation of evolutionary biologists (see Salthe, 1972). If evolutionary trends seemed to be accessed by more than one lineage, they could roughly be explained away as the results of selection on similar genotypes in response to similar selective pressures from approximately similar components of environments. Such trends are only awkwardly treated in a discourse based in difference.

But convergent evolution, as seen, for example, when comparing the eyes of squids and vertebrates, really has had no adequate Darwinian interpretation. Surely these eyes are adaptations for similar needs, but how could they come to be virtually identical? It certainly could not be because of any identity of genes! (unless one believes in the transfer of large chunks of genomes as proposed, e.g. by Williamson (1992), a perspective which must be the acme of mechanicism).

Well, it was the widespread occurrence of parallel, convergent and iterative evolution, producing ecological vicarage at the level of both organisms and whole biotas in both contemporary and sequential biotas, that led me toward structuralism, and so away from Darwinism. On first glance, such similarities seem to suggest the great power of natural selection in being able to mold anything from any kind of beginning. That is until you realize that the forms constructed in organic nature are such a small sample of possible forms. This realization alone must show you that Darwinism is insufficient to explain the limited variety of organic forms, past and present, however well it may explain diversity as such (for which, however, Brooks and Wiley (1988) have shown us that it is unnecessary — even non-parsimonious — to invoke selection).

LIMITS ON SELECTION

Finally, I would like to clarify something that might be detected as incoherent in my approach. I have gone to some trouble here to characterize the products of natural selection as being "semiotic" rather than material. At the same time I have argued elsewhere that semiotics needs to be brought into evolutionary systems discourse, even while I also suggest here that selection will not contribute much to this discourse, which I characterize as primarily a materialist pursuit.

1. I do not claim that selection will have no effects on the subjects examined in evolutionary systems. In fact, these effects would tend to be characteristic of the highest integrative levels (e.g. Salthe, 1993) of the systems observed just because they are in the nature of fine-tuning, and so they will be there before us prominently, like a peacock's tail feathers, to dazzle our eyes. But,

2. I believe that what evolutionary systems has really been after — similarities across systems — are generated at the deeper integrative levels of systems. In order to effectively pursue this interest we need to resist being distracted by superficial characteristics, like, e.g. the differences between two species of organisms in the same family, or, for that matter, the differences between two cities, or two systems of musical expression.

3. In evolutionary systems discourse we are primarily after similarities between systems because we are searching for clues as to how to more effectively interfere with them — for whatever reasons, ecocentric or technocentric. Semiotics will be important in this quest for the reason that the systems we are primarily interested in in this context tend to be those large scale ones enveloping us within them. The internalist perspective that we must therefore assume means that we need to explicitly take into account the nature of the observer — ourselves — and I believe semiotics will be a great aid in helping us to do that (where some beginning has already been made in second cybernetics — e.g. Van de Vijver, 1992). Following the lead of Charles Peirce (e.g. Deely, 1990; Rescher, 1978), we need to see that semiotics can become a kind of scientific epistemology.

4. Some form of semiotic will be necessary in evolutionary systems because of the widely appreciated need to subsume information theory under some system capable of dealing with the meaning of information. Especially in biological and higher integrative levels, forms are seen to be *for* something — to have functions — and are not taken to merely exist willy-nilly. This means we require some approach capable of representing functions, purposes and values.

For these reasons, then, I believe that the primarily materialistic enterprise of evolutionary systems could not proceed without semiotics to deal with the immaterial aspects of complex systems. We need to characterize ourselves explicitly as both disinterested observers (for some purposes) and as interested actors (for others). These characterizations must explicitly accompany even the primary observations of nature, so that semiotics must inform experimental design as well as subsequent interpretations. Even though synthetic evolutionary theory is deeply semiotic (as I have tried to suggest here), like most of the rest of western science in its search for "objectivity", it has been at great pains to conceal not only this fact, but also the semiotic connections of its practice to the rest of modern society (see Depew and Weber, 1995, for such connections of Darwinism). For this reason the fact that the disinterested observer is also an interested actor has been obliterated, giving us only, in the event, a partially effective scientific practice.[2]

Natural Systems
New York, USA

NOTES

[1] Materialism is the viewpoint that we are confronted with a frictional world. Friction gives rise to regularities working as constraints, some of which are usefully recognized as laws. The materialist seeks to exploit these laws.

[2] If we could not justify our actions as realists, we would need to deliberate more carefully before acting. We need, as it were, more doubt.

REFERENCES

Berg, L.S., 1926, *Nomogenesis, Or Evolution Determined By Law*, London, Constable and Company.

Brooks, D.R.; Wiley, E.O., 1988 (2nd ed.), *Evolution As Entropy: Toward a Unified Theory of Biology*, Chicago, University of Chicago Press.

Calder, W.A. Jr., 1984, *Size, Function, and Life History*, Cambridge, MA, Harvard University Press.

Csányi, V., 1989, *Evolutionary Systems and Society: A General Theory*, Durham, NC, Duke UP.

Deely, J., 1990, *Basics of Semiotics*, Bloomington, IN, University of Indiana Press.

Depew, D.J.; Weber, B.H., 1995, *Darwinism Evolving: Systems Dynamics and the Genealogy of Natural Selection*, Cambridge, MA, MIT Press.

Dubbeldam, J.L.; Goodwin, B.C.; Kortmulder, K. (eds.), 1993, *Evolution, Development and Behavior*, Special edition of *Acta Biotheoretica,* vol. 41, no. 4, pp. 275-494.

Goodwin, B.C., 1994, *How the Leopard Changed Its Spots*, New York, Charles Scribner's Sons.

Goonatilake, S., 1991, *The Evolution of Information: Lineages in Gene, Culture and Artefact*, London, Pinter Publishers.

Hultmark, D., 1994, Ancient relationships, *Nature*, vol. 367, pp. 116-117.

Kauffman, S., 1993, *The Origins of Order: Self-organization and Selection in Evolution*, New York, Oxford University Press.

Laszlo, E., 1987, *Evolution: The Grand Synthesis*, Boston, Shambala Publications Inc.

Laszlo, E., 1994, An introduction to general evolutionary theory, *J. of Biol. Syst.*, vol. 2, pp. 105-110.

Livingstone, F.B., 1971, Malaria and human polymorphisms, *An. Rev. of Genetics*, vol. 5, pp. 33-64.

Maynard Smith, J., 1993, *The Theory of Evolution*, Fourth Edition, Cambridge, UK, Cambridge UP.

Morss, J., 1992, Against ontology, in *Trees of Life: Essays in Philosophy of Biology*, P. Griffiths (ed.), Dordrecht, Kluwer Academic Publishers.

Niklas, K.J., 1994, *The Scaling of Form and Process*, Chicago, University of Chicago Press.

Pattee, H.H., 1972, Laws and constraints, symbols and languages, in *Towards a Theoretical Biology*, C.H. Waddington (ed.), vol. 4: Essays, Chicago, Aldine.

Peters, R.H., 1983, *The Ecological Implications of Body Size*, Cambridge, UK, Cambridge UP.

Rescher, N., 1978, *Peirce's Philosophy of Science*, Notre Dame, IN, University of Notre Dame Press.

Salthe, S.N., 1972, *Evolutionary Biology*, New York, Holt, Rinehart & Winston.

Salthe, S.N., 1993, *Development and Evolution: Complexity and Change in Biology*, Cambridge, MA, MIT Press.

Schmidt-Nielsen, K., 1984, *Scaling: Why Is Animal Size So Important?*, Cambridge, UK, Cambridge UP.

Simpson, G.G., 1949, *The Meaning of Evolution*, New Haven, Yale University Press.

Sober, E., 1984, *The Nature of Selection: Evolutionary Theory in Philosophical Focus*, Cambridge, MA, MIT Press.

Stamatoyannopoulos, G., 1972, The molecular basis of hemoglobin disease, *An. Rev. of Genetics*, vol. 6, pp. 47-70.

Templeton, A.R., 1982, Adaptation and the integration of evolutionary forces, in *Perspectives on Evolution*, R. Milkman (ed.), Sunderland, MA, Sinauer Associates.

Turesson, G., 1922, The species and the variety as ecological units, *Hereditas*, vol. 3, pp. 100-113.

Van de Vijver, G. (ed.), 1992, *New Perspectives on Cybernetics. Self-Organization, Autonomy and Connectionism*, Synthese Library, vol. 220, Dordrecht, Kluwer Academic Publishers.

Williamson, D.I., 1992, *Larvae and Evolution: Toward a New Zoology*, London, Routledge, Chapman & Hall.

DAVID J. DEPEW

DARWINISM AND DEVELOPMENTALISM: PROSPECTS FOR CONVERGENCE

INTRODUCTION

The Darwinian tradition in evolutionary biology, whose most distinctive feature is the idea of natural selection, contrasts with an alternative approach that, for lack of a better phrase, I will call "evolutionary developmentalism." Developmentalists treat biological phenomena at various levels of scale by analogy to individual ontogeny, a view that Darwinism abandoned fairly early in its career. In *Darwinism Evolving*, Bruce Weber and I were unable to say as much as we might have about how the admittedly non-canonical view of Darwinism we espouse (sometimes dubbed "ecological Darwinism") is related to recent statements of evolutionary developmentalism (Depew & Weber, 1995). I will try to redress that omission in this essay by arguing that there are several strands in contemporary (genetical) Darwinism; that developmentalism (which has its own diversity) is flatly inconsistent with one prominent strand of Darwinism; but that developmental evolutionists (such as Stanley Salthe or Rod Swenson) need not on that account despair of Darwinism altogether, since other strands in the Darwinian tradition are showing signs of increasing integration with some forms of evolutionary developmentalism. In particular, many of the core insights of Theodosius Dobzhansky, a central figure in what Weber and I call the American wing of the Modern Evolutionary Synthesis, have been developed, whether wittingly or not, by authors as diverse as William Wimsatt, Michael Conrad, and Stuart Kauffman. They might well be developed further if evolutionary phenomena are placed more deeply within the ecological, and hence thermodynamical, context in which self-organization and natural selection can most perspicuously be seen to work together.

THE TWO WINGS OF THE MODERN SYNTHESIS

The Modern or "Neo-Darwinian" Evolutionary Synthesis began to take shape when R.A. Fisher first reconciled Mendelian genetics with natural selection by applying statistical mechanical models taken from Maxwell and Boltzmann to genetic gradients in populations. Fisher could deal with more phenomena falling within the scope of evolutionary biology than Darwin (or Galton or Morgan) could, and could deal more persuasively with the same phenomena, for the same reason that Boltzmann could deal better and with more phenomena falling within the scope of physics than Newton could, such as the behavior of ideal gases and the inability of natural systems to convert all energy into work. In both cases, the simplifying assumptions of their models made greater complexity tractable. In turn,

G. Van de Vijver et al. (eds.), Evolutionary Systems, 21–32.

the changes in Fisher's dynamical models introduced by Sewall Wright and Theodosius Dobzhansky — especially the highly heuristic notion of an "adaptive landscape" — made it possible for the Darwinian tradition to deal plausibly with still more complex phenomena, such as speciation. The long-standing confidence of the Modern Synthesis springs from these advances. According to Weber and myself, this progress was made possible by following Darwin's own lead in transferring dynamical models from physics to evolutionary biology: Darwin's Darwinism, we hold, was informed by Newtonian dynamics; Fisher's and his critics' by Boltzmann's (Depew & Weber, 1995). Recent challenges to the Modern Synthesis arise from the perception that it is unable to deal with even more complex phenomena, such as the relationship between evolutionary dynamics, ecological succession, and developmental genetics — phenomena that complex systems dynamics are now beginning informatively to model. Having pointed out that Darwinism has risen to just such occasions before, Weber and I have put forward inductive and dialectical arguments suggesting that it might be able to do so again (Depew & Weber, 1995).

In pursuing debates about the continued explanatory fecundity of the Darwinian tradition, it is important to recognize that the Modern Evolutionary Synthesis itself has only a specious unity, and that the inability of some Darwinians to comprehend certain kinds and degrees of complexity does not entail that all Darwinians are so retrograde. Close historical examination shows, in fact, that from the outset genetical Darwinism has had at least two distinct, though often interpenetrating, wings or tendencies. One, dominantly British, derives from Fisher and is currently exemplified, by his own testimony, by Richard Dawkins. The other, dominantly American, derives from Wright and Dobzhansky, and is currently exemplified, by his own testimony, by Richard Lewontin, among others (Depew & Weber, 1995). They are both versions of genetical Darwinism. What, then, is the difference between them?

Roughly put, it is that Fisher's wing thinks that the point of diminishing returns on simplifying assumptions for analyzing complexity falls at a lower point than the American wing does. This manifests itself in a sometimes over-reified commitment to the context-independent fitness values of individual genes, and in a tendency to work so exclusively within genotype state-space (where Fisher's analogy to statistical mechanics works well) as to deny the independent and interacting effect of causality at the phenotypic surface. By contrast, the American strain of genetic Darwinism has stressed what Ernst Mayr had dubbed the "principle of genetic relativity," which asserts that context-independent fitness values cannot generally be assigned to genes, because in changing environments what counts as fit is constantly changing, in part through the activities of organisms themselves. Dobzhansky wrote: "The environment does not remain constant, either in terms of geological periods or even from one year to the next. Selection and mutation rates, and hence genetic equilibria, are therefore in a state of perpetual flux. The nature of the genetic mechanism is therefore such that the composition of the species population is probably never static. A species that would remain long quiescent in the evolutionary sense is likely to be doomed to extinction" (Dobzhansky, 1937, p.179).

This being so, Dobzhansky came to believe that there is selection pressure for mechanisms that maintain enough variation to deal with a constantly changing environment, and that this end is achieved in large measure through heterosis, which Dobzhansky came to see as inherently adaptive. On this basis, he hypothesized that a higher-level trait called adaptability (conceived as the creation and maintenance of variation for changing environments) is itself an evolutionary adaptation, perhaps the most important one: "Environmental instability presents challenges to the organism (...) To maintain itself in harmony with a changing environment, the organism must not only be adapted, but also adaptable (...) A species should not only possess genetic variety, but must also be able to generate variety. It may then respond to changing environments by genetic changes" (Dobzhansky, 1962, p. 289).

In the wake of the molecular revolution, Dobzhansky's students and heirs have developed his insights. The circle of Lewontin's concerns are indicative of this Dobzhanskyan inheritance. Lewontin has extended the permissiveness with which nature tolerates variation by revealing massive amounts of natural polymorphism in proteins, and by arguing against Motoo Kimura's "neutralism" that this width of variation is either directly or indirectly functional (Lewontin, 1974). He has also attempted to separate natural selection, defined as the differential retention of heritable variation, from its historically contingent connection with the crypto-teleology of what Stephen Jay Gould and he call "the adaptationist program," making possible an "expanded" Darwinism in which selection can occur at any number of levels — genes, cell-lineages, organisms, demes, and (for Gould at least) whole species (Lewontin, 1970; Gould & Lewontin, 1979; Gould, 1982). In addition, Lewontin has antecedently blunted the reductionism of Dawkins' Fisher-inspired genic selectionism by showing that interdemic selection is genuine group selection, casting further doubt on the very possibility of assigning context-independent fitness values to genes (Lewontin, 1970). Finally, recognizing that natural selection's tendency to select for adaptations that enhance adaptability includes a tendency to select for adaptive behaviors that make organisms flexible and competent agents in their environment, Lewontin has stressed the importance of viewing organisms as active creators of their world rather than passive collections of disaggregated traits and genes coding for them (Levins & Lewontin, 1985). In the light of these and other themes, Lewontin has been able to proclaim even more loudly than Dobzhansky that "The fitness at a single locus ripped from its interactive context is about as relevant to real problems of evolutionary genetics as the study of the psychology of individuals isolated from their social context is to an understanding of man's sociopolitical evolution" (Lewontin, 1974, p. 318).

These reflections let us see that as the Darwinian tradition has matured it has moved in two opposed directions: a gene-centered, reductionist direction that, by according causal efficacy to "selfish genes," extends Fisher's vision and preserves many of the ideologically suspect trappings of the older Darwinism; and an opposite, if less popularly well-known, direction, in which it is acknowledged that nature is full of variation, activity, complexity, and scalar hierarchical levels (Depew &

Weber, 1995). I have little doubt that if Darwinism were restricted to the first orientation, one might be well advised to be an anti-Darwinian. Fisher's wing of the Modern Synthesis remains, in my view, uncomfortably close to the mechanistic ontology that Salthe imputes to the Darwinian tradition as a whole (Salthe, 1993). My abbreviated history of genetical Darwinism's two orientations should suggest, however, that anti-Darwinians cannot win an easy victory simply by assimilating Darwinism to the reductionist ideals of Fisher's lineage. That the greater creative thrust and explanatory fecundity of the Darwinian tradition lies along the second, Dobzhanskyan path is suggested by the fact that Lewontin himself is no less incensed at the mechanistic and reductionistic distortions of Darwinism than Salthe (Levins & Lewontin, 1985). Even if it can be shown that such distortions residually attach to Lewontin's own work — a point I am ready to concede — developmental evolutionists do not seem to me fully justified in dismissing the Darwinian tradition unless they can show that Darwinism does not have, and cannot acquire, the resources to deal with phenomena that the revolution in complex systems dynamics is plainly teaching us to understand. In what follows, I will try to suggest that movement in this direction on the part of Darwinians is not only possible, but already actual.

ADAPTABILITY AS A THEME IN AMERICAN DARWINISM

A telling indication of how deep are the fractures within the Darwinian tradition is their differing responses to the advent of complex systems modeling, and in particular the phenomena of self-assembly and self-organization that seem to attend the trajectories of every system with sufficiently large number of interacting elements (over a sufficiently large number of reiterations). John Maynard Smith, a student of J.B.S. Haldane and current elder statesman of British Darwinism, has reacted strongly against applying self-organizational models to evolutionary problems, maintaining that every concrete case that might at first have appeared to be a product of self-organizational dynamics (apparently even Turing's reaction-diffusion model of morphogenesis) has turned out (so far) to be a case of piecemeal natural selection (Maynard Smith, 1995). Evolutionary developmentalists there are who have risen (perhaps needlessly) to this bait. Brian Goodwin, for example, speaks (often well) for a largely British school of evolutionary developmentalists who assign to the (chemically underwritten) self-ordering properties of morphogenetic fields a causal power that directly rivals, and so is designed to exclude, the power of natural selection (Goodwin, 1994). Goodwin and his colleagues are developmentalists because their primary attention is on ontogenetic pattern. Like other evolutionary developmentalists, they see evolution as marking predictable yet irreversible changes in the self-generating morphogenetic fields that cause, indeed are, developing organisms. Accordingly, whereas Darwinians say that species are historical particulars, assembled by the lawless "higgledy piggledy" of natural selection, with its opportunistic conjunctures of chance and necessity, and so are explainable only through narrative reconstructions of unique events (Gould, 1989; Lewontin, 1991), Goodwin and other "structuralists" say that species are instantiations of the universal "laws of form" to which D'Arcy Thompson referred, and so are natural kinds rather than historically unique particulars (Thompson, 1917, 1942; Goodwin, 1994). The laws that govern these kinds, they argue, are becoming increasingly clear through complex systems modeling, and have little or nothing to do with natural selection.

Given such a categorically anti-Darwinian view, it is small wonder that defenders of natural selection like Maynard Smith have circled the wagons. My aim in reporting this is not, however, to take sides in this matter. It is to point instead to the fact that for those familiar with or working in the American Darwinian tradition the choice between natural selection and self-organization does not appear nearly as stark, nor does the argument verge so perilously on becoming a purely metaphysical one. Maynard Smith's way of conducting his dispute with his long time friend Stuart Kauffman (and indeed his reaction to Weber's and my *Darwinism Evolving*, in which Kauffman's views are summarized) intimates that he simply does not grasp that Kauffman's models of self-organizational phenomena are not hostile to natural selection, even adaptive natural selection (Maynard Smith, 1995). If so, it is probably because Maynard Smith does not see that Kauffman has inherited a whole set of evolutionary questions about natural selection and adaptabil*ity* that were never on the agenda in Britain, but, ever since Dobzhansky, have virtually defined serious Darwinian debates in the United States. Self-organization may well be a direct threat to the main line of British Darwinism. It is likely to extend, however, rather than undermine, the American wing of the Modern Synthesis.

It is true that Kauffman has long argued that self-organization will put constraints on natural selection when connectivity among the nodes of a complex system, such as ontogenetic programs or ecological communities, is relatively high. It is also true that Kauffman's early work runs almost entirely along these lines (Kauffman, 1985). The main thrust of Kauffman's *The Origins of Order*, however, is to suggest that when connectivity drops to a level at or near the "edge of chaos", natural selection will be at its most effective as an evolutionary agent because, in that region of phase space, poised between too much order and too much chaos, it selects among self-organized variation-maintaining structures that foster future evolvability, that is, the property of adaptability (Kauffman, 1993). Kauffman has even suggested that natural selection arises as a natural process precisely in systems poised on the "edge of chaos," and that its continued power as an evolutionary driver derives from its ability to keep populations tuned to this region of phase space: "Ordered systems encounter relatively correlated landscapes, chaotic systems adapt on very rugged landscapes. Landscape structure governs both evolvability and sustained fitness ... Parallel processing systems lying in this interface region between order and chaos may be the best able to adapt and evolve. Further, *natural selection* itself may be the force which pulls complex adaptive systems into this boundary region ... It seems plausible that the most complex, integrated, and also the most evolvable behavior might occur in this boundary region near the edge of chaos" (Kauffman,

1993, pp. 218-219, original emphasis).

In this view, natural selection itself has an evolutionary history. Rather than simply being a model of a natural process, it is an evolutionary phenomenon in its own right and, in the widest sense of the term, may well be a prominent part of evolutionary dynamics just because it has an evolutionary function (Depew & Weber, 1995).

The continuity with, and radicalization of, Dobzhansky's concerns in Kauffman's work is clear. Kauffman is not, however, the only person who has worked along these lines. I want in this connection to call attention to Michael Conrad's insightful 1983 book *Adaptability* and to work of his that has appeared subsequently (Conrad 1983, 1990, this volume). Conrad argues that even when degrees of connectivity between interacting nodes of complex systems are weaker than those postulated by Kauffman (indeed are almost as uncorrelated as Fisher assumes) the problem of getting trapped on a single peak in the fitness landscape — a central issue bequeathed to Dobzhansky and his students by Wright — is solved by selecting adaptability-enhancing structures that smooth out non-adaptive valleys of the sort Wright thought were passable only by genetic drift. Conrad has written: "My purpose (...) is to show that so far as evolvability is concerned the (...) structuralist [that is, constraint-centered] view is just as Darwinian, if not twice as Darwinian [as a purely adaptationist view]. The picture is that a very special class of structures is particularly amenable to evolution, and that these are themselves selected through the Darwinian mechanism of variation and selection. This special class of structures must satisfy at one and the same time two conflicting conditions. The first is that the organism be stable, that it sit in a developmental basin of attraction. This is more likely as the number of components in the organism and the number of interactions among them decreases, on the simple grounds that the chance of a valley occurring in the phase space of a system decreases with its dimensionality (May, 1973). The second condition is that the adaptive peaks corresponding to these basins of attraction should be close enough together to be connected by single genetic changes. This is more likely as the number of components and interactions increases, since the pathways between peaks in the adaptive peak space correspond to the pathways between valleys in the basin space. The only way for a system to satisfy both conditions is to have many redundant components with multiple weak interactions (...) The amenability-increasing structural features inevitably hitchhike along with the advantageous traits whose evolution they facilitate" (Conrad, 1990, p 62).

As the language of this passage suggests, Conrad is working within the ways the problem of evolution was set up by Wright and Dobzhansky. Like Kauffman, he intends to bring this inheritance to bear on problems at the interface between developmental and evolutionary biology that were first formulated by C.H. Waddington. How does evolution bring about an ontogenetic order strong enough to protect developing embryos from perturbation, but at the same time open to evolutionary innovation? Kauffman and Conrad, as well as Wimsatt, agree that the answer involves a combination of natural selection and self-assembly, although they may disagree about the proportions of each, and the precise nature of the relationship between them (Wimsatt, 1986; Conrad, 1983; Kauffman, 1991).

Two aspects of discussions of this sort must especially be borne in mind. First, from a generalized (or perhaps reconstructed) Dobzhanskyan perspective, what makes the Darwinian tradition a "progressive problem-solving research tradition" (*sensu* Laudan, 1977) is its increasing distance from treating simple morphological features (the etiology of which may in fact involve a number of non-selective factors, including constraint and chance) as paradigms of adaptive natural selection. The focus is shifted instead to the adaptive significance and etiology of fundamental life processes, such as securing greater reliability and efficiency in autocatalytic chemical cycling, the emergence of different life-cycle strategies correlated to different ways of partitioning resources in an ecological system, and the emergence of "generative locks" that find the correct ontogenetic balance between reliability and potential for, that is tolerance of, change. ("The evolution of uncertainty and of the structure of uncertainty," writes Conrad, "is just as important as the evolution of reliable complexity" (Conrad, 1984, p. 339)).

Second, in the emergence of these evolutionary phenomena, natural selection and self-organization need not be viewed as competing causes, but as complementary processes, or even aspects of a single complex process (Weber & Depew, 1996). Self-organization is undoubtedly at work in creating at least the possibility space for the modularized and hierarchical structure of the biosphere (and the architecture of the genome) of which "expanded" Darwinians like Gould speak, but whose emergence they seldom pause to explain. Yet natural selection also plays a role in fixing and maintaining these structures, and does so in several ways. It runs along lines that are pre-formed by the mathematically accessible trajectories of complex systems. It selects among self-organizational structures made accessible by such trajectories and their basins of attraction. And it holds selected systems, with their balance between fixity and flux, in place so that further evolution is possible. I cannot help but conclude that the highly disjunctive (either-or) discussions about evolutionary causality in which British selectionists and structuralists have engaged must sound almost as strange to post-Dobzhanskyan ears as the emergence of something called "adaptability" must sound to Fisherian ears.

Developmental evolutionists such as Salthe and Swenson may mildly commend ideas like these while still maintaining that even "expanded" Darwinians must fail to countenance the main theme of the evolutionary developmentalist tradition: that autonomous ontogenetic, ecological, and phylogenetic dynamics, rather than selection, are the creative forces in the hierarchical complexification of life. Natural selection can be allowed to play supplementary roles in this process, weeding out the unfit and affecting evolutionary rates (Brooks & Wiley, 1986, 1988). Its influence can even be quantified (Collier, this volume). Developmental evolutionists will brook no version of Darwinism, however, no matter how expanded, contracted, or contrite, that does not return to the integral vision of self-moving evolutionary complexification at various levels of scale (Salthe, 1993; Swenson, 1989). This is especially true at the present time. For in recent years, evolutionary developmentalists have been emboldened by the emergence of complex systems dynamics and far-from-equilibrium thermodynamics to restate their case. Why cave in when belated triumph is just around the corner?

In *Evolving Hierarchical Systems*, for example, Salthe uses a mix of hierarchy theory, systems ecology, and far-from-equilibrium thermodynamics to reaffirm isomorphisms between ontogeny, ecological succession, and (perhaps) phylogeny (Salthe, 1985). In both ecology and ontogeny, we see an average decrease over time in the intensity of energy flow (flow per unit of mass) through the system, accompanied by a corresponding increase in complicatedness (Salthe, 1985,1993). This produces greater stability in the face of perturbation. On the other hand, while immature systems can generally recover their developmental trajectories, senescent systems, precisely because of the exponential growth of complexity, grow brittle, subject to unrecoverable insult, death, and eventual recycling (Conrad (1983) characterizes evolution as the inability to forget or discard information). These regularities mean that analogies between developing organisms and maturing ecosystems, a key theme of the developmentalist tradition, remain robust. In *Development and Evolution*, Salthe completes his argument by asserting that phylogenetic diversity is like that too (Salthe, 1993). The only kinds of systems that can evolve, Salthe argues, are those that develop. But anything that develops (including in his [and Swenson's] view the entire planetary system) might also evolve. This is because, on Salthe's view, evolutionary change occurs when perturbations in developmental processes are reset by incorporating new information (Salthe, 1993). For his part, Swenson sees this process as instantiating a universal law of development-by-differentiation of the sort that was anticipated, but not explained, by Spencer's expansion of Von Baer's laws of ontogeny (Swenson, personal communication). (Charles Lyell's old argument against this view — that [nearly] as many lineages have simplified as have become more complex — is not, I concede, relevant, let alone decisive, in spite of its popularity among contemporary Darwinians. For it assumes that the units of complexity in question are species, rather than ecologically defined patterns and processes of which species are simply transient parts.)

Darwinians (of most stripes) are unlikely to accept even modest versions of such claims, in large part because these themes represent an historically touchy issue for them. Early Darwinians were more amenable to this unitary vision than later ones. Herbert Spencer and Ernst Haeckel, for example, saw evolution as a three-level isomorphism between individual development, ecological succession, and phylogenetic differentiation. Yet the subsequent history of Darwinism can be summarized in smallest compass by saying that it consisted of the progressive dismantling of this unitary, if not entirely satisfactory, vision. T.H. Morgan put developmental biology aside when he and his students founded transmission genetics. The fathers of population genetics developed a theory of evolutionary change that ratified this choice. Developmental biology was thereupon relegated to what Mayr calls "functional" biology. Its expulsion from "historical" biology meant that developmental processes were viewed as irrelevant to evolution. In virtue of their interest in such questions, workers like Waddington, who wanted to explain the evolution of development and development's role in evolution, were considered barely

Darwinian. Similarly, in the population-genetic theory of evolutionary change, energetics was considered irrelevant. Thus parallels between ontogeny and ecological succession were regarded as bogus, except for bits and pieces of community ecology, such as the Lotka-Volterra equation, that could be coopted into "population ecology." The fate of phylogenetic pattern was even more stark. It was denied that any such thing existed, and hence that anything needed to be explained in the first place. In sum, the unitary vision is the Ghost of Darwinisms Past. To this day, accordingly, informal and tacit, but strictly enforced, sociological tests of whether one is a Darwinian center far too much on acceptance of these revisionist terms. Such tests have little or nothing to do, however, with whether the Darwinian tradition itself can repossess some of what it has given up. My hope is that at least some reform-minded, anti-mechanistic Darwinians will make an attempt to do so. They might be surprised at how much common ground can be found between the expansive sort of Darwinism that I have described above and the sort of developmentalism that, instead of restricting itself to mathematical structures, important as these are, pays close attention to the ecological and thermodynamical setting of living systems.

TOWARD AN ECOLOGICAL DARWINISM

Weber and I have tried to formulate (though not to validate) a sort of "ecological Darwinism" that embeds post-Dobzhanskyan views of selection, such as those described above, in an ecological and thermodynamical matrix. Taking heart from a long line of evolutionary biochemists ranging from Alfred Lotka to Jeffrey Wicken, as well as from systems ecologists ranging from Howard and Eugene Odum to Robert Ulanowicz, we have defined living things as physically bounded, internally informed, self-replicating, autocatalytic, dissipative systems (Depew & Weber 1994, p. 467; see Lotka, 1924; Wicken, 1987; Odum & Odum, 1982). We have inferred from this definition that it is far more explanatorily fecund to treat organisms as ecological systems that have achieved higher integrity and responsiveness to perturbation by internalizing information than to treat ecosystems as superorganisms (Depew & Weber, 1995). From this vital ecological and energetic center, we descend to the level of ontogeny and ascend to the level of phylogeny. We suspect that the relevant account of the role of genes in ontogeny has been worked out by a rising school of Developmental Systems Theorists, who hold that genes are a way of stabilizing information flow, but from a causal point of view are simply one of many "resources" in a self-organizing process whose aim is the replication of life-cycles (Griffiths & Gray, 1994, p. 283; see Oyama, 1985). At the other extreme, our picture of phylogeny concedes that the record will be littered with much evolutionary wreckage. Nonetheless, it also entails that there will be a modest amount of direction and ordered differentiation in the overall shape and pattern of evolution that comes into view from the larger perspective of earth history (Depew & Weber, 1995).

Arguments for these conjectures depend on asking (as Salthe and Swenson do) whether systems whose behavior over time is well described by nonlinear dynamical models have any distinctive physical characteristics. A good answer is that many of the systems whose trajectories are well described by these mathematical models are, or are closely linked to, physical systems that maintain themselves away from the state where, under given conditions, energy can no longer be converted into work. They can do this because they are "open systems," systems that maintain their structure by pumping energy and matter into themselves, using it to grow and dissipate it in a more disordered state to their surroundings. Open systems include entities as different as dust devils, tornadoes, ecological communities, economic systems considered as ecologies, cells, and organisms. It is precisely by building internal structure that open systems succeed in dissipating energy. As Salthe points out, their tendency to maximize dissipative rates in immature stages and to minimize them as they develop is linked to their ability to build better dissipative pathways in the form of internal structures, which make them vulnerable in senescent stages (Salthe, 1985, 1993).

There is physical selection for systems that by these criteria are stable and persistent (Swenson, 1995, this volume). In our view, however, the sub-class of open systems that feed their own by-products back into the next cycle of energy use and dissipation, that is, autocatalytic systems, are selected in a different way from other systems. Whether one thinks of the resultant forms of selection as instantiating the most basic laws of physical selection under certain conditions, as Swenson does, or as the emergence of something that transcends those laws, as Weber and I have tended to do, is of less moment (once meta-theoretical problems about theory reduction are put aside) than the fact that entities with differentially superior autocatalytic powers will be favored by processes we call "chemical selection." In turn, systems that have found ways to enhance their reproductive rates by internalizing information will be favored over those that have not (Wicken, 1987; Depew & Weber, 1995; Weber & Depew, 1996). The sub-class of autocatalytic systems that find ways to retain information that guides autocatalysis, that use it to help create a new generation of things born with the competitive advantages of their progenitors, and that build novel adaptations by taking advantage of the informational variation that inherently attends such systems — these are the only systems that Weber and I are willing to call living. Only in such systems, we contend, can natural selection of the fit, as distinct from the wider class of physically and chemically selected entities, arise (Depew & Weber, 1995; Weber & Depew, 1996).

It is just here perhaps, if we understand him correctly, that Weber and I may veer off from Salthe. Salthe's phenomenological rules for open systems, together with his insightful study of scalar hierarchies, help him to restore the parallels among ontogeny, ecological succession, and phylogenetic differentiation that are deeply entrenched aspects of the developmental tradition. At the same time, they may lead him to blur the fairly sharp line that Weber and I draw between living systems and other dissipative structures. Salthe's stress on the fact that organisms and communities, and indeed the whole earth system, obey similar energetic trajectories minimizes differences between these distinct sorts of entities, even when it is conceded that organisms are more well bounded and integrated than communities. This in turn leads Salthe to screen off the notion of fitness, on reformed versions of which we, together with Kauffman and Conrad, continue to rely, as well as the elements of contingency and opportunism that come with it. We suspect that in this matter Swenson is on Salthe's side. As a reaction against orthodox Darwinism's reductionist stress on pairwise competing organisms, and its neglect of the thermodynamic context that makes dyadic competitions of this sort so artificial and ideologically suspect, Salthe's and Swenson's move is perfectly understandable. But to argue that natural selection has little to do with creative evolution is to pay too great a compliment to the defective conceptions of natural selection from which we have been trying to dissociate ourselves, and so is insufficient (yet) to sustain the general conclusion that Darwinism is inherently too mechanistic to account for the history of life.

That bounded, informed, autocatalytic, dissipative systems are by definition parts of ecological communities, and that the information which they store and use is subject to dynamics that are inseparably both competitive and cooperative, are facts that Darwinians have ignored at their peril. Their neglect has led genetic Darwinians to overlook what Weber and I take to be a most salient conjecture: that with the appearance of informational macromolecules natural selection of the fit would predictably have emerged from chemical selection for the energetically efficient (Depew & Weber, 1995). This failure has in turn inclined Darwinians to advance from the true claim that the components of fitness cannot be reduced to energetics to the false conclusion that the role played by populations in transducing energy flow through ecological systems constitutes no part of their fitness. The consequent stress on individual advantage to the exclusion of systemic cooperation that has characterized many genetical Darwinians is a direct consequence of the fateful decision to free Darwinism from an ecological and energetic context in which it once flourished, and back toward which it is slowing groping. It comes as no surprise to me that Kauffman and Conrad assert confidently that all evolution is co-evolution, and that co-evolution is an inherently, and irreducibly, ecological phenomenon (Conrad, 1983; Kauffman, 1993). For natural selection can play the deep, essential, and above all creative roles suggested by their theories only when organisms are treated ecologically.

Department of Philosophy, University of Iowa
Iowa, USA

REFERENCES

Brooks, D.R.; Wiley E.O., 1986, rev. 1988, *Evolution as Entropy: Toward a Unified Theory of Biology*, Chicago, University of Chicago Press.

Conrad, M., 1983, *Adaptability*, New York, Plenum Press.

Conrad, M., 1990, The geometry of evolution, *Biosystems*, vol. 24, pp. 61-81.

Depew, D.J.; Weber B.H., 1995, *Darwinism Evolving: Systems Dynamics and the Genealogy of Natural Selection*, Cambridge MA, Bradford, The MIT Press.

Dobzhansky, T., 1937, *Genetics and the Origin of Species*, New York, Columbia University Press.

Dobzhansky, T., 1962, *Mankind Evolving*, New Haven, Yale University Press.

Goodwin, B., 1994, *How the Leopard Changed Its Spots: The Evolution of Complexity*, New York, Charles Scribner's Sons.

Gould, S.J., 1982, Darwinism and the expansion of evolutionary theory, *Science*, vol. 216, pp. 380-387.

Gould, S.J., 1989, *Wonderful Life: The Burgess Shale and the Nature of History*, New York, Norton.

Gould, S.J.; Lewontin R.C., 1979, The spandrels of San Marco and the panglossian paradigm: A critique of the adaptationist paradigm, *Proceedings of the Royal Society London*, B, vol. 205, pp. 581-598.

Griffiths, P.; Gray P;, 1994, Developmental systems and evolutionary explanation, *Journal of Philosophy*, vol. 91, pp. 277-304.

Kauffman, S.A., 1985, Self-organization, selective adaptation, and its limits, in D.J. Depew & B.H. Weber, (eds.), *Evolution at a Crossroads: The New Biology and the New Philosophy of Science*, pp. 169-207, Cambridge, MA, Bradford Books, MIT Press.

Kauffman, S.A., 1993, *Origins of Order: Self-Organization and Selection in Evolution*, New York and Oxford, Oxford University Press.

Laudan, L., 1977, *Progress and Its Problems: Toward a Theory of Scientific Growth*, Berkeley and Los Angeles, University of California Press.

Lewontin, R.C., 1970, The units of selection, *An. Rev. of Ecology and Systematics*, vol. 1, pp. 1-18.

Lewontin, R.C., 1974, *The Genetic Basis of Evolutionary Change*, New York, Columbia UP.

Lewontin, R.C., 1991, Facts and the factitious in natural science, *Critical Inquiry*, vol. 18, pp. 140-153.

Levins, R.; Lewontin, R.C., 1985, *The Dialectical Biologist*, Cambridge, MA, Harvard University Press.

Lotka, A.J., 1924, *Elements of Physical Biology*, Baltimore, Williams and Wilkins, Reprinted as *Elements of Mathematical Biology*, 1956, New York, Dover.

May, R.M., 1973, *Stability and Complexity in Model Ecosystems*, Princeton, Princeton University Press.

Odum, H.T; Odum, E., 1992, *Energy Basis for Man and Nature*, 2nd ed., New York, McGraw Hill.

Oyama, S., 1985, *The Ontogeny of Information*, Cambridge, Cambridge University Press.

Salthe, S.N., 1985, *Evolving Hierarchical Systems*, New York, Columbia University Press.

Salthe, S.N., 1993, *Development and Evolution: Complexity and Change in Biology*, Cambridge MA, Bradford, MIT Press.

Smith, M.J., 1995, Life at the edge of chaos, *New York Review of Books*, March 2, 1995, pp. 28-30.

Swenson, R., 1989, Emergent attractors and the law of maximum entropy production: Foundations to a theory of general evolution, *Systems Research*, vol. 6, pp. 187-197.

Swenson, R., 1995, *Spontaneous Order, Evolution and Natural Law: An Introduction to the Physical Basis for an Ecological Psychology*, Hillsdale, NJ, Lawrence Erlbaum Associates.

Thompson, D.W., 1917 (rev. ed. 1942), *Growth and Form*, Cambridge, Cambridge University Press.

Ulanowicz, R.E., 1986, *Growth and Development: Ecosystems Phenomenology*, New York, Springer-Verlag.

Weber, B.H.; Depew, D., 1996, Self-organization and natural selection, *Biology and Philosophy*, vol. 11, pp. 33-65.

Wicken, J.S., 1987, *Evolution, Thermodynamics and Information: Extending the Darwinian Program*, New York and Oxford, Oxford University Press.

Wimsatt, W., 1986, Developmental constraints, generative entrenchment, and the innate-acquired distinction, in *Integrating Scientific Disciplines*, W. Bechtel (ed.), pp. 185-208.

MICHAEL CONRAD

TOWARDS HIGH EVOLVABILITY DYNAMICS

INTRODUCTION

Some organizations are more friendly to evolution than others. Biological systems, since they are the products of evolution, must fall into the friendly class. Computer programs are at the opposite extreme. They are too fragile, at the level of base code, for random changes to produce acceptable new code with any reasonable degree of probability.

My purpose in this paper is to address this difference, and thereby to elucidate the characteristics that confer evolvability on biological organizations (Conrad, 1972, 1979a, 1979b, 1983, 1990a). This issue has come to the fore in recent years in connection with debates over the relative importance of developmental constraints and natural selection (Goodwin, 1985; Kauffman, 1993; Salthe, 1993; Depew & Weber, 1995; Weber & Depew, 1995). The main points to be developed in the present paper may be summarized in the following four statements:

1. Self-organizing (non-programmable) dynamics are prerequisite for evolution friendliness.
2. The dynamics, to be evolution friendly, must allow for topological distortability and for redundancies that buffer the effect of structural changes on system function (e.g. buffer the effect of genotypic changes on the phenotype).
3. As the complexity of the dynamics increases the evolution friendliness increases, due to the fact that the probability of isolated adaptive peaks decreases and the probability of saddle points increases.
4. The evolution friendliness of organizations is itself largely a result of the evolutionary process, i.e. evolution facilitates itself.

The implication is that physical self-organizing dynamics and high evolutionary adaptability go hand in hand. In order for a system to have a high degree of evolutionary adaptability (through variation and selection) it must have a high capacity for self-organization based on the interactions among its components. In order for a system to develop such a high capacity for physical self-organization it must have a high degree of evolutionary adaptability.

SELF-ORGANIZING DYNAMICS DEFINED

First let us consider more carefully what it means for a system to have self-organizing dynamics. A number of types of self-organization are possible: self-organization driven by energy minimization (e.g. crystallization), self-organization driven by entropy production (e.g. dissipative structures); procedural or algorithmic self-organization; and self-organization guided by

G. Van de Vijver et al. (eds.), Evolutionary Systems, 33–43.

reproduction, variation, and selection.

We will refer to the first two types (energy and entropy driven) as physical self-organization. The essential feature of physical self-organizing systems is that they have non-programmable structure-function relations. The rule (or map) describing the time development of the system is not encoded in the states of the components and in the connections among them according to a simple user's manual. We will say that such systems are structurally non-programmable.

The relation between structural programmability and physical self-organization can be understood by considering the extreme opposite case: digital computers. The program, or map, that generates the state to state transitions in this case is encoded in the states of the components and connections among them according to a simple user manual. If a digital computer had self-organizing dynamic properties at the level of the base machine it would have a "mind" of its own so to speak. It would be impossible to program it in the sense of prescriptively imparting desired rules to it using a simple (finite) user manual. We would have to tune it, cajole it, negotiate with it, just as we do with a horse, a plant, or another human being. Of course a digital computer could be used to simulate physical self-organizing dynamics. The structure-function relations are then non-programmable at the level of the simulation, but not at the level of the base machine.

DEPENDENCE OF EVOLVABILITY ON SELF-ORGANIZATION

The first major claim is that physical self-organizing dynamics are necessary in order for evolution to occur. The argument is roughly as follows (for details see Conrad, 1978, 1988, 1990b).

1. Evolution can occur on an acceptable time scale only if it is possible for reasonably probable genetic changes to yield acceptable phenotypic changes. Reasonably probable means that two or more genetic steps (mutations, cross-overs, recombinations, ...) should not be simultaneously required to move from one phenotype to another. This does not mean that multiple genetic changes are precluded — for a big enough genome they are inevitable. The crucial point is that if specific multiple genetic alterations are required, the rate of evolution will then scale as a product of probabilities. This always leads to emergence times that are far longer than the historical times available.

2. Now consider a system without physical self-organizing dynamics, that is, a structurally programmable system. For the system to be eligible for significant evolutionary change we require that it be possible to produce many possible functions by making structural alterations. Digital computers are again the example. When we change the structure of such a system we change the rule (map or program) that generates its behavior. In general it is not possible to put any metric on the amount of function change that results from such a structure change. This is tantamount to an unsolvable problem, known as the halting problem.

More specifically, let us suppose that the program encoded in the structure is such that the system never comes to a halt. There is in general no way of telling, apart from running the system and waiting to find out, whether a mutated version of the system will also continue to run indefinitely, or whether it will go to a "dead" state. If it is not even possible to put a metric on such a dramatic change in behavior it certainly would be impossible to put a metric on more graded changes in behavior.

The conclusion, which corresponds to experience, is that computer programs are fragile. In general single alterations in the structure of the program do not produce acceptable computational functions. Thus, in general systems with programmable structure-function relations (i.e. without self-organizing properties) will fail the basic requirement that must be satisfied for evolution to occur on an acceptable time scale.

3. Now suppose that the system has non-programmable structure-function relations, and that physical self-organizing dynamics are operative. When we change the structure of a such a system we can assume that we change the rule (or map) that describes its state to state behavior. But in this case we can't in general ascertain what the changed rule is without doing experiments or performing calculations. In the case of the digital computer we may not know what the rule will bring; in the case of the organism we do not even know what rule is brought into play.

The importance of this point is that the argument in 2. above — that computer programs are too fragile to allow for evolution by variation and natural selection — no longer applies. The structure of the system does not encode the program, and consequently changing the structure of the system does not mean changing the map that describes its behavior in an arbitrary way. For example, changing the structure may produce an isomorphic or homomorphic version of the system, i.e. essentially a different instantiation of the same map. A mouse and an elephant, for example, might be essentially topological distortions of the same developmental plan. Graded changes in functional performance in response to likely genetic changes are therefore supportable. Organizations of this type would be evolution friendly.

As noted in the previous section it is possible to simulate non-programmable structure-function relations with digital computers. It is conceivable, though by no means certain, that all types of physical self-organizing dynamics are in principle simulatable to an adequate degree of approximation by digital computers given enough time and space resources. But this just means that the issue of computational efficiency must be added to the tradeoff relationship between evolutionary adaptability and structural programmability. The correct statement is that structurally programmable systems allow for significant evolution to occur on an acceptable time scale only to the extent that computational resources are expended to simulate non-programmable structure-function relations (Conrad, 1988). In practice the computational expenditure required to simulate biological structure-function relations is much too great to fully duplicate the evolutionary capacities of biological systems. (For the record I should add that I do not in fact believe that all natural processes are simulatable by digital computers, but it is unnecessary to consider this issue here.)

MUTATION-BUFFERING PARADIGM

Now let us consider what characteristics the dynamics must have in order to afford evolution-friendliness. The first point to note is that self-organization does not by itself ensure this. Thus a crystal self-organizes as it grows, due primarily to energy minimization. But in many cases crystal growth is extremely sensitive to defects, which may lead to radical changes in the growth pattern. But the rule that governs the growth process in these cases is inappropriate for gradual transformation of form.

Kauffman type Boolean networks (Kauffman, 1993) provide another example that is not very amenable to evolution. These are collections of elements, each of whose state transitions are controlled in a highly specific way by the states of other elements in the collection (usually a small number of other elements). Such networks tend to form attractors that are isolated from each other, somewhat like isolated peaks on a multi-dimensional mountain range.

By contrast, let us consider a system of balls and springs, with balls of different sizes and springs of different strengths (Conrad, 1979a). For simplicity we can imagine that the different balls represent different amino acids, strong springs represent covalent bonds, and weak springs represent the various weak interactions among the amino acids that lead to the folded shape. The different arrangements of the balls represent different shapes. We can suppose that some of the spatial relations are particularly crucial for some function, while others are not. In a protein the function might be recognition, catalysis, binding, or some control function.

Mutation corresponds to replacing a ball by one of different size, deleting a ball, or adding a ball. If there are a large number of balls and springs in the system the effects of such mutations will be distributed over the whole network. A significant fraction of the mutations are likely to produce only slight alterations in the critical spatial relations. This will of course depend on the specific arrangement of balls and springs and on the specific mutation. If a small ball at a crucial site is replaced by a large ball the transformation of form certainly will not be gradual. But if balls are available in a graduated series of sizes they will be more replaceable. And, if there are mechanistically superfluous regions of the network that can serve to absorb some of the effects of the mutation and hence buffer the effect of the mutation on the critical spatial relations, the transformation of form is likely to be acceptable.

Mechanistically superfluous regions and availability of balls with a high degree of replaceability (like amino acids with similar hydrophobicities) serve to buffer the effects of the mutation, and hence to increase the graceful evolutionary transformability of the structure. The opposite type of effect is also possible; mutations at certain sites, or replacement with very different size balls allow for bigger evolutionary jumps. Such larger jumps may serve to explore the space of possibilities in the large, somewhat like the gross control on a microscope. The smaller ones serve to explore in more detail, somewhat like the fine control. But this analogy should be treated with caution. Large jumps are likely to preclude any useful function, especially when the system becomes complex. The smaller, gradual jumps have a much greater likelihood of being acceptable, and therefore of allowing a large space of possibilities to be explored without requiring simultaneous mutations.

Now let us consider what distinguishes the balls-and-springs abstraction from crystal growth on the one hand and from the Boolean abstraction on the other. The main point in both cases is the role of interactions with different strengths. The balls and springs model converges to a definite form as long as the effects of balls on the springs and the effect of the springs on the balls form a self-consistent field. The collective effect of multiple weak interactions (the weak springs) plays the crucial role. When a system grows according to an addition rule that is highly history-dependent this collective aspect is much less important. The dynamics are therefore much more likely to be divergent. Of course, both types of processes play a role in biology. Evolution as a whole has both divergent and convergent aspects. The decomposition of an enzyme-substrate complex is a divergence process. The important point is that the two types of dynamics are woven together in a way that allows mutation-buffering, and therefore, that allows genetic variations to have a good chance of yielding acceptable phenotypic variations.

The Boolean type abstraction is clearly different from the balls-and-springs model in that it completely fails to distinguish strong and weak interactions. Since the models are entirely logical it is not even possible to put a metric on the difference between the states of a component. In the case of the balls and springs model we can quantify such differences by specifying their sizes, average locations, and motions. Any abstraction that fails to represent these features would be much too limited to provide useful intuitions about evolutionary transformability (however useful it may be for other purposes).

EXTENSION TO DEVELOPMENTAL ARCHITECTURES

Let us now consider how the balls and springs paradigm can be used as a metaphor for multi-cellular development. Collective and serial growth effects become mixed in this more complicated case. The problem is to add growth rules in a way that allows for high evolvability. The idea, already alluded to, is that mutation-buffering allows for the generation of many alternative realizations of essentially the same growth rule (i.e. images that are homomorphic so far as the syntactics of development are concerned).

We can suppose that the balls are now cells and the springs now correspond to affinities between different types of cells. We should also imagine a rule for duplication of balls, differentiation into different types in response to the milieu, and migratory movements. The new feature is that the number of possible variations on each ball, or cell, is much larger than the rather small number of variant amino acids (limited to twenty), and even larger than the rather enormous number of variations possible at the protein level.

In the case of proteins the genotypic description consists of a specification of the linear sequence of amino acids. This is captured in the balls and springs analogy by the sequence of balls as connected by strong springs (analogous to covalent bonds). The sequence of events in folding is implicit in this initial specification for proteins, since the weak bonds are implicit in the various weak interactions among the amino acids (including van der Waal's interactions, hydrogen bonds, coordination bonds, and so forth). The genotypic specification of a multi-cellular system is more complicated. The structural genes specify the proteins. But any given trait may be affected by many different genes (fan in) and any given gene may affect multiple traits (fan out or pleiotropy). Redundant genes may have the same effect, allowing for gradual variation in the production of some enzyme, or regulatory genes may be available that control the extent to which structural genes are read out, or whether they are activated at all. Many other regulatory mechanisms of a non-genetic nature (such as biochemical control mechanisms dependent on the milieu sensitivity of enzymes) are also available.

In the multi-cellular situation the balls and springs should thus be pictured as much more plastic and adaptive than in the protein example. Plasticity means that the specifications of the balls and of their number may be varied in a graded manner through single changes in a (genotypic) description. Adaptivity means that the specifications may change in a graded manner the effect of environmental influences. Plasticity and adaptivity together allow the whole system to form a self-consistent field that allows for plastic variations in the component characteristics and at the same time preserves the sequential aspect of the growth rule.

Thus suppose that the genotypic description is altered, and that this alters the specifications of the cells. This will in turn alter the growth rule, i.e. when and how cells differentiate or migrate. At any given stage the form of the system will largely be determined by the forces the cells exert on each other and at any given stage the forces exerted change as a result of influences from other cells (i.e. how the cell responds to the various forces and signals impinging on it). If changes in the genotypic description that affect the components must be simultaneously accompanied by changes that preserve the growth rule the basic condition that must be satisfied for evolution to occur on an acceptable time scale will fail, since it is then most likely that the form of the system will be unacceptably unstable to slight variations in the components. Here is where the adaptivity of the cells enters. If the genotypic alteration that affects the components also perturbs the growth rule to some extent the cells will be exposed to alterations in milieu that are tantamount to error signals. Their internal structural organization will be altered, thereby altering their dynamic responses. If the altered responses serve to correct the deviation in the growth rule the system will converge; if not the system will diverge, most probably leading to unacceptable consequences. In short, the growth rule is brought into the framework of the self-consistency dynamics.

At this point we will have obtained a homomorphic image of the original system, since a given growth rule will have been realized with different components. The question is, how much variation is possible in such a self-consistency scheme? Might not the scheme lock in the components in a rather tight way, allowing only for a very small degree of variation?

Here is where the buffering mechanisms discussed above play a key role. If the variation of the components is acceptable and useful it will become fixed in the population. A subsequent genetic change can then modify the components so that they preserve the growth rule without any significant error. This sets the stage for another variation of the components. The process will continue until some point in the evolution is reached where the system becomes locked in. Exploiting a new environment or keeping up with a changing environment will then be impossible due to the fact that all the pathways for evolution require two or more simultaneous mutations. However, it is always possible to introduce extra redundancies at this point that allow for more graded changes. The evolution will occur in a larger number of genotypic steps, each of which is associated with a smaller degree of phenotypic change, but it will occur inordinately faster than if multiple genetic changes are specifically required to occur simultaneously (for specific calculations demonstrating this point see Conrad, 1972, 1978, 1979b).

Such extra redundancies are an expense to the individual organism, apart from the contribution they may make to the reliability of function (assuming such extra reliability more of an advantage than a cost). For the sake of argument let us suppose that the extra redundancy has no value to the individual organism. It will still hitchhike along with the advantageous traits whose appearance it facilitates (Conrad, 1982, 1983). In this way mutation-buffering redundancies will accumulate in the course of evolution, somewhat in the fashion of parasitic DNA. The difference is that the accumulation in this case has an evolution facilitating effect. In short, evolvability evolves (point 3 of the introduction).

We cannot here review the vast array of biological mechanisms that provide for self-consistency dynamics and its buffering. Redundant genes are the simplest example. By varying the number of genes it is then possible to vary a gene product in a graded way. A greater or less quantity of a particular hormone might be produced. Any given property affected by multiple genes in this way can be varied more gradually. If cells in different tissues or organs respond in coherent ways to changes in the number of such redundant genes, or in the activity of the enzyme produced by any one of them, gradual allometric transformations will be possible (i.e. the gene will have coherent pleiotrophic effects). Suppose that the different cells do not respond in a coherent way — for example, they might all have similar responses, but it might happen that similar responses are functionally unacceptable. An error signal will be produced that disturbs the internal dynamics of the cells. The response will change. If it changes in an appropriate way the pleiotrophic effects will be coherent. Subsequent genetic changes may then alter the system to allow for preservation of the growth rule with a smaller degree of error, thereby setting the stage for further changes in the characteristics of the components.

Many other examples could be considered. On the error correction side we can include all the adaptive control mechanisms. The structural organization of cells and their interaction with dynamic processes probably plays an especially important role. Redundant features at all levels contribute to buffering. Non-coding regions of DNA can serve to buffer the effect of structural and regulatory gene changes on each other (Conrad, 1985). Redundant features of biochemical reaction networks, such as so-called futile cycles (Sel'kov, 1975) and functionally overlapping enzymes, also serve as mutation-buffers. Similar types of redundancies are characteristic of hormonal signaling systems (for a more extensive discussion see Conrad, 1983).

The reverse picture is also possible. A genotypic change might alter the growth rule, thereby generating a very different form. The change would be acceptable only if the adaptive properties of the cells allowed them to change their input-output behavior in a way consistent with the new form. Subsequent genetic changes affecting the properties of the cells could then serve to reduce the amount of error feedback they would have to experience as a result of the change in the growth rule, thereby opening up further possibilities for alterations in this rule. Our framework suggests that this type of "self-consistent divergence" plays an important role in the origin of the different phyla and in other macro-evolutionary processes.

STRUCTURE OF THE ADAPTIVE SURFACE

Now let us rephrase the mutation-buffering picture in the language of adaptive surfaces. The image, which runs back to Sewall Wright (1932), is that of a multi-dimensional surface that plots fitness against genetic properties. It is also possible to construct molecular versions of such landscapes, where the genetic axes represent DNA sequences (Conrad, 1979b, 1983). The whole concept of fitness is not, in the present author's view, well defined (especially when defined in terms of genes given to the next generation). For the present purposes we will regard it as a construct which, if used with due cognizance of its limitations, can lead to useful insights.

Recall our criterion for evolvability: particular genetic variations must not be required to occur simultaneously if evolution is to occur on an acceptable time scale. This means that the region of the adaptive surface occupied by evolving systems must be dominated by saddle points rather than isolated peaks. If the peaks are separated by rather shallow valleys this might not be so important, since genetic variants that fall into the valleys are then likely to be acceptable (less fit, but not lethal). If the valleys are deep or wide evolution will not occur. If peaks merge into each other during the dynamic changes in the surface as it develops in time (it should not be thought of as static) pathways between them will appear and disappear. If the peaks are really saddle points when all dimensions are taken into account, then upward running pathways are sure to be available. If the peaks are really saddle points in most dimensions then it will be relatively easy to find these pathways, or at least to keep pace with a changing landscape.

Now suppose that we increase the dimensionality of the adaptive surface by adding mutation-buffering redundancies. If the surface has a random topography the chance of isolated peaks decreases as the number of dimensions increases, while the chance of saddle points increases. If the dimensionality is high enough a significant proportion of likely genetic variations will be acceptable. This actually corresponds to the facts. Most recombinations are acceptable. The vast amount of neutral or quasi-neutral variation associated with protein evolution is a concomitant to the saddle point feature. If the saddle point property is associated with buffering many of the genetic variations that occur will be associated with gradualistic phenotypic changes. The existence of neutral phenomena is thus a concomitant to the essential requirement for the efficacy of natural selection, namely, that evolution can proceed through the stepwise accumulation of genetic changes.

The self-facilitating aspect of evolution can also be formulated in the language of adaptive surfaces. It simply means that biological systems inevitably move to regions of the adaptive surface that are most friendly to evolution. They do so by accumulating redundancies through the hitchhiking mechanism. The opposite process could conceivably also occur. In principle systems could lock in adaptations by dispensing with redundant features, or by converting them to functionally significant features. The probability of discovering pathways that lead away from a given high point in the space would then decrease. Some proteins that are highly conserved in evolution, such as the histones, may have evolved initially in a high dimensional space and then locked in their essential functional attributes by dispensing with evolvability conferring redundancies.

CONCLUSIONS

We can now summarize the main characteristics of an evolution friendly organization. First it must have self-organizing dynamics, otherwise it will be overly sensitive to structural variations. Second the self-organizing dynamics must be hierarchically organized. For multi-cellular organisms this means that some of the contributions must come from intracellular interactions and others from intercellular interactions. This allows for a developmental dynamics of the self-consistency type. The intracellular dynamics adapts to changes that affect the intercellular interactions and conversely. This makes it possible to alter the characteristics of the two levels somewhat independently, in effect separating the evolution of the component characteristics from the evolution of the growth rule. The third feature is mutation-buffering. Redundant components and multiple weak interactions serve to buffer the effect of genetic variations on critical aspects of form and function, thereby increasing the number of feasible pathways for evolutionary development. Weak interactions are essential for such buffering; if all interactions were strong genetic variations would be restricted to having either no effect or major effects.

The same principles apply to ecosystems, except that basins of attraction replace adaptive peaks (Conrad, 1995). They should also apply to pre-biological systems that are incapable of supporting reproduction per se. All that is necessary is that components be present that are modifiable and that error signals from inter-component interactions induce the modifications. Systems that randomly accumulate some redundant features are more likely to discover a stronger basin of attraction; at the same time the basin is less likely to be a basin in all dimensions, and therefore less likely to preclude further evolution.

Evolutionary processes in all these cases have two faces. The first is a search procedure, the main one in biology being search through reproduction, variation, and selection. The second is the material organization on which this search procedure acts. Our analysis demonstrates that the two faces are inseparable. The idea that variation and selection can mold passive systems to produce arbitrary biological forms and functions is incompatible with the manifest fact that biological organizations are not passive, with the fact that passive systems are too fragile to exhibit evolutionary transformations at all comparable to those that occur in biology, and incompatible with classification (which reflects homomorphisms). The idea that physical self-organization is the main determinant and that variation-selection search just does some minor tuning is incompatible with the immense variety of biological forms and with the fact that these forms so frequently defy unique classification schemes. The real situation, according to our analysis, is that self-organizing dynamics become increasingly evolution amenable in the course of evolution and therefore increasingly malleable; there are no archetypes that are not themselves subject to evolution.

ACKNOWLEDGMENT

This material is based upon work supported by the National Science Foundation under Grant No. ECS-9409780.

Department of Computer Science, Wayne State University
Detroit, Michigan, USA

REFERENCES

Conrad, M., 1972, Information Processing in Molecular Systems, *Currents in Modern Biology* (now *BioSystems*), vol. 5, no. 1, pp. 1-14.

Conrad, M., 1978, Evolution of the Adaptive Landscape, in *Theoretical Approaches to Complex Systems*, R. Heim & G. Palm (eds.), Heidelberg, Springer, pp. 147-169.

Conrad, M., 1979a, Mutation-absorption Model of the Enzyme, *Bull. Math. Biol.*, vol. 41, pp. 387-405.

Conrad, M., 1979b, Bootstrapping on the Adaptive Landscape, *Biosystems*, vol. 11, nos. 2-3., pp. 167-180.

Conrad, M., 1982, Natural Selection and the Evolution of Neutralism, *BioSystems*, vol. 15, pp. 83-85.

Conrad, M., 1983, *Adaptability*, New York, Plenum.

Conrad, M., 1985, The Mutation-Buffering Concept of Biomolecular Structure, *J. of Biosciences*, vol. 8, pp. 669-679.

Conrad, M., 1988, The Price of Programmability, in *The Universal Turing Machine: A Fifty Year Survey*, R. Herken (ed.), Oxford, Oxford University Press, pp. 285-307.

Conrad, M., 1990a, The Geometry of Evolution, *BioSystems*, vol. 24, pp. 61-81.

Conrad, M. 1990b, Molecular Computing, in *Advances in Computers,* vol. 31, M.C. Yovits (ed.), San Diego, Academic Press, pp. 235-324.

Conrad, M., 1995, The Ecosystem as an Existential Computer, in *Complex Ecology*, B.C. Patten & S. Jørgensen (eds.), Englewood Cliffs, NJ, Prentice-Hall, pp. 609-622.

Goodwin, B.C.,1985, Developing Organisms as Self-Organizing Fields, in *Mathematical Essays on Growth and the Emergence of Form,* P.L. Antonelli (ed.), Edmonton, Canada, University of Alberta Press, pp. 185-200.

Kauffman, S.A., 1993, *Origins of Order: Self-Organization and Selection in Evolution,* Oxford, U.K., Oxford University Press.

Depew, D.J.; Weber, B.H., 1995, *Darwinism Evolving: Systems Dynamics and the Geneology of Natural Selection,* Cambridge, MA, Bradford/The MIT Press.

Salthe, S.N., 1993, *Development and Evolution: Complexity and Change in Biology*, Cambridge, MA, Bradford/The MIT Press.

Sel'kov, E.E., 1975, Stabilization of Energy Charge, Generation of Oscillations and Multiple Steady States in Energy Metabolism as a Result of Purely Stoichiometric Regulation, *Eur. J. Biochem.*, vol. 59, pp. 151-157.

Weber, B.H.; Depew, D.J., 1995, Natural Selection and Self-Organization, *Biology and Philosophy*, vol. 4, pp. 1-33.

Wright, S., 1932, The Roles of Mutation, Inbreeding, Cross-Breeding, and Selection in Evolution, *Proc. Sixth Int. Cong. Genet.*, vol. 1, pp. 356-366.

ERIC MINCH

THE BEGINNING OF THE END: ON THE ORIGIN OF FINAL CAUSE

INTRODUCTION

Over the last decade or so we have seen a proliferation of models and simulations, largely inspired by biological evolution and cognitive development, in which complex behavior emerges from the local interactions of large populations of simple units. This includes Wolfram's revival of work based on cellular automata (Wolfram, 1986), the promotion of the neural network paradigm by the Parallel Distributed Processing Research Group (Rumelhart et al., 1986), Holland's genetic algorithms (Holland, 1975) and classifier systems (Holland, 1986), the logic nets and catalytic reaction systems investigated by Kauffman (1993), and many variations and combinations of these. This work has been done under the various designations of machine learning, artificial life, complexity studies, and evolutionary (or emergent) computation.

It is exciting that the computational sciences have developed sufficiently to support such ambitious efforts, and encouraging that the approaches are avowedly anti-reductionist and emergentist in disposition. A good many of these models exhib-it surprisingly complex and/or useful behavior. Still, those of us whose main interest is in understanding the essential nature of biological or cognitive systems may feel that they have yet to demonstrate truly emergent properties or *de novo* capabilities. Some critics (Rosen, 1991; Kampis, 1991; Cariani, 1991; Penrose, 1989) even contend that the formal basis of these paradigms precludes their ever doing so.

With respect to their evolutionary and developmental capabilities, my own appraisal was that most of these models incorporate too much of the desired systems' structure into the architecture. In particular, it is quite common to find in them two or more disjoint dynamics and two or more levels of entities represented. The genetic algorithm, for instance, applies to its chromosomes various kinds of noise and distortion in the form of genetic operators: mutation, inversion, crossover, etc. The phenotype of a given chromosome is then computed according to a second dynamics. And yet a third dynamics is used to evaluate the resulting structures according to a performance function. In neural networks we also find three disjoint dynamics: the behavioral dynamics (the activation rules governing changes in neuronal activation levels), structural dynamics (the learning rules governing changes in synaptic weights), and environmental dynamics (the performance rule governing feedback on actions to new inputs). In cellular automata, and in cellular automaton models of reaction-diffusion systems, the boundaries and neighborhoods of cells are fixed. And in ecological models we may find species, individuals, and aspects of their metabolism all hard-coded into the simulation, all with their own behavioral rules. The problem here is that structural distinctions embedded in the model's architecture are immutable. An architectural distinction between organism and environment precludes the emergence of new types of organism. An architectural distinction

G. Van de Vijver et al. (eds.), Evolutionary Systems, 45–58.

between dynamics at different levels precludes the emergence of higher levels of dynamics.

Be that as it may, it is not my intention here to discuss the architecture problem or to convince anyone of this assessment. These matters have been argued elsewhere (Minch, 1988, 1992). I mention them only as motivation for the simulation study to which they led, and whose results and implications I wish now to discuss.

THE OBSERVER'S PERSPECTIVE

In order to avoid the architecture problem, I designed a simulation (Minch, 1988) with a fixed set of primitives with fixed characteristics, interacting according to a single dynamics. This in itself is nothing new; there are many such simulations, from Conway's Game of Life (1982) to lattice gas models of hydrodynamics. A flat, non-hierarchical view of the model universe, though, only lets us see translational changes, i.e. the motion of primitives or particles. We can't see any function, goals, models, life, etc. (unless these are installed as primitives of the universe, in which case they cannot evolve). Functional properties, then, can only arise from an interpretation, in which some aspects are aggregated into an organism object or a mind object, etc., whose characteristics emerge when viewed as an aggregate. So the novel aspect of this simulation was an interpreter which monitored the activity in the simulation, constructed a hierarchical model, and reported the emergence of higher level objects.

How does one go about deriving a hierarchical model from a flat, non-hierarchical view? Allen and Starr (1982) have provided an admirable treatment of the methodology of developing hierarchical models, and of the complexity of behavior to be observed from them, and Pattee (1973) has tackled the very same problem we are now addressing, in terms of hierarchical control. Objects at a given level may be composed of objects at lower levels; objects at different levels may be loosely coupled through yet higher levels. The interpretation which provides access to higher level phenomena must be capable of handling this tangled network of strong and weak interactions.

The solution arises from the fact that, within this universe, there exist a fixed number of types of primitive objects (atoms), each having fixed characteristics, i.e. they have no internal structure, only properties. The atoms are subject to conservation, so they can neither be created nor destroyed. Accordingly, the only way in which the atoms can interact with each other is through combination, or, if already in combination, through separation. In the particular simulated chemistry of this universe, this appears as the joining and splitting of strings. Some of these interactions, like elastic collision, are transitory and do not affect the participants, while some result in more or less stable combinations of atoms (molecules). All possible molecules can occur, but the expected durations — both the expected time until occurrence of a molecule and the expected stability of a molecule — can range over many orders of magnitude. The state of this universe at any instant of time can then be specified as the distribution of quantities of each type of molecule. Since the molecules are continually interacting, however, an augmented representation of the universe can be constructed by monitoring the changes in quantity and deriving from these a weighted, directed graph of the interactions which have occurred, in which the arc weights correspond to the frequency with which a given transformation has been observed. This reaction graph is a form of the classical Newtonian or dynamical systems theoretic view of the system as a set of first order differential equations.

The *constructive aliasing* method I used for interpretation relies on this reaction graph. It involves detecting the strongly-connected partial subgraphs (the *components*) of transformations among molecular objects, characterizing them by their relaxation times, and assigning each component (along with, inevitably, its component objects) to a *level* according to its relaxation time. Each level then includes a collection of components, each of which reaches equilibrium instantaneously relative to higher levels, and each of which is composed of strongly-connected transformations among objects at lower levels.

One prototypical example of emergent function is enzymatic catalysis. Catalytic function is not primitive to a reaction network, as Farmer et al. (1986) recognize, but rather "(...) there is no clear separation of the molecular species into products, enzymes, and substrates: A given species can play all three of these roles in different reactions". In the hierarchical interpretation provided by constructive aliasing, enzymatic catalysis and higher levels of control appear as components, as shown in Figure 1b.

1a: flat view

1b: hierarchical view

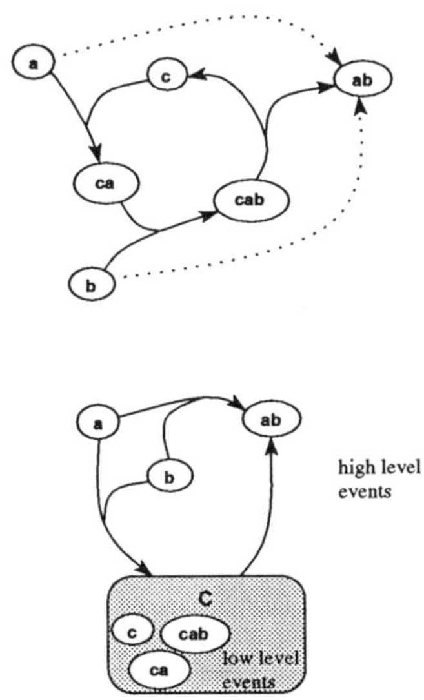

Figure 1. A catalytic loop. Figure 1a. Flat view. The catalyst, **c**, first binds to one of the substrate molecules, **a**, to form a complex, **ca**. This then binds with the other substrate molecule, **b**, to form the complex **cab**, which then dissociates into the product, **ab**, and the original catalyst. The reactions within the loop occur at a faster rate than the spontaneous reaction **a+b → ab** (dotted arrows). Figure 1b. Hierarchical view. The component **c-ca-cab** has been condensed into a virtual object, C, which equilibrates instantaneously relative to higher, slower levels.

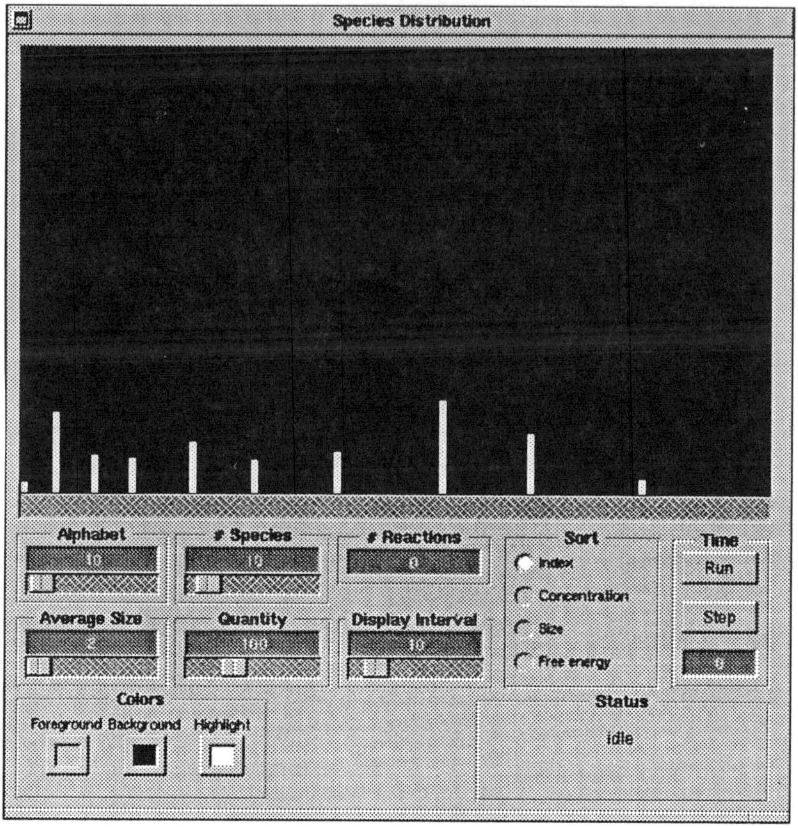

Figure 2. Screen shot of a simulated chemical system after initializing with an average of one hundred molecules each of ten chemical species with an average size of two atoms, constructed from an alphabet of ten atoms. The height of the histogram bars shows the concentration of each species.

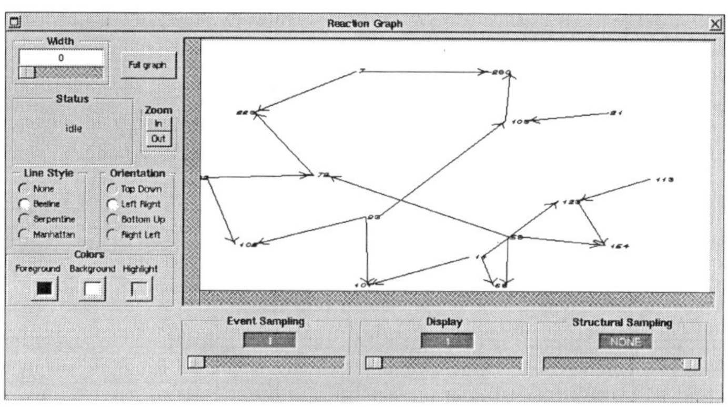

Figure 3. Screen shot of the reaction graph inferred from activity in the system over a period of several time steps, after many new species have appeared. Sixteen species have been detected as participating in reactions, but no strongly connected components have appeared.

Figure 2 is a screen shot of a simulated chemical system in its early stages. Here we see a histogram of the initial concentrations of ten chemical species, constructed from an alphabet of ten atoms. In Figure 3 is portrayed the reaction graph inferred from activity in the system over a period of several time steps, after many new species have appeared. This is a non-hierarchical view which reports every reaction detected. Figure 4 shows the same system after several hundred time steps, showing only the strongly connected partial subgraphs. Figure 5 shows a hierarchical view (from an earlier implementation) in which a catalytic component has been detected.

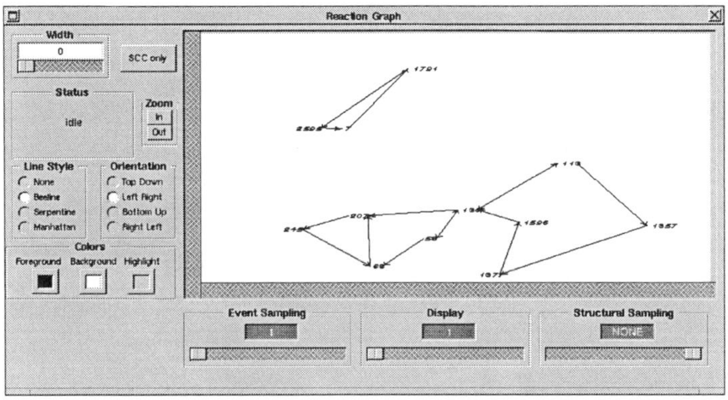

Figure 4. Screen shot of the reaction graph inferred from activity in the system over a period of several hundred time steps. Sixteen species have been detected as participating in reactions, but no strongly connected components have appeared.

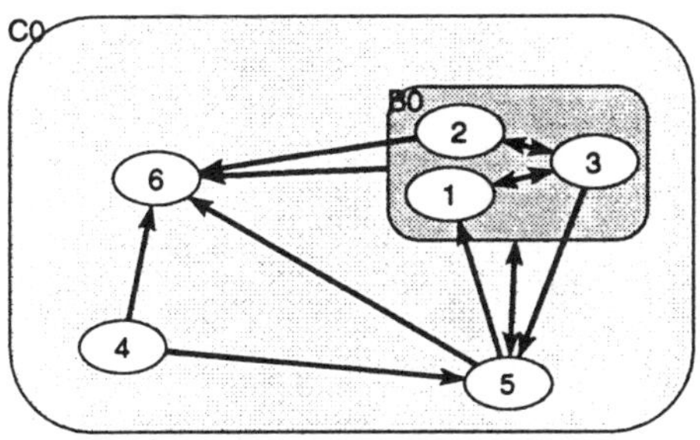

Figure 5. Structural snapshot of a catalytic system, at equilibrium in rapid-flow conditions. In the absence of the catalyst molecule (**1**), the substrate molecules **2** and **4** react slowly to form the product **6**. In the presence of sufficient catalyst, **1** and **2** react to form **3**, which reacts with **4** to form **5**, which breaks down into **1** and **6**. Due to the vagaries of sampling, the catalytic component **BO** includes substrate **2** and does not include the third member of the catalytic loop, **5**.

Although none of the simulation runs has spontaneously produced a system with multiple levels and modes of control, this is due, I believe, to shortcomings in the chemical model rather than in the interpretive mechanism. Detailed data from an actual biochemical metabolism, or from a sufficiently complex chemical model, should yield interpretations such as those of the (M, R)-systems and forward activation systems seen in Figures 6 and 7. The principle, after all, is clear: any strongly connected set of events will appear to higher-level objects as a single complex object, with multiple inputs and outputs.

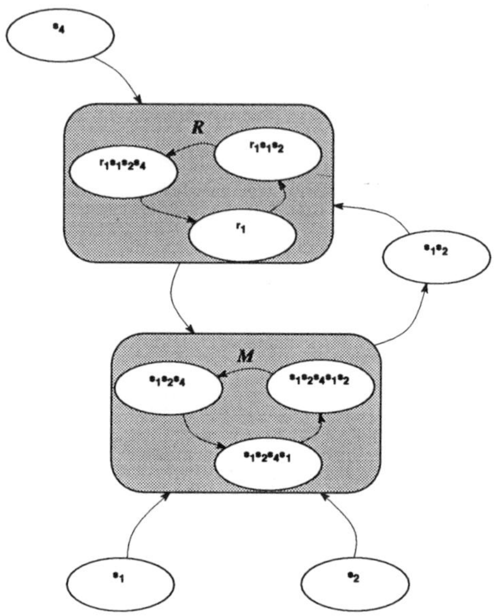

Figure 6. The minimal (**M**,**R**)-system, including four distinct substrings ($\mathbf{s}_1$, $\mathbf{s}_2$, $\mathbf{s}_4$, and $\mathbf{r}_1$) and six reactions. If the combined operational lag of *R* and the transport lag of $\mathbf{s}_1\mathbf{s}_2$ from *M* to *R* is $\mathbf{T}^R$, and *M* is somehow inhibited for a time, $\mathbf{T}^I > \mathbf{T}^R$, then the system "dies"; if $\mathbf{T}^I < \mathbf{T}^R$, the system stays "alive".

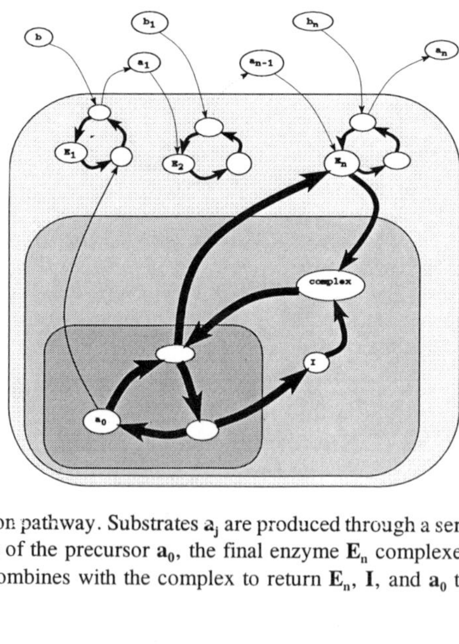

Figure 7. A forward activation pathway. Substrates $\mathbf{a}_j$ are produced through a series of catalyzed reactions $\mathbf{a}_i + \mathbf{b}_i \rightarrow \mathbf{a}_j$. In the absence of the precursor $\mathbf{a}_0$, the final enzyme $\mathbf{E}_n$ complexes with inhibitor **I**. When the precursor is present it combines with the complex to return $\mathbf{E}_n$, **I**, and $\mathbf{a}_0$ to their pools.

RECOVERING CAUSALITY

After working for some time with these simulations and the interpretations provided by constructive aliasing, it occurred to me that, in deriving descriptions in terms of higher functional units from observations in terms of lower-level units to which no apparent functional properties can be imputed, the result was a reconstruction of complex causes from simple causes. In the Aristotelian terms which Robert Rosen has been using (1971, 1985, 1991), it is possible to recover the formal, efficient, and final causes of events which are observed only as material changes. The meaning of these terms is somewhat imprecise, however, and has varied over time and among commentators, so to see how they might be recovered let us look at how they appear within a hierarchical interpretation, i.e. what does it mean for one event to cause another when each event occurs within some level of a hierarchical model, but may have effects both within its level and in higher and lower levels?

The components at each level can be interpreted as *virtual objects*. From within a given focal level, they have the same status as the atoms and molecules; the difference is that they are not restricted to elementary reactions, nor are they subject to conservation (though their constituent material still is). The resulting hierarchy has a causal structure which may best be parsed into within-level vs. between-level interactions. These can be further categorized by their topology (components vs. chains) and, for between-level causes, by their direction (ascending vs. descending):

within-level: transformations among objects at the focal level which take place at a rate no slower than the relaxation time characteristic of objects at that level,

- *components*: virtual objects which have been assigned to this level,
- *chains*: when virtual objects are assigned, they may carry along with them transformations between their components which however do not strongly connect them,

between-level: events between objects at different levels,

- *descending chains*: events in which a component of a high-level object is transformed into a lower-level object,
- *ascending chains*: events in which a low-level object is transformed into a component of a higher-level object,
- *components*: closed chains of descending and ascending events; these are of course also represented as virtual objects at their apex.

We have now enriched our view of the causal structure by adding new types of objects and of events. The within-level events still have material causes, but now the characteristics of the participating objects are no longer defined as they were at the primitive level, instead, they must emerge with the objects. The between-level events are yet more interesting; now a primitive object which was bound up in a lower-level component, and hence invisible to the focal level observer, may be transformed into a component of a focal-level object via an ascending chain; if the low-level object is present in sufficient quantity, this can radically change the stoichiometry of the focal level, as though an object has suddenly been switched into existence. Likewise with the descending chains: an object at the focal level can disappear relatively instantaneously if its rate-limiting component is drawn into a lower-level vortex. As a result, the apparent structure of the entire hierarchy can change over time. So this approach allows us to generate from elementary observations of material change a series of *structural snapshots*, which can be viewed in succession as a *morphological movie* of the structural changes through which the system progresses over time.

The direct, monotonic ascending and descending chains correspond to a kinetic control of the network of transformations. It is also possible, via non-monotonic chains, to realize something corresponding to thermodynamic control, in which the equilibrium state at a given level is changed by the relatively gradual increase or decrease of objects. This appears as an imposition of boundary conditions on the behavior of lower level objects, these boundary conditions arising as emergent properties of higher level objects.

Recovering the classical types of causality from this picture entails answering the following questions about an event: What is the instrument for the event? What properties are responsible for the event? What is the purpose of the event? The first question requires an identification of objects which play the role of instrument, the second question requires characterization of emergent properties, and the third question requires identification of objects which play the roles of representation and control system.

Efficient cause

In the universe described above, all events as seen from the lowest level involve recombination of primitives, or alternatively, joining and splitting of combinations. Taking a high-level object and analyzing it into its components and their recurrent transformations at the lowest level will give us a picture of the material cause of the object (though the object itself will no longer be visible *qua* object). The ascending and descending chains of transformations, though, have an instrumental effect: they cause objects to appear in or disappear from a given level, and by pulling them up or down from level to level they can cause these objects to become involved in new sets of interactions, thus changing their properties. Identifying objects which play the rôle of instrument, then, involves searching for components with afferent descending or efferent ascending chains. Such objects act as nonlinear constraints, effectively switching changes in the structure.

Formal cause

To answer the second question means deriving the rules specific to a level that characterize the interactions of objects at that level. Only the primitives have their laws specified; all higher-level objects have emergent properties which must be discovered and interpreted. The empirical approach is simple: observe an object in its interaction with other objects and describe what regularities can be detected. This is essentially what is done in extracting the reaction graph from observations of quantitative changes. The problem is that it confounds the constitutive parameters with the initial conditions, i.e. the same reaction graph can be produced either by a set of initial conditions (material cause) or by a set of free and activation energies (formal cause). Let me sketch here the context of a more principled approach. The properties of the primitives are explicitly specified, and include only sufficient degrees of freedom to determine their tendency to combine with one another and with other combinations. In the artificial chemistry model I have implemented, for example, each primitive and each combination of primitives is characterized by a variable-length vector I call its *profile*, a matching vector of *weights*, and a scalar corresponding to its *free energy*. Every possible reaction is then characterized by the free energies of the reactants and of the products, and by an *activation energy* of the unstable intermediate, derived by matching the weighted profiles of the reactants. The point of my describing this is to show how simple the dynamics are at the primitive level. Once we begin constructing virtual objects, though, we add new virtual degrees of freedom to our description and remove the ones we were working with at the primitive level. A virtual object, which is fundamentally a closed set of transformations, will have many more ways of behaving than any one of its elementary components, and it does not have the ability to combine (at least in the sense that its components do) with other virtual objects. In particular, some of these virtual degrees of freedom may be associated with *states* of the emergent objects. The notion of state is absent from the primitive level view, since primitives have no internal structure. When this interpretation is appropriate, objects will appear to have multiple *modes* of behavior while maintaining their identity. Deriving the emergent properties of one of these virtual objects means discovering the constraints on its virtual degrees of freedom. This is not trivial: you may recognize its real-world counterpart to be solving both the folding problem and the design problem, i.e. deriving the folded conformation(s) of a protein from knowledge of its primary sequence and then deriving the function(s) of the protein from its conformation(s).

Final cause

I stated above that answering the third question involves detecting representations and control systems. This follows from my assumption that goal-directed anticipatory behavior requires a standard against which the current situation can be compared, together with a strategy for effecting change in the situation. Both of these can vary tremendously in sophistication and in efficacy, from the simplest example of a single enzyme, which arguably makes "measurements" to detect substrate and "controls" their reaction, to the most complex examples of mental models, the legal codes of societies, or the digital architecture of embedded computer systems. I am personally uncomfortable calling an enzyme a model or a control system; I would prefer that the representation includes at least one or more of the objects involved in its metabolic support, as well as one or more of the inputs and/or outputs of the whole system (however this is delineated), as Pattee (1977) describes.

Just in case you are worried about the idea of identifying aspects of the wetware world with the discrete hardware/software world of machine logic, I would like to point out the real "switchiness" of metabolic function vs. machine function. Looked at in sufficient detail, any gating effect is really a continuous transition. The only question is: how "switchy" does a transition have to be before it makes sense to regard it as discontinuous, i.e. what is the difference in the flow with the switch on vs. the switch off? The current drawn by a standard TTL gate in the high and low states is about 1.6 mA vs. 0.04 mA, or 40:1. Being very conservative and more than doubling this, it is apparent that a separation of two or more decimal orders of magnitude is good enough for building digital computers. But enzymatic catalysis of metabolic pathways typically changes reaction rates by a factor of at least 10^5-10^6, i.e. it is at least 1000-10,000 times "switchier" than digital computers are. This has been recognized at least since Sugita's (1963) paper on molecular automata; much of Conrad's (1985, 1992, 1993) recent work is based on it also, and the more recent implementations of Adleman (1994) and Hjelmfelt (1995) are actual prototypes of chemical computation.

So it is possible to build any degree of logical sophistication into a metabolic network, and this is naturally enough what has happened. If, as Pattee (1977, 1987) has pointed out, the "unreasonable effectiveness of mathematics" (Wigner, 1960) has always been puzzling to scientists, it is not so surprising to find that the arbitrariness, the rate-independence, and the stability of symbolic representations (Pattee, 1986) have been exploited by opportunistic Nature. The use of DNA and RNA is the inevitable outcome of the "unreasonable effectiveness" of symbol systems. The problem now is to *detect* such systems. This means detecting persistent objects with multiple thermodynamically degenerate *modes* (states or messages), i.e. objects which serve as vehicles for many possible signals, all of which require approximately the same amount of energy. There must be multiple modes, because the alternative is that at each level there exists only a single component, in which case the control hierarchy would consist only of switching signals flowing up and down a linear array. The modes must be thermodynamically degenerate, because the alternative is that some states or messages are energetically favored over others, and this is quite obviously a dangerous bias for a system which is required to respond appropriately to a variety of different situations.

It is not possible to predict or design these multimodal objects from knowledge of the primitive level until a principled approach to emergent formal cause is completed. It *is* possible, though, through the empirical approach, to search entirely automatically for persistent objects whose ascending and/or descending chains vary over time (this corresponds to taking measurements and issuing controls), and for objects with inputs and outputs to multiple objects at some lower level (this corresponds to the potential existence of multiple states of comparable stability).

To put this in context, what I'm saying here is that the final cause of an event is a loop in which the highest level is occupied by a representation. The necessary implication of this is that the final cause of all objects in the loop, including the representation itself, is the loop itself. Indeed, it is entirely possible that the loop may be embedded in even larger loops, and that it, and all of the objects in it, are therefore finally entailed by the models at the top of *these* loops. Furthermore, in a truly complex system, these loops are not strictly hierarchical or tree-like, but intersect and conflict with one another.

A couple of examples will illustrate these interpretations. First, the most elementary example of an efficient cause, the catalytic loop, shown in Figure 1. Here, the flat primitive model (Figure 1a) shows only material and formal cause; the material cause of **ab** is the concentration of **cab**, and the formal cause is the free energies of **cab** and **ab**, and the activation energy of the unstable intermediate **c+ab**. This view shows no efficient cause (and of course no final cause). In the hierarchical model (Figure 1b), the material cause of **ab** is still the concentration of **cab**, but only as seen through the concentration of **C**. The formal cause of **ab** is found in **C**: {**a**,**b**} → {**ab**}, i.e. the mapping provided by the emergent object **C** from substrate to product. The efficient cause of **ab** is **C**. The material cause of **C** is found in its components, {**c**, **ca**, **cab**}. Other causal aspects of **C** are inaccessible without knowledge of further levels and objects, but the catalyst's presence, too, may be switched on or off by other catalysts, in a chain of efficient causes (activators and inhibitors, agonists and antagonists, etc.). If this chain closes, we have an autocatalytic loop, i.e. a closed loop of efficient cause.

As a second example, consider Rosen's (1981) analysis of the three mechanisms of morphogenesis: *differentiation*, in which the properties of the objects change over time "due to intrinsic autonomous dynamics, or the influence of environmental forces (including other subunits)"; *morphogenetic movement*, in which the number and properties of the objects are conserved so that "the only forces involved in generation of structures are the intersubunit affinities, and the differential movements of these subunits in space imposed by these forces"; and *differential birth and death*, in which "the properties and affinities of the subunits may remain constant, but their relative numbers may be changing in time". Using as his paradigm the reaction-diffusion network, he continues: " (...) differentiation is represented by the cell states, considered as ordered *n*-tuples of morphogen concentrations, becoming more and more distinct from each other as time proceeds. By construction, differential changes in morphogen concentrations occur through the combination of intracellular *chemical reactions* between the morphogens, and the *diffusion* of morphogens between cells. Now the process of chemical reaction can be regarded as a differential birth-and-death process; intuitively, molecules of reactants are removed from the system ("death") and new molecules of product appear ("birth"). Likewise, chemical diffusion can be regarded as a process of morphogenetic movement. Therefore, a reaction-diffusion model for *differentiation* at the cellular level can be interpreted as a combination of *morphogenetic movement* and *differential birth-and-*

death at an underlying chemical level" (p. 206).

The cause of the reactions can be ascribed to a combination at least of material and formal properties at the primitive level; if found within an organism, we would have to include also the efficient causes of enzymatic catalysis and the final cause of genomic instruction. The causes of the diffusion are likewise the material and formal properties of the chemical species, together with the formal properties of the intercellular membranes which restrict their flow.

According to this view, what emerges at the focal level as material cause (appearance and disappearance of objects), as formal cause (change in properties of objects), as efficient cause (constraint on behavior), or as final cause (goal-directed control) arises from a skein of other causes at lower and higher levels. Beyond the "circular causality" introduced fifty years ago by cybernetics, this points to a self-similar, self-generating causal structure which we could term fractal or recursive causality.

ORIGINS OF FINALITY

Given that the observed dynamics permit such an interpretation, the method discussed above will always deliver a picture of the system in terms of hierarchical control, i.e. in terms of instruments, representations, and agents. There is no consideration of whether the semantics of the system support such an interpretation. For some systems, of course, such explanations and such imputations of efficacy or finality will be inappropriate. Whether to accept such explanations always was and always will be the choice of the inquiring agent. Terms such as function, control, and purpose, are appropriate only with respect to that large class of most complex, interesting, and refractory *organic systems*, namely, organisms, ecosystems, societies, economies, languages, and so forth. The appropriateness of final cause as an explanation then turns out to be the defining characteristic of these systems, and the study of their origin has as its shadow the study of the origin of final cause, or the beginning of the End.

The above analysis allows us to draw a few conclusions about the conditions under which organic systems, and hence final cause, might arise. Measurement, representation, and control require persistent objects with multiple modes, and chains of efficient cause coupling these objects together. Reciprocal interaction between objects at a given level can be either via weak, higher-level arcs (which do not threaten their stability), or via lower level couplings whose effect on the objects is stabilized by active maintenance processes. Throughout this hierarchy, then, the topology at each level has to consist of a number of disjoint objects, i.e. it is "lumpy" rather than uniform. We know from random graph theory (Palmer, 1985) that this occurs only just below the threshold for the emergence of the giant component, i.e. when the average number of arcs per vertex is between one and two. This means that the average object — whether primitive or emergent — participates in one or two reactions with objects at or below its own level.

This distribution of reaction rates must hold for all functional levels of an organic system. How many such functional levels are there? From the theoretical end, we know that a primitive (M,R)-system requires only three functional levels. Adding the capacity for self-replication requires four levels. Adding the capacity for modelling requires five or more. From actual biology, we know that the span of relaxation times to be considered lies between the shortest metabolically significant event (about one femtosecond) and the length of a generation for an organism (about 12 days for protista, or 30 years for large plants and vertebrates). Even assuming an inter-level separation of two decimal orders of magnitude, this still allows only 12 to 13 functional levels. Since every molecular species participates in at least one reaction with every other species — and many more in chemistries more complex than that simulated and described above — this implies that the relaxation time of almost all reactions is effectively infinite, and that where multiple reactions occur within a level, they occur only within a single object. This can occur if most components are not simple cycles but rather richly interconnected nets, but this is not expected to occur in uniformly random graphs. The implication is that the spontaneous emergence of organic structures is much more unlikely than the study of random graphs would indicate.

Department of Genetics, Stanford University
California, USA

REFERENCES

Adleman, L., 1994, Molecular computation of solutions to combinatorial problems, in *Science*, vol. 266, no. 5187, pp. 1021-1023.

Allen, T.F.H.; Starr, T.B., 1982, *Hierarchy: Perspectives for Ecological Complexity*, Chicago, University of Chicago Press.

Berlecamp, E.R.; Conway, J.H.; Guy, R.K., 1982, *Winning Ways for Your Mathematical Plays*, vol. 2, New York, Academic Press.

Cariani, Peter, 1991, Emergence and Artificial Life, in *Artificial Life II*, C. Langton, C. Taylor, J.D. Farmer & S. Rasmussen (eds.), Redwood City (CA), Addison-Wesley.

Conrad, M., 1985, On design principles for a molecular computer, *Communications of the ACM*, vol. 28, no. 6, pp. 464-480.

Conrad, M., 1992, Molecular computing: the lock-key paradigm, *Computer*, vol. 25, no. 11, pp. 11-20.

Conrad, M., 1993, Integrated precursor architecture as a framework for molecular computer design, *Microelectronics Journal*, vol. 24, no. 3, pp. 2632-2685.

Farmer, J.D.; Kauffman, S.A.; Packard, N.H., 1986, Autocatalytic replication of polymers, *Physica*, vol. 22D, pp. 50-67.

Hjelmfelt, A.; Ross, J., 1995, Implementation of logic functions and computations by chemical kinetics, *Physica D*, vol. 84, nos. 1-2, pp. 180-193.

Holland, J.H., 1975, *Adaptation in Natural and Artificial Systems*, Ann Arbor (MI), University of Michigan Press.

Holland, J.H., 1986, *Induction*, Cambridge (MA), MIT Press.

Kampis, G., 1991, *Self-modifying Systems in Biology and Cognitive Science: A New Framework for Dynamics, Information, and Complexity*, Oxford, Pergamon Press.

Kauffman, S.A., 1993, *The Origins of Order*, New York, Oxford University Press.

Minch, E., 1988, *Representation of Hierarchical Structure in Evolving Networks*, Doctoral dissertation, State University of New York, Binghamton, New York.

Minch, E., 1992, Anima y Animus: Una perspectiva sobre la vida artificial y la inteligencia artificial, *Revista de la Real Academia de Ciencias Exactas, Físicas Y Naturales*, Madrid, vol. 86, no. 3, pp. 389-421.

Palmer, E.M., 1985, *Graphical Evolution*, New York, Wiley.

Pattee, H.H., 1973, The physical basis and origin of hierarchical control, in *Hierarchy Theory*, H.H. Pattee (ed.), New York, George Braziller.

Pattee, H.H., 1977, Dynamic and linguistic modes of complex systems, *International Journal of General Systems*, vol. 3, pp. 259-266.

Pattee, H.H., 1986, Universal principles of measurement and language functions in evolving systems, in *Complexity, Language, and Life: Mathematical Approaches*, J.L. Casti & A. Karlqvist (eds.), Berlin, Springer-Verlag.

Pattee, H.H., 1987, Simulations, realizations, and theories of life, in *Artificial Life*, C. Langton (ed.), Redwood City (CA), Addison-Wesley.

Penrose, R., 1989, *The Emperor's New Mind: Concerning Computers, Minds, and the Laws of Physics*, Oxford, Oxford University Press.

Rosen, R., 1971, Some realizations of (M,R)-systems and their interpretation, *Bulletin of Mathematical Biophysics*, vol. 33, pp. 303-319.

Rosen, R., 1981, Pattern generation in networks, *Progress in Theoretical Biology*, vol. 6, pp. 161-209.

Rosen, R., 1985, *Anticipatory Systems: Philosophical, Mathematical, and Methodological Foundations*, Oxford, Pergamon Press.

Rosen, R., 1991, *Life Itself: A Comprehensive Inquiry into the Nature, Origin, and Fabrication of Life*, New York, Columbia University Press.

Rumelhart, D.E.; McClelland, J.L.; the PDP Research Group, 1986, *Parallel Distributed Processing* (vol. I-II), Cambridge (MA), Bradford/MIT.

Sugita, M., 1963, Functional analysis of chemical systems in vivo using a logical circuit equivalent, II. The idea of a molecular automaton, *Journal of Theoretical Biology,* vol. 4, pp. 179-192.

Wigner, E.P., 1960, The unreasonable effectiveness of mathematics in the natural sciences, *Communications on Pure and Applied Mathematics*, vol. 13, pp. 1-14.

Wolfram, S. (ed.), 1986, *Theory and Applications of Cellular Automata*, Singapore, World Scientific.

BRUCE H. WEBER

EMERGENCE OF LIFE AND BIOLOGICAL SELECTION FROM THE PERSPECTIVE OF COMPLEX SYSTEMS DYNAMICS

IN THE BEGINNING ...

All cultures, including the scientific, have creation stories. These reflect the knowledge and values of the tellers of such narratives. Until relatively recently, most such scenarios posited a divine creator or engineer-like god responsible for the creation of the cosmos and the origin of life.

Modern naturalistic explanations for the origin of life could be said to begin with the "active materialism" of early nineteenth-century French biologists such as Lamarck and Geoffroy, which was introduced into England by "philosophical anatomists" such as Robert Grant (one of Darwin's early teachers) and taken up avidly by English socialists of the 1820's and 1830's (Desmond, 1989). This view assumed that matter was capable of spontaneous self-organization. The firestorm that greeted such views formed the background to the development of Darwin's thinking (Desmond & Moore, 1991). In his public writings Darwin ducked the issue of the origin of life and spoke of life being "breathed into a few forms or into one" (Darwin, 1859, p. 490). Privately, however, he let his materialistic inclinations fuel speculation on a possible scenario for the origin of life in a letter to Joseph Hooker in 1871 in which he wrote, "But if (and oh what a big if) we could conceive in some warm little pond with all sorts of ammonia and phosphoric salts, light, heat, electricity and etc., present, that a protein compound was chemically formed, ready to undergo still more complex changes" (DAR 94, pp. 188-189).

It would be some time before further speculation about the origin of proteins and metabolism would occur and indeed could even be reasonably undertaken. In the period after World War I, biochemistry was in a period of rapid growth, which allowed for more informed models of life's origin. J.B.S. Haldane and Alexander Oparin suggested that there could be an abiotic synthesis and accumulation of organic compounds of increasing complexity from which would emerge a proto-metabolism (Haldane, 1929; Oparin, 1924, 1938). Within this approach, J.D. Bernal suggested that clays could have provided the catalytic and template properties that would have facilitated the emergence of such a proto-metabolism (Bernal, 1951). Concurrent with the solution of the structure of DNA, the abiotic synthesis of amino acids was demonstrated (Miller, 1953). The subsequent demonstration of the abiotic synthesis of proteinoids (amino acid polymers) suggested that not only might polypeptides be catalysts of protometabolism, but that their formation might even be prior (Fox, 1965). The demonstration of the abiotic synthesis of purine and pyrimidine bases led to the suggestion that RNA might be the original catalytic agent in the origin of life (Woese, 1967; Orgel, 1968). This suggestion became more plausible with the demonstration of catalytic activity of present-day RNA (Cech, 1986, 1987; Doudna & Szostak, 1989; Wang, Downs & Cech, 1993; Lorsch &

G. Van de Vijver et al. (eds.), Evolutionary Systems, 59–66.

Szostak, 1994; Wilson & Szostak, 1995). Recently the emphasis has been on an RNA-first model for the origin of life in which RNA becomes template, replicator, and ribozymic catalyst (Orgel, 1992; Joyce & Orgel, 1993). Indeed, it has become quite popular among biochemists and molecular biologists to speak of an "RNA world" that preceded the emergence of true life (Gesteland & Atkins, 1993). This view is particularly attractive if the fundamental fact about life is taken to be the self-replication of information-bearing nucleic acids. Indeed, in Dawkins' vision, life began with the highly improbable accident of the first replicating macromolecule that subsequently decorated itself with proteins, lipids, carbohydrates, and cellular structure, which provided protection for the genes and exploited catalysis and metabolism for their benefit, such that first cells and then later organisms emerged as "gene machines" (Dawkins, 1976, 1989).

THIS (SYSTEMS) VIEW OF LIFE

Each of the approaches to thinking about the origin of life emphases some aspect of contemporary living systems (metabolism, enzymic catalysis, nucleic acid replication) as the primary process through which life is brought into being. Some proponents of the RNA world are well aware that such a simple picture of RNA replication and catalysis is incomplete and that life emerged, and continues to exist, far from thermodynamic equilibrium, with the consequence that resultant autocatalytic cycles allow the emergence of information and systems complexity (Eigen, 1992). Also, there are some serious problems in the RNA-first model, such as the fact that no convincing chemical scenario has been advanced as to how the purine and pyrimidine bases might be attached to ribose abiotically, even under the catalytic activity of agents such as clays (Joyce & Orgel, 1993). Unfortunately, in a more general sense, while RNA can do some limited catalysis of reactions on itself, and even has been induced *in vitro* to have a kinase activity and an alkyl transferase activity (Lorsch & Szostak 1994; Wilson & Szostak 1995), it is highly unlikely that it has the catalytic capacity that would match that of a set of polypeptides.

Alternatively, life can be viewed as emerging with its *general organization* of catalytic and metabolic functions prior to, or at least along with the appearance of genetic replication (Fox, 1980; Cairns-Smith, 1982; Conrad, 1982; Wicken, 1987; Weber *et al.*, 1989, Salthe, 1991; Morowitz, 1992). J.D. Bernal commented that DNA did not just wash up on the beach from the primordial sea (Bernal, 1967). Life, in this perspective, emerges as a whole *system* that over time becomes more coherent and articulated. Fox has argued for life emerging within proteinoid microspheres that could give significantly greater chemical concentrations than the dilute solution chemistry assumed in many accounts (Fox, 1965, 1988). More recently, Morowitz has argued that amphophile bilayers could have formed prior to other events and that within such membranes the emergent chemistry could occur (Morowitz, 1992). This idea that life arose within a protocellular context is particularly attractive since this also suggests that primitive chemosmotic functions could have been involved in providing energy transduction to keep such protocellular systems far from equilibrium (Mitchell, 1961, 1979; Harold, 1986).

Morowitz postulates that vesicular protocells spontaneously form from amphiphilic bilayers of primitive phospholipids (formed by terrestrial processes or of meteoric origin). This bilayer is a barrier to the diffusion of polar solutes, thus allowing molecules once brought into the protocell to be kept at higher concentrations than in the ambient fluid and for concentration gradients to form. The lipid phase in between the aqueous phases can partition non-polar solutes, some of which could be chromophores capable of capturing and transducing light energy, thus permitting chemical, electrical and osmotic work. Given such a source of free energy, the protocellular system can be driven (or constrained) further from equilibrium by such work. "Nutrients" could accumulate and more amphiphilic molecules would be made, allowing growth of the protocell. Mechanical and thermodynamic constraints would provide physical selection for optimal sizes and drive the breakdown of larger vesicles or the fusion of smaller vesicles, just as has been observed for Bénard convection cells under similar dissipative conditions (Swenson, 1989, 1998). This gives a model of: protocell + nutrients + energy → protocells + waste products + entropy (Morowitz, Deamer & Smith, 1991).

There is empirical evidence to support such a model. Amphiphiles have been isolated from a number of carbonaceous chondrites and repeated drying and wetting of such amphiphiles (mimicking the action of a tide pool) give rise to membrane-like structures, including bilayered micelles and vesicles (Deamer & Pashley, 1989). Such structures show autocatalytic self-replication (Bachmann, Luisi & Lang, 1992). Alternative lipid-like molecules of fully terrestrial origin, such as terpenoids, could also have formed the membranes of primitive protocells (Ourisson & Nakatani, 1994). Polycyclic aromatic hydrocarbons derived from carbonaceous chondrites will "dissolve" in the non-polar phase of amphiphile bilayer vesicles and upon illumination translocate protons across the membrane generating a pH shift and hence a proton-motive force that could energize chemical and osmotic work (Deamer, 1992).

Such phase-separated vesicles would not yet be living entities, but they would provide good cradles within which life could emerge. With a free-energy source producing a proton gradient, for example, polyphosphate could be made (Morowitz, 1992; Deamer & Harang, 1990). The formation of polyphosphate is important because such compounds have a unique mix of thermodynamic instability and kinetic stability (Westheimer, 1987) in water and will participate in phosphorylation of amino acids and nucleotides, which were accumulated by the cell also through the use of the electrochemical potential of the proton gradient. The phosphorylation of monomers in the protocell would help drive thermodynamically their polymerization to polypeptides and polynucleotides. At some point polyphosphate was replaced by ATP. Interactions of such macromolecules would not be random, even if their sequences at this point were random (Carter & Kraut, 1974) and at the very least their interaction would provide some mutual stability against hydrolysis (Wicken, 1987). Both polypeptides and polynucleotides could catalyze reactions, though the potential for polypeptides is much larger. Indeed, Kauffman estimates that a complete set of catalytic peptides of random sequence could "cover" the entire possible catalytic task space of the hundred million or so chemical reactions that could be catalyzed (Kauffman, 1993). Poised away from equilibrium, nonlinear interactions and autocatalytic cycles will spontaneously arise. In such instances, the set of N catalytic peptides (and/or RNAs) would be capable of catalyzing, though rather poorly, a subset of chemical reactions. As Kauffman has shown by his modeling, such systems will develop toward catalytic closure in which there is a tightening of the number of interactions between the reactions of the catalysts of the set that gives rise to a protometabolism. In such catalytic closure, every member of the autocatalytic set has at least one of the possible steps of its formation catalyzed by some member of the set, and connected sequences of catalyzed reactions lead from the maintained food set to all members of the autocatalytic set. Computer modeling of such sets of autocatalytic peptides or ribozymes shows the system exhibiting increasing complexity over time until there is a phase-transition-like change when catalytic closure obtains. With closure there will be increased rewards to the catalysts for catalytic efficiency. Thus there would be a selection pressure for a better fit of the sequence space of a given macromolecule to the catalytic task space with which it roughly overlaps. Under such conditions there would be polypeptide catalysts available that could facilitate difficult reactions, such as adding purine and pyrimidine bases onto ribose. Even without coding of protein sequences by nucleic acids, there will be selection of peptide and protein sequences that are more chemically efficient and which contribute to the stability of the set as a whole (Kauffman, 1993).

Even though there is no information in the genetic sense in such a system, there will be competition with other such autocatalytic systems for energy fluxes. As successful patterns of dissipation emerge under these competitive conditions, we would expect progressive tightening of the nucleic acid-protein relationship as the protocells evolve into true cells. It is thermodynamic or chemical selection of the efficient, rather than physical selection of the stable or biological selection of the fit, where the relevant units of selection are energy-capturing and energy-utilizing cycles. Self-organization under thermodynamic constraints provides much order without selection in such autocatalytic cycles; chemical selection can act upon these cycles to produce increased efficiency.

The emerging ensemble of proteins that catalyzed the reactions constituting the primitive metabolism of the protocell would have been "generic" proteins (Wicken, 1987). Over time they would have acquired more specific catalytic functions as mentioned above. There would have been selection pressure for any entity that could increase the efficiencies of its autocatalytic cycles by storing the information needed for autocatalysis and for expanding autocatalytic prowess by using these information-storing capacities in new ways. The close coupling of replicating nucleic acids (with some residual catalytic activity) and catalytic proteins involved in autocatalytic processes would have been highly advantageous in this context. It would have been of enormous competitive advantage to such catalytic units if they were able to "remember" information that enhanced autocatalytic activity by encoding it in the polymers of nucleic acids formed by chemical selection itself. This information would thus not dissipate as quickly as before and could be passed on in time to the next generation. So genetic information would have accumulated under thermodynamic selection for stable and efficient patterns of entropy production. There is no reason to think that the properties that were required for the emergence of life would not be maintained by living systems thereafter. Thus living things can be viewed as bounded, self-replicating, *informed* autocatalytic dissipative systems that sustain themselves by efficient environmental energy exchanges (Wicken, 1987; Depew & Weber, 1995), yet vary under the drive to configurational randomness (Brooks & Wiley, 1986,1988) in such a way that new information, guiding new catalytic functions, can be selected from the variation.

Natural selection of the reproductively fit is emergent from chemical selection of the autocatalytically efficient; it is a process that can be ascribed only to the autocatalytic dissipative structures that capture information within strongly defined boundaries and use it to guide efficient autocatalysis. Such entities would have to be able to pass the information they have to successor entities, for without that additional property, the whole point of internalizing information would be lost. Among such entities themselves, therefore, what would have been even more crucially contested than storing and deploying metabolic information would have been the ability to reproduce themselves and the information they possess. Fitness is a measure of that ability. It cannot be reduced to the efficiency of chemical selection, any more than the efficiency of chemical selection can be reduced to the stability of physical selection, for the relevant processes and entities capable of engaging in them do not exist at those levels. To say that these processes cannot be reduced to lower-level ones is not, however, to say that they are not part of a single, coherent process. It is to say rather that evolution exhibits emergent levels and emergent properties and phenomena.

From this perspective, biological information would emerge as a probable, indeed expected, result of deep physical and chemical principles governing self-organization in complex webs of catalysts rather than as a "fortuitous, frozen accident" (Kauffman, 1993; Salthe, 1991). These principles, however construed or modeled, represent an interplay of self-organizational and selective processes. The physical imperatives of self-organization and dissipation require that the particular sort of selection process leading to the emergence of living systems was at first the selection of the stable (physical selection) and of the efficient (chemical selection) rather than of the reproductively fit (biological selection). So construed, the problem of the origin of life is the emergence of the *phenomenon* of natural selection out of these more basic forms of selection (Weber & Depew, 1996). From this perspective, it may be more fruitful to regard primitive protocellular systems as the sites of the dynamics leading to life involving a coevolution of proteins and replicating nucleic acids over a "replicators first" strategy. The more basic forms of selection that obtain in such sites are inseparable from the amplification of stochastic events by the self-organizational tendencies and propensities of open systems and dissipative structures.

Kauffman's NK Boolean models allow exploration of the interplay of selective and self-organizational principles at a more abstract and level-independent manner that emphasizes the relational rather than causal aspects of complex systems. In the ordered regime of NK space, there is either no or too little variation in regions described by simple point or limit-cycle attractors, not a very interesting dynamics. In the chaotic regions, described by large numbers of strange attractors, there is insufficient stability or coherence; over time such a system will settle onto and cycle through attractors with vast numbers of states, taking billions of times longer than the history of the universe to traverse these enormous attractors. Such systems do not do anything interesting either. But, under the right conditions (supplied by the programmer in virtual systems or by the nonlinear interactions of real, autocatalytic systems), at or near the membrane-like edge of order and chaos, there is a region where interesting things can and do happen. The behavior of systems in this range is complex. It might move toward one or several different attractors. Transient islands of ordered structure can arise in a sea of chaos, only to melt away as new order appears elsewhere. In this region is also seen the distinctive dynamics of self-organization and adaptability (Langton, 1986, 1992; Kauffman, 1993; Ulanowicz, 1998). Adaptation, considered as a process, becomes an exploration of parameter space near the edge of chaos in search of a better fit to a given fitness landscape onto which Kauffman maps his NK model, or even to find better attractors. The structure of this region of phase space and its attractors are properties of the self-organization of the system itself. Kauffman goes on to speculate that it is selection that pulls such complex adaptive systems into this fruitful region. This suggests that complex adaptive behavior, at any hierarchical level (molecules, cells, organisms, ecosystems, phylogenies), develops only in systems whose range of elements and connections is such that wide variation due to stochastic processes, a feed-back-driven selection process, and self-organization are integral aspects of a single process. The interplay of chance with physical and chemical self-organization and selection led ultimately to the emergence of natural selection concomitant with the emergence of life itself, but life and natural selection remain embedded in this context from which it arose, even as they exhibit novel, emergent behavior.

Rather than cut natural selection off from its physical and chemical roots, as biological autonomists would, it may be more useful to consider that there is an on-going intimate tie between chance, self-organization and selection, including natural, or biological, selection. Such a conception, along with the use of dynamical models based upon complex systems more generally, could vitalize the Darwinian tradition and facilitate dialogue between its adherents and the proponents of the developmentalist tradition (Depew & Weber, 1995; Weber & Depew, 1996).

Department of Chemistry and Biochemistry, California State University Fullerton, USA

REFERENCES

Bachmann, P.A.; Luisi, P.L.; Lang, J., 1992, Autocatalytic self-replication micelles as modes for prebiotic structures, *Science*, vol. 357, pp. 57-58.

Bernal, J.D., 1951, *The Physical Basis of Life*, London, Routledge & Kegan Paul.

Bernal, J.D., 1967, *The Origin of Life*, Cleveland, World.

Brooks, D.R.; Wiley, E.O., 1986, *Evolution as Entropy: Toward a Unified Theory of Biology*, Chicago, University of Chicago Press.

Brooks, D.R.; Wiley, E.O., 1988 (2nd ed.), *Evolution as Entropy: Toward a Unified Theory of Biology*, Chicago, University of Chicago Press.

Cairns-Smith, A.G., 1982, *Genetic Takeover and the Mineral Origins of Life*, Cambridge, Cambridge University Press.

Carter, C.W. Jr.; Kraut, J., 1974, A proposed model for interaction of polypeptides with RNA, *Proc. Nat. Acad. Sci. USA*, vol. 71, pp. 283-287.

Cech, T.R., 1986, A model for the RNA-catalyzed replication of RNA, *Proc. Nat. Acad. Sci. USA*, vol. 83, pp. 4360-4363.

Cech, T.R., 1987, The chemistry of self-splicing RNA and RNA enzymes, *Science*, vol. 236, pp. 1532-1539.

Conrad, M., 1982, Bootstrapping model of the origin of life, *BioSystems*, vol. 15, pp. 209-219.

Darwin, C.R., 1859, *On the Origin of Species by Means of Natural Selection or the Preservation of Favored Races in the Struggle for Life*, London, John Murray.

Dawkins, R., 1976, *The Selfish Gene*, Oxford, Oxford University Press.

Dawkins, R., 1989, *The Selfish Gene: New Edition*, Oxford, Oxford University Press.

Deamer, D.W., 1992, Polycyclic aromatic hydrocarbons: Primitive pigment systems in the prebiotic environment, *Adv. Space Res*, vol. 12, pp. 183-189.

Deamer, D.W.; Harang, E., 1990, Light-dependent pH gradients are generated in liposomes containing ferrocyanide, *BioSystems*, vol. 24, pp. 1-4.

Deamer, D.W.; Pashley, R.M., 1989, Amphiphilic components of the Murchison carbonaceious chondrite: Surface properties and membrane formation, *Origin of Life and Evolution of the Biosphere*, vol. 19, pp. 21-38.

Depew, D.J.; Weber, B.H., 1995, *Darwinism Evolving: Systems Dynamics and the Genealogy of Natural Selection*, Cambridge MA, Bradford/The MIT Press.

Desmond, A., 1989, *The Politics of Evolution*, Chicago, University of Chicago Press.

Desmond, A.; Moore, J., 1991, *Darwin*, London, Michael Joseph.

Doudna, J.A., Szostak, J.W., 1989, RNA-catalysed synthesis of complementary-strand RNA, *Nature*, vol. 339, pp. 519-522.

Eigen, M., 1992, *Steps Towards Life: A Perspective on Evolution*, Oxford, Oxford University Press.

Fox, S.W., 1965, Simulated natural experiments in spontaneous organization of morphological units from proteinoid, in *The Origins of Prebiological Systems and Their Molecular Matrices*, S.W. Fox (ed.), New York, Academic Press, pp. 361-382.

Fox, S.W., 1980, The origins of behavior in macromolecules and protocells, *Comparative Biochemistry and Physiology*, vol. 67B, pp. 423-436.

Fox, S.W., 1988, *The Emergence of Life: Darwinian Evolution from the Inside*, NY, Basic Books.

Gesteland, R.F.; Atkins, J.F., 1993, *The RNA World: The Nature of Modern RNA Suggests a Prebiotic RNA World*, Cold Spring Harbor, Cold Spring Harbor Laboratory Press.

Haldane, J.B.S., 1929, The origin of life, *Rationalist Animal*, reprinted in *The Origin of Life*, J.D. Bernal (ed.), 1967, Cleveland, World, pp. 242-249.

Harold, F.M., 1986, *The Vital Force: A Study of Bioenergetics*, New York, Freeman.

Joyce, G.F.; Orgel, L.E., 1993, Prospects for understanding the origin of the RNA world, in *The RNA World*, R.F. Gesteland & J.F. Atkins (eds.), Cold Spring Harbor, Cold Spring Harbor Laboratory Press, pp. 1-25.

Kauffman, S.A., 1993, *The Origins of Order: Self-Organization and Selection in Evolution*, New York, Oxford University Press.

Langton, C.G., 1986, Studying artificial life with cellular automata, *Physica (D)*, vol. 22, pp. 120-149.

Langton, C.G., 1992, Life at the edge of chaos, in *Artificial Life II*, C.G. Langton, C. Taylor, J.D. Farmer & S. Rasmussen (eds.), Reading MA, Addison-Wesley, pp. 41-91.

Lorsch, J.R.; Szostak, J.W., 1994, In vitro evolution of new ribozymes with polynucleotide kinase activity, *Nature*, vol. 371, pp. 31-36.

Miller, S., 1953, A production of amino acids under possible primitive earth conditions, *Science*, vol. 117, pp. 528-531.

Mitchell, P.D., 1961, Coupling of phosphorylation to electron and hydrogen transfer by a chemiosmotic type of mechanism, *Nature*, vol. 191, pp. 144-148.

Mitchell, P.D., 1979, Keilin's respiratory chain concept and its chemiosmotic consequences, *Science*, vol. 206, pp. 1148-1159.

Morowitz, H., 1992, *Beginnings of Cellular Life: Metabolism Recapitulates Biogenesis*, New Haven, Yale University Press.

Morowitz, H.; Deamer, D.W.; Smith, T., 1991, Biogenesis as an evolutionary process, *Journal of Molecular Evolution*, vol. 33, pp. 207-208.

Oparin, A.I., 1924, *Proiskhozhdenie zhizy*, Moscow, Moskovski Rabochii.

Oparin, A.I., 1938, *The Origin of Life*, London, Macmillan.

Orgel, L.E., 1968, Evolution of the genetic apparatus, *Journal of Molecular Biology*, vol. 38, pp. 381-393.

Orgel, L.E., 1992, Molecular replication, *Nature*, vol. 358, pp. 203-209.

Ourisson, G.; Nakatani, Y., 1994, The terpenoid theory of the origin of cellular life: The evolution of terpenoids to cholesterol, *Current Biology*, vol. 1, pp. 11-23.

Salthe, S.N., 1991, Formal considerations on the origin of life, *Uroboros*, vol. 1, pp. 45-65.

Swenson, R., 1989, Emergent attractors and the law of maximum entropy production: Foundation to a theory of general evolution, *Systems Research*, vol. 6, pp. 187-197.

Swenson, R., 1998, *Spontaneous Order, Evolution, and Autocatakinetics: The Nomological Basis for the Emergence of Meaning*, Hillsdale, NJ, Lawrence Erlbaum Associates.

Ulanowicz, R.E., 1998, Boundaries on the complexity of evolving networks: A window of vitality, in *Between Order and Chaos: Studies in Nonequilibrium Biology*, J.D. Collier & S. Causey (eds.).

Wang, J.-F.; Downs, W.D.; Cech, T.R., 1993, Movement of the guide sequence during RNA catalysis by a group I ribozyme, *Science*, vol. 260, pp. 504-508.

Weber, B.H.; Depew, D.J.; Dyke, C.; Salthe, S.N.; Schneider, E.D.; Ulanowicz, R.E.; Wicken, J.S., 1989, Evolution in thermodynamic perspective: An ecological approach, *Biology and Philosophy*, vol. 4, pp. 373-405.

Weber, B.H.; Depew, D.J., 1996, Natural selection and self-organization: Dynamical models as clues to a new evolutionary synthesis, *Biology and Philosophy*, vol. 11, pp. 33-65.

Westheimer, F.H., 1987, Why nature chose phosphates, *Science*, vol. 235, pp. 1173-1178.

Wicken, J.S., 1987, *Evolution, Information and Thermodynamics: Extending the Darwinian Program*, New York, Oxford University Press.

Wilson, C.; Szostak, J.W., 1995, *In vitro* evolution of a self-alkylating ribozyme, *Nature*, vol. 374, pp. 777-782.

Woese, C., 1967, The evolution of the genetic code, *The Genetic Code*, New York, Harper & Row, pp. 179-195.

JUAN M. ALVAREZ DE LORENZANA

SELF-ORGANIZATION AND SELF-CONSTRUCTION OF ORDER

INTRODUCTION

As noted by many people (e.g. Salthe, 1993), self-organization is increasingly becoming the bedrock of the so-called *complex systems* which, themselves, span the whole spectrum of reality: from the physical universe to human beings and, beyond that, to constructs of human beings such as technological systems. This volume is a good example of variety within self-organization (from here abbreviated as SO). This is so because evolving systems are the quintessential expression of SO. But, why SO? Why is it so pervasive? In other words, what is it that SO responds to? What is it that SO solves in each system where it is present? These questions are in need of examination. We tend to acknowledge the existence of SO and, from that acknowledgement, go straightaway to the business of describing it as best we can, usually leaving aside the task of identifying, fully and deeply, what the problem dealt with by SO is. I shall attempt to deal principally with the question: why SO? The outcome will be used to determine what is needed to implement formal representations of SO.

WHAT IS SELF-ORGANIZATION?

Let us undertake a first attempt to answer "Why SO?". *SO is a specific characteristic of open systems*. "Open systems" are systems that are open (in space, time and matter-energy) to the environment. In order to have and maintain the condition of "systemness" when being open to the environment, an *inner source* of organizational ability has to be at play: we call it self-organization (SO). Therefore, if we are to have "open systems" (and we indeed do), it is because of systemic SO. From the assumption that SO is a key characteristic of open systems, we can infer that only systems within which SO takes place can exist in unbounded, finite, environments. If proven, and considering how pervasive open systems are (technological systems are included, considering culture, in this case, as part of the social system's environment), this assertion could account for the universality of SO.

When speaking about "open systems" we should understand:

1. systems that exist (i.e. they are viable) in an environment that is bigger than the system (unbounded, yet finite);
2. the environment is invariant in some essential way with respect to the system's initial conditions (See 3. below);
3. the environment is the source for whatever matter-energy the system is in need of.

G. Van de Vijver et al. (eds.), Evolutionary Systems, 67–78.

SELF-ORGANIZING AND SELF-ORDERING SYSTEMS

We do know a little bit about open systems: they exist within environments subject to conditions 1 to 3. We have also stated that open systems are characterized by SO. In other words, we assume, as a working proposition, that SO is a sufficient indication that open systems occur. With that understanding and if the proposition is correct, we should seek the overlap between "the conditions" and whatever SO is, in order to develop a definition of systems based on such an overlap; we name those systems *self-organizing systems* (SOS). With that established, we should determine the proper context from which to draw a formal representational counterpart to SOS. Such a formal counterpart will be named *self-ordering systems* (sos).

Let us run again through the conditions 1 to 3 so that we can give a description of a SOS. We change the order in the sequence for convenience. Condition 2: "the environment is invariant in some essential way with respect to the system's initial conditions". This means that the parameters of the environment that are fixed with respect to the system's dynamics play the role of initial conditions for that system; that is, the definitional characterization of the system is established *from the environment* and by fixing certain of its parameters (see Alvarez de Lorenzana, 1993). We can say that the relation between environment, E, and system, SOS, is that of interacting under a set of constraints (the set of "fixed" initial conditions). Those environmental constraints are *definitionally embedded* into SOS (in the same way that axioms are embedded into theorems). Those constraints are *necessary conditions* for SOS to exist within E. They are necessary because without them there could be no system defined; they are not sufficient because any systemic realization would need, in addition to the initial conditions, systemic laws such as a process dynamics (an operation or inference rule in the case of axiomatics).

Condition 3: "the environment is the source for whatever matter-energy the system is in need of". This condition deals with *boundaries* (initial as well as any other within the system's spatio-temporal web). This condition originates and regulates the dynamical unfolding of the system through E. It plays an important role in estimating the *potential* for systemic expansion. The conjunction of both conditions, 2 and 3, establishes the potential of a particular system on a particular environment. Beyond that there is still the element of novelty coming from outer-E conditions, that is, phenomena not encompassed by or within, E. Condition 1: "systems that *exist* (...) in an environment that is bigger than the system..." We cannot entertain the possibility of SO within an environment that is smaller than the system itself (it is like thinking in physics about structure within a black hole). Given, therefore, an environment E that is bigger than the system, its existence is equated to the system's viability. The viability of a system within a given E ("given" in the sense of conditions 2 and 3), can be equated to the ability to systematically apprehend its field of interactions based on its initial conditions. The field of interactions for the system is none other than its E. Systemic apprehension of E is the implementation of an internal (i.e. systemic) representation of E. An internal representation of E by SOS is the construction of a model of E by SOS.

In order for a model to be "sufficient" it has to represent E with respect to each and every definitional constraint of SOS, maintaining invariant what could be termed its definitional closure. The other element of condition 1 is the fact that "E is (much) bigger than SOS". So, it is not just a matter for the modeling to be relevant and accurate with respect to the definitional import of the system. That has to be attained under the added restriction of having to compress E within the system's modeling resources (e.g. amount of memory) without losing the previous attributes (relevance and accuracy). How can E be compressed into SOS when the environment is much larger than the system (E> >SOS)? Before any attempt is made to answer this question we should take stock of Ross Ashby's advice about modeling with systems, the components of which totally interact with each other: "(...) when a designer attempts to design (...) a system in which all the parts interact fully, complexity at the outputs can often be ignored: it is complexity at the inputs of the system that is to be feared" (Ashby, 1972, p. 90).

From the other side of the issue there are the critics' views saying that only oversimplified models can be implemented because any approximation to the intricate nature of the relationships of any "real" system, instantly brings about a combinatorial explosion (the one thing that Ashby, and any model maker, abhors). If we try to put together the severe warnings about input complexity and the task of any SOS which is to model E> >SOS, we are in the presence of a considerable challenge. And yet, the challenge is met and solved each time a SOS comes into being. If the problem is input complexity, the solution has to be along the lines of maximum reduction of input complexity. That is, the very foundation of the modeling process of E by SOS, under the condition of E> >SOS, is *reduction of input complexity*.

What, then, determines input complexity? At the risk of surprising some people, the answer is: input complexity is uniquely determined by the number of distinct system's components and not at all by the richness of the information to be embedded by E into SOS. Let us deal first with the latter part of the issue (i.e. information embedded by E into SOS) and once this is clear we will go into the role played by the "initial number of distinct system's components".

The information to be exchanged between E and SOS, has no role in the crucial aspect of input complexity. Reduction of input complexity due to oversimplification of E by SOS is not the answer. I said that the "embedded" information, which I have named "structural information" (Alvarez de Lorenzana & Ward, 1985), has no bearing on input complexity. Structural information I equated with semantics, as opposed to "signal information" which I identified with syntax.[1]

Now we deal with the issue of how to reduce input complexity in SOS. Let us follow Ashby's advice: "(...) how much will the informational quantities be increased if the system is changed from one having no interaction between its parts to one having full interaction between them?" (Ashby, 1972, p. 83). His answer: "(...) when full interaction is occurring, when the system is a 'whole' in the fullest sense, no mere doubling or trebling of the resources, or even a multiplying by a millionfold, is likely to be of any use" (Ashby, 1972, p. 84; more on this in Alvarez de Lorenzana, 1989, 1991). Ashby's "full interaction" or "the system is a 'whole' in the fullest sense" is the mathematical equivalent of defining a linear or total ordering over a set of elements. To have such an ordering relation is equivalent to saying that any element (of the given set) is related to (is distinct from) any other. A total ordering over a collection of components (Ashby's "parts") provides for a maximum in input complexity. This is the worst case scenario according to Ashby.

The opposite to a total ordering on a system is no initial ordering relations among system components (Ashby's "no interaction between its parts"). Such an unordered collection can only be mathematically represented as a singleton. SOS should maintain a cap on the combinatorial explosion while, at the same time, allowing system's components to interact among themselves and with the environment. Also, it should be clear that, given the fact that E>>SOS, the combinatorial explosion will always take place, the choice being only between sooner or later. The only way to minimize the *pace* toward combinatorial explosion is to make the interactions *scale bounded* and to start from the minimal possible scale. Once the system is ordered in relation to the smallest relevant scale, a new, also scale-bounded, relational space initiates its construction with, again, as short a range as possible. The process of relational scale construction continues, until the system reaches a last relational scale that encompasses all possibly feasible interactions between SOS and E, for the given system's potential. In this manner, the difficult task of modeling E with ostensibly limited resources by SOS, is achieved. The set of scales of interactions of increasing range, establishes a hierarchy of system dynamics, from the smallest to largest scale. Through the process of constructing a hierarchy of scales of system dynamics, an ordering among systemic components is implemented.

So, the material instantiation of such systemic construction of order is the self-organization of the system itself. *The system constructs its own order in the process of organizing itself.* Moreover, it is through this process of order construction among system components that the system increases its ability to model the environment, which in turn increases its viability.

USING ARISTOTLE'S CAUSAL CATEGORIES

What has been said so far should allow me to answer, at least heuristically, the initially posed question: why self-organization? I am going to answer by enumerating Aristotle's causal categories as interpreted by John Casti (1989) and Robert Rosen (1991).

Material cause: "the substrate, substance out of which a thing comes to be and which persists; that *in which* a change takes place" (Angeles, 1992, p. 42). We start with what I have been calling (Alvarez de Lorenzana & Ward, 1987) an undifferentiated system; that is, a system made out of a collection of indistinguishable (non-distinct) components. No matter how many initially given components there are, from the point of view of ordering relations the collection is a singleton. Mathematically this amounts to the unorthodox situation whereby a finite collection of components, with cardinality Card $= |M|$, $M > 1$, has ordinality Ord $= 1$. So we have a finite collection with cardinality $\neq$ ordinality which is something not permitted in set theory. This collection (of indistinguishables) is produced (defined, determined) by the imposition of a set of constraints on the environment (set of initial conditions for the system).

Efficient cause: "that *by which* some change is brought about; that which initiates activity" (Angeles, 1992, p. 42). This is component dynamics, that is, the dynamics between components and between components and the environment. It is a dynamics subject to spatial closure over a particular scale of interaction. It renders scale-bound discrimination (scattering) among components and/or environment, establishing a certain manifestation of diversity within a given scale.

Formal cause: "that shape (pattern, configuration) *into which* something is changed. The essence (the essential characteristic) being the process of becoming" (Angeles, 1992, p. 42). This category refers to the system's dynamics at the next level upscale from efficient cause. The components of this next scale are groupings of components from where efficient cause takes place. As opposed to efficient cause, where a dynamics of discrimination (differentiation) takes place, formal cause displays a dynamic of integration with respect to the entities that exist within the realm of efficient cause. It could be seen as the frame within which efficient cause acts (Casti, 1989), because the scale determined by formal cause encompasses the full range of discrimination within efficient cause. The level of formal cause is a dynamics subject to temporal closure over a particular scale of sequentiality. Mathematically, each component at the scale of formal cause is an element of the group of permutations of all the discriminations produced by efficient cause. It could be said that there exists a complementarity relation between efficient and formal causes (e.g. particle-wave complementarity in physics, the complementarity between functions in the Fourier transform in mathematics).

An interesting twist in the case of SOS is that, due to the fact that there are several levels, any two consecutive ones have efficient-formal causal relationship of complementarity between them. So in fact we have several instantiations within the same SOS of efficient-formal causal categories.

Final cause: "that *for the sake of which* an activity takes place; that end (purpose, goal, state of completion) *for which* the change is produced, or for which the change aims (strives, seeks)" (Angeles, 1992, p. 42). In our scheme final cause is the pursuit of complete ordering within the system or at least, order up to total separability of system components. In other words, final cause generates the construction of a partition over the initial collection of (indistinguishable) systemic components: a partition over the material cause. It could also be said that final cause is the act of inducing the maximum amount of information over the (initially uninformed) material cause.

SYNTHETIC AND ANALYTIC APPROACHES TO MODELING

Let us break-off from our train of thought for a moment in order to bring in informed views on modeling. I am thinking of Robert Rosen who has an extensive work on the subject (Rosen, 1991, 1993, 1995). In his seminal book of 1991, Rosen divides the area of modeling into two main approaches: synthetic and analytical.

1. Synthetic models: The very essence of this modeling scheme is the underlying assumption that a system, X, can be comprehensively conveyed by the union of disjoint (non-overlapping) subsystems or fractions of X. Moreover, those "fractions" have to be context-independent so that a particular definition is valid in any environment (Rosen, 1995, p. 33). The mathematical representation of such a system takes the algebraic form of direct sums (in set theory, union of disjoint sets). This is a clear reductionist view. By accepting fractionation it is implied that none of the connectivity among systemic components has any relevance for the system as a whole.

2. Analytic models: This approach starts with the whole system as a given and then proceeds to tear it to pieces. The system is defined as a cartesian product or direct product. Any piecewise solution will be implemented by means of quotient sets. Because in this case what is given is the totality of the system, we cannot assume disjointness among subsystems and, any severing of the whole in terms of quotient sets, will not permit us to recover the system in full (at least not necessarily). According to Rosen, the inability to recover all the systemic information from the totality of the pieces into which the system has been cut, is a manifestation of the irreducibility of semantics into syntax.

The big dilemma in the synthetic-analytic dichotomy is the following:

1. If we start with a disjoint union of subsystems ("fractions"), we can construct the system. But, by assuming fractionality, we dismiss connectivity between subsystems as irrelevant. This amounts to saying that whatever systemic information there is, it is all inside the subsystems (it is from that inner information within the "fractions" alone, that a system can be built).

2. If we start with the system as a whole, it cannot be partitioned (analyzed) without losing essential information (the connectivity among parts); so much so that once severed into pieces of analysis, the system as such is lost. Rosen's views on fractionality are that only simple systems, i.e. machines, can be fractionated and built from pieces. In this particular class of systems, syntax and semantics coincide. When that is not the case, we are in the presence of complex systems. In Rosen's own words: "The identity of these two quite different ways of talking about 'states' of X is a direct consequence of supposing that 'analysis' into fractions, and 'synthesis' from these same fractions, are inverse operations" (Rosen, 1995, p. 35).

APPROACHING MODELING AS SELF-CONSTRUCTION OF ORDER

The basic consideration about what was said in the last section is that the two options given by Rosen do not cover all the possibilities for modeling; in fact, it

leaves out a major class of systems, precisely the ones that go under the rubric of SOS. Let me use an analogy. Imagine for a moment the following scenario: We have a two-dimensional spatial map of a particular geographic zone in which the flora of that region is plotted. We are faced with the task of determining the kind of ecological dynamic that is present between the different species of flora. We have three different map-options that we give to three different people: A, B, C. Person A gets a map as a whole picture. Person B gets a map cut into irregular small pieces as is done with puzzles. Person C gets a map that can only be looked at by means of a *zooming device*; such a device imposes the constraint of an inverse relation between its resolving power and the range of view, that is, if we want to look at the map in the utmost detail (a point or pixel), then the area that we can cover is the point itself and possibly a few other points within a short-range radius. If we want to "see" a broader area we will have to do so by proportionally diminishing the degree of resolving power, i.e. we would have to give up detail in order to attain a broader view. Now we ask the first two persons, A and B, to causally explain the relations among the different species of flora present in the map. In this scenario:
1. person A will have to do an analytic modeling; person B, a synthetic modeling.
2. Although person B has no whole picture of the map, each and every piece (element) is *distinct*, that is, is distinguishable from and among all others. The fact that the elementary pieces are distinct is a remarkable property and it should be looked at in some detail.

Let us assume that the number of pieces is n. If we were to have n disjoint, non-distinct, pieces, we could only define n unary relations (a unary relation is a property of the object to which that relation applies, in this case to a 'fraction' or subsystem). If, on the other hand, we are considering (as Rosen seems to do) n disjoint and distinct pieces, we would have defined one n-dimensional relational space. The dimensionality of this relational space turns out to be the same as the dimensionality of the analytic option, the reason being that if we can distinguish any element from any other it is because those elements are embedded within a relational space of the same dimensionality as the number of elements, that is, n. So, as long as we are working with n distinct elements, whether analytically or synthetically, we are de facto in an n-dimensional relational space. Both, analytical and synthetical approaches, although very different in some respects, are implemented or take place in one and the same relational space: at least up to the total number of fractions where, after all, the connectivities among systemic components are defined.

3. Dimensionality is of the utmost importance because it tells us that what is giving ground to the condition of "distinctiveness" among initial building blocks (or pieces, or "fractions") of the puzzle is the fact that the representation takes place within one macro scale only (that is, a scale that encompassed all the possible connectivities among pieces, "fractions", etc.). This is equal to assuming that the same ordering relations are present throughout, that is, in both local and global phenomena from the outset (it amounts to assuming that every fraction has relational access to every other "fraction").

4. Person A has access to the map in a way that B does not. A shares B's total ordering (which comes from the distinctiveness of the pieces) and also carries an ordering that determines the relative position (coordinate) of all the pieces within the map as a whole. This latter ordering is conveyed through the cartesian product and it amounts to making it very easy to figure out what the relation is between any point and the rest of the picture. B on the other hand has to make the puzzle (i.e. construct an ordering) before reaching that "easiness".

5. Person A has no ability in establishing local and global relationships within the picture because, as with B, A's map is defined on one macro scale: the scale given by the initial distinctiveness (total ordering) among pieces which, in the case of A, are labeled as coordinates. From this example and back to Rosen's options we can conclude the following: although the two options for modeling (analytic and synthetic) are very different in the way they label the definitional set representing natural systems, and although the analytic one does carry some extra ordering referring to the system as a whole, both methods share the inherent limitation of not having a *zooming device* (such as in option C) that would allow them to locate specific classes of system states (i.e. classes of coordinates for A or classes of pieces for B, in Rosen's examples) to specific spatio-temporal systemic levels of a developmental dynamics. Without this "device" in the representational language, we are forced to renounce modeling any natural system carrying the ability of constructing its own systemic order (going from local to global connectedness). The "device" is an *encoding device* that would allow us to transfer a set (or hierarchy) of scale-bound processes taking place within the natural system (SOS) into a set of levels in the formal system (sos); without such an ability to transfer information between natural and formal systems, there is no modeling relation possible (that is, the modeling diagram would not commute).

To summarize: the monoscaled (one and only macro scale) nature of both analytic and synthetic modeling, overshadows their differences and reduces their applicability to systems where self-construction of order cannot be formally implemented in a genuine way. We need to formally represent inner systemic scale dependency in order to model systems which are open to the environment and that come into being as a restriction of that environment, and become viable because of their ongoing modeling process of that environment. Another way to state the critical question that we are trying to grapple with is the following: does a process of (systemic) self-construction of order necessarily carry information about the genesis of the process itself? If so, should its formal counterpart carry it too and how? Rosen's two options for modeling preempt the previous question in the sense that both assume that all the information is present from the outset. That being the case, nothing is generated and we only need to pick up the information that is already there. And the reason why we only need to pick up what is already there is because we are looking from the outside. Rosen's analytic and synthetic perspective are at all times what, in terms of Otto Rössler (1987), would be called *exomodeling*: modeling from the point of view of an external observer. That is why, when Rosen talks about the synthetic building blocks (see synthetic model above) he defines them as a direct sum of (disjoint) pieces: he is in front of a *pile of pieces* of a puzzle.

Perhaps the most essential point of this discussion is to clarify why it is that Rosen interprets direct sums as a family of disjoint *but distinct* elements. The answer is again: because he is in the position of an external observer looking from a mono (macro) scale. From the SOS perspective, the only feasible position for an observer, has to be an internal view, that is, the observer is inside the system. What SOS does is *endomodeling* (Rössler, 1987; see also Salthe, 1993, pp. 166-173), that is, modeling from inside the system. Still we have to add (to the endomodeling) a "zooming device" in order to render SO and SOS. By defining the system as a set of constraints imposed on the environment (condition 2) we are doing two things: 1. we fix the environment in terms of a set of necessary and initial conditions (equivalent to axiomatic closure in mathematics), that allow the system to be defined; 2. we define the system internally, devoid of any (external) connectivity among initial components and, therefore, at its maximum degree of potentiality, that is, at its point of indistinguishability: in space, in time and in class membership of components (see Alvarez de Lorenzana, 1991, 1993, pp. 305-308). This second condition translates, mathematically, into saying that the direct sum of system components is, in fact (being defined over a collection of elements that belong to one and only one equivalence class), a singleton. In other words, we don't have a pile of distinct pieces of the puzzle but a *bag of indistinguishable* (equivalent) *components*. The term "bag" is used in computer science (Scheurer, 1994), or, equivalently, the term "multiset" (Blizard, 1989).

SELF-ORGANIZATION AND SELF-CONSTRUCTION OF ORDER

The previous survey on SO and SOS, should provide a general view of basic requirements for their formal representation. The key element to point out is the role played by the systemic construction of ordering relations: it is the very core of SO. In other words, *self-organization is for natural systems what self-construction of order is for formal systems*. SO is a realization (in the mathematical sense, see Rosen, 1991) of self-construction of order. If we agree on that, then knowledge about the way in which such a construction takes place is essential. The deeper and sharper our formal representation is, the better chance we have of developing a good understanding of SOS and the better corresponding ability to make models for simulation and experimentation.

It is my assessment that, up to now, a sound and clear formal scheme for systemic multiscaled phenomena is lacking. This is so whether in statistical or analytical mechanics. Even non-linear approaches seem to be in difficulty when trying to represent causality within systems which span over a hierarchy of scales of interaction. If my assessment is correct, then, none of those options would give us the necessary degree of conceptual sharpness and overall control required for multi-scaled SO. Only in the use of algebraic approaches have I found a power of description capable of matching the intricacy and subtlety of the phenomena at play. Let us be clear and say it again, I am referring to SO of complex systems, that is, systems that carry and develop SO and therefore implement self-construction of order to the fullest, in other words, make use of the full range of the hierarchy of scales (that is what the condition E> >SOS means). It is precisely the multi-scaling nature of systemic phenomena that makes, in my view, those formalisms mentioned before either exceedingly difficult or artificially cumbersome or ad hoc, or not sufficiently specific. There is a serious problem with any realm of phenomena that takes place across several scales. Physics, the "hard" and mature natural science par excellence, is a good example of such a problem; the many formalisms attempting unification of the four basic forces is a sample of the unfinished struggle in the formalization of multi-scale phenomena. Non-linear dynamics and chaos work well within the micro-macro paradigm. This two-level framework does not allow careful formalization of more than two levels. Remember that scales of interaction exist in SOS as a way to curtail the ever-present risk of combinatorial explosion (resulting in, if not avoided, the breakdown of the system or, at best, partial collapse into previous levels). At the same time the viability of the system is based on good modeling of the environment (which means the intrinsic necessity of increasing the scale of system interaction across the environment).

The systems' existence lies in between two opposing needs: the need to expand and the need to control. If it does not expand to favour control, it becomes vulnerable because the control is exerted over too small an area. If it loosens control, favoring expansion, the system also becomes vulnerable because it diminishes its grip over the covered environment. The only way to negotiate between those two opposing needs while attaining SO is by means of scaling, not any arbitrary set of scales, but the set that both *minimizes interaction growth and maximizes control*. There is a very critical balance that has to be maintained and optimized. It is my view, that the algebraic road is the one more readily available. It is also my belief that once such a formalism is well established it could make incursions into analytical and statistical mechanics if needed.

DIFFICULTIES IN FORMALIZING SELF-ORGANIZATION

Before ending the description of this viewpoint, we should be aware of the fact that algebra is also marred with implicit assumptions that are questionable. The advantage with algebra is that it has ways to circumvent or overcome, to a considerable degree, its shortcomings. Category theory is an important, relatively new, tool (see Golblatt, 1979; Gago, 1986) that allows to formalize construction of order. The questionable assumptions that I just mentioned are present across the whole realm of mathematics. In particular, I am referring to taking as a given the existence of natural numbers. Kronecker's well publicized phrase declaring that only the whole numbers came from God, all else is the work of Man, expresses the rational grounds for such a claim. According to mathematicians, *N* is the most elementary entity to be encountered and therefore has to be assumed:

"We cannot expect that the natural number sequence can be defined in terms of anything essentially more primitive than itself..." (Stoll, 1979, p. 57). Such statement is shocking coming from the SO side of things, where order of any kind is not trivial and therefore has to be constructed. We have seen how difficult the construction of order is in the world of the real where the threat of a combinatorial explosion never fades away. From our vantage point, to assume *N* as a minimal proposition is not only not necessary but very damaging for any formal conceptualization of self-organization. Mathematics, as we know it today, is to a considerable degree based on this assumption. The call for an inquiry into the foundations of the construction of order is pervasive.[2]

ACKNOWLEDGEMENTS

I want to acknowledge long and engaging conversations with Stan Salthe while visiting him in June 1995 and thank him for his involvement in proof reading my draft. I also want to thank Gertrudis Van de Vijver for her comments on my paper and all her help before and after the ISES conference.

Institute of International Relations, University of British Columbia
Canada, USA

NOTES

[1] There is, of course, an important relationship between semantics and syntax (as portrayed in that paper and to which I still subscribe). Syntax is a carrier of semantics in a process of increased refinement of discrimination over an initial, undifferentiated, semantically closed system. A good analogy is an axiomatic system in mathematics, where, starting with axiomatic closure, theorems are constructed. Clearly, the theorems of such a system are syntactical constructs. They also convey a certain math-ematical "meaning" that escapes syntax but which is specific to that particular system of axioms (e.g. Euclidean triangles have a different "meaning" than hyperbolic triangles). The "meaning" of the theorems comes from the specific nature of the set of axioms, but the consistency of the theorems has nothing to do with that particular set of axioms (save axiomatic closure) and everything to do with the rules of inference (syntactic rules), the stuff of which theorems are constructs. So, the "meaning" of the axiomatic structure does not have any immediate impact on the theorems. The axiomatic structure could be relevant or irrelevant (e.g. in relation to possible usefulness of the theorems), but that aspect of the axiomatic structure does not have an effect on the theorems. This analogy (for a more in-depth look see Rosen, 1991) should clarify any doubts about the necessarily separate realms on which both semantics and syntax operate. Conversely, it also shows the impossibility of reducing semantics to syntax alone because the roots of semantics lie, ultimately, outside the system in question (are meta-systemic).

[2] The author is presently attempting to develop an implementation of these foundations.

REFERENCES

Alvarez de Lorenzana, J.M., 1989, *Las Relaciones Internacionales en el contexto de la historia. Un caso de sistemas evolutivos jerarquicos*, PhD Dissertation, Madrid, Universidad Complutense de Madrid.

Alvarez de Lorenzana, J.M., 1991, Informal Comments on Indistinguishability, the Combinatorial Hierarchy and Evolutionary Systems Framework, in *Proc. 7th Ann. Mtg. ANPA West*, F. Young (ed.), Stanford, Palo Alto, CA.

Alvarez de Lorenzana, J.M., 1993, The constructive universe and the evolutionary systems framework, in *Development and Evolution. Complexity and Change in Biology*, S.N. Salthe, Cambridge, Mass., The MIT Press.

Alvarez de Lorenzana, J.M.; Ward, L.W., 1985, Semantic and Syntactic Information, *Proc. 29th Ann. Mtg. Society for General Systems Research*, Louisville, Kentucky, pp. 78-86.

Alvarez de Lorenzana, 1987, On Evolutionary Systems, *Behavioral Science*, vol. 32, pp. 19-33.

Angeles, P.A., 1992, *HarperCollins Dictionary of Philosophy*, New York, HarperCollins Publishers.

Ashby, R.W., 1972, Systems and Their Informational Measures, in *Trends in General Systems Theory*, G. Klir (ed.), New York, John Wiley & Sons.

Blizard, W.D., 1989, Multiset Theory, *Notre Dame Journal of Formal Logic*, vol. 30, pp. 36-66.

Casti, J., 1989, Newton, Aristotle, and the Modeling of Living Systems, in *Newton to Aristotle. Toward a Theory of Models for Living Systems*, J. Casti; K. Anders (eds.), Boston, Birkhuser, pp. 47-89.

Gago, F., 1986, *Topos Theory, Why?*, Santiago de Compostela, Servicio de Publicaciones Universidad Santiago de Compostela.

Goldblatt, R., 1979, *Topoi, the categorical analysis of logic*, New York, North-Holland Pub. Co.

Rosen, R., 1991, *Life Itself. A Comprehensive Inquiry into the Nature, Origing, and Fabrication of Life*, New York, Columbia University Press.

Rosen, R., 1993, On Models and Modeling, *Applied Mathematics and Computation*, vol. 56, pp. 359-372.

Rosen, R., 1995, The Mind-Brain Problem and The Physics of Reductionism, *Communication & Cognition-Artificial Intelligence*, vol. 12, pp. 29-43.

Rössler, O.E., 1987, Endophysics, in *Real Brains. Artificial Minds*, J. Casti and K. Anders (eds.), New York, North-Holland.

Salthe, S.N., 1993, *Development and Evolution. Complexity and Change in Biology*, Cambridge, Mass., The MIT Press.

Scheurer, T., 1994, *Foundations of Computing. System Development with Set Theory and Logic*, Reading, Mass., Addinson-Wesley.

Stoll, R.R., 1979, *Set Theory and Logic*, New York, Dover Pub. Inc. (republication from W.H. Freeman and Co., 1963).

WERNER CALLEBAUT

SELF-ORGANIZATION AND OPTIMIZATION: CONFLICTING OR COMPLEMENTARY APPROACHES?

The attraction of optimality models is clear — they allow one to avoid history and genetics. Years ago in a discussion about number theory, Bertrand Russell said, "The method of 'postulating' what we want has many advantages; they are the same as the advantages of theft over honest toil. Let us leave them to others and proceed with our honest toil." (1919, p. 71). These are exactly our thoughts with respect to optimality models and the rigorous test of adaptationism"

- Brandon & Rausher, 1996

"In a naturalistic world view of the operation of functionally organized evolving systems, it is heuristics all the way down as far as that perspective is meaningfully applied — i.e. as far down as entities or configurations of them can be products of selection"

- Wimsatt, 1997a

INTRODUCTION

In the current phase in the long-standing debate over structuralism vs. functionalism in biology (Riedl, 1975, 1983; Gould, 1977, 1980, 1983; Mayr, 1982; Richards, 1992; Asma, 1996) attempts are made to revive the Aristotelian doctrine of the four causes to further the structuralists' case. For the purposes of this paper, structuralism may be defined very broadly as any "attempts to explain entities — biological or social — in terms of relationships or organizing principles, rather than as a sum of parts" (Mae-Wan Ho). Self-organization as "a process in which pattern emerges at the global (collective) level through interactions among the components of the system at the individual level, without these interactions explicitly specifying the global pattern" (Callebaut & Deneubourg, 1995) is an obvious candidate for structuralist explanation. In this paper I assess pleas for causal pluralism from the perspective of a thoroughly naturalized philosophy of science (section 2) by linking the causality debate in biology to the more general issue of the status of optimality principles in the sciences: physics and engineering, biology, and social science, in particular economics (section 3).

Optimization arguments are not just the magic cloak the economist puts on in order to demonstrate the methodological if not theoretical equivalence of the "queen of the social sciences" with Physics;[1] they have also become "the foundation upon which a great deal of theoretical biology now rests" (Oster & Wilson, 1978, p. 292). Although "adaptationism" or "Panglossian functionalism" has been the target of systematic criticism in biology (Gould & Lewontin, 1979; Levins & Lewontin, 1985; Riedl, 1995; Rose & Lauder, 1996b; Seger & Stubblefield, 1996; cf. the counterclaims of Williams, 1985, 1992, and Dennett, 1995), psychology (Sober, 1985; contrast Rachlin et al., 1981), and economics (Hodgson, 1993) for almost two decades now, many scientists in these and related

G. Van de Vijver et al. (eds.), Evolutionary Systems, 79–100.

fields continue to be unaware of the rationalist and perfectionist pitfalls of optimization as conventionally conceived (section 4). Qua scientific idealization, optimality is both descriptively and normatively inadequate or, as Hooker (1994b) puts it, "degenerate". An alternative, *bounded rationality* approach along the lines of Herbert Simon's theory of *satisficing behavior* is more in tune with the naturalistic Zeitgeist (section 5).[2] Optimization tools need not be abandoned altogether: if put on a different epistemological footing, viz., a general, thoroughly naturalized (i.e. fallibilist) theory of heuristic problem solving that accords with the biological and cultural nature of human reason (Hooker, 1994a; Callebaut, 1995c; Wimsatt, 1997a, 1998), optimizing becomes a limiting case of satisficing, and its real potential and limitations can be appreciated better. I conclude by pondering how self-organization theorists can benefit from taking the optimizing/satisficing controversy seriously.

THE PHILOSOPHER OF BIOLOGY AS PARTICIPANT OBSERVER

Scientists, like most ordinary folk, "live forward" (Dewey): the past is important to them mainly as a basis for future action (Nickles in Callebaut, 1993, pp. 31-32). The current concern among biologists and systems theorists with the Aristotelian doctrine of the four causes (e.g. Riedl, 1978/79; Rieppel, 1990; Alvarez de Lorenzano, this volume) is such a case of *reculer pour mieux sauter*: a historical resource is appropriated opportunistically, because one hopes it may help solving current research problems.[3] To the professional historian's dismay, scientists involved in this exercise rarely take into account such things as the wider metaphysical context in which Aristotle's view was embedded (e.g. the importance of the dichotomy of temporality and timelessness),[4] or the implications of his identification of *change* with *movement* (e.g. how his 'biology' relates to his 'physics'), or, more generally, the very possibility of incorporating classical philosophical ideas into the (post-)modern scientific world view. Recall that what inspired one of the central tenets of Thomas Kuhn's (1962/1970) model of science, the "incommensurability" of paradigms, were his difficulties to come to grips with the Aristotelian world view as a modern physicist. Good science tends to be bad history (Nickles, 1990), and one useful thing philosophers can do (one hopes) is investigate why scientists tend to write their own history whiggishly — through present-day spectacles — and how they do so in specific cases. In the next section I thus try to make sense of the often voiced claim that modern science has focused on efficient causation to the detriment of the three other Aristotelian causes. The mere fact that most of us relish in Whig history, even when we are aware of its dangers, suggests that Whig history serves a function in scientists' looking forward (Wimsatt in Callebaut, 1993, pp. 37-38). In a different mind set, philosophers of science or scientists acting as philosophers can also stand back and try to make sense of the larger picture, asking retrospectively, say, what the Darwinian revolution was all about in terms of societal context, methodology, epistemology, metaphysics, or even ethics. Very rarely, the backward- and forward-looking perspectives are happily married (e.g. Depew & Weber, 1995).

On one influential view, philosophy is the *anti-chambre* of science in the sense that today's (good) philosophy prepares the ground for tomorrow's (revolutionary) science.[5] The work of most scientists in this volume is philosophical at least in the sense that they deal with issues for which no well-established approaches, let alone solutions, exist to date. Conversely, one consequence of the recent naturalization of the philosophy of science is the blurring of the distinction between science and philosophy, sometimes to the point where the two become inseparable — whether one likes this (Hull, 1969; Wimsatt, 1997a) or not (Fuller, 1994; Reif, 1995).[6]

The philosophy of biology has come a long way since David Hull's rather cynical assessment of the state of the field almost two decades ago (Hull, 1969). One attractive dimension of the naturalistic turn in current science studies, in which the life sciences feature prominently along with cognitive science, is the careful attending to *matters of fact* (Giere, 1988), *model building* as "the cutting edge of theory construction" (Wimsatt, 1997a; e.g. Darden, 1991), *experimental design*, as well as other material aspects of the scientific enterprise.

The old dichotomies on which the formalistic and reductionistic view of science of the logical empiricists was based — analytic vs. synthetic statements, theory vs. observation, epistemologically relevant justification vs. epistemologically irrelevant discovery, normative philosophy vs. descriptive history, psychology and sociology — have all been given up in favor of more sophisticated distinctions (Callebaut, 1995a,b,c). *Pace* what some social constructivists want us to believe, the room vacated by positivism has not been invested by a new hegemonic view of science and its place in the world. The kind of *pragmatic pluralism* (Wimsatt, 1997b) in matters epistemological and methodological that emerges from a genuine symbiosis of philosophy with science must not be mistaken for an 'anything goes' relativism. The radically relativistic or solipsistic tenets of some of the currently fashionable varieties of constructivism (see Niiluoto, 1991 and Kitcher, 1994 for critical reviews) may be properly diagnosed as consequences of the continued dependency on long superseded rationalist or empiricist philosophies and, even more importantly, *metaphilosophies*, as Hooker (e.g. 1987, 1994a) and Wimsatt (1995, 1998) have argued independently.[7] Not only does a properly understood and qualified (constructive, local) *evolutionary realism* remain a feasible option, but there are good reasons to regard it as the epistemological position that resonates best with a principled, overarching naturalism (e.g. Hooker, 1987, 1994a; Giere, 1988; Wimsatt, 1994, Azevedo, 1997).[8] (Note also that there are intricate connections between structuralism in biology and realist philosophy of science; see, e.g. Verhoog et al., 1995, p. 3.) In Wimsatt's words: "We need to be much less absolutist and much more contingent, contextual, and historicist in our analyses of science. But we must do this by recognizing the real complexities we are increasingly able to study in natural systems whose simplicity we have been taking for granted for decades or centuries. A major fraction of these complexities are not a function of our conceptual schemes, language, or interests, but products of the way the world is.

Realism lives! But any wise realist must recognize that the social, cultural, and ideational entities of the 'social relativists' are real, too, and imbed them with the natural entities we theorize about. We do this via the idealized models we construct and the carpentered, 'natural' entities, tools, practices, procedures, and phenomena we experiment with and on. We recognize we are guided, regulated, and maintained by the social structures, languages, and values of science as part of the appropriate (panrealist) world picture." (Wimsatt, 1995, p. 494)

I take for granted here that the pragmatically oriented new naturalists have good grounds for being dissatisfied with the excessive formalism that was typical of the positivist and Popperian approaches to scientific rationality and remains the hallmark of most work in rational decision theory and game theory today (Wimsatt, 1976, 1979; Hooker, 1981, 1987, 1994a). But I cannot round off this bird's eye survey of the new philosophy of biology without including a caveat on the interrelations between naturalism, reductionism, and scientism (I offer a more substantial analysis in Callebaut, 1998).

Is naturalism reductionistic? The kind of reductionism I take to be defensible and desirable, viz. *explanatory reduction* or *reductive explanation*, and which differs in important respects from the subsumptive intertheoretical reduction of the logical empiricists, for instance, by making room for *emergent phenomena* (Wimsatt, 1976, 1979; Callebaut, 1995b) is compatible with a principled naturalism. Roughly, the argument goes like this: If we accept the fact of biological evolution, it is not far-fetched to claim that (only) in humans has evolution become self-conscious or reflexive. From this ontological premise, the naturalist's methodological agenda follows directly, viz., "to develop relevantly unified theories" (Hooker, 1987, p. 262; Azevedo, 1997, ch. 7). The naturalistic ideal of a *methodological unity of the sciences of nature and culture*, the view of *philosophy as continuous with science*, and the requirement of *testability*, viz. the claim that *matters of fact are as relevant to philosophical theory as they are to science* (which signalizes a more general concern about the productive integration of theoretical and empirical levels of inquiry) are successive ramifications of the naturalist's methodological agenda (Callebaut, 1993, 1997). Toward the end of the perceptive "Interpretive Afterword" to their first book, *Evolution at a Crossroads*, Depew and Weber expressed a double concern: "(...) the persistence of natural theological discourse, and perhaps even of residual tendencies to vitalism, together with the desirability of confuting these, might suggest that it is advisable to retain some attachment to reductionistic ideals. These ideals, whether they are given methodological, theoretical, epistemological, or ontological import, are still powerful tools for deconstructing attempts at non-natural explanations. But this may be to take these residual phenomena too seriously, if the effort diminishes our recognition that our knowledge is most perspicuously encoded in a fully naturalistic, but definitely non-reductionistic, conceptual matrix. Such a preoccupation might even retard efforts to attack the most pressing problems in evolutionary biology (Depew & Weber, 1985, p. 255).

In my view it is one thing to be wary of the scientistic proclivities inherent in many varieties of naturalism — Quine's would be a prime example (Callebaut,

1995c, 1998; Code, 1996), another to reject reductionism and/or mechanism wholesale, as, e.g. Goldsmith (1990) or, for that matter, quite a few systems theorists are willing to do. A brief glance at what the average bookstore in Europe or the United States offers under the label of "Life Sciences" also makes me less optimistic as to the negligibility of Depew and Weber's "residual phenomena" than I would like to be. In what follows, I will try to substantiate this claim by investigating the 'four causes' debate.

FROM TELEOLOGY TO CAUSALITY AND BACK?

In his *Causality and Modern Science* (1979), Mario Bunge described the historical development of Western science in terms of its progressive naturalization in the following sense:

(N1) a progressive restriction of causation to *natural* causation,

(N2) the further restriction of all varieties of natural causation to *efficient* causation,

(N3) the endeavor to reduce efficient causes to *physical* ones (*mechanism*),

(N4) the requirements of *testing* causal hypotheses by means of repeated observations and, whenever possible, through reproduction in controllable experiments, and

(N5) *parsimony*, understood as both caution in the assignment of causes and minimization of the number of 'ultimate' natural causes.

One could add

(N6) the current naturalization of the philosophy of science itself (self-reflexivity of science) to this list.

In this context several remarks are in order. First, the historical movement described by Bunge still continues today, and many aspects of it are to be welcomed as genuine scientific progress. Consider as an example the distance covered in the last three decades in social-scientific thinking about the 'development' of 'Third World' countries. Crude thinking in terms of an *evolutionist*, i.e. linear and teleological, logic — typically cast in terms of stage laws such as Rostow's (1960) "stages of economic growth"[9] — has given way to *evolutionary*, i.e. multi- or non-linear, context-sensitive, dynamical, and teleonomic thinking (e.g. Riggs, 1970).[10,11] Or consider Dennett's "killjoy", i.e. a purely causal physical mechanism (say, oleic acid secretion in bees) explaining apparently intentional behavior. Whether simple or complicated, finding a killjoy is a way to "pay the debt" in the sense that killjoys "allow us to replace the intentional stance by a plausible mechanistic picture" (Atlan, 1994, p. 78). Gradually replacing pieces of "the metaphorical web that surrounds theories of selection in biology and the social sciences" by point-at-able mechanisms — for instance, by redirecting the focus of researchers "from genes to levels of organization that may well serve as causal units in selection", as Bradie (1994, p. 612) provocatively puts it — still seems to me an important (if not the only) way for science to go.

Evolutionary naturalist realists among the philosophers of biology of the current generation like William Bechtel, Robert Richardson, and William Wimsatt articulate accounts of explanation and reduction in which the 'bottom up' aspect — mechanisms and causal factors, which may be 'single shot' events — feature prominently, and in which the 'top down' role of (exceptionless) laws of nature is far less important than on the traditional subsumption account of explanation due to Hempel and Oppenheim (Wimsatt, 1976, 1997a; Bechtel & Richardson, 1991; Salmon, 1993; additional references in Callebaut, 1993, ch. 4, 1995b).

Second, talking about "a reduction of the four Aristotelian causes to one" (Rieppel, 1990, p. 293), viz. *causa efficiens*, is ambiguous. It is correct that at the *foundational* level, attempts to rephrase functional analyses and explanations in terms of causal explanations and analyses have been legion (e.g. Nagel, 1977; Rosenberg, 1985, 1994), and that a similar reduction of intentions, reasons etc. to causes is being attempted since a decade or so (e.g. Nelson, 1987; Dretske, 1988; Atlan, 1994; Juarrero, this volume).[12] Now quite obviously, these foundational efforts have not kept *practicing scientists* from invoking functions, intentions and what-have-you ad libitum, to the extent that Elster (1983) was able to conclude that explanation in the physical sciences is essentially causal, explanation in (evolutionary) biology is essentially functional (cf. Hull, 1982 and Rosenberg, 1985 on the ubiquity of functional explanation, even in molecular biology), and social-scientific explanation is essentially intentional. One can also think here of Gould's quip that teleology is the evolutionary biologist's mistress: he can't live without her but does not want to be seen in public in her presence. Confusion of these two levels explains at least part of the persistent misunderstanding or even non-communication among scientists working on the levels of proximate ('how?') and ultimate ('why?') causal analysis, respectively (see, e.g. the biochemist Schoffeniels' (1976) arguments against Monod), or, what amounts to the same thing, efficient-cause analysis vs. final-cause analysis — the latter being exemplified by optimality analysis: Oster and Wilson (1978); Alexander (1982); Kitcher (1987); Rachlin (1993).[13] At the pragmatic level, both efficient causation and purpose appear to be psychological primitives that we naturally use to make sense of our surrounding world (Schoemaker, 1991).[14] Schoemaker's contention that it remains unclear which of these two crucial concepts is *epistemologically* more objective or valid cannot be extended to the foundational level, however, where the only plausible evolutionary scenario is one in which matter and energy transfer preceded function or purpose. As to the odds that the naturalistic reduction program (in particular, Bunge's N2 and N3) succeeds or fails, it is probably fair to say that the genesis of functions, including the 'highest' ones (intentionality), is explainable in purely causal terms *in principle*,[15] though not necessarily *in practice*.[16] (cf. Wimsatt, 1976, 1979, 1997a on the misleading character of 'in principle' talk.)

My third and last remark concerns the relation between the 'four causes' issue and the debate concerning adaptationism and self-organization. Aristotle invoked the four causes as complementary rather than competing explanations, and thought all four were necessary and jointly sufficient to capture all kinds of phenomena, including complex living things (Wimsatt, 1989, p. 1; Rieppel, 1990). This does not rule out the possibility of certain reductions; e.g. in organisms, the formal cause (cf. *Bauplan*, equation) and the final cause may be (but mustn't be) seen as identical; and in human action, efficient, formal, and final causes can coincide (Wimsatt, 1989, p. 1). More importantly, although this is sometimes suggested in the literature on structuralism or self-organization, it would be quite misleading to oppose 'full-fledged' structuralist explanations in terms of the four causes (with generative rules as the formal cause[17]) to 'empoverished' functionalist (adaptationist) explanations with natural selection as the (only) efficient cause. For if natural selection is the creative force neo-Darwinians such as Mayr claim it to be, the functionalist research program may claim to address the formal cause of evolution (viz., adaptation) as well, as Rieppel (1990, p. 311) admits. Cf. also Amundson (1994, p. 572) on the difference between *hard adaptationism* ("Any developmental constraints can be [and have been] overcome by the forces of natural selection") and *soft adaptationism* ("A constraint on form must result in a constraint on adaptation").

ADAPTATIONISM AND OPTIMIZATION

Optimality analysis in biology has roots in the physics envy of biologists; with Fisher's (Not So) Fundamental Theorem of Natural Selection, "teleology had been given mathematical respectability" (Oster & Wilson, 1978, p. 293). Going back further, natural theologians read divine intent into physical laws interpreted as extremum principles.[18] Through the widespread but wrong assumption that "evolutionary processes lead generally in the direction of optimality and efficiency" (Hodgson, 1993, p. 197), optimality is linked quite directly to the adaptationist program.[19] Adaptationism may be loosely defined as the view that "adaptation to the environment is the guiding force [sic] of evolutionary change" (Dobzhansky, 1956, p. 339).[20] More specifically, adaptationists maintain that natural selection is the sole evolutionary force responsible for the adaptation (in the sense of 'resulting in a local optimum'[21]) of the traits of individuals in a population[22] (Oster & Wilson, 1978, p. 292; Campbell, 1989; Hodgson, 1993, p. 197; Ridley, 1993, pp. 326-350; Brandon & Rausher, 1996). Although adaptationists need not — and nowadays seldom do — insist that *all* evolutionary changes in nature are adaptive[23], the adaptationist fallacy, their critics contend, is to take for granted that "nature is an optimizer" (Hodgson, 1993, p. 197), and to "systematically exaggerate the prevalence of adaptations in biology" while being "insensitive to possible non-adaptational explanations of biological phenomena" (Amundson, 1994, p. 556). It may be useful in this context to stress that in their famous attack on adaptationism, Gould and Lewontin (1979) did not accuse the adaptationist research program of unfalsifiability, but rather claimed that when individual adaptationist hypotheses are falsified, they are immediately replaced by other ones (cf. Callebaut, 1993, pp. 400-403). "What seems never to be falsified is the belief that the trait is an adaptation of some kind" (Amundson, 1994, p. 557). Whereas Darwin invoked imperfections in nature as evidence for his naturalistic explanation of life, the latter-day 'Panglossian' adaptationists regard the existence of perfectly adapted structures and behaviors as evidence of evolution by natural selection. In Levins and Lewontin's words: "The modern adaptationist program, with its attempts to demonstrate that organisms are at or near their expected optima, leads to the consequence that although species come into existence and go extinct, nothing really new is happening in evolution. (...) There is no progress because there is nothing to improve. Natural selection simply keeps the species from falling too far behind the constant but slow changes in the environment" (Levins & Lewontin, 1985, p. 84).

The problem, however, is that natural selection will be a sufficient explanation of all, or most, of evolution only if all possible variants eventually come to exist. If there are significant biases in the production of variation, something beyond natural selection will be needed (Brandon & Rausher, 1996).

Not unlike the biological adaptationists, economists such as Milton Friedman and Friedrich Hayek have postulated that the kinds of behaviors that are selected in competitive market environments are *necessarily* superior[24] and thus justify their profit- or utility-maximizing assumption, i.e. the economic equivalent of neo-Darwinian fitness maximization (cf. Dawkins, 1995).[25] (For critical assessments, see, e.g. Alden Smith, 1987; Hodgson, 1992, 1993; Campbell, 1993.[26])

With the notable exception of Oster and Wilson (1978), skepticism as to the feasibility and usefulness of optimization approaches to the study of complex biological systems has been voiced by the anti-adaptationists among students of animal form and behavior (the locus classicus being Gould & Lewontin, 1979; cf. Dupré, 1987; Campbell, 1989; Horan, 1992). In the emerging specialism of evolutionary economics, criticism of optimization was a main focus from the outset (Winter, 1964; Nelson & Winter, 1982; Witt, 1991).[27] In the philosophy of science, systematic criticism of the 'unnatural' assumptions underlying standard approaches to rationality is even more rare (Callebaut, 1983; Cherniak, 1986; Giere, 1988; Oaksford & Chater, 1993; Hooker, 1994b; Wimsatt, 1997a). In practice, mathematical modeling in biology, psychology, and the social sciences typically continues to rely heavily on optimization or maximization as if it were unproblematical (see, e.g. Rachlin et al., 1981 and Schoemaker, 1991 for recent vindications).

NATURALIZED RATIONALITY IS BOUNDED RATIONALITY

Against the background of his general model of generative entrenchment (Wimsatt, 1997a; for a summary, see Wimsatt in Callebaut, 1993), Wimsatt (1996a) talks of the *ossification* of our cultured selves "reflecting increasing dependencies on and commitments to earlier pivotal decisions and experiences, and to social forms that we acquire early, and deviate from later with increasing difficulty if at all". When it comes to stubbornly sticking to optimality despite overwhelming evidence to the contrary, both economics and biology abound with examples of how "the narrowly trained mind often operates as if it were an unreconstructed Kantian tied to a priori categories", as Freedman (1993, p. 14) puts it. "Faced with any deviation from these few well travelled paths they resolve this potential dissonance by reshaping alien ideas to fit a more familiar outline" (*ibid.*). The non-occurrence of Herbert Simon's 'behavioral revolution' in economics illustrates Wimsatt's and Freedman's point perfectly well.

Simon developed his theory of *bounded rationality* or *satisficing behavior* with the avowed aim to replace the unrealistic rationality assumption of conventional economic theory with an empirically adequate principle: "The capacity of the human mind for formulating and solving complex problems is very small compared with the size of the problems whose solution is required for objectively rational behavior in the real world (...) or even for a reasonable approximation of such objective rationality." (Simon, 1957, p. 198)

Evolutionary considerations are at the basis of Simon's satisficing theory: the survival of human beings "depends on attending selectively to [their] environment, and on finding satisfactory behavioral alternatives for handling these problems to which they attend" (Simon, 1977, p. 174). The *complexity* of their environment and their *uncertainty* about the world and about the consequences of their possible or effective actions compel humans to learn to live with *satisfactory* rather than optimal solutions to their problems. In the dynamical version of the theory, *levels of aspiration are adjusted* in terms of the success or failure of previous attempts to reach satisfactory solutions. Trivial as the principle of bounded rationality may seem, it has revolutionary consequences for economic theorizing: "If the principle is correct, then the goal of classical economic theory — to predict the behavior of rational man without making an empirical investigation of his psychological properties — is unattainable" (*ibid.*, p. 199). Simon was acutely aware that his "behavioral approach" would meet with considerable resistance among his fellow economists: "The reluctance of economic theory to relinquish its classical model of economic man is understandable. When even a small concession has been made in the direction of admitting the fallibility of economic man, his psychological properties are no longer irrelevant. Deductive reasoning then no longer suffices for the unique prediction of his behavior without constant assistance from empirical observation" (*ibid.*, p. 198).

In his article, "Rational Choice and the Structure of the Environment" (Simon, [1956] 1979a), Simon included the object pole of the relation of adaptation of subject to environment in the scope of his theory: "If an organism is confronted with the problem of behaving approximately rationally, or adaptively, in a particular environment, the kinds of simplifications that are suitable may depend not only on the characteristics — sensory, neural, and other — of the organism, but equally upon the structure of the environment. Hence, we might hope to discover, by a careful examination of some of the fundamental structural characteristics of the environment, some further clues as to the nature of the approximating mechanisms used in decision making" (Simon, 1979a, p. 21).

More recently Simon has likened the significance of his naturalization of rationality to that of some of the major historical scientific revolutions: "Behavioralism does not hold forward the hope of a theory that can be arrived at in an armchair. The new theory can only be built through extensive empirical research — on the scale, let us say, of the empirical research of the last century or two in field biology or geology" (Simon, 1984, p. 52).

Right from the start, the temptation has been to diminish the threat posed by satisficing by interpreting satisficing as a case of optimization subject to constraints (e.g. Maynard Smith, 1983; references in Elster, 1979, 1983 and Van Parijs, 1981). But as Winter (1964) has cogently argued, this won't do: "Since the decision problem confronting a firm is not fully structured for it *a priori*, as the traditional theory assumes it to be, the search for an optimum position would be of indeterminate length and the notion that the firm searches for such a position is either meaningless or hopelessly impractical. It is this consideration which constitutes the basic objection to any attempt to subsume satisficing behavior as a special type of maximizing behavior that is appropriate for certain 'optimum search' problems. Of course, any behavior can in one way or another be rationalized as maximizing behavior. *But to identify satisficing behavior with optimum search behavior is merely to push the crucial problem back a stage*: How does it happen that the task of learning about the relevant probability distributions, the ability to identify the problem as one of requiring a certain type of search procedure, and the ability to determine the precise procedure required, are within the limited information-obtaining-and-processing resources of the decision maker?" (Winter, 1964, p. 228, italics mine).[28]

Because the neoclassical approach leads to an infinite regress, then, it is much more natural to opt resolutely for the satisficing approach: "Unless one could prove (and I do not see how one could prove) that the deviation from the 'real' optimum converges to zero or at any rate becomes smaller for each new level in the hierarchy of information structures, there must be a cut-off point [in every decision] where calculation stops and you simply have to make an unsupported choice, but also *this point might as well be as close to the action itself as possible*. Why, indeed, seek for precision in the second decimal if you are uncertain about the first?" (Elster, 1979, p. 59).

"The classical program has failed", Simon (1984, p. 52) observes, in the sense that "it patches the rationality principle with ad hoc assumptions of bounded rationality, but it has no way to choose among alternative patches without obtaining new data". Maximizing behavior, one may safely conclude, is "just what Simon has long sought to reject" (Freedman, 1993, p. 14). Now, how do mainstream economists deal with this? Freedman quotes from a paper by Jensen and Meckling: "Unfortunately, Simon's work has often been misinterpreted as a denial of maximizing behavior, and misused, especially in the marketing and behavioral science literature. His later use of the term 'satisificing' has undoubtedly contributed to this confusion because it suggests rejection of maximizing behavior rather than maximization subject to costs of information and of decision making." (Jensen & Meckling, 1976, p. 306n, quoted in Freedman, 1993, p. 14). Comment superfluous.

Meanwhile, "the principle of rationality has already been abandoned and abandoned in fundamental ways, by classicists, neoclassicists, Keynesians, and rational expectationists alike" (Simon, 1984, p. 52). Another irony is that neoclassical and mathematical economics are being defended qua *rhetoric*, i.e. "by appeal to the tools of the tender-minded philosophers: the hermeneuticists, the deconstructionists, the post-modernists" (Rosenberg in Callebaut, 1993, p. 92).[29]

Callebaut (1983), Giere (1988), Wimsatt (in a series of papers including Wimsatt, 1986 and culminating in Wimsatt, 1997a), and a few others have extended the satisficing approach to the study of science itself (cf. also Simon, 1977, 1979a). Writes Wimsatt: "I seek a philosophy of science whih can be pursued by *real* scientists in *real* situations in real time with the tools that they *actually* have at their disposal. This must be a central requirement for any naturalistic account." (Wimsatt, 1997a, p. 2). Superficially it may look like Wimsatt's radicalism in rejecting transcendent reason ("Things are now changing but we still have a long way to go to realize the full impact of Simon's revolution"), which has made him explore the deep structural and functional similarities between heuristics (as human problem-solving tools) and biological adaptations,[30] brings him close to the postmodernist position of someone like Richard Rorty (who occasionally calls himself a naturalist as well). Rorty is convinced that the human need which is "gratified by the attempt to stand outside all human needs — the need for what [Thomas] Nagel calls 'transcendence'" is "culturally undesirable to exacerbate" and "eliminable by means of a suitable moral education — one which raises people up from the 'humility' which Nagel recommends" (Rorty, 1991, p. 8). Hooker could not disagree more: "Humans are finite, ignorant, error-prone, and bear other imperfections. Without these imperfections, there would be no point to a drive to transcend imperfection and hence no role for active reason." (Hooker, 1994b, p. 181). However, being principled naturalists who endorse realism as a consequence of our evolutionary constitution, both Wimsatt and Hooker agree contra Rorty that the transcendence which reason provides is a transcendence *within* natural life, not *of* natural life. This insight then allows them to shift foreground and background in the debate over the limitations of human reason and to see optimizing itself as a heuristic (non-algorithmic) tool. The upshot for our discussion of optimizing vs. satisficing is that their views — which are quite sophisticated and cannot even be adequately summarized here — confirm Simon's own interpretation, in his Nobel Prize acceptance speech, of optimizing behavior as a special case of satisficing behavior: "The salient characteristic of the decision tools employed in management science is that they have to be capable of actually making or recommending decisions, taking as their inputs the kinds of empirical data that are available in the real world, and performing only such computations as can reasonably be performed by existing desk calculators or, a little later, electronic computers. For these domains, idealized models of optimizing entrepreneurs, equipped with complete certainty about the world — or, at worst, having full probability distributions for uncertain events — are of little use. *Models have to be fashioned with an eye to practical computability, no matter how severe the approximations and simplifications that are thereby imposed on them.*

Model construction under these stringent conditions has taken two directions. The first is *to retain optimization, but to simplify sufficiently so that the optimum (in the simplified world!) is computable*. The second is to construct satisficing models that provide good enough decisions with reasonable costs of computation. By giving up optimization, a richer set of properties of the real world can be retained in the models. Stated otherwise, decision makers can satisfice either by finding optimum solutions for a simplified world, or by finding *satisfactory solutions for a more realistic world*. Neither approach, in general, dominates the other, and both have continued to co-exist in the world of management science." (Simon, 1979b, p. 498; italics mine).

CONCLUSION: OPTIMIZATION AND SELF-ORGANIZATION

Developmental biology was all but excluded from the Modern Synthesis of evolutionary theory (Gould, 1983; Riedl, 1983; Rieppel, 1990). As a result of the "hardening" of the Synthesis (Gould), the explanatory and methodological interests of mainstream evolutionary biologists on the one hand and of the structuralists among the embryologists and other developmental biologists on the other now strongly diverge (Amundson, 1994; Verhoog et al., 1995). Adaptationists (e.g. Maynard Smith, 1995) explicitly justify the use of optimality modeling in terms of its allowing to "black-box" ontogeny. One influential anti-adaptationist thinks, to the contrary, that "as a methodology the adaptationist program has been most detrimental in the subtle ways it discourages any emphasis on structure and its consequences" (Gould, 1985, p. 47).

As a result of their marginalization, many a structuralist is being tempted by a confrontation strategy, as most of the contributions to this volume amply document. Another strategy has been to stress the complementarity of the adaptationist and structuralist research programs, with the adaptationists addressing *why*-issues (ultimate or final causation) and the structuralists trying to find out *how* form emerges mechanically (Rieppel, 1990; Amundson, 1994). Thus Callebaut and Deneubourg (1995, p. 1) suggested that self-organization (SO) as "one particular script of the dynamics of biological organization" is very often "erroneously presented as a 'heretic approach' ignoring natural selection or being a substitute to it", and specified that "SO is only one of the mechanisms subjected to natural selection and one actor of natural selection".

Numerous misunderstandings occur between the two groups; e.g. the 'self-organizers' are being perceived by 'the others' as physicists ignoring biology, while 'the others' have sometimes been regarded as unaware of relevant advances in certain scientific disciplines. The gap is likely to persist for some time to come, for it is far from obvious that the 'Synthesists' will be easily convinced of the *need* for a developmental complement to their own endeavors. In Amundson's words, "Pre-Synthesis Darwinians at least realized the need for a theory of inheritance, although they doubted that Mendelism was that theory. Most post-

Synthesis neo-Darwinians do not require developmental biological contributions to evolutionary theory. Developmentalism may or may not be able to demonstrate that a knowledge of the processes of ontogenetic development is essential for the explanation of evolutionary phenomena" (Amundson, 1994, p. 576). In this *cul-de-sac* situation, appeals to the effect that causal-mechanistic accounts of developmental processes are required are legion (surveys in Callebaut, 1989, 1993, pp. 128-147). Often, testability is at stake. Thus, without a one-to-one mapping of mechanism onto result, a method that looks only at results could not possibly test a hypothesis about mechanisms, Brandon and Rausher (1996) write *contra* optimality. The human physiologist Cabanac concurs: "One shortcoming of the theories of the optimization of behavior proposed by ethologists and behavioral ecologists is that the mechanism, by which behavior is optimized, is never mentioned. In other words, they do not explain how the subject 'decides'. Most often they work at a level of ultimate causation, emphasizing the evolution of traits that contribute to lifetime reproductive success. This is reasonable when the object studied is animal behavior. With humans, however, it is possible to obtain verbal reports and to analyze in cognitive terms the proximate physiological and psychological mechanism and, thus, to go further than the mere measurement of behavior, especially when the behavior studied is serving a physiological regulation." (Cabanac, 1992, p. 173). Examples like this could be multiplied (cf. Rachlin et al., 1981, p. 406). The black-boxer/optimizer's defense is that the bracketing of scientific domains is scientifically respectable. Amundson (1994, p. 576) even invokes the patron saint of evolutionary biology himself: "Evolutionary biology was built on a huge black box — Darwin could never have written the *Origin of Species* if he had not wisely bracketed the mechanism of inheritance". True as this may be, Darwin was also much more sensitive to structuralist arguments than his post-Synthesis descendants (Richards, 1992; Asma, 1996).

It would be inconsistent with their self-image for naturalistic philosophers of science to make normative pronouncements on the structuralism vs. functionalism debate that would go way beyond the arguments that are actually being used by the scientists in both camps. As I hope to have shown in this paper, however, with respect to the optimality issue, at the level of the *justification* of the approach at least, the burden of proof has shifted from the satisficers' to the optimizers' camp. Writing about the value of game-theoretic models in biology, two decades ago, Rapoport (1975) argued that it is "conceptual rather than specifically predictive": "The use of specific models to predict, say, distributions of genotypes would be very difficult for lack of definitive independent measures of 'fitness', that is, of 'utilities'. On the other hand, game-theoretic models have been shown to provide plausible explanations for such observable biological phenomena as the stability of polymorphic populations, anomalous sex ratios, and the wide occurrence of restraints in intraspecific combat. Actually, this function of game-theoretic models in their better known applications (namely, in the social sciences as models of conflict and of conflict resolution) is quite analogous. It manifests itself not in specific predictions or observations (which is the usual function associated with mathematical models in the physical sciences) but rather in plausible explanations of a wide variety of phenomena, and as a source of ideas that generate novel hypotheses and enrich the conceptual repertoire of those sciences" (Rapoport, 1975, p. 58). In view of the *in principle* non-achievability of the normative standards of optimality by cognitively limited human agents as well as of the sheer adhocery of the optimizers' self-immunization strategies, it may be doubted that optimality models can even have a modest "conceptual rather than specifically predictive" function. Unless, that is, one takes at heart Schoemaker's (1991) admonition that each optimality principle seems to beg for an associated process explanation that describes causally, within the constraints of an organism or system, how it operates. Thieves ought to pay the price.

ACKNOWLEDGEMENTS

Drafts of various parts of this paper were discussed at ISES, at ISHPSSB in Leuven, Belgium, on 20 July 1995, and in the Evolution and Complexity Group at Ghent University on 23 October 1996. I thank the participants at these occasions, and in particular Michael Conrad, David Depew, Jean-Louis Deneubourg, René Lefever, and Isabelle Stengers for their stimulating criticisms. I have also benefited from discussions with or remarks by Joel Brown, Greg Cooper, Dick Nelson, Rupert Riedl, Karola Stotz, and Bill Wimsatt. Error clause as usual.

Konrad Lorenz Institut für Evolutions- und Kognitionsforschung
Altenberg/Donau, Austria
Limburgs Universitair Centrum, Diepenbeek, Belgium
Universiteit Maastricht, The Netherlands

NOTES

[1] From the naturalist's point of view, it is ironic that in his Nobel Prize acceptance speech, "Maximum principles in analytical economics", Samuelson (1975) actually paid lip service to his teacher Joseph Schumpeter's evolutionary epistemology ("man's ability to operate as a logical animal capable of systematic empirical induction was itself the direct outcome of the Darwinian struggle for survival" (p. 323)) before developing his main, quite ingenious argument to the effect that neoclassical maximization is formally equivalent to what the physicist does: "Often the physicist gets a better, a more economical, description of nature if he is able to formulate the observed laws by a maximum principle. Often the economist is able to get a better, more economical, description of economic behavior from the same device." (p. 325). Samuelson's *pièce de résistance* is the reformulation of Le Chatelier's principle as a maximum principle (pp. 330-333).

[2] Elsewhere (in Callebaut, 1993, 1995b,c, 1997) I have distinguished 4 semi-independent dimensions of naturalization: (1) *anti-transcendence*: the denial that there exist entities or events which lie, in principle, beyond the scope of scientific explanation, (2) *anti-transcendentality* or *reflexivity*: the theory of science can be developed on a par with science only, (3) *testability*, which is intimately related to the preference for causal modeling for explanatory purposes (cf. section 6), and (4) the replacement of a *categorical*

(substantive) by an *instrumental* (procedural) conception of rationality.

[3] "Opportunistic" is not meant pejoratively here. That science, considered as a material activity, consists to a large extent in opportunistic tinkering is what evolutionary epistemology predicts (cf. Wimsatt, 1997a on "Nature as backwoods mechanic and used parts dealer"), and what empirical science studies (e.g. Darden, 1990) confirm.

[4] For Aristotle, "efficient and final causes were extensions to dynamic systems (movements and processes) of static material and formal causes. Just as any static substance must have *both* matter and form, Aristotle said, so any movement or process must have *both* efficient and final causes." (Rachlin, 1993, p. 623).

[5] This view was, of course, held by many positivists and by Popper; yet competing conceptions existed within the positivist tradition, such as Duhem's or Wittgenstein's demarcationism. Conversely, people who have little or nothing to do with positivism have expressed a similar idea (e.g. Varela, Thompson, and Rosch, 1992). In principle, the 'maieutic' role of philosophy of science is compatible with the 'underlaborer' role as defined by Locke ("clearing the ground a little, and removing some of the rubbish that lies in the way to knowledge") and vindicated by many contemporary naturalists (Callebaut, 1993, p. 200; Fuller, 1994). In practice, forward-looking scientists are often disappointed by the "anally retentive" ways of philosophers (Ruse in Callebaut, 1993, p. 465).

[6] Among the various new roles Hull envisaged for philosophers of biology in his visionary (1969) paper, in addition to uncovering and explicating problems in biological theory and methodology, were *solving* biological problems (i.e., the philosopher of biology acting as theoretical biologist) as well as various *go-between* functions between philosophy, biology, and other sciences (cf. Callebaut, 1993, pp. 239-242, 442-452 on the current state of the art).

[7] In my view, Hooker's (1987, 1994a) and Wimsatt's (1979, 1995, 1997a, 1998) departures from the empiricist metaphilosophy are at least as important as Quine's original contributions to the naturalization of epistemology. In the context of this volume it may also be useful to stress that Hooker (1994a) argues that the formalist logical/AI model, as a *fundamental* or *basic* philosophical and scientific model of cognition, is "multiply defective", and that as far as the study of heuristic rules is concerned, Wimsatt (1998) urges to go from "fussy too-formalistic disembodied AI engines" to "embodied multi-level theories of practice and cognition with strong biological and AI/AL flavors".

[8] One difficulty such a 'global' vindication of realism poses but that cannot be dealt with here concerns its *testability* (Callebaut, 1997), which may be even more difficult to assess than the testability of adaptationism (Gould & Lewontin, 1969; Campbell, 1989; Orzack & Sober, 1994; Brandon & Rausher, 1995), or of aprioristic theories of human (ir)rationality (Stich, 1985; Smokler, 1990) for that matter. The challenge for philosophical naturalists is to move beyond the current situation in which it is doubtful that "anything that might count as data or evidence could possibly influence the debate over instrumentalism and realism", as these positions have become "way too sophisticated" (Hull, 1989, p. 320).

[9] An arbitary sample, from an author who is held in high esteem as a historian and theoretician of technological evolution to this day: "Resistance to science on the basis of philosophical and religious beliefs must be overcome and replaced by positive encouragement of scientific research" (Basalla, 1967).

[10] As Campbell (1993) has correctly noted, not just Spencerian evolutionism ("Let things alone and everything will optimize") must be avoided, but the interventionist's optimism as well. For when one intervenes by changing laws or discount rates, etc., one is changing the selective system in many ways, most often unanticipated, and often destructive.

[11] Van Parijs (1981) has elaborated the contrast between "evolutionist" and "evolutionary" approaches in some detail.

[12] Contemporary varieties of naturalism are usually explicitly construed as anti-foundationalist. Yet a new, *dynamical foundationalism* may be envisaged on the basis of a generative evolutionary perspective: roughly, the idea is that in a generative structure, changes in earlier developmental features have more effects 'downstream', and so often don't survive compared to changes that occur later in development (Wimsatt, 1996b).

[13] Wimsatt (1996a) hints at the same distinction when he notes that questions as to the reducibility and scientific acceptability of the reduction of functional analyses and explanations to causal ones may be interesting questions to the philosopher, but that there are questions relating to the concept of function which are of far greater interest and import to the biologist. (The questions he envisages concern

biological organization and the grounds for judgments of homology and functional analogy.)

[14] What this boils down to in terms of methodology is this: "If an efficient-cause analysis fails, psychologists would not (or should not) therefore assume that their subjects are not basically machines, that their behavior is not explicable as a mechanical process. Rather, they have to go to work and try to discover a better mechanism. Only a large series of failures might cause them to abandon the efficient-cause method just as only a large series of failures might lead to an abandonment of final-cause analysis. Fortunately, neither method has definitively failed." (Rachlin, 1993, p. 624).

[15] See, for instance, Juarrero (this volume) on understanding the emergence of mental properties in terms of "the workings of contextual constraints operating bottom up".

[16] Thus for Van Parijs (1981, 1987), who aims to rehabilitate functional explanation as a social-scientific "pattern of intelligibility" on a par with "actional explanation", any explanation presupposes both causal linkages between two or more phenomena, and reference to a mechanism recognized as plausible by the scientific community to explain the causal connection(s). For Van Parijs, functional explanations are *consequence* explanations that "causally ascribe the presence of an item, not to its consequences, but to a dispositional property of the situation in which it appears" (Van Parijs, 1981, p. 26). So-called "hard core" function statements relate to *persistent* features, the consequences of which are *good* (and the feature persisting because the consequences are good) (p. 45); they presuppose a mechanism of *optimization* as found in evolutionary theories. Natural selection, Van Parijs holds, can account for a tiny fraction of social-scientific functional explanations only, but *reinforcement* — the direct selection of the features to be explained within the entity they characterize, cf. operant conditioning — he takes to be more promising (pp. 53-54). From the perspective of the working social scientist, one problem with an evolutionary reconstruction of social-scientific explanation such as Van Parijs's is that it does not seem to have any implications for explanatory practice (Verdon, 1985).

[17] For the structuralist, developmental biology's explanandum is not adaptation but *form*; the embryological mechanisms (causa efficiens!) they propose are understood, then, as constraints on form not adaptation (e.g. Gould, 1980; Amundson, 1994, pp. 563-564).

[18] Cf. Kant (quoted in Kitcher, 1987, p. 77): "In the natural constitution of an organized being, i.e., one suitably adapted to life, we assume as an axiom that no organ will be found for any purpose which is not the fittest and best adapted to that purpose".

[19] Note that optimality is not necessarily desirable. Thus Conrad (1983, p. 314) observes that because of limits on the optimizing power of the evolutionary process, most ecological communities must be very far from the optimal situation in terms of investment and return energy. "But peculiarly these limits afford important protection to the community. If the community were at the optimum, there would remain for it no option other than a catastrophic response to worsening conditions."

[20] Compare and contrast the contemporary, more stringent reconstruction of evolutionary theory as a "theory of forces" by Sober (1984a). Sober's influential causal analysis has not gone unchallenged itself — one reason being the unsatisfactory state of current philosophical accounts of causation in general. Although this issue is relevant to the vindication of causal-mechanistic explanation in biology in this paper, space limitations prevent me from addressing it here; see, e.g. Callebaut (1993, chs. 4 and 6), Brandon et al. (1994), Sober & Wilson (1994), and Wilson & Sober (1994, pp. 642-43).

[21] Thus Oster and Wilson (1978, p. 292): "[Our] use of the concept of optimality does not (...) require that natural selection create phenotypes 'better' than their predecessors in any absolute sense. Rather, we suppose only that physiological design features providing a mortality or fecundity advantage in the local environment will be amplified". Or Williams (1985, p. 12): "given that all organisms throughout their history have been subject to (...) genetic bookkeeping, any given organism is expected to have certain properties and not have others. The properties are expected to form a closely-optimized strategy for the maximal proliferation of the genes that directed the development of the organism. (...) A strategy can be said to be optimized only in relation to the rules of the game being played and to other necessary constraints. An important constraint arises from the gradualist view of adaptive change. Optima will never be more than local optima." In this context it is crucial not to conflate the selection *process*, which is one of *relative* advantage only (selection of the fitter, not fittest) with its alleged *results*; at least since the days of Haldane and Fisher, neo-Darwinists have maintained that small relative increments can have a large cumulative effect, and held that "it is plausible that one of those effects (given sufficient genetic variation

and consistence of selection pressures) will often be not just a trajectory of improvement in designs, but a resulting design that can fairly be characterized as 'optimal' with respect to the fitness currency" (Alden Smith, 1987, pp. 222-223).

[22] Since "traits have no adaptive significance in isolation from the whole developmental pattern of the organism which exhibits them at certain stages of its life cycle", a trait must not to be viewed statically but rather as "an aspect of all or of a certain portion of the path or the trajectory of the development"; it is the trajectory as a whole which confers upon the genotype of the individual its fitness to survive and to reproduce (Dobzhansky, 1956, p. 346).

[23] "Given the complexity of the eucaryotic genome, the finiteness of population sizes, and the limited number of generations in which evolution can occur, it is simply impossible for selection to independently optimize every aspect of an organism's genotype." (Oster & Wilson, 1978, p. 314).

[24] The locus classicus of this position is Friedman (1953, p. 35): "Unless the behavior of businessman in some way or other approximated behavior consistent with the maximization of returns, it seems unlikely that they would remain in business for long. Let the apparent immediate determinant of business behavior be anything at all — habitual reaction, random change, or whatnot. Whenever this determinant happens to lead to behavior consistent with the rational and informed maximization of returns, the business will prosper and acquire resources with which to expand; whenever it does not, the business will tend to lose resources and can be kept in existence only by the addition of resources from outside. The process of 'natural selection' thus helps to validate the [maximization of returns] hypothesis — or rather, given natural selection, acceptance of the hypothesis can be based largely on the judgment that it summarizes appropriately the conditions for survival."

[25] 'Optimizing' and 'maximizing' are not synonymous. Thus, in the context of adaptation, Williams (1985, p. 14) distinguishes "'strategies', which organisms try to optimize, from 'winnings', which they try to maximize". For the purposes of this paper such distinctions are irrelevant, however. Other complications, having to do with the egoistic, individually focused view of human social behavior yielded by economics-based optimizing theories (Alden Smith, 1987; Campbell, 1989), and with the relations between game theory and optimization, will be disregarded here as well.

[26] Hodgson (1993) identifies the following "problems for Dr. Pangloss": (1) routinized behavior within firms and other institutions will not generally be on an optimal track; (2) matters hindering market entrance; (3) path dependency; (4) lock-in and chreodic development (Waddington: a chreod is a relatively stable trajectory of development for a species, caused in part by the evolution of hierarchical control sequences in the genotype); (5) context- and frequency-dependence; (6) multiple or shifting adaptive peaks; and (7) critical mass and intransitivity. For a discussion of the biological equivalents of most of these constraints see, e.g. Ridley (1993, pp. 327-348).

[27] Hodgson (1992, p. 22) notes: "Strangely, the invocation of the 'natural selection' argument by Friedman and others [was] not used to bolster an evolutionary approach to the understanding of economic problems, but profit maximization along traditional lines. (...) Usefully, modern evolutionary theory immediately suggests a variety of circumstances in which the validity of the maximization idea is under strain."

[28] If anything, Hodgson's (1992) detailed reexamination of Winter's critique of Friedman *strengthens* the case against Friedman- or Alchian-type maximization.

[29] Murphy's (1992) plea to "deepen" Simon's position by making it "socially sensitive by understanding facts and reason to originate from the *Lebenswelt*" (p. 293) actually goes in the same direction. As Murphy sees it, "the absence of perfect rationality should not be understood to stem from inadequate cognitive skills, but the inability of [actors] to overcome the influence of interpretation" (p. 301). The argument (inspired by Lyotard) is that because interpretation antedates reasoning (any system of formalized reasoning originates in the *Lebenswelt*), "reason can never be completely formalized or axiomatic" (*ibid.*). It is not altogether clear to me what the actual target of Murphy's criticism is: for one thing, Simon's theory of satisficing can be — and actually has been — formulated without recourse to complete formalization of axiomatization. For another, the naturalistic accounts of reason that are now being articulated (Hooker, 1994a; Wimsatt, 1997a, 1998) take the inadequacy of axiomatized formal reason as their very point of departure.

[30] "If we stand back and look at ourselves as biological social cognitive beings, we see that our responses to the problems of adaptation in a complex world are crafted with heuristics, and the biological and cultural nature of our reason — insofar as these are products of selection processes — must also be heuristic — *they are re-tuned, re-modulated, re-contextualized, and sometimes newly reconnected piecemeal re-arrangements of existing adaptations or exaptations*" (Wimsatt, 1997a, p. 2, italics in original).

REFERENCES

Alden Smith, E., 1987, Optimization theory in anthropology: Applications and critiques, in J. Dupré (ed.), 1987, pp. 201-49.

Alexander, R. McN., 1982, *Optima for Animals*, London, Edward Arnold.

Amundson, R., 1994, Two concepts of constraint: Adaptationism and the challenge from developmental biology, *Philosophy of Science*, vol. 61, pp. 556-78.

Asma, S.T., 1996, Darwin's causal pluralism, *Biology and Philosophy*, vol. 11, pp. 1-20.

Atlan, H., 1994, Intentionality in nature. Against an all-encompassing evolutionary paradigm: Evolutionary and cognitive processes are not instances of the same process, *Journal for the Theory of Social Behaviour*, vol. 24, pp. 67-87.

Azevedo, J., 1997, *Mapping Reality: An Evolutionary Realist Methodology for the Natural and Social Sciences*, Albany, NY, State Un. of New York Press.

Basalla, G., 1967, The spread of Western science, *Science*, vol. 156, pp. 612-621.

Bechtel, W.; Richardson, R.J., 1991, *A Model of Theory Development: Localization as a Strategy in Scientific Research*, Princeton, Princeton Un. Press.

Bradie, M., 1994, Metaphors and mechanisms in vehicle-based selection theory, *Behavioral and Brain Sciences*, vol. 17, p. 612.

Brandon, R.N.; Antonovics, J.; Burian, R.; Carson, S., Cooper, G.; Davies, P.S.; Horvath, C.; Mishler, B.D. & Thrall, P., 1994, Sober on Brandon on screening-off and the levels of selection, *Philosophy of Science*, vol. 61, pp. 475-486.

Brandon, R.N.; Rausher, M.D., 1996, Testing adaptationism: A comment on Orzack and Sober, *American Naturalist*, vol. 148, pp. 189-201.

Bunge, M., 1979 (3rd ed.), *Causality and Modern Science*, New York, Dover.

Cabanac, M., 1992, Pleasure: The common currency, *J. of Th. Biology*, vol. 155, pp. 173-200.

Callebaut, W., 1983, Bijdrage tot een algemene rationaliteitstheorie op evolutionaire grondslag, met een toepassing op de organisatie van wetenschappelijke kennis, unpublished Ph.D. thesis, University of Ghent.

Callebaut, W., 1989, Post-Positivist views of scientific explanation, in *Explanation in the Social Sciences: The Search for Causes in Demography*, J. Duchêne, G. Wunsch & E. Vilquin (eds.), Brussels, Ciaco, pp. 141-196.

Callebaut, W., 1993, *Taking the Naturalistic Turn, Or How Real Philosophy of Science is Done*, Chicago, University of Chicago Press.

Callebaut, W., 1995a, Science Dynamics: The difficult birth of a metascience, in *Pragmatik: Handbuch des pragmatischen Denkens*, H. Stachowiak (Hrsg.), vol. 5, *Pragmatische Tendenzen in der Wissenschaftstheorie*, Hamburg, Felix Meiner, pp. 1-29.

Callebaut, W., 1995b, Réduction et explication mécaniste en biologie, *Revue Philosophique de Louvain*, vol. 91, pp. 33-66.

Callebaut, W., 1995c, The future of naturalistic philosophy of science, *Ludus Vitalis*, vol. 3, no. 5, pp. 19-52.

Callebaut, W., 1997 (forthcoming), Neonaturalism in contemporary theories of science, *Proceedings of the Workshop on Current Trends in the History, Philosophy and Social Studies of Science*, H. Nowotny & G. Kampis (eds.), Collegium Budapest, November 6, 1995.

Callebaut, W. (ed.), 1998 (forthcoming), The New Scientism: A Post-modernist Chimera?, Special issue of *Philosophica*, Ghent.

Callebaut, W.; Deneubourg, J.-L., 1995, Self-organization vs. optimization in non-human behavioral biology. Introduction to the theme of the session, Conference of the International Society for the History, Philosophy and Social Studies of Biology, Leuven, Belgium, 19-23 August 1995.

Campbell, D.T., 1989, Uncritical adaptationism criticized, *Cont. Psychology*, vol. 34, pp. 128-129.

Campbell, D.T., 1993, How individual and face-to-face group selection undermine firm selection in organizational evolution, in *Evolutionary Dynamics of Organizations*, J.A.C. Baum & J.V. Singh (eds.), Oxford/New York, Oxford University Press.

Cherniak, C., 1986, *Minimal Rationality*, Cambridge, Mass., MIT Press.

Code, L., 1996, What is natural about epistemology naturalized?, *Am. Phil. Quart.*, vol. 33, pp. 1-22.

Conrad, M., 1983, *Adaptability: The Significance of Variability from Molecule to Ecosystem*, New York/London, Plenum Press.

Darden, L., 1990, Diagnosing and fixing faults in theories, in *Computational Models of Scientific Discovery and Theory Formation*, J. Shrager & P. Langley (eds.), San Mateo, California, Morgan Kaufmann, pp. 319-346.

Darden, L., 1991, *Strategies for Theory Change: The Case of the Gene*, Oxford, Oxford Un. Press.

Dawkins, R., 1995, God's utility function, *Scientific American*, vol. 273 (November), pp. 62-67.

Dennett, D., 1995, *Darwin's Dangerous Idea: Evolution and the Meanings of Life*, New York, Simon & Schuster.

Depew, D.J.; Weber, B.H., 1985, Innovation and tradition in evolutionary theory: An interpretive afterword, in *Evolution at a Crossroads: The New Biology and the New Philosophy of Science*, D.J. Depew & B.H. Weber (eds.), Cambridge, Mass., MIT Press, Bradford Books, pp. 227-260.

Depew, D.J.; Weber, B.H., 1995, *Darwinism Evolving: Systems Dynamics and the Genealogy of Natural Selection*, Cambridge, Mass., MIT Press.

Dobzhansky, T., 1956, What is an adaptive trait?, *American Naturalist*, vol. 40, pp. 337-347.

Dretske, F., 1988, *Explaining Behavior: Reasons in a World of Causes*, Cambridge, Mass., Bradford Books, MIT Press.

Dupré, J. (ed.), 1987, *The Latest on the Best: Essays on Evolution and Optimality*, Cambridge, Mass., MIT Press, Bradford Books.

Elster, J., 1979, *Ulysses and the Sirens: Studies in Rationality and Irrationality*, Cambridge, Cambridge University Press.

Elster, J., 1983, *Explaining Technological Change*, Cambridge, Cambridge University Press.

Freedman, C., 1993, Why economists can't read, *Methodos*, vol. 3 (June), pp. 6-1.

Friedman, M., 1953, *Essays in Positive Economics*, Chicago, University of Chicago Press.

Fuller, S., 1994, Underlaborers for science, *Science*, [review of Callebaut 1993], vol. 264, pp. 982-983.

Giere, R.N, 1988, *Explaining Science: A Cognitive Approach*, Chicago, University of Chicago Press.

Goldsmith, E., 1990, Evolution, neo-Darwinism and the paradigm of science, *Ecologist*, vol. 20, no. 2, pp. 67-73.

Gould, S.J, 1977, *Ontogeny and Phylogeny*, Cambridge, Mass., Harvard University Press.

Gould, S.J., 1980, The evolutionary biology of constraint, *Daedalus*, vol. 109, pp. 39-52.

Gould, S.J., 1983, The hardening of the modern synthesis, in M. Grene (ed.), 1983, pp. 71-93.

Gould, S.J.; Lewontin, R.C., 1979, The spandrels of San Marco and the Panglossian paradigm: A critique of the adaptational programme, *Proceedings of the Royal Society of London B*, vol. 205, pp. 581-598, reprinted in E. Sober, 1984b.

Grene, M. (ed.), 1983, *Dimensions of Darwinism: Themes and Counterthemes in Twentieth-Century Evolutionary Theory*, Cambridge, Cambridge University Press.

Hodgson, G.M., 1992, Optimization and evolution: Winter's critique of Friedman revisited, Report no. 11/92 of the Research Group on Biological Foundations of Human Culture, Center for Interdisciplinary Research (ZiF), University of Bielefeld, FRG.

Hodgson, G.M., 1993, *Economics and Evolution: Bringing Life Back into Economics*, Cambridge, Polity Press.

Hooker, C.A., 1981, Formalist rationality: The limitations of Popper's theory of reason, *Metaphilosophy*, vol. 12, pp. 247-266.

Hooker, C.A., 1987, *A Realistic Theory of Science*, Albany, State University of New York Press.

Hooker, C.A., 1994a, *Reason, Regulation and Realism: Toward a Naturalistic Regulatory Systems Theory of Reason*, Albany, State University of New York Press.

Hooker, C.A., 1994b, Idealisation, naturalism, and rationality: Some lessons from minimal rationality, *Synthese*, vol. 99, pp. 181-231.

Horan, B.L., 1992, What price optimality? Review of Dupré (1987), *Biology and Philosophy*, vol. 7, pp. 89-109.

Hull, D.L., 1969, What philosophy of biology is not, *Journal of the History of Biology*, vol. 2, pp. 241-268 and *Synthese*, vol. 20, pp. 157-184.

Hull, D.L., 1982, The naked meme, in *Learning, Development, and Culture*, H.C. Plotkin (ed.), New York, Wiley, pp. 273-327.

Hull, D.L., 1989, A function for actual examples in philosophy of science, in *What the Philosophy of Biology is Not*, M. Ruse (ed.), Dordrecht, Kluwer, pp. 309-321.

Kitcher, P., 1987, Why not the best?, in J. Dupré (ed.), 1987, pp. 77-102.

Kitcher, P., 1994, How the road to relativism is paved, paper presented at the annual meeting of the History of Science Society, New Orleans, October 12-16, 1994.

Kuhn, T.S., 1970 (1962), *The Structure of Scientific Revolutions*, Chicago/London, University of Chicago Press.

Levins, R.; Lewontin, R.C., 1985, *The Dialectical Biologist*, Cambridge, Mass., Harvard Un. Press.

Maynard Smith, J., 1983, Adaptation and satisficing, *Behavioral and Brain Sciences*, vol. 3, pp. 370-371.

Maynard Smith, J., 1995, Life at the edge of chaos? Review of Depew and Weber (1995), *New York Review of Books*, vol. 42, no. 4, pp. 28-30.

Mayr, E., 1982, *The Growth of Biological Knowledge: Diversity, Evolution, and Inheritance*, Cambridge, Mass., Harvard University Press, Belknap Press.

Murphy, J.W., 1992, Reason, bounded rationality, and the Lebenswelt: Socially sensitive decision making, *American Journal of Economics and Sociology*, vol. 51, pp. 293-305.

Nagel, E., 1977, Teleology revisited, *Journal of Philosophy*, vol. 74, pp. 261-301, reprinted in E. Nagel, *Teleology Revisited and Other Essays in the Philosophy and History of Science*, 1979, New York, Columbia University Press.

Nelson, R.J., 1987, *The Logic of Mind*, Dordrecht, Kluwer.

Nelson, R.R.; Winter, S.G., 1982, *An Evolutionary Theory of Economic Change*, Cambridge, Mass., Harvard University Press, Belknap Press.

Nickles, T., 1990, Good science is bad history, in *Social Aspects of Science*, E. McMullin (ed.), Notre Dame, University of Notre Dame Press.

Niiluoto, I., 1991, What's wrong with relativism?, *Science Studies*, vol. 2, pp. 17-24.

Oaksford, M.; Chater, M., 1993, Reasoning theories and bounded rationality, in *Rationality: Psychological and Philosophical Perspectives*, K.I. Manktelew & D.E. Over (eds.), London, Routledge, pp. 31-60.

Orzack, S.H.; Sober, E., 1994, Optimality models and the test of adaptationism, *American Naturalist*, vol. 143, pp. 361-380.

Oster, G.F.; Wilson, E.O., 1978, A Critique of Optimization Theory in Evolutionary Biology, Ch. 8 of *Caste and Ecology in the Social Insects*, Princeton, Princeton University Press, reprinted in E. Sober (ed.), 1984b.

Rachlin, H., 1993, Optimality and Aristotle's concept of final cause, *Behavioral and Brain Sciences*, vol. 16, pp. 623-624.

Rachlin, H.; Battalio, R.; Kagel, J.; Green, L., 1981, Maximization theory in behavioral psychology, *Behavioral and Brain Sciences*, vol. 4, pp. 371-417.

Rapoport, A., 1975, Uses of game theory in models of biology, *General Systems*, vol. 20, pp. 49-58.

Reif, W.-E, 1995, Review of Callebaut 1993, *Zentralblatt für Geologie und Paläontologie*, Pt II, nos. 3-4.

Richards, R.J., 1992, *The Meaning of Evolution: The Morphological Construction and Ideological Reconstruction of Darwin's Theory*, Chicago, University of Chicago Press.

Ridley, M., 1993, *Evolution*, Oxford, Blackwell.

Riedl, R., 1975, *Die Ordnung des Lebendigen: Systembedingungen der Evolution*, Hamburg/Berlin, Paul Parey, English: *Order in Living Organisms: A Systems Analysis of Evolution*, trans. R.P.S. Jefferies, New York, Wiley.

Riedl, R., 1978-1979, Über die Biologie des Ursachen-Denkens-Ein evolutionistischer, systemtheoretischer Versuch, *Mannheimer Forum*, no. 7, pp. 8-70.

Riedl, R., 1983, The role of morphology in the theory of evolution, in *Dimensions of Darwinism: Themes and Counterthemes in Twentieth-Century Evolutionary Theory*, M. Grene (ed.), 1983, pp. 205-238.

Riedl, R., 1995, Deficiencies of adaptation in human reason: A constructivistic extension of evolutionary epistemology, *Evolution and Cognition* (N.S.), vol. 1, pp. 27-37.

Rieppel, O., 1990, Structuralism, functionalism, and the four Aristotelian causes, *Journal of the History of Biology*, vol. 23, pp. 291-320.

Riggs, F.W., 1970, The 'Sala' model: An ecological approach to the study of comparative administration, in *Comparative Perspectives on Formal Organizations*, H.A. Landsberger (ed.), Boston, Little, Brown, pp. 151-167.

Rorty, R., 1991, *Objectivity, Relativism, and Truth*, Philosophical Papers, vol. 1, Cambridge, Cambridge University Press.

Rose, J.; Lauder, G.V. (eds.), 1996a, *Adaptation*, New York, Academic Press.

Rose, J.; Lauder, G.V., 1996b, Post-spandrel adaptationism, in J. Rose & G.V. Lauder, 1996a, pp. 1-8.

Rosenberg, A., 1985, *The Structure of Biological Science*, Cambridge, Cambridge University Press.

Rosenberg, A., 1994, *Instrumental Biology, Or the Disunity of Science*, Chicago, University of Chicago Press.

Rostow, W.W., 1960, *The Stages of Economic Growth: A Non-Communist Manifesto*, Cambridge, Cambridge University Press.

Salmon, W.C., 1993, The value of scientific understanding, *Philosophica*, vol. 51, pp. 9-19.

Samuelson, P.A., 1975, Maximum principles in analytical economics, *Synthese*, vol. 31, pp. 323-344.

Schoemaker, P.J.H., 1991, The quest for optimality: A positive heuristic of science?, *Behavioral and Brain Sciences*, vol. 14, pp. 205-245.

Schoffeniels, E., 1976, *Anti-Chance: A Reply to Monod's Chance and Necessity*, Oxford, Pergamon.

Seger, J.; Stubblefield, J.W., 1996, Post-spandrel adaptationism, in J. Rose & G.V. Lauder (eds.), pp. 93-123.

Simon, H.A., 1957, *Models of Man, Social and Rational*, New York, Wiley.

Simon, H.A., 1977, *Models of Discovery*, Dordrecht, Reidel.

Simon, H.A., 1979a, *Models of Thought*, New Haven, Yale University Press.

Simon, H.A., 1979b, Rational decision making in business organizations, *American Economic Review*, vol. 69, pp. 493-513.

Simon, H.A., 1984, On the behavioral and rational foundations of economic dynamics, *Journal of Economic Behavior and Organization*, vol. 5, pp. 35-55.

Smokler, H., 1990, Are theories of rationality empirically testable?, *Synthese*, vol. 82, pp. 297-306.

Sober, E., 1984a, *The Nature of Selection: Evolutionary Theory in Philosophical Focus*, Cambridge, Mass., MIT Press, Bradford Books.

Sober, E. (ed.), 1984b, *Conceptual Issues in Evolutionary Biology: An Anthology*, Cambridge, Mass., MIT Press, Bradford Books.

Sober, E., 1985, Panglossian functionalism and the philosophy of mind, *Synthese*, vol. 64, pp. 165-193.

Sober, E.; Wilson, D.S., 1994, A critical review of philosophical work on the units of selection problem, *Philosophy of Science*, vol. 61, pp. 534-555.

Stich, S., 1985, Could man be an irrational animal?, *Synthese*, vol. 64, pp. 115-135.

Van Parijs, P., 1981, *Evolutionary Explanation in the Social Sciences: An Emerging Paradigm*, Totowa, NJ, Rowman & Littlefield.

Van Parijs, P., 1987, The evolutionary explanation of beliefs, in *Evolutionary Epistemology: A Multiparadigm Program*, W. Callebaut & R. Pinxten (eds.), Dordrecht, Reidel, pp. 381-401.

Varela, F.; Thompson, E.; Rosch, E., 1992, *The Embodied Mind: Cognitive Science and Human Experience*, Cambridge, Mass., MIT Press.

Verdon, M., 1985, Midwife or toad? Philosophy and the social sciences, review of Van Parijs, 1981,

Philosophy of the Social Sciences, vol. 15, pp. 53-63.

Verhoog, H.; Maas, M.; Sprey, T., 1995, Position paper: The philosophy of biological form, Conference of the International Society for the History, Philosophy and Social Studies of Biology, Leuven, Belgium, 19-23 August 1995.

Williams, G.C., 1985, A defense of reductionism in evolutionary biology, *Oxford Surveys in Evolutionary Biology*, vol. 2, pp. 1-27.

Williams, G.C., 1992, *Natural Selection: Domains, Levels, and Challenges*, New York, Oxford University Press.

Wilson, D.S.; Sober, E., 1994, Reintroducing group selection to the human behavioral sciences, *Behavioral and Brain Sciences*, vol. 17, pp. 585-654.

Wimsatt, W.C., 1976, Reductive explanation: A functional account, in *PSA 1974*, Dordrecht, Reidel, reprinted in E. Sober (ed.), 1984b.

Wimsatt, W.C., 1979, Reduction and reductionism, in *Current Research in Philosophy of Science*, P.D. Asquith & H.E. Kyburg, Jr. (eds.), East Lansing, Mich., Philosophy of Science Ass., pp. 352-377.

Wimsatt, W.C., 1986, Heuristics and the study of human behavior, in *Metatheory in Social Science: Pluralisms and Subjectivities*, D.W. Fiske & R.A. Shweder (eds.), Chicago, University of Chicago Press, pp. 294-314.

Wimsatt, W.C., 1989, Genetics in an Evolutionary Perspective, Syllabus, University of Chicago.

Wimsatt, W.C., 1994, The ontology of complex systems: Levels of organization, perspectives, and causal thickets, *Canadian Journal of Philosophy*, suppl. vol. 20, pp. 207-274.

Wimsatt, W.C., 1995, Lewontin's evidence (that there isn't any), in *Questions of Evidence*, J. Chandler, A.I. Davidson & H. Harootunian (eds.), Chicago, Un. of Chicago Press, pp. 492-503.

Wimsatt, W.C., 1996a, The Evolution of Generative Structures: The New Contextualism and a New Dynamical Foundationalism, Unpublished ms., University of Chicago.

Wimsatt, W.C., 1997a (forthcoming), *Approximations to Reality: A Realist Philosophy of Science for Limited Beings*, Cambridge, Mass., Harvard University Press.

Wimsatt, W.C., 1997b (forthcoming), Functional organization and the logic of functional analogy, *Evolution and Cognition*, (N.S.), vol. 4.

Wimsatt, W.C., 1998 (forthcoming), Emergence as non-aggregativity, in *Natural Contradictions: Perspectives on Ecology and Change*, Y. Haila & P. Taylor (eds.), Stonybrook, SYNU Press.

Winter, S.G., 1964, Economic "natural selection" and the theory of the firm, *Yale Economic Essays*, vol. 4, pp. 225-272.

Witt, U., 1991, Reflections on the present state of evolutionary economics, in *Rethinking Economics: Markets, Technology and Economic Evolution*, G.M. Hodgson & E. Screpanti (eds.), Aldershot, Edward Elgar, pp. 83-102.

MARTIN BAATZ

PLEIOTROPY AND THE EVOLUTION OF ADAPTABILITY

INTRODUCTION

The systems approach to the investigation of organismic evolution is a conceptual framework introduced by Riedl in the 1970s (Riedl, 1977, 1978). One aim was to elucidate the relationship between certain evolutionary implications of morphological concepts, such as homology, and the population genetic approach of Modern Synthesis. At about the same time an intense scientific discussion began anew regarding the interpretation of macroevolution. Some scientists doubted the usefulness of the population genetic program to the area of macroevolution (Gould, 1980; Stanley, 1975). In fact, there are reasons in the mathematical structure of evolutionary theory: the formal apparatus of population genetics was created to describe elementary mechanisms of evolution, i.e. the distribution and change of gene frequencies, but not the process of evolution itself, i.e. the transformation of genotypic and phenotypic patterns of organization. Two statements, which at first glance seem to be incompatible, mark the positions in this discussion between microevolutionists and macroevolutionists:

1. There are no elementary mechanisms other than these offered by population genetics: mutation, duplication, selection, migration and random drift; every neo-Darwinian explanation must be based on these mechanisms.
2. Phenomena of macroevolution follow their own laws which cannot necessarily be derived from the principles of microevolution.

Until now this dilemma has not been solved in a convincing way. One is inclined to overcome it by denying one of the two statements. Nevertheless, there are good reasons and arguments for each of them. If one takes both positions for true, one can come to the following conclusion: because there are no elementary mechanisms other than mutation, selection etc., laws of macroevolution can only have the status of systemic laws, i.e. rules which depend on the existence of certain systemic structures (Wagner, 1985). In fact, the system approach to evolution offers interesting perspectives. A type is characterized by a relatively stable combination of phenotypic characters. The environment can only partly be made responsible for this stability, inner systemic correlations and interdependencies must restrict the degrees of freedom of potential change into a surprising degree. These conditions, under which the adaptation of organs or organsystems happens, are called constraints. In contrast to other definitions they are understood here in a general, structural sense. In the same way in which organisms adapt to an outer environment, there exists a selective pressure to adapt to the inner functional conditions; the dependencies of functional organization provoke the constitution of a certain inner environment with important consequences for evolutionary dynamics. Epigenetic constraints influence and organize phenotypic plasticity and, thus, the paths of

G. Van de Vijver et al. (eds.), Evolutionary Systems, 101–112.

evolution. However, until now no attempts have been successfully made to formulate the evolution of these constraints in population genetic models.

An important implication of a systemic approach to evolution is also that the ability of organisms to evolve doesn't follow the principle of mutation and selection in a trivial sense. Different analyses (Rechenberg, 1973; Wagner, 1985) lead to the expectation, that only if the ability to evolve itself is object to a process of optimization, the evolutionary principle introduced by Darwin explains the evolution of complex organisms. The optimization is assumed to take place in the interpretation and canalization of mutational effects by the epigenetic system. Furthermore, one of the major challenges of organismic evolution theory is to explain how complex organisms were able to evolve by random mutations in spite of the severe functional constraints that canalize their route of change. In complex systems almost any random change can be expected to disturb the functional balance rather than to improve it. It has therefore been repeatedly proposed that functionally coupled characters have to develop genetic correlations in order to permit their coevolution (Alexander, 1962; Frazzetta, 1975; Hoy et al., 1977; Riedl, 1977; Wagner, 1982). Fundamental to the understanding of couplings or inheritable covariances on the phenotypic level is the phenomen of pleiotropy, i.e. the fact that a gene (or: geneproduct) typically influences not only one, but several characters or physiological functions. Pleiotropic effects are among the major causes of genetic covariance among quantitative characters (for references see Falconer, 1981). Every mutation of a pleiotropic structure gene product will result in changes in different dimensions of the phenotypic phase space. The "achondroplasia"-mutation in the rat, for example, causes an inability to suck, defective lung circulation, an occlusion of the incisors and inhibited development, all arising from an abnormal development of cartilage early in the ontogeny. In contrast to this direct pleiotropy, indirect (relational) pleiotropy is characterized by a series of different after-effects caused by a single primary gene activity. Just as each structure gene influences different characters, every change in character is influenced by different genes, which is called polygeny. In an antagonistic sense, pleiotropy and polygeny are describing the fact that genotype-phenotype interactions are polycausal and multi-dimensional.

Much research has been done to investigate different aspects of pleiotropy. Pleiotropic effects for example can contribute to the genetic composition of a population by a balance between pleiotropic mutations and stabilizing effects (Lande, 1980), or genetic covariances cause correlated responses to directional selection (Falconer, 1981; Lande, 1979) and indirect effects of stabilizing selection (Lande & Arnold, 1983). In the focus of a research program concerning a system theory of evolution of the Center of Computational Ecology at the Yale University (G.P. Wagner) and the Konrad Lorenz Institute, Altenberg, are the following questions:
— How can pleiotropy evolve, and what are the underlying evolutionary dynamics?
— In which way are pleiotropic effects controlled by the epigenetic system?
— In which way can such an organization contribute to an evolution of adaptivity?

Without becoming too mathematical, I would like to summarize in the following two sections some results which attempt to answer these questions. It is demonstrated that evolving epigenetic constraints adapt to the direction of functional dependencies. Thus, they bundle the distribution of pleiotropic effects and canalize the degrees of freedom of potential phenotypic change. Further adaptation is accelerated. All results were gathered in close collaboration with G.P. Wagner.

PLEIOTROPY

The model

The approach introduced here contains two basic assumptions:

1. Phenotypic change is best understood in a situation in which the vast majority of characters remains stabilized. It typically takes place in just a few of all dimensions of the phenotypic parameter space. If one character comes under selective forces which cause its change, all other features which already are functionally optimized need to be maintained; they remain under stabilizing selection. The simplest way to model such a situation is a two dimensional character space, in which one character comes under directional selection and the other under a stabilizing selection regime, representing all other stable characters. A corresponding fitness landscape was introduced by Wagner (1988a) and called "corridor". As malthusian fitness it reads

$$m(\underline{z}) = \sigma z_1 - \frac{1}{2w_2^2} z_2^2 \tag{1}$$

where z_1 and z_2 are the values of the quantitative characters 1 and 2; σ gives the strength of directional and $1/2w_2^2$ that of stabilizing selection. It can be visualized as a rising ridge in direction of character 1 with its flanks defined by the stabilizing selection and the top of the ridge at $x_2 = 0$ (see Figure 1).

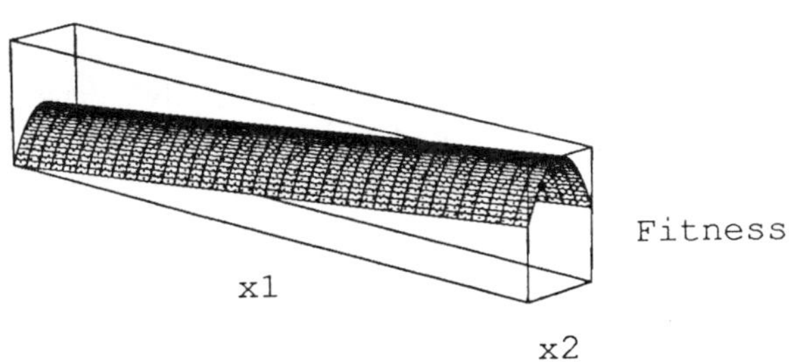

Figure 1. Corridor in the malthusian fitness landscape: directional selection on character 1, stabilizing selection on character 2

2. Genotype and phenotype are connected to each other in complex branched interdependencies, i.e. pleiotropy and polygeny are ubiquitous phenomena. This

assumption is transformed in a deterministic two locus/two allele model. The two loci have pleiotropic effects which connect the character under directional selection (first character) with the one under stabilizing selection (second character). The influence of the two genes on the second character is such that they are additive and can compensate for each others effects, i.e. they are hidden pleiotropic effects.

Table I: Allelic effects

	x1	x2
A, B	0	0
a	b	b
b	b	-b

Each positive change in direction of character 1 is pleiotropically connected with a detrimental change of character 2 by definition of the fitness function. Two alleles *a* and *b*, which represent the pleiotropic correlation, compete against two alleles *A* and *B*, which have no effects. The four alleles can be combined to form the four different gametes *AB*, *Ab*, *aB* and *ab*. A system of differential equations describes the evolution of these gametes.

We consider the usual decomposition of the phenotype z_i into its genotypic and environmental components:

$$\underline{z} = \underline{x} + \underline{e} \tag{2}$$

$$\overline{\underline{z}} = \overline{\underline{x}}$$

$$P = G + E$$

where $\underline{z}$ is the vector of phenotypic $\underline{e}$ values, $\underline{x}$ the vector of genotypic values and the vector of the environmental effects. *P*, *G* and *E* are the covariance matrices of the phenotypic values, the additive genetic and the environmental effects. Epistasis is excluded. The environmental effects are stochastically independent and normally distributed. In one of the basic laws of evolutionary theory, Lande (1976, 1980, 1982) predicts for the rate of change of a character under selective pressure

$$\dot{\bar{z}} = \sigma V_g \tag{3}$$

where V_g is the underlying genetic variance of the character in the population. The rate of change of a character depends in a linear way on the underlying genetic variance and the strength of selection. With our model, we want to compare the rate of change of a character pleiotropically connected to others with that predicted by Lande. Furthermore, the intention is to analyze the evolutionary dynamic of pleiotropy.

Results

For the rate of change of the character under directional selection (z_1), we get a general and an explicit result. The explicit result is

$$\dot{z}_1 = \sigma V_{g1} - \frac{1}{2w_2^2} cov(x_1; x_2^2) \tag{4}$$

(Baatz, 1991; Baatz & Wagner, 1996) and has some interesting consequences:
— It shows that the rate of change predicted by Lande has to be completed with a term which describes the covariance of the character under investigation with the effects on fitness of the second character;
— as long as the growth of the character 1 is connected with an increase of variance of character 2, the change will occur more slowly. If with a growth of character 1 is connected a decline of fitness effects of character 2, the growth will be accelerated. Compared to a situation without pleiotropic side-effects, the dynamic is similar to a bottleneck: as a consequene of the success of alleles *a* and *b*, recombination increases the genetic variance of character 2. This slows down the dynamic until the variance of character 2 reaches a maximum. From that moment the decline of variance contributes to an acceleration of evolution of character 1. In any case, the time which an advantageous mutant need to invade a population is considerably longer if this mutant has pleiotropic side effects.
— An analysis of the possible attractors in the simplex shows that, no matter where one starts, in most cases gamete *ab* succeeds, because the antagonistic effects of *a* and *b* on character 2 compensate for each other. However, if stabilizing selection is strong in relation to directional selection, $1/2w_2^2 > \sigma$, a further attractor can arise, the point $\rho AB = 1$. It leads to a bifurcation of the systems dynamic. If the frequency of the alleles *a* and *b* is low, a growth of their distribution produces too much variance on character 2 caused by recombination and pleiotropy. In this case, they will be selected out of the population and the dynamic will end at the point $\rho AB = 1$. This is an important result: strong stabilizing forces can hinder the success of a new mutation with disadvantageous pleiotropic effects in a population. Such a situation causes a kinetic constraint on the evolutionary dynamic: the condition $\sigma < 1/2w_2^2$ is a principle criterion for the possibility of suppression of evolutionary change.

In an enlarged model with more loci, polygenic influence, i.e. additional loci which influence the same phenotypic traits under the same selection regime in the same or a similar way as loci *a* or *b*, reduces the expected constraint due to cooperative effects between all heterozygote gametes. Thus, pleiotropy and polygeny have antagonistic effects on phenotypic evolution: as pleiotropy causes inhibition, polygeny reduces this effect of pleiotropy to a certain degree (Baatz & Wagner, in press). Nevertheless, the principle characteristic of pleiotropic effects remains unchanged.

With a simple 45 degree-transformation of the coordinates, the model describes the evolution of two functionally coupled characters: stabilizing selection defines the functional correlation, directional selection lies on the evolution of the function. The consequences correspond with previous analysis: as long as the evolution of the functional complex is connected with an increasing variance orthogonal to the function, the development of the function will be slowed down. This occurs especially for low frequencies of the active alleles *a* and *b*. If stabilizing selection on functional coupling is strong compared with the directional selection on the function itself, evolution can be totally inhibited.

The conclusion of this section is: assumptions 1. and 2. together directly lead to the consequence of an inner systemic inertia against phenotypic change.

EVOLUTION OF ADAPTABILITY

As the systems approach suggests, organisms in changing environments are expected to develop mechanisms of genome organization which facilitate and accelerate adaptation. The epigenetic system itself should be able to change and to adapt to frequent periodical changes of the environment, with the consequence of an evolutionary flexibility of the concerned characters. In the previous section it was shown that pleiotropic side effects handicap the evolution of an advantageous mutant. Thus, a sensible canalization of the distribution of pleiotropy should support evolutionary flexibility in a certain direction of the parameter phase space, and should suppress it orthogonal to this direction.

A developmental mapping function, the B-matrix

A useful formal concept and tool to handle pleiotropy and polygeny with is the B-matrix, introduced by Wagner (1989). The B-Matrix is an attempt to include developmental constraints on the phenotypic expression of genetic variation in quantitative genetic models. The structure of the model is not based on a mechanistic model of development. Let us consider a set of N quantitative phenotypic characters $z = (z_1, \ldots, z_N)$ which are influenced by n genes. Each locus $j = 1, \ldots, n$ is assumed to produce one gene product which has one physiological property relevant for the genotypic value of the characters x. Let us denote the value of this physiological property with y_j, where the index j refers to the locus that produces this gene product. Assuming additivity among alleles at the same locus, the genetic value of this underlying variable is the sum of the properties of the gene products of paternal and maternal alleles

$$y_j = y_j' + y_j''$$

The way in which the underlying variable y influences the genotypic values x of the phenotypic characters z is determined by a "developmental function" $f(y)$ which maps the space of underlying variables y, R^n, onto the phenotype space, R^N. A linear transformation now is sufficiently characterized by a $N * n$ matrix, say B.

$$x = By$$

The coefficients b_{ij} of the B-matrix determine the phenotypic effect of locus j on character i

$$x_i = \sum_{j=1}^{n} b_{ij} y_j$$

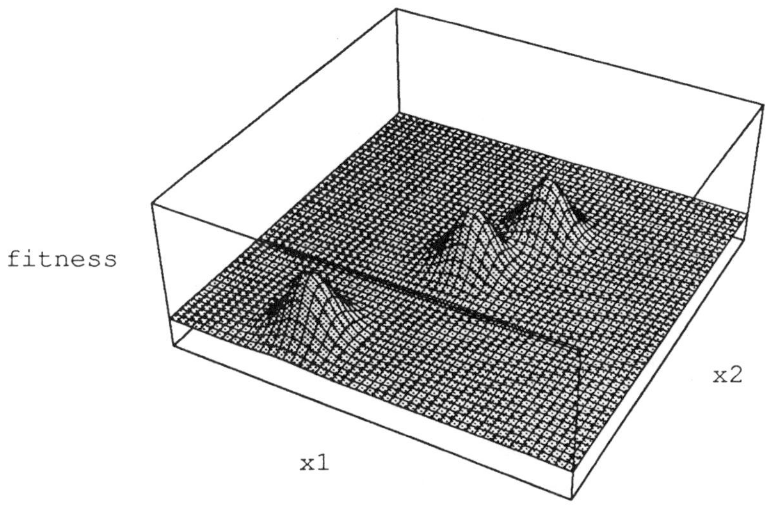

Figure 2. Three states of a randomly moving optimum in a 2 character space: the direction of the move determines a linear functional coupling between character $x1$ and character $x2$.

This equation reveals the biological meaning of the B-matrix more clearly than the matrix notation. A row vector of the B-matrix determi nes how much the corresponding gene product contributes pleiotropically to every genotypic value of x. On the other hand, a column vector determines to what extend a genotypic value is influenced in a polygenic way by each locus. Replacing the developmental function $f(y)$ by a matrix B is identical to the assumption of additivity of allelic effects among loci widely used in quantitative genetic theory.

Functional dependencies, moving optimum in the fitness landscape

In all simulations, we are using fitness functions with a moving optimum, i.e. a peak in the fitness landscape, which randomly moves in predetermined directions. The direction of the move in the parameter space defines a certain functional dependency among the concerned characters, it describes a functional constraint, see Figure 2. A population which until now has been situated on top of an optimum, but now begins to move, is supposed to follow and to adapt on the changing environment. At this point the situation can be compared with a population in the corridor fitness landscape: disadvantageous pleiotropic effects slow down the evolution of a character. A suppression or canalization of such effects orthogonal to the direction of the moving optimum should lead to a higher adaptability.

Evolvable epigenetic system

A certain B-matrix implies a certain distribution of pleiotropy. In different types of simulations, we are testing the ability of control genes to modify the B-matrix and, thus, to canalize disadvantageous pleiotropic effects. Together with these modifiers the B-matrix gives an evolvable epigenetic system which organizes and canalizes the multitude of possible interactions, the 'sea of pleiotropic and polygenic effects'. In an environment in which certain parameters are stable and other parameters are changing frequently, such an evolvable epigenetic system suppresses pleiotropic effects on characters which remain stable and supports pleiotropy in direction of characters, which need to be more flexible.

Results

Deterministic simulations

In a deterministic model similar to that in section 1 but under a moving optimum selection regime, an additional modifier allele *m* suppresses disadvantageous pleiotropic effects orthogonal to the direction of the moving optimum. It competes against a neutral allele *M*. Additivity of alleles is assumed, so that *mm* has the full modifying effect, *mM* the half and *MM* no effect. The rate of evolution of the modifying allele under all conditions is positive, the modifier always succeeds against allele *M*. The rate of evolution of the modifier is one to two orders of magnitude lower than that of the structure genes, which continuously are adapting on the changing environment. It directly depends on the amount of disadvantageous genetic variance caused by pleiotropic effects. From that moment in which the modifier is put through, it remains stable and further adaptation of the system to similar environmental change happens more quickly (Baatz, 1996).

Stochastic simulations

In stochastic simulations the possibilities of modifiers are more obvious: in contrast to a deterministic model, the development of a stochastic model is in principle open, allowing relatively complex modifications to evolve. In our simulations two separate modifiers evolve: the first determines the direction in which canalization takes place, the second gives the amount of canalization. The

B-matrix remains constant throughout the whole process, but is in each generation interpreted anew by the modifiers in the following way: each row j of the B-matrix determines the direction in which locus j intervenes pleiotropically into the phenotypic parameter space. Dependent on the direction given by the first modifier, the part of vector B_j orthogonal to this direction is suppressed. The extent of the suppression is given by the second modifier. Similar to the concept of the B-matrix, this kind of modification is not based on a mechanistic model of development. Each simulation starts with an arbitrary chosen composition of the population, an arbitrary chosen direction of the moving optimum and an arbitrary chosen B-matrix. After an initial evolutionary process under constant stabilizing selection to reduce the initial high genetic variance and to bring the population into a state of equilibrium, the movement of the optimum starts.

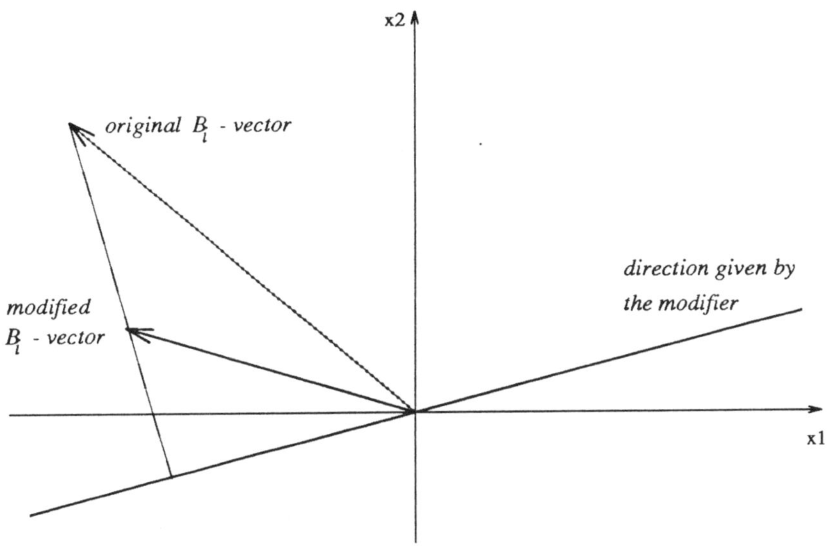

Figure 3. Modification of the pleiotropic effects: the part of vector B_i which is orthogonal to the direction given by the first modifier is suppressed to a certain degree, given by the second modifier.

In all investigated simulation series, the modifier loci evolve and clearly increase the epigenetic canalization in direction of the functional constraint. The effect of increased adaptability is demonstrated by the significantly higher number of selected individuals in each generation.

In contrast to that, we investigated situations with an optimum which moves in randomly changing directions. In these cases, the modifier doesn't evolve and the genetic variance of the modifiers remain high.

DISCUSSION

Two basic assumptions lead this investigation:
— Genotype and phenotype are connected to each other in complex branched interdependencies. Pleiotropy and polygeny are ubiquitous phenomena.
— Evolutionary change is best understood in a situation in which the vast majority of characters remains stabilized.

As described in the second section, the evolution of a character under directional selection is slowed down or even inhibited if it is pleiotropically connected to others under stabilizing selection. Thus, the two assumptions directly lead to the consequence of an internal systemic inertia against easy phenotypic change. From an evolutionary perspective this is a highly senseful result, as organisms are functionally integrated systems and most phenotypic changes can be expected to bring the organism out of this functional balance. The result implies also, that phenotypic change is supposed to take place in directions of the phenotypic phase space which are connected with relatively low pleiotropical side effects. Thus, the distribution of pleiotropy canalizes potential evolutionary change. The organization of this distribution must be an important domain of the epigenetic system.

"Expressed the other way around, functional complexes should be characterized by a high amount of internal pleiotropical correlations and lower correlations to other characters or functional units.

Although epigenetic constraints are quite well described phenomena especially in developmental biology and also morphology, their evolution until now has eluded formal analysis. Our simulations introduced in the third section demonstrate that it is possible to describe auto-catalytic evolutionary processes by means of population genetics. The approach opens the possibility to investigate the dynamics of the evolution of the epigenetic system. Thus, it establishes a new tool in population genetics.

The concept of the B-matrix as well as the described kinds of modifiers are abstract concepts which help to devise and formulate different aspects of the epigenetic system. Nothing is said about the biological mechanisms of epigenetic organization. The interpretation of the results presented in the last section is: given a certain evolvable epigenetic plasticity, the epigenetic system will adapt to the constraints of internal functional dependencies. It causes evolutionary stability of functional integration, whereas functional units as such further on are able to evolve. The nature of this adaptive canalization is important: with gaining evolutionary flexibility in a certain direction, i.e. a higher degree of pleiotropic influence in this direction, the organism loses flexibility in others, because the increased pleiotropic effects slow down the change into other directions. This elucidates the principle concept of epigenetic constraints, as it is implicitly contained in our models: A constraint is determined by the relation between the degrees of freedom in the epigenetic system to intervene pleiotropically in different directions of the phenotypic parameter space. Let us emphasize the relation between the pleiotropic effects: high flexibility in one direction itself necessarily causes low flexibility in other pleiotropically connected directions.

The process of this epigenetic adaptation takes much longer than that of phenotypic adaptation. Epigenetic organization is characterized by a high conservatism.

Thus, the consequences of the evolution of epigenetic constraints are far-reaching (Baatz, 1996):

— They bundle the multitude of pleiotropic effects in the directions of functional constraints — as a direct consequence, further adaptation is accelerated.

— The same functional unit can be found throughout a long process of speciation in always new expressions, whereas the internal integration of the unit remains stabilized.

A developmental function, which characterizes the mechanical process better, should be one which itself is determined by the genetic information completely, and not only the modification of an otherwise predetermined B-matrix. Typically, epigenetic processes are non-linear processes in contrast to the linear transformation given by the B-matrix. In a corresponding model, the degrees of freedom of potential canalization would be higher, which especially is important for models with a higher number of characters. Nevertheless, our approach is well suited to investigate the principle aspects and dynamics of the evolution of constraints.

At the end it should be mentioned that this contribution of course raises more questions than it answers. There is still much investigation to be done to describe the conditions for stability of constraints, the relations and dependencies of different epigenetic constraints to each other, or the conceptual implications for an understanding of evolution. Nevertheless, it seems to be a new step towards a systemic synthesis of microevolutionary and macroevolutionary levels of description. The approach and the results of this investigation support an attitude, which takes neo-Darwinism as well as population genetics for open, still evolvable concepts, which indeed already now have a higher explanation power than any other known theory.

Delphi Systemsimulation
München, Germany

REFERENCES

Alexander, R.D., 1962, Evolutionary change in cricket acoustical communication, *Evolution*, vol. 16, pp. 443-467.

Baatz, M., 1991, *The influence of pleiotropy on the rate of phenotypic evolution*, Diplomarbeit, 1991, University Vienna.

Baatz, M., 1995, The Evolution of Epigenetic Constraints, Phd thesis, University of Vienna.

Baatz, M., Wagner, G.P., 1996, Adaptive Inertia caused by Hidden Pleiotropic Effects, *Theor. Pop. Biol.*, in press.

Falconer, D.S., 1981, *Introduction to Quantitative Genetics*, Ed. 2. Langmann, New York.

Frazzetta, T.H., 1975, *Complex adaptations in evolving populations*, Sinauer Ass., Sunderland, MA.

Gould, S.J., 1980, Is a new and general theory of evolution emerging?, *Paleobiology*, vol. 6, pp. 119-130.

Hoy, R.R.; Hahn, J.; Paul, R.C., 1977, Hybrid cricket auditory behavior: Evidence for genetic coupling in animal communications, *Science*, vol. 195, pp. 82-84.

Lande, R., 1976, Natural selection and random drift in phenotypic evolution, *Evolution*, vol. 30, pp. 314-334.

Lande, R., 1980, Quantitative genetic analysis of multivariate evolution, applied to brain: Body size allometry, *Evolution*, vol. 33, pp. 402-416.

Lande, R., 1982, A quantitative genetic theory of life history ecology?, *Ecology*, vol. 63, pp. 292-305.

Lande, R.; Arnold, S.J., 1983, The measurement of selection on correlated characters, *Evolution*, vol. 37, pp. 1210-1226.

Maynard Smith, J., 1981, Macroevolution, *Nature*, vol. 289, pp. 13-14.

Rechenberg, I., 1971, *Evolutionstrategie: Optimierung technischer Systeme nach Prinzipien der biologischen Evolution*, Friedrich Frommann Verl., Stuttgart-Bad Cannstadt.

Riedl, R., 1977, A systems-analytical approach to macroevolutionary phenomena, *Q. Rev. Biol.*, vol. 52, pp. 351-370.

Riedl, R., 1978, *Order in living organisms*, New York, J. Wiley Stanlay.

Wagner, G.P., 1984, Coevolution of functionally constrained characters: Prerequisites for adaptive versatility, *Biosystems*, vol. 17, pp. 51-55.

Wagner, G.P., 1985, Ueber die populationsgenetischen Grundlagen einer Systemtheorie der Evolution, in *Evolution, Ordnung und Erkenntnis*, J.A. Ott, G.P. Wagner and F.M. Wuketits (eds.), Parey, pp. 97-111.

Wagner, G.P., 1988, *The significance of developmental constraints for phenotypic evolution by natural selection Population Genetics and Evolution*, Publ. G. de Jong, pp. 222-229.

Wagner, G.P., 1988a, The influence of variation and of developmental constraints on the rate of multivariate phenotypic evolution, *J. Evol. Biol.*, vol. 1, pp. 45-66.

Wagner, G.P., 1989, Multivariate mutation-selection balance with constrained pleiotropic effects, *Genetics*, vol. 122, pp. 223-234.

DANIEL R. BROOKS

THE UNIFIED THEORY AND SELECTION PROCESSES

INTRODUCTION

The patterns of evolutionary diversification and the distribution of biodiversity result from complex interactions between ecological components of evolution, explaining *maintenance* of biological systems, and genealogical components, explaining their *origins*. Evolutionary theory is under-developed with respect to questions of origin, and of integration among processes derived from "intrinsic" and "extrinsic" factors operating on different temporal and spatial scales. Biology has also resisted efforts to reconcile its general principles with basic natural laws of physics and chemistry, despite persistent indications that thermodynamics and statistical mechanics might provide the key (e.g. Boltzmann, 1877; Lotka, 1913, 1925; Lindeman, 1942; Prigogine & Wiame, 1946; Newman, 1970; Brooks & Wiley, 1986, 1988; Wicken, 1987; Demetrius, 1992; Salthe, 1993). The unified theory of evolution (Wiley & Brooks, 1982; Brooks & Wiley, 1986, 1988; Brooks et al., 1989; Brooks & McLennan, 1990; Maurer & Brooks, 1991; Brooks, 1992) asserts that 1. orderliness and organization in biological systems result from the interaction of historical uniqueness, cohesive tendencies among subunits of biological systems, and functional integration of those subunits, *in addition to* natural (environmental) and sexual selection; 2. current evolutionary theory lacks general explanations for the existence and expected effects of these three elements; and 3. finding such explanations requires extending some principles from general physico-chemical laws to complex biological systems.

The *historical uniqueness* of biological systems manifests itself in historical contingency and temporal irreversibility. Events that take place in the past may have an effect on the subsequent behavior and fate of the system, and conditions affecting a system subsequently may also contain traces of the past. In addition, many biological processes are tied to the directional nature of time, albeit on different time scales. Biological processes such as reproduction, development, aging and death, speciation, and extinction, are temporally irreversible on any time scale. These processes involve growth, increasing complexity and manifestations of at least some of the systems' history.

Cohesive properties of biological systems range from cell-cell adhesion and recognition, to sexual reproduction and specific mate recognition systems, to common phylogenetic history. They limit the ways in which, and the extent to which, populations can respond to environmental selection, and are the "glue" of functional integration and hierarchical organization (Wake & Roth, 1989). Many irreversible biological processes manifest changes as a result of interactions among cohesive and diversifying factors. Speciation in sexually reproducing species results when the developmental and reproductive constraints acting as cohesive forces keeping a species intact are overridden by environmental forces

G. Van de Vijver et al. (eds.), Evolutionary Systems, 113-128

acting to split it into descendants.

Functional integration and hierarchical structure are hallmarks of biological organization (e.g. Salthe, 1985, 1993; Wake & Roth, 1989). A major component of functional integration is the interdependence of parts. Evolutionary changes in biological systems do not occur all at the same time; thus, when such changes occur, only part of the system changes. All changes, whether they are point mutations on a chromosome or alteration of part of a complex mating ritual, must integrate with the rest of the system, which has not changed, in such a way that viable organisms result. Hierarchical structure, from genealogical relationships of individuals and species to complex social systems and trophic interactions among species in ecosystems, provide stability, reinforce boundaries between system and surroundings, allow increasing amounts of complexity without losing organizational coherence, and provide a way in which causation and control can be tied together. Salthe (1985, 1993) decomposed hierarchical structure into sets of "triads", comprising 1. upper level (causal or initiating), 2. lower level (control or boundary) elements impinging on 3. a focal level, from which emerges a particular form of structure and/or organization. Complex hierarchical systems are combinations of linked triadic units. Within a given hierarchical system, relatively random lower level effects are screened off by the cohesive properties of the higher level effects, and various levels in the hierarchy have diminishing effects on other levels in proportion to the remoteness of their interactions.

Environmental and genealogical phenomena are intimately connected in biology. Pre-biotic environmental conditions established the boundary conditions within which life could originate. Conversely, genealogical processes that characterize life are autonomous enough from environmental conditions to be capable of overrunning available resources and of changing the environmental conditions substantially. For example, the evolution of photosynthetic prokaryotes from anaerobic ancestors resulted in increased oxygen content in the atmosphere. This increase then altered the diversity and changed the distribution of anaerobic organisms, limiting them to relatively rare environments, and paving the way for the evolution of a new array of species. Most of the environment relevant to biological systems consists of products of genealogical processes (Maynard Smith, 1976). This contributes greatly to the far from equilibrium nature of the environment in which evolution takes place and environmental pressures under far from equilibrium conditions can lead to self-organization mediated by selection (Wicken, 1987).

CONCEPTUAL FOUNDATIONS OF THE UNIFIED THEORY

Brooks and Wiley (1986, 1988) examined different kinds of biological processes that show direction in time, growth and increasing complexity, coupled variation and constraint, and hierarchical and self-organization, to see if there were any underlying generalities. The search for regularities in spontaneously irreversible behavior in natural systems can be approached from two perspectives, an "externalist" or "total system (TS)" perspective, in which emergent order in a subsystem of the total system is imposed by the rest of the system (the "surroundings"), and an "internalist" or "bound matter (BM)" perspective, in which emergent order results from dynamical behavior of the system itself. Brooks and Wiley felt that evolution does not result from an interaction between a random system and an organized environment, in which case all order and organization in biological systems results from environmentally-induced changes (an extreme TS perspective), or strictly from intrinsic dynamics of biological (essentially genomic/developmental) systems (an extreme BM perspective), but from an interaction between a self-organized system (a BM component) and an organized environment (a TS component), each with their own "rules" of behavior. They used a general mathematical formalism summarizing key elements of complexity. This simple "entropy function" has four potentially illuminating formulations: 1. the "actual" entropy (H_{obs}) calculated on the basis of the observed distribution of components of the system; 2. the "maximum possible" entropy (H_{max}), estimated by calculating the entropy value for the components of the system at any given time if they were all randomized; 3. an absolute difference (H_{max}-H_{obs}) (Information or Macroscopic Information: Gatlin, 1972; Layzer, 1975; Landsberg, 1984a,b); and 4. a relative difference (H_{max}-H_{obs}/H_{max}) (Redundancy or Order: Gatlin, 1972; Landsberg, 1984a,b).

Heuristic simulations emulating reproduction, ontogeny, mutation and speciation (Brooks, LeBlond & Cumming, 1988) produced three generalities: 1. H_{obs} is an increasing function of time; 2. H_{obs} is a concave function of time, as historical constraints retard the rate of entropy increases; and 3. the difference between H_{max} and H_{obs} is an increasing function of time, a function of the growth of organization in the system. Entropy and information/organization/order (redundancy) increased over time (Brooks, LeBlond & Cumming, 1984; Brooks, Cumming & LeBlond, 1988; Smith, 1988). If H_{max} is a function of the capacity, or potential, of a system and H_{obs} is a function of the content, or expression, of that potential, the difference between information capacity and information content is a function of the constraints on the system, both inherent (BM) and extrinsic (TS). The BM perspective is appropriate for studying macroscopic properties manifested in genealogical phenomena, because genealogical processes stem from the dynamics of physical information systems, which are BM systems (Collier, 1986).

Frautschi (1982, 1988; see also Layzer, 1975; Landsberg, 1984a,b) contrasted two classes of processes that generate entropy. The first is equilibration of temperatures between system and surroundings. Biological systems exhibit this kind of entropic behavior through processes that result in the dissipation of energy (heat loss) to their surroundings. The second is expansion of the phase space (the realm of all possibilities) in which the system resides, increasing its number of accessible microstates (possible configurations). System organization increases so long as equilibration (equiprobable distribution of the system over its microstates) occurs at a slower rate than the expansion of the phase space, allowing a lag between the increase in realized entropy (H_{obs}) and the increase in the maximum possible entropy of the system (H_{max}), which is a function of the size of the phase space. So long as the phase space expands faster than the system can fill it, increasing entropy can be accompanied by the emergence of organized structure. In cosmology, this argument explains the spontaneous and irreversible emergence of stars, solar systems, galaxies, and other organized structures, in which fundamental forces linking material bodies slow down the entropic expansion of matter in the universe to such an extent that organized structures emerge as a result of and not at the expense of increasing entropy. A similar phenomenon accompanies the expansion of energy from the sun into space (Frautschi, 1982, 1988). In biological systems, mutations act to expand the genetic phase space within which life occurs (Layzer, 1978, 1980), while genetic and genealogical bonds linking all living systems, as well as the environments in which they exist, play roles analogous to the fundamental forces linking material bodies (Brooks et al., 1989; Brooks & McLennan, 1990; Brooks, 1992).

Both biological and cosmological evolution exhibit long time scale regularities. Both show positive correlation between the spatial and the temporal scale of emergence of regularities (e.g. Brooks, 1988; Maurer & Brooks, 1991). Both are characterized by entropy production, which requires a non-equilibrium state, and by the emergence of low-entropy structures, the sources of continuing entropy production. This seemingly paradoxical behavior occurs not as a result of the degradation of a static environment by the entropy production of the system (although biological systems do degrade their environments as a result of using energy and matter from their surroundings), leaving the system in an absolute negentropic state, but from the behavior of a dynamic environment in which maximum possible entropy increases faster than the systems which produce the entropy can equilibrate a relative negentropic state. In cosmological models, the dynamic environment is produced by the expansion of the universe and symmetry-breaking into distinct forms, such as matter and radiation. The resulting order is the energy in the universe available to do work.

The unified theory treats biological evolution as a microcosm of cosmological evolution in which the energy/entropy relationship is highly relative. From the perspective of autotrophic biological systems solar heat and light energy is free energy. From the perspective of the sun, however, that heat and light is entropy, relatively low-grade energy lost as a result of thermonuclear reactions. Some solar energy is picked up by autotrophs on this planet and converted into other manifestations of entropy, heat lost from the plant and organic structure. The biological phase space is the genetic system, which expands over time as mutations add new dimensions to it. This increases the maximum possible entropy of the system, making it a dynamic environment for evolving systems. The effects of common ancestry and of the environment on mate finding, reproduction, and development, provide constraints on the increases in the entropy (diversity) of the biological information system.

Energetics and the externalist perspective

Lotka (1913, 1925) recognized that biological systems persist in space and time by transforming energy from one state to another in ways that generate and maintain organized structure. Maurer and Brooks (1991) recognized two classes of such energy transformations. Heat-generating transformations involve a net loss of energy from the system, usually in the form of heat. Conservative transformations involve changing free energy into states that can be stored and utilized in subsequent transformations; for example, the use of ATP to control the burning of glucose. All conservative transformations in biological systems are coupled with heat-generating transformations, but the reverse is not true; there is a heavy energetic cost to maintaining structure. Lotka (1913) suggested that the inevitable structural decay that must accompany such costs could be delayed, although not reversed, by the system's accumulation of bound energy from conservative transformations. Or, the interplay between flow and partitioning of energy in biological systems acts to slow the rate at which energy stored by conservative transformations is degraded by heat-generating transformations.

Entropy changes (dS) in a system include changes in the surroundings due to exchanges with the system (d_eS) and changes within the system due to production by internal irreversible processes (d_iS). Exchanges between biological systems and their surroundings are accompanied by a great deal of waste; hence, d_eS is very large compared with d_iS. However, if biological systems are able to maintain their structural integrity, they must produce entropy internally ($d_iS > 0$). Or,

$$dS = d_eS + d_iS, \; d_iS > 0$$

Therefore, internal production is critically important in biological evolution, though it represents a small portion of the total energy budget for living systems. Biological production rules are those processes for which there is an energetic "cost" or "allocation". Following Prigogine and Wiame (1946) and Zotin and co-workers (e.g. Zotin & Zotina, 1978), Brooks and Wiley (1988) denoted such allocations using the symbol ψ, denoting a specific dissipation function. The function includes at least two major classes of processes: 1. those involved in dissipation from the system, called the external dissipation function (ψ_α, thermal entropy) and 2. those involved in dissipation within the system, called the bound dissipation function (ψ_μ, structural entropy). In biological systems, ψ_μ can be further subdivided into allocations for accumulating biomass ($\psi_\mu{}^b$) and allocations for accumulating genetic diversity ($\psi_\mu{}^i$). Brooks and Wiley (1988) suggested that all three components of biological production (ψ) should be included in the thermodynamic production term, shown heuristically as

$$d_iS = \psi_\alpha + \psi_\mu{}^b + \psi_\mu{}^i$$

Energy used in the uptake of raw materials from the surroundings into the system produces entropy which is dissipated into the surroundings (d_eS: Prigogine & Wiame, 1946; Prigogine, 1980). Different manifestations of entropy production resulting from irreversible processes within the system (d_iS: Prigogine & Wiame,

1946; Prigogine, 1980) are associated with each of the classes of transformations. Heat-generating processes occur when energy and entropy flow in opposite directions, entropy production tending to move the system towards disordered states. Conservative transformations are characterized by energy and entropy flowing in the same direction, entropy production being retained within the system and tending to move the system towards more structured states. As entropy and energy flow through biological systems at different rates, structure accumulates at different levels of organization; furthermore, the structure at any given level is constrained by energy and entropy flows at other levels. Different types of constraints emerge from rate gradients in entropy production.

Organisms allocate a very small proportion of their total energy budget for the critical function of transmitting information from one generation to the next. This component is tiny compared to the total thermal entropy production of the system, most of which is lost from the system, so the entropy of the surroundings increases much more than the entropy of the system. Biological systems remain low-entropy (far from equilibrium) systems relative to their surroundings; however, there is a net increase in entropy within the system.

Products of the genealogical hierarchy are maintained through time by the exploitation of "entropy" or "resource" gradients in the surroundings (Wicken, 1987; Weber & Depew, 1995; Depew, this volume; Weber, this volume; Swenson, this volume; Salthe, 1993, this volume), determined by interactions between abiotic and biotic factors. The abiotic portion of the ecological hierarchy can be structured in part by the genealogical hierarchy. For example, metabolic processes are involved in the degradation of high grade energy sources into lower grade forms of energy, including heat. Both the capture of incoming solar energy by biological systems, and the mass re-radiation of heat by these organisms affects the thermal profile of the earth. Additionally, the production of oxygen as a byproduct of photosynthesis or of carbon dioxide as a byproduct of aerobic metabolism affects the composition of the earth's atmosphere. This means that the production term (d_iS) can influence the exchange term (d_eS). The biotic portion of the ecological hierarchy is subject to the influences of the structural portion of the genealogical hierarchy (ψ_μ). The ecological hierarchy has a propensity to move the products of the genealogical hierarchy in the direction of minimizing energy gradients in the environment, to the extent permitted by the phylogenetic, developmental, and genetic constraints for each species.

Entropy is produced at different rates in biological systems because energy stored by conservative transformations is degraded at different rates. Thus, biological systems manifest organized structures on different temporal and spatial scales. If we examine cellular or sub-cellular structure over short time intervals, processes such as metabolism and respiration dominate explanations of observed structure. Most entropy production is dissipated into metabolic heat loss, and biological systems behave as classical dissipative structures. At more intermediate levels of organization, space or time, most entropy production is dissipated into accumulation and maintenance of biomass. On the largest and longest scales, the patterns relevant to biological explanations are formed mainly by accumulation and maintenance of genetic diversity. From the perspective of the environment, such patterns of biodiversity tend to be organized with respect to energy gradients, whereas from the perspective of the genealogical system, biodiversity is organized with respect to sister-group relationships and patterns of geographical distribution that mirror geological evolution occurring on similar temporal and spatial scales.

Infodynamics and the internalist perspective

Communications theory and physical measurement theory share the belief that information is an abstraction rather than a material part of the system transmitted from a "source" through a "channel" to a "receiver". In communications theory, the amount of information sent from a source is calculated using a statistical entropy function. Errors in transmission can result from poor encoding at the source or from noise in the transmission channel. The meaningful information is that subset of information transmitted which is actually recorded by the receiver. All processes affecting the transmission and reception of information thus decrease the entropy of the message from its maximal value at the source. Physical entropies are expected to increase as a result of work done on the system, so the communications view of entropy is non-physical. Physical measurement theory provides a second formalism. Brillouin (1962) distinguished "free information", an abstraction involved in descriptive exercises, and "bound information", referring to material properties of the system (but not stating that information is a material part of the system *per se*). Bound information is determined with respect to the "complexions" (microstates) of the system. Hence, it is also calculated using a statistical entropy function but, contrary to communications theory, is expected to exist only in systems for which there is a non-arbitrary microstate/macrostate distinction. Bound information is defined as

$$I = H_{max} - H_{obs}$$

where H_{max} refers to the totally relaxed state of the system (usually estimated by a randomization of the observed components of the system). Brillouin defined I as "negentropy", which is converted into bound information by measurement (measuring devices are receivers), so negentropy = information. Information thus has a physical basis, but is not material.

Biological information has both communications functions and a physicochemical (material) basis. What is thus needed is an account of biological information that is 1. physically realistic (can be shown to have an objective basis), 2. intrinsic to the system (is a material part of the system) rather than to devices for measuring the system, and 3. can grow spontaneously over time. Collier (1986, 1988, 1990, in press; Smith, in press) proposed that physical information systems occur as structural arrays in which there is 1. a physical (material) basis for the information, 2. an energetic cost in producing the information, and 3. a real (non-arbitrary) macrostate/microstate distinction. Since the discovery of the chemical structure and function of DNA, there has been a material basis for biological information, satisfying 1. above.

Energy dissipated within the system as a result of work done on the system (including heat-generating transformations, or ψ_α of d_iS) is *intropy*, which stands for internal entropy (Ulanowicz, 1986). Energy that is converted into structure (conservative transformations, or ψ_μ of d_iS) is *enformation*, standing for intrinsic information (these distinctions originated in Collier, 1990). Conservative processes within biological systems are coupled with heat-generating processes, so there is an energetic cost associated with the production and maintenance of biological information. Intropy and enformation are interconvertable (e.g. energy brought in from the surroundings can be converted into structure, say glycogen, which can then be converted into heat). Intropy is converted into enformation by cohesive properties of the system. Thus, there is an energetic cost to making and maintaining informational structures. Cohesion is analogous to inertia, which provides resistance to change. Cohesive properties, ranging from molecular affinities to cell-cell adhesion to genetic compatibility, mate recognition, and genealogy, also provide resistance to intrinsic or extrinsic fluctuations, allowing macroscopic properties to emerge.

Macrostate/microstate distinctions are determined objectively by part/whole associations. The number of accessible microstates is increased by the production of new components, either at a given level or through the opening up of new levels of organization. Biological systems accomplish this by conservative transformations. For example, auto-catalytic processes producing monomers make "monomer space" available for molecular evolution. Some monomers have high chemical affinities for each other, and will spontaneously clump into polymers. Once polymers begin to form, "polymer space" becomes available to the evolving system. At this level, polymers are macrostates and monomer and distributions are microstates. Causal interactions among polymers create new levels of organization in which polymer distributions are the microstates and new levels of organization are the macrostates, and so on. Each new functional level creates a hierarchy of increasing structural intricacy, manifested by increasing allocation of the entropy production in structure. Therefore, the allocation of d_iS to ψ_μ might be nearly proportional to entropy increases due to the expansion of phase space resulting from the creation of new possible microstates. A protein coding unit might be considered a macrostate, while all the actual sequences that code for that protein would be the microstates; a locus could be a macrostate, and all alleles corresponding to that locus the microstates; phenotypes could be macrostates, and all genotypes corresponding to a given phenotype would be microstates (Layzer, 1978, 1980; Collier, 1986). This formulation answers objections that biological informational entropies do not include a non-arbitrary macrostate/microstate distinction. Encoded information is also the carrier of the cohesive properties, so production of biological information involves simultaneous production of variation and constraints, ensuring that genealogy will be a combination of continuity and change.

The primary implication of the above discussion is that no single objective level of organization, spatial scale, or temporal scale has causal primacy in biological evolution. Neither intrinsic nor extrinsic processes are primary, much less exclusive, over all spatial or temporal scales, for all types of biological interactions, or over all groups of organisms. A variety of processes operate on many different scales, and all contribute to evolution; however, they do not all play equally important roles at all levels, at all times, or for all groups. Therefore, the macroscopic manifestations of evolutionary principles will differ depending on the group of organisms and the window of observation. A second implication is that transmission of information from one generation to the next uses a tiny fraction of the total energy budget of an organism, so evolution is more an informationally than an energetically limited process. Biomass is produced by genealogical rules but functions in the ecological realm; only genetic information functions solely within the genealogical hierarchy, and it does not cost very much energetically. Many genealogical processes appear negentropic from the perspective of the environment, because so much of the entropy production (i.e. so large a proportion of the total energy budget) of genealogical systems is returned to the surroundings; hence, the surroundings tend to decay more rapidly than the systems and this could be (and has been) misinterpreted as negentropic behavior on the part of the biological system. In addition, those selection processes which originate externally to the genetic systems (abiotic and biotic: Wade & Kalisz, 1990), may favor entropic influences on biological systems (Demetrius, 1992; Collier, in press). Organisms do not simply degrade their immediate environments, they serve as environmental sources of energy for other organisms; in fact, the largest portion of the environment for organisms is other organisms (Maynard Smith, 1976).

THE UNIFIED THEORY AND SELECTION PROCESSES

Darwin's original intuition about natural selection came by way of analogy with artificial selection informed in part by the writings of Malthus on human economic theory. If natural selection is a real process, however, there must be a theoretical framework from which an expectation of the reality of natural selection flows. Darwin based the theory of natural selection on the intuition that production of offspring overruns production of resources. This leads to a "struggle for existence" in which only the "fittest" survive, fitness being defined by an individual's ability to compete for limited resources and measured by survival. How can production of offspring overrun available resources if biological systems are adapted to their environments? First, the rules that lead to the production of organisms must be independent of, or at least highly insensitive to, fluctuations in the environment. Otherwise, the number of organisms produced would conform to a fluctuating equilibrium number determined by the availability of environmental resources, and we would not find different species exhibiting different reproductive rates in the same environments. Second, those rules must be characterized by a high degree of replication (redundancy), because if intraspecific competition is a driving force in natural selection, conspecific offspring must display overlapping requirements and abilities.

Biological systems are characterized by inherent production rules and a high degree of self-organization that is manifested, in part, by the maintenance of organized structure. Naturally occurring populations of highly similar (redundant) organisms exist within boundaries defined both by genealogical history and by the environment. During relatively short time periods historical effects are essentially constant, and evolutionary dynamics can be explained solely by reference to environmental changes, corresponding to standard population biological treatments of natural selection. Over longer time periods, the stability and fate of a population will depend also on its history, which will constrain the population's responses to environmental perturbations or changes. If this were not so, the environmental changes could not cause extinction.

Eigen and Winkler (1981) discussed a thought experiment involving two chemotrophs living in the same environment and being identical in all ways except for different replication rates. The chemotroph with the higher rate will always outcompete the one with the slower rate. This thought experiment was designed to show natural selection as a primitive concept, but it works only if each chemotroph is self-determining rather than environmentally-conforming with respect to replication rates. In reality, the thought experiment supports the idea that natural selection emerges from self-determination of biological systems. Thus, in order for natural selection to be a real process, there must be rules governing the production of organisms that are largely insensitive to environmental resources. If this is true, however, those rules may affect the ways in which, and the extent to which, populations respond to natural selection.

Macroscopic information (I), or organization, increases over time in evolving systems; therefore, we expect to see an increase in the value of the function Q, the macroscopic order (Landsberg, 1984), or redundancy (Gatlin, 1972) of the system:

$$Q = 1 - (H_{obs}/H_{max}) = I/H_{max}$$

If the environment is the primary source of macroscopic ordering for biological systems, then the percentage of biological information that is shared between system and environment should be high. This presents a paradoxical view of natural selection because, if the amount of this mutual information is high, the "fit" between biological systems and their environments is almost perfect; therefore, evolutionary change will be negligible or stochastic with respect to the environment. This paradox can be resolved by postulating that the shared information between biological systems and their environments is low (Collier, 1986). Natural selection thus emerges under conditions established by self-organization which, in turn, constrain the way in which and the extent to which populations respond to natural selection. Natural selection emerged as a result of biological evolution, and subsequently became part of the process.

The unified theory thus suggests that the effectiveness of natural selection will be inversely proportional to the amount of mutual information between the genealogical system and the environment. The environment decays with respect to the population average fitness, and inherent constraints limit the rate at which, the ways in which, and the extent to which the population can respond. This leads to what Maynard Smith (1976) termed *lag load* and what Ulanowicz (1986) termed *ascendency*. The environment also deteriorates with respect to the most fit genotype, and this creates an environmental "pull" to which genealogical systems respond in a manner that leaves room for the genealogical system to "experiment". Selection effects should be microscopically ubiquitous, tending to optimize populations locally in terms of the immediate environment, to the extent possible given the cohesive properties of the species (Demetrius, 1992, illustrated this for the cases of rapidly and slowly-growing populations). Such effects contribute to macroscopic structure by limiting the range of variation that participates in reproduction, eliminating some otherwise functional genetic variants. Environmental selection can thus act as a form of external cohesion. If genealogical changes do not occur in response to, and/or if they occur more slowly than, environmental changes, the genealogical system will lag behind the environment, and will always be "trying to catch up". The extent to which this is true is the extent to which natural selection will always be potentially an evolutionary force. That is, there will always be ecological ascendency or "unoccupied niche space", allowing for experimentation by the genetic system, but only so long as genealogical changes *do not* track environmental changes closely.

The degree to which sexual selection is effective is directly proportional to the degree of mutual information between males and females. From this it follows that in order for sexual selection to be a real process, males and females must be able to communicate with each other in such a way that sexual preferences influence the flow of genetic information from one generation to another. It is genealogical processes that are responsible for mutual information, so sexual selection should be an evolutionarily cohesive process.

Gatlin (1972) made one of the most determined efforts to relate information theory to biological systems. She adopted a standard view considering information systems to consist of a source of signals, a channel through which the signals are transmitted, and a receiver to translate the portion of the signals that made it through the channel into information. In biological systems, she argued, the genetic system is the source, reproduction and ontogeny are the channel, and the environment is the receiver. Genetic data thus become phenotypic signals as a result of reproduction and ontogeny, and become biological information as a result of causal interactions between the phenotype and the environment. The environment cannot be a receiver in a physical sense, because its only causal interaction with biological information is the possible elimination of some of it; it does not measure or interpret the information (Brooks & McLennan, 1990). Rather, the environment acts as noise in the channel, eliminating relatively less fit organisms in a population. This begs the question of just what is the receiver. The answer lies in establishing that biological information is a material part of biological systems. If this is so, then it is possible for the system to be its own source and receiver (see also Csányi, 1989). Current standard evolutionary theory stresses external causality for orderliness in biological systems, so the receiver has been construed as part of the surroundings, i.e. localized in space. While it is true that biological systems are localized in space, they are also localized in time. Therefore, the receiver can be a "time" (Brooks & McLennan, 1990). The source is a genetic system at time t_0, the channel is reproduction and ontogeny, and the receiver is the same genetic system at any given time $t_{1...n}$; thus, the receiver is temporally distinct from the source (Figure 1). If the source precedes the receiver in time, it can produce the system that acts as receiver and then become a source itself. This reinforces the biological analogy, because DNA has significant self-repair capabilities and sexual reproduction may enhance those capabilities (Bernstein et al., 1989).

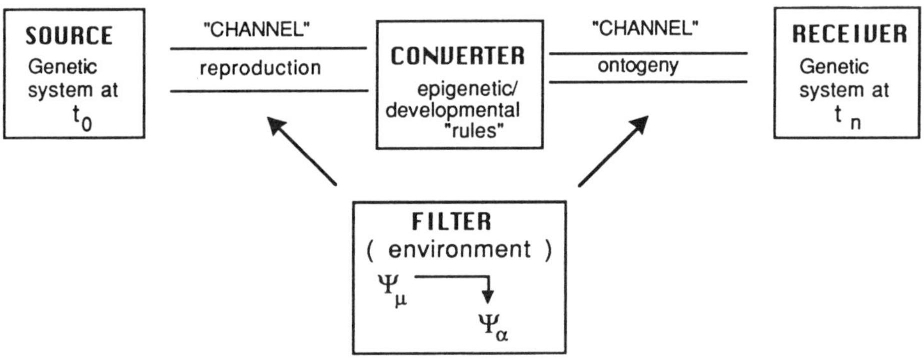

Figure 1. Biological information flow. (1) Source (gene pool at t_0), encoder (reproduction), channel (ontogeny), noise (environment), receiver (genetic system at t_1).

Figure 2 portrays the differences between natural and sexual selection using this formalism. Sexual selection operates at the source of information transmission, which is reproduction, and tends to increase the efficiency of encoding information in such a way that members of the next generation express heritable mate preferences. Natural selection operates at the endpoints of information transmission, the products of reproduction, decreasing the capacity of the channel through which information is transmitted so that a smaller fraction of all possible bits of information is transmitted to the next generation. Or, natural selection emerges as a result of low mutual information between the genetic system and the environment. Sexual selection emerges as a result of high mutual information between members of the genetic system. Natural selection changes population means and reduces variance by restricting the capacity of the channel through which information can be transmitted successfully. Sexual selection changes population means and reduces variance by increasing the efficiency of encoding information at the (reproductive) source. For natural selection the probability of survival is the result of interfering events, whereas for sexual selection the probability of survival is the result of non-response to interfering events. Natural and sexual selection thus arise in different ways, and affect different parts of the information flow system, so it is possible for them to be in conflict (Fisher's runaway sexual selection theory, in which sexual selection could lead to efficiently encoded information in such a way that less and less of it could get through the channel permitted by the environment), or to complement each other ("truth in advertising" views of sexual selection, in which sexual selection could lead to efficiently encoded information in such a way that it flowed through the channel with reduced distortion).

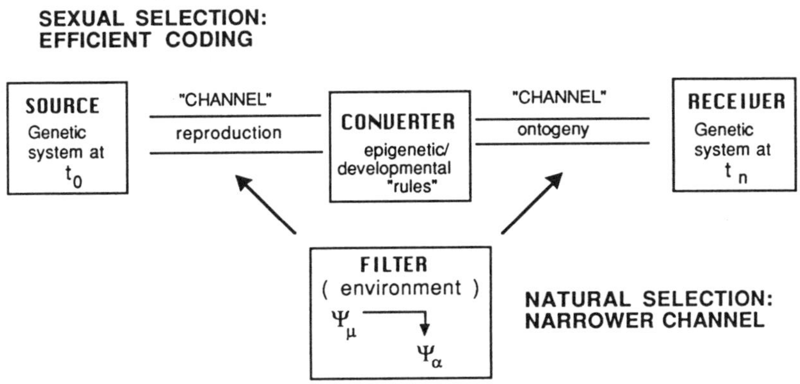

Figure 2. Natural and sexual selection in biological information flow.

CONCLUSIONS

The unified theory postulates that genealogical systems always "mean something" or "talk to" themselves (Csányi, 1989; Brooks & McLennan, 1990), and also "mean something" or "talk to" their surroundings. Evolution is fundamentally a Darwinian process because whenever there is a conflict between what an organism means to itself (including sexual selection) and what it means to its environment, self-meaning always takes precedence, with a resulting decrease in fitness of the individual. The juxtaposition of evolutionary ordering due to the conservative and relatively autonomous nature of the genetic system with environmental selection returns us to the holistic view of evolution originally propounded by Darwin, and in doing so allows us to begin to integrate biological evolution with the evolution of the cosmos at large. Substantial recent convergence of perspective by a number of research groups gives renewed hope that we are converging on a truly unified view of the origin and evolution of biological systems. These perspectives agree substantially on the following points: 1. biological information is material and not just an abstraction; that is, living systems are fundamentally auto-catalytic material information systems and not just epiphenomenal eddies on the edge of chaos; 2. biological information systems can internalize at least some of their environments directly (especially in the origin of life) or in abstract form (especially in the case of various forms of selection); 3. the transmission of biological information also involves auto-catalytic and self-organizational phenomena ("internal dynamics") that produce, and are characterized by, macroscopic properties (macrostates) manifested in genealogical phenomena such as ontogenies, demes, populations and species; and 4. internal (i.e. self-organizing) dynamics render biological systems relatively insensitive to their immediate environments, so natural selection emerges within the living system, and becomes intimately and unreducibly tied to it.

The internal dynamics produce scalar hierarchies of biological organization as a result of emergent downward causation/constraints, microscopic randomizing effects, mediated by cohesion/friction phenomena (described by Salthe, 1993), and resulting in the coupled production of variation and constraint through time. This leads to a propensity for progressive complexification of biological systems that emerge in various ways and is invested in various portions of scalar hierarchies (Salthe, 1993). Physical stability as well as fitness are important for understanding and explaining biological systems and their evolution. When stability considerations and selection pressures agree, evolution will look like a mechanical optimizing process even though it is supposed to be blind; under these conditions evolutionary rates will be high and it will appear that natural selection is favoring the evolution of energetically most efficient systems. When stability considerations and selection pressures disagree, evolutionary rates will be slow and the system might reorganize along novel and unpredictable lines. The relative insensitivity of biological systems to their environments has three important consequences: 1. these macrostates are the units of selection; genes are never the units of selection, although they can effectively be seen as tokens of selection for the purposes of fruitful research; 2. there will be constraints on the ways in which and the extent to which biological systems can respond to particular environments over particular (especially short) temporal and spatial scales; and 3. biological systems will rarely attain, and therefore be trapped on, any particular adaptive peak, thus retaining a high degree of adaptability.

Finally, global macroscopic features of biological systems manifested in ecosystems structure emerge from the internal dynamics played out over this finite planet (e.g. Salthe, this volume; Swenson, this volume). When the environmental filter becomes loaded with other species (Maynard Smith, 1976), top-down constraints can become a significant source of stabilizing selection, acting in concert with global energy-flow stability constraints to produce predictable ecosystem structure (e.g. Swenson, this volume). Drastic environmental changes will lead to mass extinction and convergent re-emergence of structure following those mass extinctions. This is the emergence of significant convergent evolution; the same kind of macroscopic properties pertaining to ecosystem structure re-emerges after mass extinctions, even though the taxa involved in "filling the niches", and the "niches" themselves, differ each time, and exhibit different evolutionary modifications that deal with different functions.

Department of Zoology, University of Toronto
Toronto, Canada

REFERENCES

Bernstein, H.; Hopf, F.A.; Michod, R.O., 1988, Is meiotic recombination an adaptation for repairing DNA, producing genetic variation, or both?, in *The Evolution of Sex*, R.E. Michod and B.R. Levin (eds.), Sunderland Mass, Sinauer Assoc, pp. 139-160.

Boltzmann, L., 1877, Über die Beziehung eines allgemeine mechanischen Satzes zum zweiten Haupsatzes der Warmtheorie, in *Sitzungsber. Akad. Wiss.*, Wien, *Math. Naturwiss. Kl.*, vol. 75, pp. 67-73.

Brandon, R., 1990 (2nd ed.), *Adaptation and Environment*, Princeton, Princeton Univ. Press.

Brillouin, L., 1962, *Science and Information Theory*, New York, Academic Press.

Brooks, 1988, Scaling effects in historical biogeography: A new view of Space, Time and Form, *Syst. Zool.*, vol. 37, pp. 237-244.

Brooks, D.R., 1992, Incorporating origins into evolutionary theory, in *Understanding Origin: Contemporary Ideas on the Genesis of Life, Mind and Society*, F. Varela and J.P. Dupuy (eds.), Amsterdan, Reidel/Kluwer Associates, pp. 191-212.

Brooks, D.R.; Collier, J.; Maurer, B.A.; Smith, J.D.H.; Wiley, E.O., 1989, Entropy and information in evolving biological systems, *Biol. Philos.*, vol. 4, pp. 407-432.

Brooks, D.R.; Cumming, D.D.; LeBlond, P.H., 1988, Dollo's law and the second law of thermodynamics: analogy or extension?, in *Entropy, Information and Evolution: New Perspectives on Physical and Biological Evolution*, B.H. Weber, D.J. Depew and J.D. Smith (eds.), MIT Press, Cambridge, pp. 189-224.

Brooks, D.R.; LeBlond, P.H.; Cumming, D.D., 1984, Information and entropy in a simple evolution model, *J. Theor. Biol.*, vol. 109, pp. 77-93.

Brooks, D.R.; McLennan, D.A., 1990, Searching for a general theory of biological evolution, *J. Ideas*, vol. 1, pp. 35-46.

Brooks, D.R.; McLennan, D.A., 1991, *Phylogeny, Ecology and Behavior: A Research Program in Comparative Biology*, Univ. Chicago Press, Chicago.

Brooks, D.R.; Wiley, E.O., 1986 (1st ed.), *Evolution as Entropy: Toward a Unified Theory of Biology*, Univ. Chicago Press, Chicago.

Brooks, D.R.; Wiley, E.O., 1988 (2nd ed.), *Evolution as Entropy: Toward a Unified Theory of Biology*, Univ. Chicago Press, Chicago.

Collier, J., 1986, Entropy in evolution, *Biol. Philos*, vol. 1, pp. 5-24.

Collier, J., 1988, Supervenience and reduction in biological hierarchies, in *Philosophy and Biology: Canadian Journal of Philosophy*, M. Matthen and B. Linsky (eds.), suppl. vol. 14.

Collier, J., 1990, Intrinsic information, in *Information, Language and Cognition: Vancouver Studies in Cognitive Science*, University British Columbia Press, vol. 1.

Collier, J., (in press), Incorporating Adaptation in Evolution as Entropy, in *Between Instruction and Selection: Studies in a Unified Theory of Non-Equilibrium Biology*, J.D. Collier and D. Siegel-Causey (eds.), Baltimore, John Hopkins University Press.

Csányi, V., 1989, *Evolutionary Systems and Society: A General Theory*, Durham, N.C., Duke Univ. Press.

Darwin, C., 1872 (6th edition), *The Origin of Species*, London, John Murray (ed.).

Demetrius, L., 1992, The thermodynamics of evolution, in *Physica A*, vol. 189, pp. 417-436.

Depew, D.J.; Weber, B.H., 1995, *Darwinism Evolving: Systems Dynamics and the Genealogy of Natural Selection*, Cambridge, Massachusetts, MIT Press.

Eigen, M.; Winkler, R., 1981, *Laws of the Game: How the Principles of Nature Govern Chance*, New York, A.A. Knop.

Eldredge, N., 1985, *Unfinished Synthesis*, New York, Columbia Univ. Press.

Eldredge, N., 1986, Information, economics and evolution, *Ann. Rev. Ecol. Syst.*, vol. 17, pp. 351-369.

Eldredge, N.; Salthe, S. N., 1984, Hierarchy and evolution, in *Oxford Surveys in Evolutionary Biology*, R. Dawkins and M. Ridley (eds.), vol. 1, pp. 182-206.

Frautschi, S., 1982, Entropy in an expanding universe, *Science*, vol. 217, pp. 593-599.

Frautschi, S., 1982, Entropy in an expanding universe, in *Entropy, Information and Evolution: New Perspectives on Physical and Biological Evolution*, B. Weber, D.J. Depew and J.D. Smith (eds.),

Cambridge, Massachusetts, MIT Press, pp. 11-22.

Gatlin, L.L., 1972, *Information Theory and the Living System*, New York, Columbia Univ. Press.

Harvey, P.; Pagel, M., 1991, *The Comparative Method in Evolutionary Biology*, Oxford, Oxford Univ. Press.

Landsberg, P.T., 1984a, Can entropy and "order" increase together?, *Physics Lett.*, vol. 102A, pp. 171-173.

Layzer, D., 1975, The arrow of time, *Sci. Amer.*, vol. 233, pp. 56-69.

Layzer, D., 1978, A macroscopic approach to population genetics, *J. Theor. Biol.*, vol. 73, pp. 769-788.

Layzer, D., 1980, Genetic variation and progressive evolution, *Amer. Nat.*, vol. 115, pp. 809-826.

Lindeman, R.L., 1942, The trophic dynamic aspect of ecology, *Ecology*, vol. 23, pp. 399-418.

Lotka, A.J., 1913, Evolution from the standpoint of physics, the principle of the persistence of stable forms, *Sci. Amer. suppl.*, vol. 75, pp. 345-346, p. 354, p. 379.

Lotka, A.J., 1925, *Elements of Physical Biology*, Williams and Wilkins, Baltimore.

Maurer, B.A.; Brooks, D.R., 1991, Energy flow and entropy production in biological systems, *J. Ideas*, vol. 2, pp. 48-53.

Maynard Smith, J., 1976, What determines the rate of evolution?, *Amer. Nat.*, vol. 110, pp. 331-338.

Newman, S.A., 1970, Note on complex systems, *J. Theor. Biol.*, vol. 28, pp. 411-413.

Prigogine, I., 1980, *From Being to Becoming*, W.H. Freeman, San Francisco.

Prigogine, I.; Wiame, J.M., 1946, Biologie et thermodynamique des phénomènes irréversibles, *Experientia,* vol. 2, pp. 451-453.

Salthe, S.N., 1985, *Evolving Hierarchical Systems: Their Structure and Representation*, Columbia Univ. Press, New York.

Salthe, S.N., 1993, *Development and Evolution: Complexity and Change in Biology*, MIT Press, Boston.

Schank, J.C.; Wimsatt, W.C., 1988, Generative retrenchment and evolution, in *PSA 86*, vol. 2, A. Fine and P.K. Machamer (eds.), Philosophy of Science Association, East Lansing, Michigan, pp. 33-60.

Smith, J.D.H., 1988, A class of mathematical models for evolution and hierarchical information theory, *Inst. Math. Appl. Preprint Series*, vol. 396, pp. 1-13.

Smith, J.D.H., in press, Mathematical approaches to the unified theory of biology, in *Between Instruction and Selection: Studies in a Unified Theory of Non-Equilibrium Biology*, J.D. Collier and D. Siegel-Causey (eds.), Baltimore, John Hopkins University Press.

Ulanowicz, R.E., 1986, *Growth and Development: Ecosystems Phenomenology*, New York, Springer-Verlag.

Wade, M.J.; Kalisz, S., 1990, The causes of natural selection, *Evolution*, vol. 44, pp. 1947-1955.

Wake, D.B.; Roth, G. (eds.), 1989, *Complex Organismal Functions: Integration and Evolution in Vertebrates*, Dahlem Workshop, New York, Wiley.

Wicken, J.S., 1987, *Evolution, Thermodynamics and Information: Extending the Darwinian Paradigm*, Oxford Univ. Press, Oxford.

Wiley, E.O., 1981, *Phylogenetics: The Theory and Practice of Phylogenetic Systematics*, Wiley-Intersci., New York.

Wiley, E.O.; Brooks, D. R., 1982, A non-equilibrium approach to evolution, *Syst. Zool*, vol. 31, pp. 1-24.

Wiley, E.O.; Mayden, R. L., 1985, Species and speciation in phylogenetic systematics, with examples from the North American fish fauna, *Ann. Mo. Bot. Garden*, vol. 72, pp. 596-635.

Zotin, A.I.; Zotina, R.S., 1978, Experimental basis for qualitative phenomenological theory of development, in *Thermodynamics of Biological Processes*, I. Lamprecht & A.I. Zotin (eds.), Berlin, deGruyter, pp. 61-84.

JOHN D. COLLIER

INFORMATION INCREASE IN BIOLOGICAL SYSTEMS: HOW DOES ADAPTATION FIT?

INTRODUCTION

The notion of progress has been virtually banned from mainstream contemporary biology (Nitecki, 1988, p. viii), even though it still lurks in the background (Ruse, 1988). This is a marked change from earlier views, which saw evolution to be distinctly progressive. The complete explanation of this change is no doubt complex, involving factors dear to historians and sociologists of science, such as the breakdown of the Enlightenment view of science and society, and the education and class commitments of prominent authorities. It is probably too early to historically evaluate these forces with any sort of objectivity. I will focus instead on the internal logic of the shift, and the extent to which it is justified: progress has no theoretical role in contemporary neo-Darwinian orthodoxy. As John Maynard Smith (1988) points out, given two states of a biological system, there is nothing in Fisher's "fundamental theorem of natural selection" that would allow a biologist to determine which state is earlier. The hardening of the Modern Synthesis of genetics and population biology has permitted the gradual realization that directed processes of any kind, let alone progressive ones, are neither a probable consequence of nor presupposition of its core theses alone.

This situation is not entirely satisfying, since it seems evident that there has been an increase in something from the earliest life-forms to those we find today (Ayala, 1988), and the neo-Darwinian orthodoxy tells us that traits that have not been eliminated are more adaptive, and hence more *functional*, than those that are. Given that functionality is a teleological notion, and that adaptive traits are the only neo-Darwinian candidates that can support an increase in *anything*, it isn't a great step to assume that biological progress is just the increase in functionality resulting from adaptive processes. Indeed, this has been a popular view among prominent biologists (Ruse, 1988; but see Williams, 1966). This consequence can be avoided by admitting that directional changes are found in the evolutionary record, but that these changes do not involve increased overall functionality (nor, perhaps, are they directed by increases in any global property). Furthermore, directional change is arguably not global, but local only (Hull, 1988). Evolution might be locally, or even globally, directional, but it isn't progressive. There are several ways to support this move without doing (too much) violence to orthodox neo-Darwinism.

On the other hand, the problem of directionality, and other empirical problems with the Synthesis have led to calls for extending, expanding or replacing the neo-Darwinian paradigm (e.g. Eldredge, 1985; Wicken, 1987; Brooks & Wiley, 1988; Wesson, 1993; Salthe, 1993). Behind these calls is a recognition that evolution is an historical process, in which historical contingencies constrain later evolution, leaving their trace in contemporary observable traits. This implies at least an overall

G. Van de Vijver et al. (eds.), Evolutionary Systems, 129–139.

direction of evolution, if not an increase in some property over time (Ayala, 1988). In this chapter, I will show how the implications of the historical constraints on evolution can account not only for directed adaptation, but also for genuine increases in adaptation, and hence progressive evolution. The basis for this is a more precise notion of biological form that permits us to make comparisons of the "fit" of traits (and of the organisms that bear them) to their environment.

ADAPTATION PRODUCES INFORMATION

Although everything must have some form, morphology and morphogenesis have been especially important in our understanding of biological systems. In the past, our understanding of biological form has been largely qualitative. Now, information theory allows us to precisely quantify (if not measure) form. Perhaps the most salient feature of biological form is its organized complexity. Although individual subsystems might expire, the typical pattern we observe in biology (whether in evolution, ecology or development) is growth of complexity and organization across an integrated system. In principle, this can be measured as an information increase, either in terms of computational complexity or computational depth, or (more likely) both.

One important aspect of (to speak more definitely, a *component* of) biological information is adaptedness. Adaptedness is the fit of an organism with its environment that increases the survival potential of the organism's genes. Adaptation is the process by which adaptedness is produced and/or maintained through natural selection. An adaptation, in turn, is a trait that is fit, and exists because of adaptation. I will give an account of the information involved in adaptation that is based in orthodox accounts of fitness. This account involves the comparison of the information in a trait with the information in the environment, i.e. between the form of a trait and the structure of its environment. I will then discuss how that information can increase. My account is compatible with a physical interpretation of information that makes it measurable, at least in principle, through direct intervention, like any other scientifically useful physical property. My account also makes increase in the information of adaptation a likely if not inevitable outcome of evolutionary processes. We can keep the core of selection theory, while reconceiving it in a way that makes information increase a likely outcome of adaptation. This approach can, at the same time, lay the ground for the integration of adaptation with other biological processes involving information transmission and transformation.

GENETIC DETERMINATION OF INHERITANCE

Natural selection is environmentally caused differential reproductive success resulting in fitness differences.[1] Selection can act on populations, organisms or genes. I will focus on the fitness of organisms and their traits for convenience, but much of what I say can be applied, *mutatis mutandis*, to populations and genes. The fitness of a trait of an organism is a measure of the trait's adaptive value for the organism, relative to the organism's environment. A trait has adaptive value if it serves the survival of the lineage of which the organism is a member. The complete fitness of an organism is the net effect of the fitnesses of its traits. Since survival is necessary for the continuing existence of lineages, and evolutionarily significant traits are inherited, the existence of some traits can be explained by their contribution to adaptive value, or fitness (Wright, 1973; Sober, 1984). A trait is biologically functional just insofar as it contributes to the fitness of the organism that possesses the trait.

It is doubtful that fitness can be reduced to lower level physical properties (Sober, 1984). This means that fitness must be defined in terms of its effects on survival. On the other hand, it is impossible to give an adequate definition of fitness entirely operationally (Brandon, 1990). In order to avoid circularity, the fitness of a trait should be defined as a propensity for the organism that has it to produce lineages of organisms that also have the trait, rather than being measured in terms of the number of replicas of the trait actually produced (Mills & Beatty, 1979). (Roughly, propensity is an inherent tendency that something will occur with a certain probability. I will have more to say below.)

Traits that are not inherited, either directly or indirectly, cannot be fit, even though they might contribute to the survival and reproduction of the organism that has them. The reason for this is that they can play nothing but an accidental role in evolution. Only inherited traits can be the cause of a consistent trend. The effects of non-inherited traits last for only one generation, and have no regular effect on the long term survival of a lineage. Inheritance is not an all or nothing phenomenon: a trait might have an environmental and a hereditary component. Only the latter matters to patterns in evolution. The main form of inheritance is genetic. A trait is *genetically determined* if it "differs from other traits at least in part as the result of the presence of one or more distinctive genes" (Wilson, 1978, p. 19). Genetic determination does not preclude other forms of inheritance, which can be cultural, environmental or somatic. The totality of all means of inheritance is called a *developmental system* (Griffiths & Grey, 1994). In what follows I am going to simplify things by considering only the genetically determined part of inheritance. This permits a reduction of the information in traits to the information in their genetic determinants. Information in a developmental system that is stored environmentally or culturally will complicate the story considerably, but I will rely on the assumption that the pattern of explanation will be the same. Neo-Darwinism generally assumes that genetic determination is the only evolutionarily significant form of hereditary determination, so the simplification is not unusual.

INFORMATION OF ADAPTATION

Consider a population of organisms with a set of alleles over the population (the gene pool). Selection will act on members of the population in proportion to the total fitness of each organism to produce a certain statistical distribution of alleles in each of the varying environments the population inhabits. Other effects might distort this distribution, so the selection component will pick out a distribution that has a certain propensity. Given the connection between propensities and probabilities, we can define the fitness of an allele in an environment as the probability of the allele given the environment, where the probability is understood as the propensity due to selection:

$$f_{i,j} = p\,(a_i | E_j) = \frac{p(a_i \& E_j)}{p(E_j)} \tag{1}$$

The fitness $f_{i,j}$ of the ith allele in the jth environment is the propensity due to selective forces for the allele a_i to occur in the environment E_j. By probability theory, this is equal to the probability of the allele and the environment both occurring, divided by the probability of the environment. Suppose, for example, we are considering malarial resistance in an environment with malaria present. We know there are three related traits: malarial resistance, no resistance, and the sickle cell trait. We know that the probability of malarial resistance is relatively high, but we also know that — because it is heterozygotic in origin — that there will be a certain amount of the deleterious sickle cell allele. The propensity for this deleterious allele derives from the genetic conditions that produce it. The sickle cell allele is actually somewhat fit, given both the environment and genetic conditions. If we consider an environment in which malaria is not present, then malarial resistance is less fit, and the sickle cell allele is correspondingly less fit. This method of defining fitness allows us to take into consideration interactions between an allele and other factors.

It would be nice to be able to define the fitness of a trait in a similar manner, but we immediately run into a problem. A propensity distribution of traits would contain information not only about the fitness of the traits, but also about the correlation among traits due to pleiotropic effects. Equation (1) implicitly presumes that the alleles are independent, i.e. that a given allele does not contain information about the probabilities of other alleles. This is not true for traits, since they depend on two alleles (at least). To determine the fitness of traits, we must discount the fitness for the effects of correlation with other fit traits. In the sickle cell case, the fitness of the sickle cell trait can be zero, but it can still appear due to its correlation with the malaria resistant trait. In the following, for convenience, I am going to assume that the discounting for correlations has been done. A similar procedure could be done to correct (1) for gene linkages. For traits, (1) becomes

$$f_{i,j} = p\,(T_i | E_j) = \frac{p(T_i \& E_j)}{p(E_j)} \tag{1'}$$

Where the selection propensity for trait i in environment j, p(T_i & E_j), is discounted for pleiotropic effects. The propensity of a trait to appear under selection can be recovered by adding the pleiotropic component back in. The discounting move is required to avoid double counting selection effects. This sort of manoeuvre must be done very carefully to generalize (1') to traits in a full developmental system.

Interestingly, we can sum or even integrate over environments to judge the fitness of a trait under varying conditions (say, for example, the prevalence of malaria varies from location to location). This can be done either for an individual or an interbreeding population to get the fitness of the trait for the individual according to the average time it spends in each environment, or for the population, according to how it is distributed among different environments. The net trait fitness in this case is the weighted sum of the fitness in each environment:

$$f_i = \sum_j p(E_j)p(T_i|E_j) = \sum_j p(T_i\&E_j) = p(T_i) \tag{2}$$

This is just the average discounted propensity of the trait. We could also apply the same method to the fitnesses of individual alleles, depending on their physical and genetic environments.

The adaptive information content of a trait, intuitively, is a measure of how well it matches the environment. On the statistical account of information, it is a function of the logarithm of a probability. Since the fitness is a probability, the natural thing to do is to take its logarithm and take the sum over environments, weighted by the fitness:

$$\boldsymbol{I}_{T_i} = \sum_j f_{i,j} \log_2 f_{i,j} \tag{3}$$

This, however, would not give us the information of adaptation of a trait, but would give us the information about the distribution of the trait among environments. We need to subtract the information we have about the trait in general, and weight the result by the probability of the environment (in addition to weighting by the fitness) in order to get the average information the trait gives about environments:

$$\boldsymbol{I}_{T_i,E} = \sum_j p(E_j) f_{ij} \log_2 \frac{f_{ij}}{p(T_i)} \tag{4}$$

It is worth noting that the environments do not have to be actual, as long as we have some way to estimate the relevant propensities. $\mathbf{I}_T$ the information in the trait, is maximal for a trait and range of environments in a population when the fitness of the trait in each environment is optimized. If some T is improbable, but highly likely in a common environment nonetheless, it has a high information content, as might be expected. It should also be noted that a trait which is fit in many environments has a higher information content than if it were fit in only one environment, again as we might expect. It is somewhat artificial to use traits in these equations, since traits are not fit in isolation, but traits can be replaced with phenotypes. With the choice of suitable environments, the same equations apply to genic selection (Williams, 1966; Dawkins, 1979) or to group selection (Lewontin, 1970). Equations (1)-(4) follow logically from the propensity definition of fitness, and are no more and no less adequate than that definition. In particular, the definition tells us nothing of the source of the propensities.

$$\boldsymbol{I}_{\boldsymbol{T,E}} = \sum_i \sum_j p(E_j) f_{i,j} \log_2 \frac{f_{i,j}}{p(T_i)} \tag{5}$$

We can expand the procedure further to consider a range of traits in a variety of environments. This requires a summation over traits as well: Substituting the value of $f_{i,j}$:

$$I_{T,E} = \sum_i \sum_j p(T_i \& E_j) \log_2 \frac{p(T_i \& E_j)}{p(T_i) p(E_j)} \tag{6}$$

This is a measure of correlation: the information of adaptation is just the mutual information of alternative environments E and alternative traits T.

The use of propensities, especially the discounted propensities, required to derive these results might be questionable. There is another version of information theory based in computation theory that provides a better justification for the use of discounted propensities. On this account, the information content of a string is the length of the shortest computer program that can produce that string. By using 1-1 maps of more general forms to strings, we can account for the information content of anything with a definite form. One problem with this approach is that there is a computational overhead involved in the definition. This overhead, though, is small for high information strings, and it largely drops out when we are comparing string lengths. The information of x given y is the information in x, but not in y, i.e. the length of the shortest program to compute x given an input y. The information of adaptation in (6) is a measure of the information that does not need to be computed, given the information in the organism, to determine the environmental conditions. Beyond the limit of this information, the environment will be effectively unpredictable by the organism, and it will be effectively random. This justifies the use of propensities. Justifying the discounting manoeuvre follows a similar pattern: correlations between traits, or linkages among genes, amount to mutual information. As such, even if it is independently embodied in different places, the total amount of environmental prediction that can be based on it is limited to the non-redundant part of the sum of the information of all traits (or genes), so any mutual information should be counted only once.

The information of adaptation is a relation between the environment and the biological entity (gene, trait, organism or group). Since by definition it depends on selection, there is a (somewhat metaphorical) sense in which the biological entity represents the information in its environment. This sense of representation is strengthened by the observation that the origins of differences in phenotypic form within the genealogical hierarchy (Eldredge & Salthe, 1984; Eldredge, 1985) are largely (but not exclusively) genetic. For genetically determined adaptations it is not unreasonable to say that the information of adaptation is encoded in the DNA. Again somewhat metaphorically, the DNA is capable of representing information about the environment. In the same metaphorical sense, DNA can represent information about organisms and species. It is worth noting, however, that equation (6) is completely symmetrical (being a correlation), and does not itself sanction the representational metaphor. We might just as well say that the environment represents the organisms in it. The important thing is that increasing adaptation leads to increasing correlation, which is equivalent to greater mutual information between the environment and the biological entities.

DIRECTIONAL EVOLUTION

Natural selection is a powerful tool for explaining the existence of adaptive traits and their functionality. In its neo-Darwinian guise, however, it is deficient for explaining the growth of biological organization and complexity. A respected biology text (Luria, Gould & Singer, 1981, p. 647) states: "Darwinism is not a theory of intrinsic progress", and goes on to note that adaptations are local only. If that were the whole story, selection could account at best only for local increases of information. Appropriate conditions could just as easily lead to adaptive decreases in organization, and complexity. Particularly if environmental conditions go through long cycles, we might observe cyclic gain and loss of information. There is no obvious reason why increased organization and complexity should be adaptive, especially throughout large cycles in environmental conditions (Maynard Smith, 1988). Why is selection for apparently simpler forms (as, for example, in viruses, parasites and yeasts[2]) relatively rare? The information of adaptation will increase the most if there is more refined adaptation. A precondition for this is greater variety both among and within organisms. If the intuition that adaptation involves a growth of information is correct, then not only must it be temporally directed, but it must also involve the production of variety.

There are neo-Darwinian accounts of the growth of organization (Mayr, 1982; Dawkins, 1987, pp. 169-220; perhaps Schanck & Wimsatt, 1988), but these accounts *do not arise from the core of neo-Darwinism, if the core is restricted to natural selection as it is usually understood*. One approach is to maintain that the organization produced by adaptation, and the irreversibility of evolution are independent of each other. Irreversibility, on this account, comes from the constant change in the genotype: if a need arises again for some capacity that has been lost, the capacity will be generated by a different genotype than the original (Mayr, 1982, p. 609). The change in the genotype is partly due to selection, and partly due to random changes (i.e. changes that are not ordered with respect to selection). Although this process can produce increased complexity if the starting point is towards the simpler end (McShea, 1994), there is no clear reason why this process should yield any sort of directionality of the sort that Maynard Smith requires: given two (relatively complex) states of a biological system, there is nothing in Mayr's account of irreversibility that would allow a biologist to tell which state is earlier. Furthermore, since the irreversibility is due to chance events, and not selection, progress is out of the question.

One block to decreasing order is what Wimsatt (Schanck & Wimsatt, 1988) calls generative entrenchment. Once a particular function or structure has been established, functions or structures that evolve later are likely to depend on the earlier evolved traits. Consequently, changes in the earlier traits are less likely to produce a viable organism, since changes to the earlier traits will likely disrupt the traits that depend on them as well. Dawkins (1987, p. 169ff) pointed out that once a particular chemical pathway has become established, it has an advantage over possible competing pathways, because pathways generally involve interdependencies that must co-evolve together. It is unlikely that a competing pathway will appear, with all the required interdependencies, when an effective pathway is already present. This idea can be generalized to other evolved traits in any circumstance in which co-evolution is required. This mechanism can account for increases in the magnitude of diversity: once a particular set of traits has co-evolved, any further evolution will tend to continue to diverge in a direction that preserves these traits together. Dawkin's ecological examples involve evolutionary "arms races" that are examples of the "Red Queen" phenomenon, producing directionality, but no net gain in fitness. Together, Wimsatt's generative entrenchment and Dawkins' co-evolutionary fixation of diversity can explain why order does not decrease, and why existing diversity tends to be amplified. They do not, however, explain progress, since they do not require an increase in adaptation.

One way to explain the origins of variety is to postulate isolating mechanisms that divide otherwise unified populations. No doubt these mechanisms are important. External forces (such as geological changes) can separate a population into two or more groups that, no longer reproductively cohesive, will evolve in different directions due either to differing environmental conditions or to chance differences between the genomes of the separated populations (the "founder effect"). Such external influences will need to be invoked by any evolutionary theory, since their presence is well-established. Sexual assortment can serve as an internal mechanism to achieve isolation. It is an interesting question whether there are other internal mechanisms. But again, this mechanism implies no increase in adaptation, and hence no progress.

PROGRESSIVE EVOLUTION

The above gives us four separate phenomena that can produce directionality: the local phenomenon of adaptation on random variants, mechanisms like generative entrenchment and positive feedback that produce irreversibility, mechanisms for the production of variation, and mechanisms to amplify variants. To some extent these overlap and interact, but they are largely independent. This is hardly parsimonious, and reflects the fact that the explanations have been largely added on to the core of traditional neo-Darwinism. It would be nice to have a more systematic explanation. Furthermore, these mechanisms give directionality, but not according to any standard, so they cannot account for evolutionary progress.

One possibility involves entropic increase of information, along the lines described by Brooks and Wiley (1988a, 1988b) in the absence of adaptation. If we assume that the environment contains genetic variation, much as a bell jar contains a gas, then we have a source of cohesion that allows us to distinguish genetic macrostates (the environmentally constrained state) from genetic microstates (all of the variant states permitted by a particular environment). This distinction is similar to that I proposed (Collier, 1986) for a macrostate-microstate distinction between species information and the genetic information of all the possible varieties permitted by a given species. If genetic variation approaches equilibrium with the environment, then adaptation will cease. In fact, a strong selective force (large fitness differences) will lead to this happening quite quickly. We can expect that near to equilibrium, adaptation will be static. Furthermore, near to equilibrium processes can go in any direction, just like adaptation in traditional neo-Darwinism. But this is not what we observe. We observe a distinct direction of increase in the information of adaptation. This suggests that the genome is not in equilibrium with the environment.

One reason to expect non-equilibrium is an analogue to the Brooks-Wiley mechanism for increasing genetic phase space (Collier, 1986). As new variants are selected, the genome becomes more highly adapted, constraining the genetic phase space more severely. On the other hand, the same process often opens up new possibilities for adaptation, which constitutes a release of constraints, possibly making more genetic information potentially adaptive. This would increase the effective genetic phase space. Furthermore, the increased overhead from new possibilities allows the possibility of intermediate levels of organization that are permitted by non-equilibrium conditions, but not under equilibrium (for example, chance concentrations of particular genes in a relatively small group of closely interbreeding organisms). These spontaneously organized structures might give further chances for adaptation. On a more mundane level, mutations and recombinations continually add information to the genetic phase space which are potential mutual information with the environment. Other mechanisms that can create new possibilities for adaptation, increasing the size of the adaptive phase space are described by Wimsatt and Schanck (1988), Conrad (this volume), and Baatz (this volume). Unless new forms are maintained strongly and spread rapidly through the population (as would happen if they were especially fit), they are likely to be formed faster than they can equilibrate, thus maintaining non-equilibrium conditions. It is noteworthy that the main reason why equilibrium might be reached reflects exactly the thinking found in traditional neo-Darwinism: perturbation followed by return to equilibrium. However, among all the possibilities, it is an extreme case. These extreme cases would, of course, provide strong confirming evidence for the traditional view, at least in the short run.

Both the production of variability required for unidirectional increases in the information of adaptation, and its unidirectionality itself, can be explained by a genetic system that is well out of adaptive equilibrium with the environment. It is interesting that such conditions enhance the possibilities for adaptation. This might seem counterintuitive, but it is merely the counterpart to the stasis that results from steep peaks in the adaptive landscape. Away from equilibrium, processes involved in adaptation are both more turbulent and more gently productive. These processes permit not just directional change, but increases in the information of adaptation. Evolution in these regions is the winner in the long run.

Department of Philosophy, University of Newcastle
Newcastle, Australia

NOTES

[1] Rather than selection by chance or deliberate intervention. The former is called neutral (Kimura, 1983), while the latter is called artificial selection (but see Brandon, 1990).

[2] Even in these cases there is evidence that this decrease is compensated for by complexity increase across a clade (Brooks & Wiley, 1988).

REFERENCES

Ayala, F.J., 1988, Can "progress" be defined as a biological concept?, in *Evolutionary Progress*, M.H. Nitecki (ed.), Chicago, University of Chicago Press, pp. 75-96.

Brandon, R.N., 1990, *Adaptation and Environment*, Princeton, Princeton University Press.

Brooks, D.R.; Wiley, E.O., 1988 (2nd ed.), *Evolution as Entropy: Toward a Unified Theory of Biology*, Chicago, University of Chicago Press.

Collier, J.D., 1986, Entropy in evolution, *Biology and Philosophy*, vol. 1, pp. 5-24.

Eldredge, N., 1985, *The Unfinished Synthesis: Biological Hierarchies and Modern Evolutionary Thought*, New York, Oxford.

Eldredge, N.; Salthe, S.N., 1984, Hierarchy and evolution, *Oxford Studies in Evolutionary Biology*, vol. 1, pp. 184-208.

Griffiths, P.; Grey, R., 1994, Developmental systems and evolutionary explanation, *Journal of Philosophy*, vol. 91, pp. 277-304.

Hull, D., 1988, Progress in ideas of progress, in *Evolutionary Progress*, M.H. Nitecki (ed.), Chicago, University of Chicago Press, pp. 22-48.

Kimura, M., 1983, *The Neutral Theory of Molecular Evolution*, Cambridge, Cambridge University Press.

Lewontin, R.C., 1970, The units of selection, *Ann. Review of Ecology and Systematics*, vol. 1, pp. 1-14.

Luria, S.E.; Gould, S.J.; Singer, S., 1981, *A View of Life*, Menlo Park, Benjamin Cummings.

Maynard Smith, J., 1988, Evolutionary progress and levels of selection, in *Evolutionary Progress*, M.H. Nitecki (ed.), Chicago, University of Chicago Press, pp. 219-230.

Mayr, E., 1982, *The Growth of Biological Thought: Diversity, Evolution and Inheritance*, Cambridge, Harvard University Press.

McShea, D.W., 1994, Mechanisms of large-scale evolutionary trends, *Evolution*, vol. 48, pp. 1747-1763.

Mills, S.; Beatty, J., 1979, The propensity interpretation of fitness, *Phil. Sci.*, vol. 46, pp. 263-286.

Nitecki, M.H. (ed.), 1988, *Evolutionary Progress*, Chicago, University of Chicago Press.

Ruse, M., 1988, Molecules to men: Evolutionary biology and thoughts of progress, in *Evolutionary Progress*, M.H. Nitecki (ed.), Chicago, University of Chicago Press, pp. 97-126.

Salthe, S.N., 1993, *Development and Evolution: Complexity and Change in Biology*, Cambridge MA, MIT Press.

Schanck, J.C.; Wimsatt, W.C., 1988, Generative entrenchment and evolution, *PSA 86*, vol. 2, A. Fine and P.K. Machamer (eds.), East Lansing, Philosophy of Science Association, pp. 33-60.

Sober, E., 1984, *The Nature of Selection*, Cambridge, MA, MIT Press.

Weber, B.; Depew, D.; Smith, J.D. (eds.), 1988, *Information, Entropy and Evolution*, Cambridge, MA, MIT Press.

Wesson, R., 1991, *Beyond Natural Selection*, Cambridge MA, MIT Press.

Wicken, J.S., 1987, *Evolution, Thermodynamics and Information: Extending the Darwinian Paradigm*, New York, Oxford University Press.

Wiley, E.O., 1988a, Entropy and evolution, in *Information, Entropy and Evolution*, B. Weber, D. Depew and J.D. Smith (eds.), pp. 173-188.

Wiley, E.O., 1988b, Entropy, evolution and progress, in *Evolutionary Progress*, M.H. Nitecki (ed.), Chicago, University of Chicago Press, pp. 275-292.

Williams, G.C., 1966, *Adaptation and Natural Selection*, Princeton University Press.

Wilson, E.O., 1978, *On Human Nature*, Cambridge, MA, Harvard University Press.

Wimsatt, W.C.; Schanck, J.C., 1988, Adaptations and the means of their avoidance, in *Evolutionary Progress*, M.H. Nitecki (ed.), Chicago, University of Chicago Press, pp. 231-274.

Wright, L., 1973, Functions, *Philosophical Review*, vol. 82, pp. 139-168.

JONATHAN D.H. SMITH

CANONICAL ENSEMBLES, EVOLUTION OF COMPETING SPECIES, AND THE ARROW OF TIME

MATHEMATICS AND PHILOSOPHY

At the outset, it may prove helpful to clarify the respective roles of mathematics and philosophy in the study of evolutionary systems. Historically, the two subjects have not always been considered as distinct. (Indeed, when I was a student in Cambridge, applied mathematics was called "natural philosophy", while philosophy was called "moral science".) In order to evolve, systems need to be somewhat complex. This complexity forces a certain subtlety on the concepts used to analyze the systems — concepts like "entropy", "hierarchy", "fitness" and "species". Such concepts can easily lead to paradox and confusion if they are not defined and used with considerable care. Mathematics and philosophy share the task of ensuring that the concepts are properly formulated, that they are clearly understood. On the other hand, the distinction between the two subjects may be summarized by the observation that philosophy works with natural language, with words, whereas mathematics works objectively, with symbols. Sciences such as physics, chemistry, biology, economics and sociology begin with verbal formulations, and then move progressively towards symbolic codifications. As this move takes place, mathematics inherits from philosophy the job of validating the science's concepts and arguments (while philosophy may then assume the job of interpreting the increasingly refined science in "layman's terms", in natural language). One special benefit that mathematics has to offer is its universality: a single mathematical model may be capable of describing a wide range of phenomena by means of varying interpretations of the symbols used. The current paper presents a striking example. The canonical ensemble model, originally developed in the context of equilibrium statistical mechanics to describe a system that has cooled to an ultimate state of maximal entropy, may be reinterpreted to describe the evolution of a non-equilibrium system away from a state of maximal entropy and towards an ultimate state of certainty. Examples like this are salutary reminders of the need for the precision and clarity that mathematics and philosophy can bring.

REDUCTIONISM AND HOLISM

The two basic methodologies for tackling the study of complex systems (and indeed science in general) are usually classified as reductionism and holism. Up to now, it is reductionism that has had the most solid mathematical support, centered on Newtonian calculus and the study of differential equations. The key physical concept underlying reductionism is its infinitesimal and reversible time

G. Van de Vijver et al. (eds.), Evolutionary Systems, 141–153.

parameter, extended by the process of integration to the absolute time of Newtonian physics or to its modern variants in relativistic physics. The paradigm of reductionism is to isolate a suitably small part of a system and discover the differential equation that describes its instantaneous evolution. The longer-term behavior of the isolated part of the system is then predicted by integrating the differential equation subject to the appropriate initial conditions. Most mathematical biology as currently understood and practiced tends to conform to this general pattern. While there are certainly many situations like chemotherapy where this procedure has literally vital significance, one cannot escape the feeling that it is inadequate for a complete study of the emergent properties, such as life itself, that are truly characteristic of biology.

A serious symptom of the inadequacy of a purely reductionist methodology is the awkwardness with which it tries to reconcile the inherent reversibility of its time parameter with irreversibility in the evolution of complex systems. Neither the return-time arguments of the Ehrenfests (1907) and Smoluchowski (1915), nor Prigogine and George's invocation of the Second Law of Thermodynamics as a selection principle (1983), are entirely satisfactory. Briefly, the return-time arguments align the arrow of time with the passage of a dynamical system from an "unlikely" non-equilibrium state to "average", equilibrium states. One difficulty here is that the non-equilibrium state may be just as likely as any given equilibrium state. Another difficulty is that, in the usual models, the system is almost certain to return to the initial non-equilibrium state, thereby destroying the orientation of time that had temporarily been constructed. The use of the Second Law of Thermodynamics as a selection principle may be illustrated by the wave equation, as the differential equation

$$\frac{\partial^2 \varphi}{\partial r^2}+\frac{2}{r}\frac{\partial \varphi}{\partial r}=\frac{1}{c}\frac{\partial^2 \varphi}{\partial t^2} \tag{1.1}$$

for the spherically symmetric air pressure $\varphi(r,t)$ at time t and distance r from the center of the symmetry, c being the speed of sound. There are two basic types of solutions: expanding or "retarded" waves

$$\varphi=\frac{1}{r}f(r-ct) \tag{1.2}$$

for a suitable function f(x), and contracting or "advanced" waves

$$\varphi=\frac{1}{r}f(r+ct). \tag{1.3}$$

A solution of the retarded type might describe sound waves emanating from a whistle at the center of symmetry. On the other hand, solutions of the advanced type are generally an embarrassment to the theory, as they describe spherical waves converging on the center of symmetry. Note how time reversal interchanges the retarded and advanced waves. Invocation of the Second Law of Thermodynamics as a selection principle excludes advanced waves as being implausible, and thereby attempts to deny the time reversibility. Again, there are several difficulties here. Firstly, advanced waves as in (1.3) may actually occur, for example in the firing of the detonation lens of an atomic bomb. Secondly, superpositions of both types of solution arise in the formation of stationary waves. Thirdly, it is peculiar that retarded and advanced solutions to the one-dimensional analogue of (1.1) are physic-

$$\frac{\partial^2 \varphi}{\partial r^2} = \frac{1}{c} \frac{\partial^2 \varphi}{\partial t^2} \tag{1.4}$$

ally on an equal footing, so the Second Law does not break the symmetry in this case.

In the study of evolving systems, the desire to tackle emergent properties and the need for a naturally irreversible time parameter incline one towards the use of a holistic methodology. Unfortunately, holism does not yet have the range of mathematical tools that are available to reductionism. In a holistic approach to a complex system, one is certainly forced to concentrate on a few aspects of the system, neglecting everything else as irrelevant detail. Probability theory is the mathematical discipline that has been developed to handle such tasks. Thus the holistic approach has to begin with a satisfactory understanding of probability and randomness.

RANDOMNESS

Much of the controversy surrounding statistical mechanics has arisen from the long-standing lack of a solid basis for the underlying statistics. Measure theory provided a sound axiomatic foundation from the pure mathematics standpoint, but begged the question of the applicability of the axioms. Martin-Löf (1966) finally gave an acceptable definition of randomness. The definition may be paraphrased as follows. Consider a scientific experiment which may have any one of N possible outcomes. By convention of modern science, a "scientific" experiment has to be repeatable and reproducible. The experiment is said to be *random* if no statistical test available to the experimenter can detect any pattern in repeated outcomes. It is very important to note that this definition is contingent on the power of the apparatus available. Rolling a die under casino conditions should be random. On the other hand, if equipment such as a high speed precision camera was available, it would be possible to predict the outcome of each roll from the initial motion of the die, and the experiment would no longer be random. (There is a curious parallel with the wave equation (1.1), where the advanced solution (1.3), normally excluded by the Second Law in its guise of selection principle, may actually be realized by the elaborate technology of the detonator lens). A subtler distinction arises when rolling a very slightly loaded die under casino conditions. If the number of rolls required for even a sophisticated statistical test to reveal the bias is in excess of the number of rolls sufficient to wear the spots off the die, then the experiment would still qualify as random. (Once again there is a curious parallel, here with the issue of return times underlying the

Ehrenfests' and Smoluchowski's models).

The basic concepts of entropy and probability follow from the concept of randomness. For a random experiment with N outcomes, the (*natural*) *entropy* is

$$H = \log N \tag{2.1}$$

(using natural logarithms) or the (*binary*) *entropy*

$$H = \log_2 N \text{ bits} \tag{2.2}$$

using logarithms to base 2 and bits as units. (Occasionally logarithms to base 10 are used, with Hartleys as units). The *probability* $\pi(x)$ of any one particular outcome x of the random experiment is

$$\pi(x) = N^{-1} \tag{2.3}$$

The probability $\pi(x)$ and natural entropy H are connected by the mutually inverse relationships

$$\begin{aligned} &\text{(a) } H = -\log \pi(x) \\ &\text{(b) } \pi(x) = \exp(-H). \end{aligned} \tag{2.4}$$

Using binary entropy H, these take the form

$$\begin{aligned} &\text{(a) } H = -\log_2 \pi(x) \\ &\text{(b) } \pi(x) = 2^{-H}. \end{aligned} \tag{2.5}$$

The probability $\pi(x)$ represents a fair stake to buy into the following game: win one unit if the outcome of the experiment is x. Randomness of the experiment means that there are no winning strategies in this game. The entropies (2.1) and (2.2) measure one's ignorance about the outcome of the experiment. If you let someone else run the experiment, and instead question them afterwards as to what the outcome was, then the number of yes/no questions required to elicit the outcome would be given by (2.2).

The above formulae are too narrow to be of general use, where one wishes to deal with non-random experiments. These may be modelled using an underlying random experiment whose set of outcomes, called the *phase space*, has N elements. The phase space is completely partitioned into a set $\xi = \{C_1,\ldots, C_r\}$ of mutually exclusive subsets called *states*. The partition ξ represents the non-random experiment (also denoted ξ) of sampling an outcome x from the phase space and locating the state C_i in which it lies. If the state C_i contains n_i outcomes of the underlying random experiment, each of whose outcomes has probability N^{-1} according to (2.3), then the *probability* $p(C_i)$ of the state C_i is given as

$$p(C_i) = n_i N^{-1} \tag{2.6}$$

The state C_i may be regarded as a random experiment in its own right: select an outcome from C_i. The entropy of this random experiment, according to (2.1), is

$$H(C_i) = \log n_i . \tag{2.7}$$

If you perform experiment ξ and obtain the result C_i, then your ignorance will have been reduced by $\log N - \log n_i = -\log p(C_i)$. This happens with probability $p(C_i)$. Thus the average loss of ignorance or gain in knowledge obtained on performing experiment ξ is its *entropy*

$$H(\xi) = -\sum_{i=1}^{r} p(C_i) \log p(C_i) . \tag{2.8}$$

(Of course, one may take logarithms to base 2 and quote $H(\xi)$ in bits). The mathematical discipline of measure theory extends the definitions of probability and entropy to appropriate infinite phase spaces, where "counting outcomes" may be replaced by "measuring volumes". The entropy $H(\xi)$ satisfies the inequality

$$0 \leq H(\xi) \leq \log r . \tag{2.9}$$

Equality obtains on the left in (2.9) if and only if $p(C_i) = 1$ for some i: if you already know in advance that ξ will come up with state C_i then you gain no knowledge by performing the experiment. Equality obtains on the right in (2.9) if and only if $p(C_i) = r^{-1}$ for each i: the most informative experiments are those designed so that all their different outcomes are equally likely. In particular, randomness of an experiment is characterized by its entropy. Moreover, the three concepts of "entropy", "probability", and "randomness" turn out to be equivalent. Each leads to the other two. In complexity theory, it is an entropy measure, namely the self-delimiting algorithmic complexity, which is usually taken as basic. Probability is then obtained via (2.5) (Uspenskii, Semenov & Shen', 1990). The complexity class of the algorithms invoked corresponds to the power of the apparatus used in the statistical tests for randomness above.

STATISTICAL MECHANICS: THE CANONICAL ENSEMBLE

For the experiment $\xi = \{C_1, \ldots, C_r\}$ considered in the previous section, absolute randomness — complete ignorance about the outcome — was characterized by unconstrained maximization of the entropy $H(\xi)$, attaining the value $\log r$ according to (2.9). In practice it may be possible to assign a numerical value E_i to each state C_1, e.g. the number of spots on the face of a die or an energy in electron volts. If the expected value

$$E=\sum_{i=1}^{r} p(C_i)E_i \tag{3.1}$$

is known, then the non-negative probabilities $p(C_i)$, which have to satisfy the relationship

$$1=\sum_{i=1}^{r} p(C_i), \tag{3.2}$$

are determined by the assumption of relative randomness: no pattern is discernible in repeated outcomes except for maintenance of the fixed value E. This is equivalent to maximization of the entropy $H(\xi)$ subject to the constraints (3.1) and (3.2). Setting $g_1=E-\sum_{i=1}^{r} p(C_i)E_i$, $g_2=1-\sum_{i=1}^{r} p(C_i)$, *and* $f=H(\xi)+\beta g_1+\lambda g_2$ with Lagrange multipliers β, λ, the stationarity conditions $\partial f/\partial p(C_i)=0$ lead to $\log p(C_i)=-\beta E_i-(1+\lambda)$ or $p(C_i)=\exp(-\beta E_i)/\exp(1+\lambda)$. Substituting into (3.2), noting that λ is independent of i, one obtains

$$p(C_i)=Z(\beta)^{-1}\exp(-\beta E_i) \tag{3.3}$$

with the *partition function* or *Zustandsumme*

$$Z(\beta)=\sum_{i=1}^{r}\exp(-\beta E_i). \tag{3.4}$$

The fixed value E from (3.1) is recovered as

$$-\frac{d\log Z(\beta)}{d\beta}; \tag{3.5}$$

on the other hand this equation may yield β. The entropy (2.8) may be recovered via

$$-\sum_{i=1} p(C_i)\log p(C_i)=\sum_{i=1}^{r} p(C_i)[\beta E_i+(1+\lambda)] \ \textit{as}$$

$$\mathrm{H}(\xi) = \beta E + \log Z(\beta). \tag{3.6}$$

Note that (3.3)-(3.6) are just consequences of the assumption of randomness of ξ relative to (3.1) and the attribution of the numerical value E_i to each state C_i. There is no inherent reason for such an experiment ξ to be inappropriate as a model in certain biological circumstances. (In many cases additional numbers F_i, G_i, etc. may be assigned to each state, with known expected values F, G, etc. The Lagrange multiplier method readily extends to such cases, using vectors in place of scalars). An experiment ξ with random probabilities subject to (3.1) is called a *canonical ensemble*, since it generalizes the models of that name in (equilibrium) statistical mechanics. A correspondence with thermodynamics arises when the E_i and E are energies in suitable units. Introducing *Boltzmann's constant* K, the *thermodynamic entropy* is

$$S = KH. \tag{3.7}$$

The *temperature* is

$$\Theta = 1/K\beta \ . \tag{3.8}$$

The *thermodynamic potential* is

$$\Psi = \log Z(\beta) \ . \tag{3.9}$$

The *Helmholtz free energy* is

$$F = -K\Theta\Psi \ . \tag{3.10}$$

Equation (3.6) then reduces to

$$F = E - \Theta S \ . \tag{3.11}$$

All this formalism is available in the general context of the canonical ensemble, although the appropriate units may vary from case to case.

COMPETITION WITHOUT MUTABILITY

The simplest application of the canonical ensemble (with non-thermodynamic parameters) to an evolving system arises in Eigen's (1971) phenomenological rate equation for the case of constant overall organization and negligible mutability. Consider r species 1, ..., r. Suppose that the i-th species has a known net growth rate E_i. In other words, if a population of n_i units of the i-th species is allowed to develop without constraint, its rate of change is given by

$$\dot{n}_i = E_i n_i . \tag{4.1}$$

Note that the units of the quantity E_i in equation (4.1) are $(\text{time})^{-1}$. Now suppose that the r species are brought together in a joint population maintained at a constant count N (e.g. by control of a common food supply). Thus the birth of one individual has to be compensated by the death of another, not necessarily of the same species. If n_i now represents the number of individuals of species i present in the joint population, its rate of change is given by

$$\dot{n}_i = (E_i - E) n_i . \tag{4.2}$$

Summing the equations (4.2) over i, constancy of the joint population amounts to

$$0 = \sum_{i=1}^{r} \dot{n}_i = -NE + \sum_{i=1}^{r} n_i E_i \ . \tag{4.3}$$

Consider the experiment ξ of capturing a random individual from the population and determining its species. State C_i corresponds to the i-th species. Equations (2.6) and

(4.3) yield the constraint

$$E=\sum_{i=1}^{r} p_i E_i \tag{4.4}$$

on the experiment ξ. IF NOTHING ELSE IS ASSUMED TO BE KNOWN (including the initial values of the n_i), then the canonical ensemble is the appropriate model for ξ, and (3.4) determines the probability of C_i as

$$p_i = \exp(-\beta E_i)/Z(\beta) \tag{4.5}$$

with

$$Z(\beta)=\sum_{i=1}^{r} \exp(-\beta E_i). \tag{4.6}$$

To interpret this, set

$$t = -\beta . \tag{4.7}$$

Since the dimensions of E_i are $(\text{time})^{-1}$, the dimensions of the Lagrange multiplier t are those of time. Indeed, rewriting (4.5) and (4.6) in terms of t yields

$$p_i=\exp(E_i t)/\sum_{j=1}^{r} \exp(E_j t) \tag{4.8}$$

or

$$n_i=N\exp(E_i t)/\sum_{j=1}^{r} \exp(E_j t). \tag{4.9}$$

It is instructive to compare (4.9) with a conventional solution of (4.2) as a Bernouilli differential equation. The conventional solution uses the Ansatz

$$n_i=Nx_i/\exp\int_0^t E(\tau)d\tau, \tag{4.10}$$

which leads to the linear equations

$$\dot{x}_i=E_i x_i. \tag{4.11}$$

In deriving (4.9), nothing was assumed to be known about the initial distribution of the n_i. The canonical ensemble model, maximizing the entropy $H\xi$, thus selects the maximum entropy initial distribution

$$n_i(0)=N/r, \tag{4.12}$$

i.e. $x_i(0) = r^{-1}$. With this initial condition, the conventional solution of (4.2) is

$$n_i=N\exp(E_i t)/r\exp\int_0^t E(\tau)d\tau. \tag{4.13}$$

Comparison with (4.9) yields the partition function

$$Z(t) = \sum_{j=1}^{r} \exp(E_j t) = r \exp \int_0^t E(\tau) d\tau . \tag{4.14}$$

As a potential, the thermodynamic potential (3.9) is only determined to within an additive constant. Using (4.14), one may thus express the "thermodynamic potential" in the present context as the "energy integral"

$$\Psi = \int_0^t E(\tau) d\tau . \tag{4.15}$$

By (4.4), $E(t)$ is the cull rate necessary to hold the total population constant. Thus (4.15) represents the total number of individuals lost to population control over the first t time units. The "Helmholtz free energy" (3.10) becomes the average cull rate

$$F = \frac{1}{t} \int_0^t E(\tau) d\tau \tag{4.16}$$

over the first t time units.

COMPETITION WITH MUTABILITY

In the presence of mutations, Eigen's phenomenological rate equation for the case of constant overall organization $N = \sum_{i=1}^{r} n_i$ takes the form

$$\dot{n}_i = -En_i + \sum_{j=1}^{r} A_{ij} n_j . \tag{5.1}$$

Positive off-diagonal terms A_{ij} of the r x r-matrix A represent mutations from species j to species i. Assuming that the matrix A has r distinct real eigenvalues, the canonical ensemble technique provides a partial analysis of (5.1). To this end, it is helpful to rewrite (5.1) in vector-matrix form as

$$\underline{\dot{n}} = (A - E) \underline{n} . \tag{5.2}$$

Here, as usual, the scalar E is considered as the scalar matrix EI with I the $r \times r$ identity matrix. Let Q be an $r \times r$ matrix whose columns Q_j, for $j = 1, \ldots, r$, are eigenvectors of A, say

$$A \underline{Q}_j = \lambda_j \underline{Q}_j , \tag{5.3}$$

ordered so that $\lambda_1 > \lambda_2 > \ldots > \lambda_r$. Normalize the eigenvectors as follows. If $\sum_{i=1}^{r} Q_{ij} \neq 0$, choose $\underline{Q}_j$ with $\sum_{i=1}^{r} Q_{ij} = 1$. If $\sum_{i=1}^{r} Q_{ij} = 0$, choose $\underline{Q}_j$ with $\sum_{i=1}^{r} |Q_{ij}| = 1$ and with the first non-zero term of the sequence $Q_{1j}, Q_{2j}, \ldots, Q_{rj}$ being positive. (This normalization of eigenvectors $\underline{Q}_j$ with $\sum_{i=1}^{r} Q_{ij}=0$ is purely conventional). Since the columns of Q form a basis of the vector space $\mathbb{R}^r$, there are unique scalar functions $m_1(t), \ldots, m_r(t)$ such that

$$\underline{n}=\sum_{j=1}^{r} m_j(t)\underline{Q}_j. \tag{5.3}$$

This expression is to be contrasted with the expression

$$\underline{n}=\sum_{j=1}^{r} n_j(t)\underline{I}_j \tag{5.4}$$

of the vector $\underline{n}$ as a linear combination of the columns $\underline{I}_j$ of the identity matrix I. The column $\underline{I}_j$ represents one individual of the species j. By analogy, the column $\underline{Q}_j$ of Q will be considered to represent one individual of *virtual species j*. Consider a population determined by a vector $\underline{n}$. According to (5.4), it may be described by the presence of n_j individuals from species j. According to (5.3), it could then equally be described by "the presence of m_j individuals from virtual species j". It is this latter description which facilitates application of the canonical ensemble technique.

The virtual species j is said to be *substantial* if and only if $\sum_{i=1}^{r} Q_{ij}\neq 0$, i.e. if and only if $\sum_{i=1}^{r} Q_{ij}=1$ according to the normalization. Now substitution from (5.3) into (5.2) gives

$$\sum_{j=1}^{r} \dot{m}_j(t)\underline{Q}_j=\dot{\underline{n}}=(A-E)\underline{n}=\sum_{j=1}^{r} m_j(t)(A-E)\underline{Q}_j=\sum_{j=1}^{r} m_j(t)(\lambda_j-E)\underline{Q}_j.$$

Equating coefficients of $\underline{Q}_j$ yields, for each virtual species j,

$$\dot{m}_j=(\lambda_j-E)m_j \tag{5.5}$$

On the other hand, substitution from (5.3) into the constraint

$$N=\sum_{i=1}^{r} n_i \tag{5.6}$$

of constant overall organization gives

$$N=\sum_{i=1}^{r} n_i=\sum_{i=1}^{r}\sum_{j=1}^{r} m_j(t)Q_{ij}=\sum_{j=1}^{r} m_j(t)\sum_{i=1}^{r} Q_{ij}$$

or

$$N=\sum_{j}{}' m_j, \tag{5.7}$$

where $\sum_{j}{}'$ denotes summation restricted to substantial virtual species j. The insubstantial virtual species make no contribution to the overall organization. The reduction performed for the case of r distinct real eigenvalues of A may be summarized as follows:

> Eigen's phenomenological rate equation for the evolution of species in the presence of mutation, and subject to constant overall organization N, yields the phenomenological rate equation for the evolution of substantial virtual species in the absence of mutation, subject to the same constant overall organization N. (5.8)

The equation for the evolution of the substantial virtual species in the absence of mutation may be analyzed by the canonical ensemble technique, as discussed in the preceding section. In particular, it should be noted that the dominant virtual species 1 is substantial if the non-negative matrix $B=A-I\min\{A_{ii}|A_{ii}< 0, 1\leq i\leq r\}$ is *primitive* in the sense of Birkhoff (1967). Indeed, the matrices A and B have the same eigenvectors, and the Perron-Frobenius Theorem (Birkhoff, 1967, Theorem XVI.4.5) shows that B has a positive dominant eigenvector.

VIRTUAL SPECIES AND HIERARCHICAL SYSTEMS

The concept of virtual species, emerging from the analysis of competition with mutation given in the preceding section, is interesting from a number of different standpoints. To begin with, it would appear to be a good candidate for the somewhat tentative notion of *environ* discussed by Patten (1982) in his search for the "elementary particles" of ecology. The virtual species comprise the proportioned collections of individuals that are capable of functioning autonomously at a certain level, whereas the original species may not necessarily have that capability. For example, the giant panda *Ailuropoda melanoleuca* would not form a virtual species on its own, without a suitable proportion of bamboo shoots being included. If the bamboo shoots are not present, the panda becomes extinct. Given the difficulties involved in formulating a clear, generally valid definition of what constitutes a species (compare Brooks & Wiley (1986), Chapter 4, for example), it may prove easier to identify and classify virtual species instead.

The distinction between substantial and insubstantial virtual species is crucial to the further development of the techniques being introduced here. Consider the equations

$$\begin{aligned} \dot{n}_i &= -D_i n_i + A_i n_{s+i} - E(t) n_i \\ \dot{n}_{s+i} &= A_i n_i - D_i n_{s+i} - E(t) n_{s+i} \end{aligned} \tag{6.1}$$

for $i = 1, \ldots, s$, with $A_1 - D_1 > A_2 - D_2 > \ldots > A_s - D_s > 0$. They are simple versions (without mutation) of the phenomenological equations for self-reproduction based on complementary recognition, as considered by Eigen (1971)(§ IV. 2), and by Thompson and McBride (1974). The virtual species are $Q_i = \frac{1}{2}I_i + \frac{1}{2}I_{s+i}$ and $Q_{s+i} = \frac{1}{2}I_i - \frac{1}{2}I_{s+i}$ for $i = 1, \ldots, s$. The Q_i are substantial, while the Q_{s+i} are not. For the purposes of canonical ensemble analysis, (6.1) reduces to

$$\dot{m}_j = (A_j - D_j)m_j - E(t)m_j \tag{6.2}$$

for $j = 1, \ldots, s$. The canonical ensemble technique describes the evolution of the combined populations of complementary pairs, in particular the eventual dominance of the master pair $\{1, s + 1\}$, but does not concern itself with the differences between n_i and n_{s+i} represented by the insubstantial virtual species Q_{s+i}. In general, insubstantial virtual species are functioning at a higher level within a hierarchy than the level of organization being addressed by the canonical ensemble. The distinction between substantial and insubstantial virtual species identifies the point at which a new hierarchical level arises. The discovery of mathematical techniques to analyze the development of complex systems functioning at more than one hierarchical level is one of the prime goals of future research.

IRREVERSIBILITY AND THE NATURE OF TIME

The most fundamental insight to be elicited from the mathematical models under discussion concerns the nature of time. The time parameter implicit in Newton's dot notation in equations (4.1) and (4.2) is an infinitesimal time. It is convenient to refer to it as *instantaneous time*. The classical procedure of integrating differential equations subject to initial conditions is based on Newton's concept of *absolute time*. Even in general relativity, the concept of time is essentially an integrated version of instantaneous time, albeit over curved manifolds rather than Euclidean space. The time introduced as a Lagrange multiplier by (4.7) is different. Looking back to (3.8), the corresponding Lagrange multiplier in the thermodynamic interpretation of the canonical ensemble is essentially the reciprocal of the temperature. Temperature is not a local quantity. It is a global, statistical property. The same is true of the time defined by (4.7), which may be called *historical time*. It is a statistical property of the natural process of competition between the species. In general historical time should be determined as a statistical property of natural processes whose local behavior may be described in terms of instantaneous time. And while the instantaneous time is reversible, as is the absolute time to which it integrates, historical time is irreversible in general, according to the nature of the process by which it is embodied.

Department of Mathematics, Iowa State University
Ames, USA

REFERENCES

Birkhoff, G., 1967, *Lattice Theory*, Providence, R.I., American Mathematical Society.

Brooks, D.R.; Wiley, E.O., 1986 (1st ed.), *Evolution as Entropy: Towards a Unified Theory of Biology*, Chicago, University of Chicago Press.

Ehrenfest, P.; Ehrenfest, T., 1907, Über zwei bekannte Einwände gegen das Boltzmannsche H-Theorem, *Phys. Zeit*, vol. 8, pp. 311-314.

Eigen, M., 1971, Self-organization of matter and the evolution of biological macromolecules, *Die Naturwissenschaften*, vol. 58, pp. 465-523.

Martin-Löf, P., 1966, The definition of random sequences, *Information and Control*, vol. 9, pp. 602-619.

Patten, B., 1982, Environs: relativistic elementary particles for ecology, *Am. Nat.*, vol. 119, pp. 179-219.

Prigogine, I.; George, C., 1983, The Second Law as a selection principle, the microscopic theory of dissipative processes in quantum systems, *Proc. Nat. Acad. Sci.*, vol. 80, pp. 4590-4594.

Smoluchowski, M.V., 1915, Molekulartheoretische Studien über Umkehr thermodynamisch irreversibler Vorgänge und über Wiederkehr abnormaler Zustände, *Wien. Ber.*, vol. 124, pp. 339-368.

Thompson, C.J.; McBride, J.L., 1974, On Eigen's theory of the self-organization of matter and the evolution of biological macromolecules, *Math. Biosci.*, vol. 21, pp. 127-142.

Uspenskii, V.A.; Semonov, A.L., Kh. Shen', A., 1990, Can an individual sequence of zeros and ones be random?, *Russian Mathematical Surveys*, vol. 45, pp. 121-189.

ROD SWENSON

SPONTANEOUS ORDER, EVOLUTION, AND AUTOCATAKINETICS: THE NOMOLOGICAL BASIS FOR THE EMERGENCE OF MEANING

INTRODUCTION

Something has meaning when it has the property of "information about" some part of the world with respect to an end or ends of some intentional system or agent. Intentional systems are end-directed systems, but not all end-directed systems are intentional. Not all end-directed systems require information about or meaning to determine their actions towards their ends. We need not invoke intentionality, or meaning, for example, to explain the flow of a river down a slope, or the flow of heat down a temperature gradient from a hotter to a cooler region because these processes are explicable in terms of local physical potentials and fundamental laws. On the other hand, when a bird flies above the Earth, a bacterium swims up a concentration or chemical gradient, or a human drives a car, takes a plane, or moves a bit of food from her plate to her mouth, the action is seen to go in directions that are different, or orthogonal, to processes taken to causally follow from local physical potentials (the "orthogonality condition"). Such behavior is determined by information about the paths to ends, not by local potentials and it is these systems we call intentional.

The orthogonality condition provides some immediate clues to the nature of intentionality and meaning. To begin with it provides a dimension that has progressively increased during the course of planetary evolution. This is interesting enough because orthodox evolutionary theory (Darwinian theory) can supply no ordinal measures with respect to the progressive nature of evolution, but it is even more interesting by virtue of the fact that it refers to the epistemic or cognitive function of terrestrial evolution writ large, again something that is well beyond the scope of Darwinian theory. It suggests a principled basis for recognizing the commensurability between psychology, biology and physics, and hence between knower and known as part of a general evolutionary or universal process.

The most significant clue to the origin and nature of intentionality and meaning is the distinctive property that defines the orthogonality condition itself. In particular, the fact that intentional dynamics work in directions and over lengths of time different from those determined by local physical potentials makes it clear that intentional systems define dimensions of space-time that are determined differently from those systems determined by local potentials. What might this mean? According to Boltzmann's interpretation of the second law of thermodynamics the transition from disorder to order, the filling of ordered dimensions of space-time is infinitely improbable. What then is the basis for the epistemic or psychological dimension which appears thus to actively work against the apparent universality of physical law? In this paper I will show that contrary to the older view which falls out of Cartesian metaphysics, the epistemic dimension is not only fully commensurable

G. Van de Vijver et al. (eds.), Evolutionary Systems, 155–180.

with universal principles, but a direct manifestation of them.

CARTESIAN METAPHYSICS, THE RISE OF MODERN SCIENCE AND THE POSTULATES OF INCOMMENSURABILITY

Although the physics of Newton eclipsed the physics of Descartes it was the latter's dualistic metaphysics that provided the ground on which the former was able to flourish, and because psychology and physics were defined at their modern origins by Descartes, he is often referred to not only as the father of modern philosophy but the father of modern psychology and physics as well. What Cartesianism effected with its dead mechanical, or clock-work, view was a means for the religious authority of Descartes' time to see science within a context it could accept. Humans were seen sitting dualistically outside the clock-like world learning the laws of physics to manipulate it towards their own, and hence, as privileged creations on Earth, towards divine ends.

The main point with respect to this paper is that Cartesian dualism literally took the active epistemic dimension out of the physical part of the world by defining physics and psychology by their mutual exclusivity (call this "first postulate of incommensurability", Swenson, 1997a, 1997b). According to this postulate the world was divided into an active, purposive, perceiving "mind" (the "free soul", "thinking I", "Cartesian ego", or "self") on the one hand (the psychological part), and passive, "dead", purposeless "matter" (the physical part) on the other. The physical part, defined exclusively by its extension in space and time, was seen to consist of reversible, qualityless, inert particles governed by deterministic causal laws from which the striving mind, seen as active boundless and without spatial or temporal dimension, was immune.

An immediate implicate of this view was that spontaneous ordering in general, and intentionality and meaning in particular were eliminated from the physical world by definition, and needed to be extra-physically imposed from the outside. Another immediate implicate, certainly an unintended one, was that the epistemic and physical parts of the world were logically forbidden from interacting — the problem of Cartesian interactionism. For one part to interact with another it would have to violate the postulate of incommensurability defining the dualism. For example, for "mind", supposedly without spatial or temporal dimension, to act on matter defined in space and time, it would have to become defined in space and time at the point of interaction. But then part of mind would be defined in space and time and part would not. How does the part that is not then interact with the part that is without itself becoming defined in space and time? The problem leads to an infinite regress until mind takes on the properties of matter and the dualism is demolished.

Leibniz recognized this central problem of Cartesianism, and dualism in general, by anticipating the law of energy conservation (the first law of thermodynamics). For one thing to interact with another requires something conserved over the interaction, and if something is conserved over the two things or processes they are, at some level, part of the same thing. Without a conservation the two would be truly incommensurable — two separate worlds without any possible relation or causal connection. Descartes' theory of perception which had mind perceiving itself when coupled with the intractable problem of interactionism, made the epistemic dimension a closed circle with no way in or out (the "Cartesian circle") — the subjective simply given with no basis for its existence or meaningful connection with anything other than or beyond itself.

Cartesian metaphysics came full-blown into modern biology with Kant who, arguing that the active, end-directed striving of living things, could not be adequately accounted for as part of a dead, reversible mechanical world, promoted a second major dualism (the "second postulate of incommensurability", Swenson, 1997a, 1997b), the dualism between biology and physics, or between living things and their environments (Swenson & Turvey, 1991). This argument for the autonomy of biology from physics is still promoted today by leading proponents of Darwinian theory (e.g. Mayr, 1985). "Darwin", in Lewontin's words, following the tradition of incommensurability, "completely rejected [the] world view (...) that what was outside and what was inside were part of the same whole system (...) and replaced it with one in which organisms and environments were totally separated". "The fundamental dichotomy of evolutionary theory", in the words of Levins and Lewontin (1985, p. 52), became "that of organism and environment", and in this way, the core problems of Cartesian metaphysics were effectively spread from the question of the nature of human minds and their relation to the world to life as a whole.

CLOSED-CIRCLE THEORY VS. EVOLUTIONARY EPISTEMOLOGY

Post-Cartesian theories of psychology or epistemology can be generally located in one of two competing paradigms — the work of the later Wittgenstein, Kuhn and others being exemplars of the first and the work of Popper, Campbell, Lorenz and others being exemplars of the second (Munz, 1985, 1987). The proponents of the first, "closed-circle theorists", who have worn incommensurability and relativism almost as a badge of enlightenment, look to sociology or social psychology as the basis for meaning and intentionality, while evolutionary or natural selection epistemologists, proponents of the second, look to biology, or, more particularly, to Darwinian evolutionary theory as the ground for the epistemic dimension.

Closed-circle theory and the sociology of knowledge

Closed-circle theory finds roots in the functionalism of Durkheim and Malinowski, and the "sociology of knowledge" of Mannheim. Earlier roots go back to Marx and Engels' work on ideology, and Spencer before. While the common thread to all these various ancestors was the idea that social ordering determines individual action, it should not be construed that all were closed-circle theorists in the extremized post-modern form of Wittgenstein and Kuhn (Swenson, 1997a).

Social systems, according to Durkheim (1895-1938), were said to have a reality outside the existence of individual humans that acts to determine their meaningful relations and intentional behavior. In explicit reaction to "psychologism", the idea that cultural systems are rational constructions of individual intentional agents, as well as the prevalent evolutionary views of history or culture, Malinowski held that human social or cultural systems were effectively closed circles where the parts all function to maintain the whole. Given that the circular relations of such a system were seen to refer back to themselves — that the function of the system is *to maintain itself* — they were said to exist *sui generis*. Everything is explained with respect to something else that happens internal to the circular relations of the system.

Wittgenstein further extremized this view by transposing it into a sociological version of the Cartesian circle. Rather than the circular relations constituting the self-referentiality of the human mind, Wittgenstein's circle was constituted through the intersubjective circular relations of humans within a cultural system. Meanings, said Wittgenstein, are formulated and stated in "language games" consisting of a set of rules that constitute closed circles of meanings. Because there is no individual language there can be no individual meanings, and because such systems are closed circles there can be no ostensive pointing or reference to anything outside the system (i.e. an objective "world"). What is more, because meaning is entirely relative to the rules of each system and thus meaning invariance is denied, such circles of meaning are incommensurable with respect to each other. Truth thus varies from one closed circle to the next, and can only be measured with respect to the rules or authority of a particular community.

In Kuhn's influential history and philosophy of science Wittgenstein's closed-circle language games were turned into paradigms, and the history of science the shift from one paradigm to another (scientific revolutions). Since reality is taken to be an ideal construction of human cognizers operating under particular paradigms, and since paradigms as closed circles are incommensurable with each other there is no progress in science or in Kuhn's view — without meaning invariance there is no way to make a comparison. In this view, Einstein's physics does not subsume or explain Newton's but rejects it. The post-modern structuralism of Foucault, Derrida, and the post-modern pragmatism of Rorty, which uses Foucault, in effect, to justify Wittgenstein (Munz, 1987), are all closed circle theories that share the common premises of the relativity of meaning to circularly closed systems, and the incommensurability of such systems with respect to each other and an external world. Closed-circle theory carries forward the anti-realist position of positivism, but in addition attacks its rationality.

Evolutionary epistemology: evolution as a knowledge acquisition process

Evolutionary or natural selection epistemologists have a point of view almost directly opposite to that of the closed-circle theorists (e.g. see Callebaut &

Pinxten, 1987; Radnitzky & Bartley, 1987). Whereas closed-circle theorists such as Wittgenstein and Kuhn are arch anti-evolutionists, evolutionary epistemologists hold that knowledge is the product of evolution, and that evolution at its core is a progressive and continuous knowledge acquisition process, in Popper's words, "from amoeba to man" following from natural selection. Every living thing has knowledge in the expectations on which its intentional behavior depends, and this knowledge as the consequence of natural selection is true (hypothetically) since if it were not the living thing in question would be dead. Thus while true knowledge to the closed-circle theorist follows from cultural authority under a particular paradigm, to the evolutionary epistemologist it is determined with respect to the performance of an epistemic agent in the world. Scientific knowledge is seen to be continuous with evolution by natural selection since it too involves a trial and error process of selection through the proposal and refutation of falsifiable hypotheses (Campbell, 1987).

Both approaches are Cartesian at their cores

Closed-circle theory is the Cartesian circle regressed

Although closed-circle theory is often given as a kind of enlightened alternative to modernism, it is itself modernism carried to a certain post-Humean, post-Kantian, extremized conclusion — the Cartesian core is still there only wrapped in sociological packaging. More precisely, it is the Cartesian circle regressed from the individual to the cultural level. Regressed because the original epistemic problem of the Cartesian circle still remains, and by extending the idea to culture, a second one is added. Regressive moves of this kind are defining symptoms of what Lakatos (1970) has called "degenerative problem shifts", and they are inherent to dualist schemes. The various problems of closed-circle theory are of four main kinds which are all related.

1. *The premises of closed-circle theory make it anti-evolutionary by definition, and it thus offers no account (nor cares to) of evolutionary dynamics, in particular, the directed and expansive nature of the epistemic or psychological dimension in evolution.* Closed-circle theory is anti-evolutionary in two ways. The first is that because closed circles are incommensurable with each other there is no way to assert that they are part of any evolutionary process — no continuity over the discontinuity, and thus no ground to assert an ordinal measure with respect to the direction of time. From this perspective there is a deep incoherence in closed-circle theory as a history of science because it must rely implicitly on principles to state its position that it theoretically denies. Kuhn, for example, relies on a historical comparison of paradigms to claim that such paradigms are incomparable. In addition, while the history of science for Kuhn is taken to be constituted by paradigm shifts there is no principled way to address when or why such transformations should occur, or that they should appear in any particular order, e.g. Einstein's theory could just as well have preceded Newton's, the

theory of oxygen that of phlogiston, or the theory of heat that of the caloric.

The second way that closed-circle theory is anti-evolutionary is by grounding meaning in the intersubjective relations of humans. The consequence is similar to the consequence of Descartes' view of perception as a rational process which, when coupled with the further claim that only humans are capable of rational processes, took the entire psychological dimension away from the rest of life. Closed circle theory, while abandoning "the thing in itself", is a further extremization of the anthropocentric "revolution" of Kant which has the epistemic acts of humans, in effect, dictating reality to the rest of the world, which in the case of closed-circle theory is ideally created anew, and entirely at random, without rhyme or reason, with each new paradigm.

2. *The formal causality postulated at the core of closed-circle ontology constitutes an illegitimate teleological principle with respect to accepted physical and biological principles.* By making the fundamental reality the circularly defined cultural system, closed-circle theory substitutes "formal cause" (the form or shape of a thing, in this case the circular relations) for the usual efficient cause that constitutes the notion of causality in modern science. The components are determined by the dynamic functional ordering of the whole rather than the other way around.

But here arises the problem that has discredited virtually every one of closed-circle theory's functionalist ancestor's before it (see Turner & Maryanski, 1979) — by the most widely accepted principles of physics and biology, the end-directed behavior it invokes entails an illegitimate teleological principle. Downward causality, or formal cause, was removed from modern science, along with final cause, at its origins. Biology, on the one hand, rejects such causality because functional behavior is taken to follow from natural selection, and the functional ordering of closed-circle systems, which are populations of one cannot follow from natural selection. Physics, on the other hand, has traditionally rejected such spontaneous ordering because the world, on the received view of the second law of thermodynamics, has been taken to be collapsing to disorder.

3. *The intersubjectivity at the core of closed-circle theory begs the old Cartesian questions and doubles the problem.* The Cartesian circle follows logically from the first postulate of incommensurability and the claim that perception is by the mind of the mind. What I know indubitably, said Descartes, in his famous *cogito ergo sum*, is my own active self-reflective mind. Claims about knowledge of an outside or objective world, on this view, are impossible, and claims for the existence of such a world, therefore, effectively irrelevant (since two incommensurable things have no causal efficacy with respect to one another), and on parsimony simply unenlightened superstition. By taking the idea of the self-reflective Cartesian circle and invoking it at the cultural level instantiated by the intersubjective relations of individual humans, closed-circle theory begs the original Cartesian questions and doubles the problem.

Briefly put, meanings for the closed circle theorist exist in the persistent and invariant relations constituted by the intersubjective relations that define the closed circle. To each individual, however, this requires persistent and invariant relations with an outside world, and this requires a non-Cartesian theory of perception. In short, intersubjectivity requires breaking the Cartesian circle since the individual mind is no longer simply perceiving itself, but something external in relation to which it comes to be determined or defined, and this requires a commensurability between knower and known which undercuts the ground of closed-circle theory — once the Cartesian circle is thus broken there is no longer any principled basis for it.

4. *Closed-circle theory simply supports the status quo, a dubious distinction in a rapidly changing world.* Closed-circle theory is sometimes taken to be enlightened because in its denial of a measure for the ordinal comparison of cultures it fosters an egalitarian view, but the idea of closed-system theory that truth is determined by the authority of a community carries a severe ideological price tag of its own since it makes no distinction between ways this epistemic authority is established. When taken to its logical conclusion there is no way to discriminate between truth obtained by scientific methods, by religious revelation, or authority achieved at gun point. Meaning and truth are simply determined by those in power, and given the incommensurability of various power structures operating under particular paradigms there is no way to compare or make a judgement about one with respect to another — pragmatism, conventionalism, instrumentalism, and so on, under this view can only serve, by definition, to reinforce hegemony, or the status quo, whatever it is.

The problem with evolutionary epistemology is Darwinian evolutionary theory

The problem with evolutionary epistemology is that it relies on Darwinian evolutionary theory, but evolutionary theory reduced to Darwinism can never provide an account of the epistemic dimension, either the active intentional striving of living things or the meaningful relations it entails (Swenson, 1997a, 1997c). The four distinct, but related, reasons for this, all follow directly from its Cartesian foundations:

1. *The situational logic of natural selection, and hence Darwinism as a theory of evolution assumes the psychological or active epistemic dimension to begin with.* The core concept of Darwinian theory is natural selection (Depew & Weber, 1995). Evolution, according to Darwinism, follows from natural selection and natural selection is entailed by a situational logic (Popper, 1985) — if certain conditions hold then natural selection will necessarily follow. These conditions are heritable variation, finite resources and the fecundity principle, a biological extremum principle that captures the active striving of living things. Natural selection, said Darwin (1959/1937, p. 152), follows from a population of replicating or reproducing entities with variation "striving to seize on every unoccupied or less well occupied space in the economy of nature". Because "every organic being", he said (Darwin, 1959/1937, p. 266), is "striving its utmost to increase, there is therefore the strongest possible power tending to make each site support as much life as possible". Paraphrasing Darwin, in Schweber's (1985, p. 38) words, this says that nature acts to "maximize the

amount of life per unit area" given the constraints. Because natural selection is thus seen as a consequence of the fecundity principle, and because the fecundity principle assumes the active intentional dynamics or psychological or epistemic dimension of living things, an explication of the psychological dimension is beyond the reach of natural selection, and hence, by definition Darwinian theory.

2. *The claim that the growth of knowledge is a progressive evolutionary process is an assertion that can neither be made nor explained from the ground of Darwinian theory because the relevant observable (fitness) is relativized to members of breeding populations.* Evolutionary epistemology makes the claim that the growth of knowledge is continuous from "amoeba to man", and the measure of the knowledge a living thing possesses is its fitness. But fitness, on the Darwinian view, and thus knowledge, is relativized to members of a breeding population (species). Thus a zebra who can run faster than another zebra, avoid predators better, and thereby produce more offspring is more fit, and can be said to possess more evolutionary knowledge, than a zebra that is slower, but a zebra cannot be compared on the same basis to a mouse, or to an amoeba. Mice can only be judged more or less fit with respect to other mice, zebras with respect to other zebras, and amoebas with respect to other amoebas, and this makes fitness an incommensurable observable with respect to any assertions about evolution writ large.

3. *Given the second postulate of incommensurability assumed by Darwinian theory there are no grounds within the theory from which epistemic or meaningful relations between living things and their environments can take place.* The fecundity principle on which evolution on the Darwinian view crucially depends assumes the active intentional dynamics of living things — it assumes the meaningful determination of the behavior of living things with respect to their intentional ends. Given the second postulate of incommensurability built into the foundations of Darwinian theory, however — the rejection by Darwinism that what is inside and what is outside are part of the same whole system (Lewontin, 1992), there is no principled basis for meaningful relations to take place. Such a basis cannot be given by a theory like Darwinism that holds biology and physics or living things and their environments to be incommensurable.

4. *Darwinism has no account of the insensitivity to initial conditions (like consequents from unlike antecedents) required to account for the reliability of intentional dynamics or the evolutionary record writ large.* On the mechanical view of the world causality is reduced almost entirely to efficient cause (like antecedents produce like consequents), but the intentional dynamics of living things require like consequents from unlike antecedents. In short, to use the felicitous terms of Dyke (1997), they must be "end-specific" and not "start specific". The dynamics of living things require giving a fundamental ontological status to "formal causes", in Aristotle's terms, that function nomologically at a macroscopic level to attract and entrain micro-components towards determinate macroscopic ends regardless of differences (within tolerance) of microscopic starting conditions. Among other things, this goes to the issues of multiple realizability, downward causation, and the fact that forms in nature, including those considered as intentional dynamics "fall

out" at critical thresholds of dimensionless ratios. Darwinian theory, based as it is on efficient cause, continuous change, and the idea, in effect, that "anything goes" cannot address this fundamental dynamic.

5. *Darwinism cannot address cultural evolution or evolution of life as a single process at the planetary level because such systems evolve as populations of one, and natural selection can only 'explain' the results following from a competing population of many.* In contrast to the closed-circle theorists who take scientific knowledge to be irrationally determined under a paradigm, evolutionary epistemology takes the production of scientific knowledge to be determined by a process of rational, critical selection. But cultural systems are circularly causal, and the production of human knowledge is culturally determined — cultural systems are defined by world views or paradigms and what is accepted as meaningful pursuits or intentions, including the pursuit of scientific questions (and thus answers) are typically those that support the underlying paradigm. What is more, evolution itself at the planetary level is a single global circularly causal system on which the ordinary objects of Darwinian study as well as cultural systems depend (e.g. for a steady and reliable supply of oxygen). Darwinism cannot address the circular causality of cultural systems or life at the planetary level on which all other terrestrial evolutionary products depend because such systems evolve as populations of one and natural selection only works on a population of many (Dawkins, 1982; Maynard Smith, 1988; Swenson, 1991a).

AUTOCATAKINESIS, THE EPISTEMIC DIMENSION, AND EVOLUTION AS SELF-ORGANIZATION

Return to cogito

Descartes had two important things right. The first was his admonition to challenge authority (what are closed-circle theorists to do!?), and the second was the general idea of his *cogito ergo sum*, his assertion that what each of us knows indubitably (viz., what certainly exists) is the fact of our own epistemic experience. What Descartes had wrong after that, however, by asserting the first postulate of incommensurability (the dualism of self vs. other, knower vs. known, or psychology vs. physics), in a sense, was everything. Being more precise, of course, Descartes did not exactly say that "what I know indubitably is the fact of my own epistemic experience", what he said was "what I know exists indubitably is the independent 'self' or 'thinking I' ('mind')", in short, the Cartesian circle, and this is where Descartes made his major error because such an independent self or mind is just what none of us knows. Whether a consequence of individual perceptions, actions, feelings, thoughts, imagination or any form of experience each of us only experiences our self in relation to that which we are not — no self without not-self, no knower without known, no subjective without objective. Even the conception is unimaginable. In short, the Cartesian circle is a myth. Self-reference, in the epistemic experience, is defined by the relation of the self to that which it is not.

What is more, the persistence or invariant properties that define the relation means there is a conservation over the relation, a third thing known that is neither self nor other that remains invariant over the two, and there are two other fundamental properties entailed by the epistemic experience (the "epistemic given") too. The first is the circular relations implied by self-reference, and the second is time-asymmetry or irreversibility. The circular relations defining self-reference take place through a one-way flow of time. Persistence of the defining relations means conservation through process — the conservation must flow for a self-other relation to persist. These fundamental properties, a defining relation between self and other, the circularity constituting the relation, a conservation over the relation, and one-way flow of the conservation provide a kind of minimal set of world properties or ontological conditions entailed by the epistemic experience. Figure 1 shows a schematic of this minimal ontology — (a) shows the conservation, that which is neither self nor other, and out of whose flow the self-other relation is constituted, and (b) shows the self-other relation circularly constituted in the directional flow.

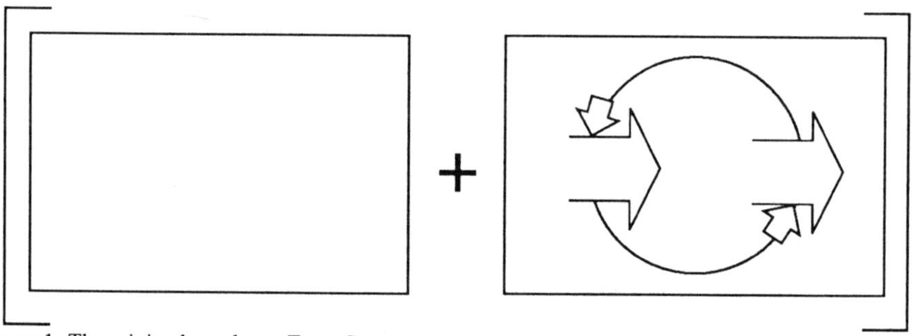

Figure 1. The minimal ontology. From Swenson, 1997c, p. 68, copyright 1997 Lawrence Erlbaum Ass., Inc., reprinted by permission.

Autocatakinetics

The non-Cartesian conjunction shown in Figure 1 (call it the "embedded circle") provides the basis for understanding the commensurability between physics, psychology and biology where it becomes possible to subsume what is right about closed-circle theory and evolutionary epistemology, while getting rid of the Cartesian ghost and the resulting degenerative problemshift to give a principled account of the emergence and evolution of meaning and intentionality or the epistemic dimension. Epistemic agents, following Figure 1, can be seen to be members of a class of systems called "autocatakinetic", a term referring to self-organizing or spontaneously ordered systems.[1] Autocatakinetic systems are flow structures that by pulling resources into themselves maintain their identities through the flux or motion of their components. In other words, invariance at one level, the form of the thing, is constituted by change or motion at the component level.

The root of the idea goes back at least to Heraclitus (536 B.C.) who characterized the world as a process of continual flow, and its objects as constituted by a generalized metabolism or combustion. Fire, as Aristotle (1947, p. 182) wrote centuries later in *De Anima*, stressing the active agency and generalized metabolism of such systems, "alone of the primary elements (earth, water, air, and fire) is observed to feed and increase itself". In this century, von Bertalanffy (e.g. 1952) developed these ideas under the name of "open systems", and Prigogine (e.g. 1978) under the heading of "dissipative structures". An autocatakinetic system is defined as one that maintains its "self" as an entity constituted by, and empirically traceable to, a set of nonlinear (circularly causal) relations through the dissipation or breakdown of field (environmental) potentials (or resources) in the continuous coordinated motion of its components (from *auto-* "self" + *cata-* "down" + *kinetic*, "of the motion of material bodies and the forces and energy associated therewith" from *kinein*, "to cause to move") (Swenson, 1991a). From this definition other examples of autocatakinetic systems, in addition to flames, can be seen to include dust devils, hurricanes, tornadoes, and all the entities typically taken to be living, including human cultural systems (e.g. tribes, chiefdoms, nation states and empires), and perhaps most importantly, the planetary system as a whole of which all the rest are component productions. Figure 2 shows a generalized drawing of an autocatakinetic system.

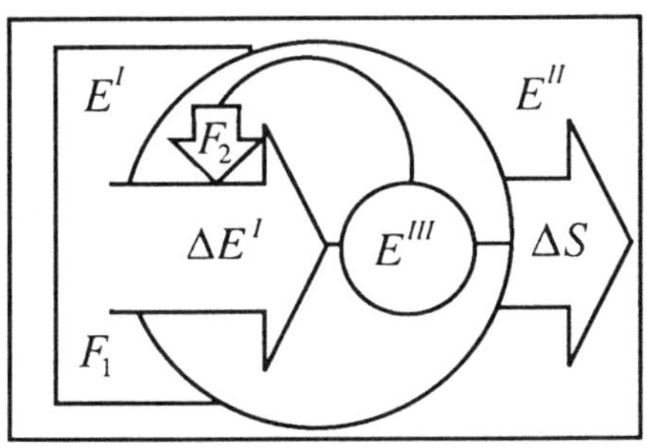

Figure 2. A generalized autocatakinetic system. E^I and E^{II} indicate a source and a sink with the difference between them constituting a field potential with a thermodynamic force F_1 (a force being the gradient of a potential) the magnitude of which is a measure of the difference between them. ΔE^I is the energy flow at the input, the drain on the potential which is transformed into entropy production ΔS at the output. E^{III} is the internal potential carried in the circular relations that define the system by virtue of its distance from equilibrium that acts back to amplify or maintain input during growth or non-growth phases respectively with an internal force F_2. From Swenson, 1991a, p. 45, copyright 1991 Intersystems Publications, adapted by permission.

Kinds of causality, autocatakinesis, and non-Cartesian circles

Part of mechanistic or Cartesian hegemony is the idea that nomological description means reduction to efficient cause, but to accept this notion is simply to proceed from the view of "normal science", in Kuhn's phrase, *within* the Cartesian paradigm, and intentional dynamics, and the minimal ontology it entails as expressed in autocatakinesis are not reachable from the Cartesian core. The Aristotelian causal system (in various versions), the system mechanistic theory replaced asserted a multiple causal framework with four distinct and kinds of causes of which efficient cause was just one. The three other kinds were material, formal, and final. Such a four-cause framework, suitably brought up to date is remarkably useful in understanding the dynamics of self-organizing systems.[2] In particular, the removal of formal and final causes, the removal of all enddirectedness from the physical world was a crucial part of the dead mechanical world view that gained ascendancy with modern science. But the world, as Aristotle recognized, is inherently active and end-directed and if the epistemic dimension is ever to be understood it is precisely this aspect of the world that must be grasped. The epistemic dimension of things, their shape, development, and intentional behavior, is not, and, as we shall see, *cannot* be micro-determined (viz., determined by efficient cause), but arise and are determined nomologically nevertheless.

Insensitivity to initial conditions, downward causation, or macrodeterminacy

Central to the mechanistic conception is that like antecedents produce like consequents — that macroscopic dynamics are micro-determined, and that to change the micro-conditions produces a change in the macro-conditions that follows deterministically from the equations of motion. If one knows the coordinates of the components at some time then, in principle, one knows all future and past states ("Laplacian determinism"). A hallmark of the intentional dynamics of living things, however, is that the same consequence or end state typically follows from different sets of initial conditions. Intentional dynamics are not Laplacian deterministic. Given the crucial requirement that precise behavioral patterns of living things must be repeatedly realized for living things to function or survive, surely means, however, that such systems, even though not Laplacian determined, must still be determined.

In fact, it is striking to realize that real-world initial micro-conditions are never the same twice. Whether one believes that quantum states are objectively probable (meaning that behavior of individual events is not predictable although the average behavior is), or the older notion that stochasticity is simply a matter of human ignorance, the micro-conditions at one time will, for all practical purposes, never be the same as at some other time. If Laplacian determinism were true then the repetition of forms not only of behavior, but of behaving things themselves would be inconceivable.[3] The real-world dynamics of autocatakinetic systems of which intentional ones are a kind show a remarkable *insensitivity to initial conditions*. They produce the same end states from different initial conditions. The significance of this remarkable fact, first used by Driesch against mechanism and in support of vitalism, recognized in a non-vitalistic form by von Bertalanffy under the name of "equifinality", and by Weiss under the name of macro-determinacy, has been repeatedly stressed by developmentally minded evolutionary theorists (e.g. see Salthe, 1994; Swenson, 1991a, 1992).[4]

Figure 3 shows two time slices from the Bénard fluid experiment, an exemplar of spontaneous order production or self-organization in a simple physical system. The left-hand photograph shows the disordered regime and the right-hand one shows the spontaneous production of order that occurs when the energy gradient or potential between source and sink is increased beyond a critical minimal threshold and the fluid spontaneously fills with Bénard "cells". The cells in the photograph are seen to be of variable size and shape. As time continues, however, a spontaneous process of selection occurs that includes the subsumption of smaller cells by larger ones, the competitive exclusion of smaller cells by larger ones, and the spontaneous division of larger cells to smaller ones (e.g. see Swenson, 1989a, 1989b, 1992, and in press, for the time series). The end result is a regular array of hexagonal cells of uniform size and shape. The variability that is seen at the beginning of the process is a consequence of the fact that order production is stochastically seeded.

In particular, in the disordered regime the dynamics are characterized by random collisions between micro-components which constitute fluctuations around an average state. When the critical threshold is crossed spontaneous order is seeded by any fluctuation anywhere in the fluid that is of a minimal amplitude. Since the location and actual amplitude of such fluctuations is stochastically determined the cells will form at different places in the fluid and grow at different rates every time the experiment is done. Seconds after the critical threshold is crossed the fluid thus fills with cells of variable size, but each and every time the experiment is run the variability in the size and shape is progressively eliminated by a process of selection to produce a final state of hexagonal cells.

In a decidedly non-Laplacian fashion, *unlike micro-antecedents lead to like macroscopic consequences*. Here we see a process of "blind variation" in the stochasticity of the micro-components in the disordered regime and a lawful process of selection leading to a macro-determinate result. Random initial conditions at the micro-level do not mean that the evolution of the system is random or undetermined.

Figure 3. Bénard cells. Two time slices from the Bénard experiment. The first time slice (left) shows the homogeneous or disordered "Boltzmann regime" where entropy is produced by heat flow from the disordered collisions of the molecules (by conduction), and the second (right) shows entropy production in the ordered regime. Spontaneous order arises when the field potential is above a minimum critical threshold and stochastic microscopic fluctuations are amplified to macroscopic levels and hundreds of millions of molecules begin moving coherently together. From Swenson, 1989c, copyright 1989 Pergamon Press, reprinted by permission.

The Interplay of micro and macro and the expansion of space-time dimensions

The establishment of autocatakinesis or the origins of order is generic across scales. In each case it involves 1. stochasticity or "blind variation" at the micro-level that "seeds" order at the macro-level, 2. circular causality to amplify the microscopic seeding and constitute the order at the macro-level, and 3. a source-sink gradient above some minimal critical level sufficient to pump up or fill out the new dimensions of space-time spontaneous ordering entails. These principles can be seen at work in the Bénard experiment with a more detailed look. Figure 4 shows a schematic of the pattern of flow that defines the autocatakinesis of a single Bénard cell.

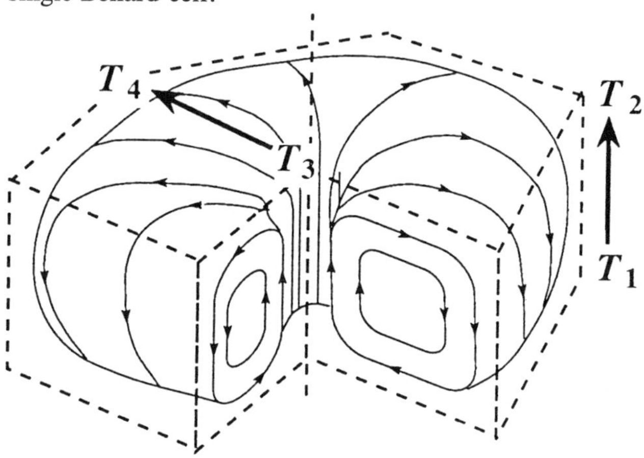

Figure 4. Autocatakinetic flow in a Bénard cell. The autocatakinetic flow of the fluid constituting a Bénard cell is shown by the small arrows. $T_1 \rightarrow T_2$ is the heat gradient between the heat source below and the cool sink above that constitutes the potential that motivates the flow. Because density varies inversely with temperature there is also a density gradient from bottom to top giving groups of molecules ("parcels") that are displaced upwards by stochastic collisions and upward buoyant force. If the potential is above the minimum threshold parcels will move upward a faster rate than their excess heat can be dissipated to their surrounds. At the same time such an upward flow of heat will increase the temperature of the upper surface directly above it creating a surface tension gradient $T_3 \rightarrow T_4$ which will act to further amplify the upward flow by pulling the hotter fluid to the cooler surrounds. The upward displacement of fluid creates a vacuum effect pulling more heated fluid from the bottom in behind it which in turns makes room for the fluid which has been cooled by its movement across the top to fall, be heated, and carry on the cycle thus establishing autocatakinesis. From Swenson, 1997a, p. 24, copyright 1997 JAI Press, Inc., reprinted by permission.

Because the intrinsic space-time dimensions for any system or process are defined by the persistence of its component relations the transformation from disorder to order increases its dimensions dramatically. In the ordered regime of the Bénard example the intrinsic dimensions are of the order of seconds and centimeters — it takes the fluid some seconds to make an autocatakinetic cycle between source and sink, and the distance covered, or the dimensions of a single cell, is in numbers of centimeters. This is in stark contrast to the disordered regime where the intrinsic spacetime dimensions are defined by mean free path distances and relaxation times (the distances and times between collisions) and are on the order of 10^{-8} centimeters and 10^{-15} seconds.

The relation between micro and macro in the filling of higher-ordered dimensions of spacetimes follows from the amplification of a micro-stochasticity to macroscopic proportions by macroscopic feedback relations, and the way macro-motion works circularly to determine micro-motion by spontaneously selecting microscopic degrees of freedom, by entraining or pulling micro-components into its formation and maintenance, is clearly seen in Figure 4. Here we see the dynamics of the minimal ontology at work, the origin and persistence of non-Cartesian circles and their inextricable dynamical relation to their environments, but what is the nomological basis for such spontaneous ordering?

THE LAW OF MAXIMUM ENTROPY PRODUCTION OR WHY THE WORLD IS IN THE ORDER PRODUCTION BUSINESS

The laws of thermodynamics are special laws of physics

To say something is conserved or remains the same over certain transformations is to describe a symmetry, and to say that there is a one-way flow of something is to describe a broken symmetry. From this we see that these two first principles of the epistemic dimension given in the conjunction shown in Figure 1 are symmetry principles. Of great interest is the fact that they are also statements, in very profound form, of the first and second laws of thermodynamics. This underscores the point that the first and second laws of thermodynamics are not ordinary laws of physics. Because the first law, the law of energy conservation, in effect, unifies all real-world processes, it is thus a law on which all other laws depend. In more technical terms, it expresses the time-translation symmetry of the laws of physics themselves — there could be no laws, no invariances, no persistences or ongoing relations without time-translation symmetry, without a conserved quantity underlying the change. The second law not only governs the ordinary laws of physics, but the first law as well. If the first law expresses the underlying symmetry principle of the natural world (that which remains the same) the second law expresses the broken symmetry (that which universally changes). It is with the second law that end-directedness, and time itself, the ordinary experience of then and now, of the flow of things, is given nomological grounding.

The classical statements of the first and second laws

Following the work of Davy and Rumford, the first law was first formulated by Mayer, then Joule, and later Helmholtz in the first half of the nineteenth century with various demonstrations of the equivalence of heat and other forms of energy. Figure 5 shows the experiment devised by Joule. The law was completed in this century with Einstein's demonstration that matter is also a form of energy.

The first law says that (a) all real-world processes consist of transformations of one form of energy into another (e.g. mechanical, chemical, or electrical energy or energy in the form of heat), and that (b) the total amount of energy in all real-world transformations always remains the same or is conserved (energy is neither created nor destroyed).

The first law was not fully understood until the second law was formulated by Clausius and Thomson in the 1850's. Some twenty-five years earlier Carnot had observed that like the fall of a stream that turns a mill wheel, it was the "fall" of heat from higher to lower temperatures that motivated a steam engine. That this work showed an irreversible destruction of "motive force" or potential for producing change suggested to Clausius and Thomson that either the first law was false (energy was not conserved), or else energy was not the motive force for change. Recognizing that the active principle and the conserved quantity could not be the same they realized that there were two laws at work and showed their relation. Clausius coined the word "entropy" to refer to the dissipated potential, and the second law states that all natural processes proceed so as to maximize the entropy (or equivalently minimize or dissipate the potential), while energy, at the same time is entirely conserved. The balance equation of the second law, expressed as $\Delta S > 0$, says that in all real world processes entropy always increases.

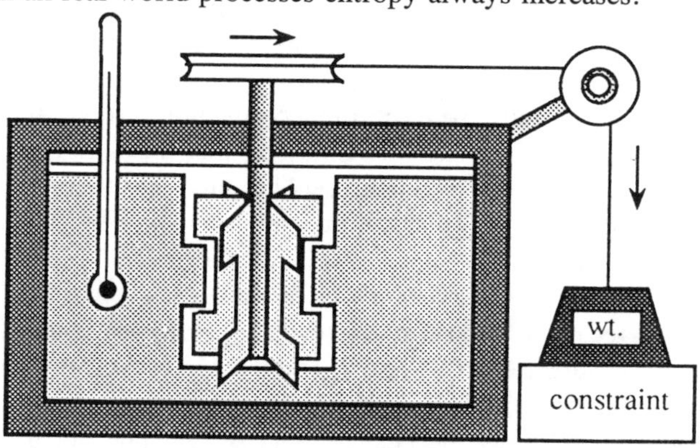

Figure 5. Joule's experiment showing the conservation of energy. When a constraint is removed, potential energy in the form of a suspended weight is converted into the mechanical or kinetic energy of a moving paddle wheel in an energy-tight container of water heating the water by an amount equal to the amount of potential energy lost by the falling weight. From Swenson, 1997a, p. 31, copyright 1997 JAI Press, Inc., reprinted by permission.

The second law provides the nomological bases for understanding the directed flow entailed by the minimal ontology. The crucial role of this broken symmetry in epistemic processes can be further grasped by returning to the experiment of Joule — although designed to show the first law it, and every other experiment designed to demonstrate the first law (e.g. Mayer and Helmholtz), demonstrated the second law as well. The act of turning the paddle wheel by the fall of the weight, and the measuring of the conservation (the energy) is an irreversible act — the measurement or epistemic act can only take place through or in the flow.

The second law as a law of disorder

The active nature of the second law presented a profound blow to the mechanical world view which Boltzmann attempted to save by reducing it to the stochastic collisions of mechanical particles — to a law of probability. Modeling the gas molecules of a gas in a box as colliding billiard balls, Maxwell had shown that non-equilibrium velocity distributions (groups of molecules moving at the same speed and in the same direction) would become increasingly disordered with each collision leading to a final state of macroscopic uniformity and maximum microscopic disorder. Boltzmann recognized this state as the state of maximum entropy (where the macroscopic uniformity corresponds to the obliteration of all field potentials). Given this, he argued, the second law was simply the result of the fact that in a world of mechanically colliding particles disordered states are the most probable. There are so many more possible disordered states than ordered ones that a system will almost always be found either in the state of maximum disorder — the macro-state with the greatest number of accessible micro-states such as a gas in a box at equilibrium — or moving towards it. A dynamically ordered state, one with molecules moving "at the same speed and in the same direction" said Boltzmann (1886/1974, p. 20), "is the most improbable case conceivable (...) an infinitely improbable configuration of energy".

Although Boltzmann himself acknowledged that his hypothesis of the second law had only been demonstrated for the case of a gas in a box near equilibrium, the science of his time (and up until quite recently) was dominated by linear, near-equilibrium or equilibrium thinking, and his hypothesis became widely accepted, and the second law came to be seen as a law of disorder. But the world is not a linear, near equilibrium system like a gas in a box, but is instead nonlinear and far-from-equilibrium, and the second law is not reducible to a stochastic collision function. As the next section outlines, rather than being infinitely improbable, a spontaneous ordering can now be seen as the expected consequence of natural law.

The law of maximum entropy production

A nomological basis for the world as active, and end-directed came with the recognition of the second law, but it did not seem to be the right kind of end-directedness for biology and psychology. Particularly with Boltzmann's interpretation, as Fisher, one of the founders of neo-Darwinism, noted, the end-directedness of the second law appeared to run directly opposite to the active, end-directedness of living things the fecundity principle and the intentional dynamics it entails has life producing as much order as it can. The problem was partly put aside in the middle of this century when von Bertalanffy (e.g. 1952, p. 145) showed that "spontaneous order (...) can appear in [open] systems" (systems with energy flows running through them) by virtue of their ability to build their order by dissipating potentials in their environments. Along the same lines, pointing to the balance equation of the second law, Schrödinger (1945) popularized the idea of living things as a streams of order which like flames are permitted to exist away from equilibrium because they feed on "negentropy" (potentials) in their environments. These ideas were further popularized by Prigogine (e.g. 1978).

Schrödinger's point was that as long as living things like flames (and all autocatakinetic systems) produce entropy (or minimize potentials) at a sufficient rate to compensate for their own internal ordering (their ordered persistence away from equilibrium) then the balance equation of the second law would not be violated. Order on the von Bertalanffy-Schrödinger-Prigogine view *can* arise spontaneously, and living things, on this view are thus *permitted* to exist, as it became popular to say, as long as they "pay their entropy debt". This worked for the classical statement of the second law per Clausius and Thomson, but on Boltzmann's view such "debt payers" were still infinitely improbable. Living things, and *a fortiori* evolution as a planetary process as a whole, were still infinitely improbable states struggling against the laws of physics — the urgency towards existence captured in the fecundity principle and the intentional dynamics it entails and in planetary evolution as a whole were entirely anomalous on this view with respect to universal law. What is more, as seen from the discussion of the Bénard experiment, simple physical systems also falsify Boltzmann's hypothesis. Order is seen to arise, not infinitely improbably, but with a probability of one, that is, whenever, and as soon as it gets the chance. The nomological basis for this opportunistic ordering was still a mystery.

Return to space-time relations, order production, and the balance equation of the second law

A return to the balance equation of the second law provides the first clue towards solving the puzzle. As discussed above and illustrated with Figure 4, transformations from disorder to order dramatically increase the space-time dimensions of a system. What von Bertalanffy and Schrödinger emphasized was that as long as an autocatakinetic system produces entropy fast enough to compensate for its development and maintenance away from equilibrium (its own internal entropy reduction or increase in space-time dimensions) it is permitted to exist. Ordered flow, in other words, to come into being or exist must function to increase the rate of entropy production of the system plus environment — must pull in sufficient resources and dissipate them — to satisfy the balance equation of the second law. This makes an important point implicitly that now will be stated explicitly: *Ordered flow must be more efficient at dissipating potentials than disordered flow* (Figure 4 shows exactly how this works in a simple physical system), and this brings us to the final piece in the puzzle.

The law of maximum entropy production

The crucial final piece to the puzzle that provides the nomological basis for dissolving the postulates of incommensurability between physics and psychology and physics and biology — between thermodynamics and evolution — is the answer to a question that classical thermodynamics never asked. The classical statement of the second law says that entropy will be maximized, or potentials minimized, but it does not ask or answer the question of which out of available paths a system will take to accomplish this end. The answer to the question is that *the system will select the path or assembly of paths out of otherwise available paths that minimize the potential or maximize the entropy at the fastest rate given the constraints*. This is a statement of the *law of maximum entropy production* the physical selection principle that provides the nomological explanation, as will be seen below, for why the world is in the order production business (Swenson, 1988, 1991, 1992, 1997a, 1997d; Swenson & Turvey, 1991). Note that the law of maximum entropy production is in addition to the second law. The second law says only that entropy is maximized while the law of maximum entropy production recognizes that it is maximized (potentials minimized) at the fastest rate given the constraints. Like the active nature of the second law, the law of maximum entropy production is intuitively easy to grasp and empirically demonstrate.

Consider the case of the warm mountain cabin sitting in cold, snow-covered woods. The difference in temperature between the cabin and the woods constitutes a potential and the cabin-woods system as a consequence will produce flows of energy as heat from the cabin to the woods so as to minimize the potential. Suppose the house is tight and heat is flowing to the outside primarily by conduction through the walls. Imagine now opening a window or a door which amounts to removing a constraint on the rate of dissipation. What we know intuitively, and can confirm by experiment, is that whenever a constraint is removed and a new path or drain is provided that increases the rate at which the potential is minimized the system will seize the opportunity. In addition, since the opened window, for example, will not instantaneously drain all the potential some will still be allocated to conduction through the walls. Each path will drain all that it can, the fastest (in this case the open window) procuring the greatest amount with what is left going to the slower paths (in this case conduction through the walls). The point is that no matter what the specific conditions, or the number of paths or drains, the system will automatically select the assembly of paths from among those otherwise available so as to get the system to the final state, to minimize or drain the potential, at the fastest rate given the constraints. This is the essence of the law of maximum entropy production. What does this have to do with spontaneous ordering, with the filling of dimensions of space-time?

Given what has already been discussed above, the reader may have already leaped to the correct conclusion. If the world selects those dynamics that minimize potentials at the fastest rate given the constraints, and if ordered flow is more efficient at reducing potentials than disordered flow, then *the world will select order whenever it gets the chance — the world is in the order-production business because ordered flow produces entropy faster than disordered flow* (Swenson, 1988, 1991, 1992, 1997a, 1997d; Swenson & Turvey, 1991), and this means the world can be expected to produce as much order as it can, to expand space-time dimensions whenever it gets the chance. Autocatakinetic systems, in effect, are self-amplifying sinks that by pulling potentials or resources into their own self-production extend the space-time dimensions of the fields (system plus environment) from which they emerge and thereby increase the dissipative rate.

INTENTIONAL DYNAMICS AND THE PHYSICAL BASIS FOR ABOUTNESS

The law of maximum entropy production and its relation to autocatakinesis provides the nomological basis for the commensurability between physics, biology, and psychology. Ecological psychologists (e.g. Gibson, 1979) have argued the necessity of a mutuality postulate, namely, the recognition that living things and their environments constitute single systems, as a prerequisite to understanding the epistemic dimension of living things, and the law of entropy production shows how this postulate directly follows from natural law (Swenson & Turvey, 1991). The active striving of living things is no longer seen as a struggle against the laws of physics, but a manifestation of them. The fecundity principle, along with the intentional dynamics it entails, is seen as a special case of the law of maximum entropy production where the substrate (material cause in Aristotle's terms) is replicative. The epistemic dimension rather than being incommensurable and improbable is instead the expected behavior of the world acting back on itself in its own becoming. This provides a principled ground for the minimal ontology, the conditions of existence of the epistemic experience itself, and for evolution itself as about something other than individual fitness, as an epistemic process, in particular the dynamical means for accessing new dimensions of space-time.

The orthogonality condition

There is one property that is implied in the description of the minimal ontology, but it needs to be explicitly stated. Namely, that the directedness towards that characterizes intentional behavior is experienced only by the fact that it is directed differently than the one way flow through which it is distinguished. In the most immediate case, I know my self by invariance through change, and through this recognition comes the deep connection of the epistemic experience to the cosmic project of building space-time dimensions captured by the "orthogonality condition" referred to in the introduction. Summarizing and expanding briefly as follows: all end-directed systems act to minimize potentials, and intentional systems as end-directed systems are no exception, but not all end-directed systems are intentional.

In particular, there is a principled distinction to be made, for example, between rivers that flow down hills, heat that flows down temperature gradients and living things whose intentional dynamics entails moving up not down gradients, and who unlike rivers and heat that stop flowing, or tornadoes and dust devils (or Bénard cells) that fall apart or "die" when the local potentials are removed, become more active instead. Intentional dynamics are not determined by local potentials. Instead the intentionality of living things functions to permit them to constitute their autocatakinesis through the minimization of (to feed on) non-local potentials — to move, to grow, to act in such a way so as to, in effect, "skate across" local potentials to access non-local potentials discontinuously located in space-time — to move in directions different from or orthogonal to those determined by local potentials. This is the essence of the orthogonality condition, and it is easily seen to have progressively increased over evolutionary time providing an ordinal measure not only for the epistemic or psychological dimension but for evolution as a single planetary process. This in turn, coupled with the nomological basis for the orthogonality condition in the expansion of space-time dimensions underscores the point that evolution itself is an active epistemic process, and one that follows from natural law.

If we remember from the discussion above that the intrinsic space-time dimensions of a system are determined by the persistence of its component relations, we see at once the dramatic increase in accessible space-time dimensions as a consequence of the orthogonality condition, the ability of intentional systems to build their autocatakinesis off of non-local potentials.

From extension to intension

If intentional dynamics are not determined by local potentials, then how are they determined? Autocatakinesis has the property of insensitivity to initial conditions and macro-determinacy, but what is the basis for this insensitivity to initial conditions so crucial to the performance of intentional acts in a world that is microscopically different at every moment? The Bénard convection, which, in effect, "solves the packing problem" by producing a regular array of hexagonal cells during the course of its evolution or development can be understood in terms of the system's proximal relation to, or embeddedness within, a field of local potentials, but how is intentional behavior determined with respect to non-local or distal potentials? We have noted that such systems surely are determined, and we have said in the introduction that they are meaningfully so, or determined by information about, but how does one get from an otherwise meaningless world of extension or usual physical descriptions to a meaningful world of information about?

The principles of thermodynamics provide some immediate clues, namely, that one needs to look for macroscopic observables that capture invariant properties with relevance to intentional ends. In addition one would need to look for symmetry and broken symmetry, for observables that capture the nomological relation between persistence and change of the distal objects of intention with respcct to their proximal or local space-time position with respect to the epistemic subject. The answer which follows from the two laws of thermodynamics has been most powerfully captured by Gibson's (1979/1986; Swenson & Turvey, 1991; Turvey & Shaw, in press) ecological concept of information. Living things are surrounded by ambient energy distributions (e.g. optical, mechanical, chemical) for which the mean energy content is extremely low relative to the energy used by living things from their on-board potentials to power their intentional acts. As a consequence of the fact that lawful or invariant relations exist between the macroscopic properties of such ambient energy distributions that are relevant to the prospective control of intentional ends, the former can specify or determine the latter.

Bacteria, for example, may move up a chemical gradient that lawfully specifies the source of their food, animals may use diffusion fields of diffusing volatiles that lawfully specify the sources of their intentional ends, and fields of mechanical waves and optical fields are used in similar ways. The ability to effect (or avoid) controlled collisions, for example, is a particularly crucial and widespread requirement for the intentional dynamics of many living things (e.g. soft collisions with little or no momentum exchange as in a bird landing on a branch, hard collisions with substantial momentum exchange as when a predator attacks a prey, and collision avoidance where the ends of an intentional agent require that it not collide with particular things). Information about such collisions is lawfully carried in the ambient energy field (the "optical flow field") that transforms itself as a living thing moves through it. Just as in the Bénard case where laws of form specify the origin of order, and its development, so too with intentional dynamics.

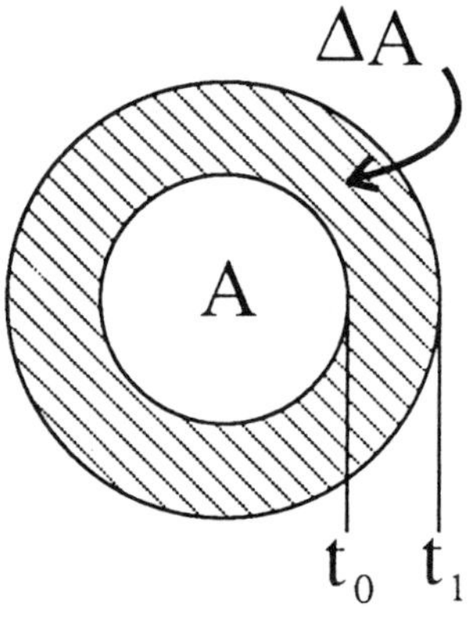

$$\tau = \left[\frac{\Delta A}{\Delta t} \cdot \frac{1}{A}\right]^{-1}$$

Figure 6. Information about is lawfully carried in ambient energy distributions. From Swenson, 1997a, p. 41, copyright 1997 JAI Press, Inc., reprinted by permission.

With respect to controlled collisions, the time-to-contact (τ), as shown in Figure 6, is determined by the inverse of the relative rate of expansion of the optical flow field, and the information about whether a collision will be hard or soft is given by the time derivative or rate of change of the relative rate of expansion ($\dot{\tau}$) (Lee, 1980; Kim et al., 1993). For an intentional act requiring a soft collision, for example, such as a bird landing on a branch the rate of change must be $\dot{\tau} \geq -.5$. This single macroscopic variable nomologically carried in the optic flow precisely determines when a particular bird, for example, must open its wings to decelerate so that it does not, in effect, crash into a branch. This is again, a deceivingly simple idea with remarkable and profound consequences. It exposes the fact that, not only are the shapes and forms things assume nomologically determined by laws of form (e.g. that there is, within tolerance, a requisite ratio between flight muscle weight and body weight, or between wing span and body weight, or brain weight and body weight (e.g. Alexander, 1971)), but that information about or meaning carried in macroscopic flow variables nomologically determines the behavior of things towards their intentional ends (See Swenson, 1997a, 1997d; Swenson & Turvey, 1991 and Turvey & Shaw, 1995 for further discussion).

CONCLUSION

The search for the nomological basis for the origin of intention and intension, the search for the origin and nature of the epistemic or psychological dimension, is a search for the nomological basis for end-directed ordering determined by meaning. Following Boltzmann's interpretation of the second law of thermodynamics it is "no surprise", in the words of Levins and Lewontin (1985, p. 19) "that evolutionists [came to] believe organic evolution to be the negation of physical evolution". Noting the opposite directions followed by physical and biological (and by implication, psychological) systems, Ronald Fisher (1930/1958, p. 39), one of the founders of neo-Darwinism, wrote that "entropy changes lead to a progressive disorganization of the physical world (...) while evolutionary changes [produce] progressively higher organization (...)" Konrad Lorenz (1973, p. 20), one of the founding fathers of evolutionary epistemology, wrote that the aspect of life "most in need of explanation, is that, in apparent contradiction to the laws of probability, it seems to develop from (...) the more probable to less probable, from systems of lower order to systems of higher order".

The law of maximum entropy production, coupled with the generic properties of autocatakinesis, and the fact that information about distal potentials is lawfully carried in the invariant properties of ambient energy flows, solves the problem of Fisher and Lorenz, and provides the ground for a nomological account of spontaneous epistemic ordering or intentional dynamics. The law of maximum entropy production says that potentials are minimized at the fastest rate given the constraints, and, following the balance equation of the second law, autocatakinetic or spontane-ously ordered systems work spontaneously to increase the rate. As a consequence, the world acts opportunistically to produce as much order as it can. The epistemic dimension, the urgency towards existence characterizing the intentional dynamics of living things expressed by the fecundity principle and the process of terrestrial evolution writ large is thus not only commensurable with universal first principles, but a direct manifestation of them. This view provides a principled basis for uniting living things and their environments, knower and known, or self and other as reciprocal parts of an active world acting back on itself in its own becoming.

Department of Psychology, University of Connecticut
Connecticut, USA

NOTES

[1] The term draws the distinction between "real-world" law-based self-organizing systems and rule-based systems (e.g., those relating to automata theory, Boolean nets and various other connectionist models) which depend on rules programmed into them by human designers. All rule-based systems are ultimately internal productions of autocatakinetic systems, but the reverse is not true.

[2] For further discussion of the limitations of efficient cause and re-examination or reinterpretation of Aristotle's causal framework see Swenson, 1990, 1992, 1996 (on all four causes), Swenson & Turvey, 1991 (on final cause), and also Rosen, 1987; Salthe, 1985, 1994 (on all four causes) and Van de Vijver (all four causes), this volume.

[3] The failure to understand this crucial point, has led particularists or historicists (e.g. Gould, 1989) to argue that if one "rewound the tape of life on Earth" that, as a consequence of chance fluctuations ("errors") of the microcomponents, the system would repeatedly branch off onto different trajectories with the result that evolution would be dramatically different every time — there would be an astronomically high number of possible states the system might assume, and thus the possibility of the repetition of similar trajectories such as those that would lead to "intelligent life" such as humans would also be astronomically improbable. But if the world were actually microdetermined in this way the world as we know it would not merely be improbable but impossible.

[4] See Depew (this volume) for some discussion of development and evolution.

REFERENCES

Alexander, R., 1971, *Size and shape*, London, William Clowes & Sons, Ltd.

Aristotle, 1947, De anima, in *Introduction to Aristotle*, R. McKeon (ed.), pp. 163-329, New York, Random House.

Boltzmann, L., 1886 (1974), The second law of thermodynamics, *Populare Schriften*, Essay 3, address to a formal meeting of the Imperial Academy of Science, 29 May, 1886, reprinted in Ludwig Boltzmann, *Theoretical Physics and Philosophical Problems*, S.G. Brush (transl.), D. Reidel Publishing, Boston, 1974.

Callebaut, W.; Pinxten, R. (eds.), 1987, *Evolutionary epistemology, A multiparadigm program*, Dordrecht, D. Reidel Publishing Company.

Campbell, D.T., 1987, Evolutionary epistemology, in *Evolutionary epistemology, rationality, and the sociologyh of knowledge*, G. Radnitzky and W.W. Bartley (eds.), La Salle, IL, Open Court, pp. 47-89.

Darwin, C., 1859 (1937), *On the Origin of Species by Means of Natural Selection or the Preservation of Favored Races in the Struggle for Life*, D. Appleton-Century Company, Inc. NY.

Dawkins, R., 1982, *The extended phenotype*, San Francisco, Freeman.

Depew, D.; Weber, B., 1995, *Darwinism Evolving*, MIT Press, Cambridge, MA.

Durkheim, E., 1895 (1938), *The rules of sociological method*, S. Solovay and J. Mueller (Transl.), G. Catlin (ed.), Chicago, Chicago University Press.

Dyke, C., 1997, The heuristics of ecological interaction, in *Advances in Human Ecology*, vol. 6, L. Freese (ed.), Greenwich, CT, JAI Press, Inc., pp. 47-72.

Fisher, R.A., 1930 (1958), *The genetical theory of natural selection*, New York, Dover.

Gibson, J.J., 1979 (1986), *The Ecological Approach to Visual Perception*, Lawrence Erlbaum Associates, Hillsdale, NJ.

Gould, S.J., 1989, *Wonderful life: The burgess shale and the nature of history*, New York, W.H. Nortron & Company.

Kim, N.; Turvey, M.T.; Carello, C., 1993, Optical information about the severity of upcoming collisions, *Journal of Experimental Psychology*, vol. 19, no. 1, pp. 179-193.

Lakatos, I., 1970, Falsification and the methodology of scientific research programmes, in *Criticism and the growth of scientific knowledge*, I. Lakatos and A. Musgrave (eds.), Cambridge, Cambridge University Press, pp. 51-58.

Lee, D., 1980, A theory of visual control of braking based on information about time-tocollision, *Perception*, vol. 5, pp. 437-459.

Levins, R.; Lewontin, R., 1985, *The Dialectical Biologist*, Harvard University Press, Cambridge, MA.

Lewontin, R., 1992, *Biology as Ideology: The Doctrine of DNA*, Harper Collins, NY.

Lorenz, K., 1973, *Behind the mirror: A search for a natural history of human knowledge*, New York, Harcourt Brace Jovanovich.

Maynard Smith, J., 1988, Evolutionary progress and levels of selection, in *Evolutionary progress*, M. Nitecki (ed.), Chicago, IL, University of Chicago Press, pp. 219-230.

Mayr, E., 1980, Prologue: Some thoughts on the history of the evolutionary synthesis, in *The Evolutionary Synthesis*, E. Mayr and W.B. Provine (eds.), Harvard University Press, Cambridge, MA, pp. 1-48.

Mayr, E., 1985, How biology differs from the physical sciences, in *Evolution at a Crossroads*, D. Depew and B. Weber (eds.), MIT Press, Cambridge, MA, pp. 43-63.

Munz, P., 1985, *Our knowledge of the growth of knowledge: Popper or Wittgenstein?* London, Routledge & Kegan Paul.

Munz, P., 1987, Philosophy and the mirror of Rorty, in *Evolutionary epistemology, rationality, and the sociology of knowledge*, G. Radnitzky and W.W. Bartley (eds.), LaSalle, IL, Open Court, pp. 345-387.

Popper, K., 1985, *Unended quest: An intellectual autobiography*, La Salle, IL, Open Court.

Popper, K., 1987, Natural selection and the emergence of mind, in *Evolutionary epistemology, rationality, and the sociology of knowledge*, G. Radnitzky and W.W. Bartley (eds.), La Salle, IL, Open Court, pp. 139-154.

Prigogine, I., 1978, Time, structure, and fluctuations, *Science*, vol. 201, pp. 777-785.

Radnitzky, G.; Bartley, W.W. (eds.), 1987, *Evolutionary epistemology, rationality, and the sociologyh of knowledge*, La Salle, IL, Open Court.

Salthe, S., 1985, *Evolving hierarchical systems*, New York, Columbia University Press.

Salthe, S., 1994, *Evolution and Development*, MIT Press, Cambridge, MA.

Schrödinger, E., 1945, *What is life?* New York, Macmillan.

Schwartzman, D.; Shore, S.; Volk, T.; McMenamin, M., 1994, Self-organization of the Earth's biosphere — Geochemical or geophysiological?, *Origins of Life and Evolution of the Biosphere*, vol. 24, pp. 435-450.

Schweber, 1985, The wider British context of Darwin's Theorizing, in *The Darwinian Heritage*, D. Kohn (ed.), Princeton University Press, Princeton, NJ, pp. 35-69.

Swenson, R., 1988, Emergence and the principle of maximum entropy production: Multi-level system theory, evolution, and non-equilibrium thermodynamics, *Proc. 32nd Ann. Meet. ISGSR*, vol. 32, p. 32.

Swenson, R., 1989a, Engineering initial conditions in a self-producing environment, in *A delicate balance: Technics, culture and consequences*, M. Rogers and N. Warren (eds.), Los Angeles, Institute of Electrical and Electronic Engineers, IEEE Catalog, no. 89CH2931-4, pp. 68-73.

Swenson, R., 1989b, Emergent evolution and the global attractor: The evolutionary epistemology of entropy production maximization, *Proceedings of the 33rd Annual Meeting of the ISSS*, vol. 33, no. 3, pp. 46-53.

Swenson, 1989c, Emergent Attractors and the Law of Maximum Entropy Production: Foundations to a theory of General Evolution, *Systems Research*, vol. 6, pp. 187-197.

Swenson, R., 1990, Evolutionary systems and society, *World Futures*, vol. 30, pp. 11-16.

Swenson, R., 1991a, End-Directed Physics and Evolutionary Ordering: Obviating the Problem of the Population of One, in *The Cybernetics of Complex Systems: Self-Organization, Evolution, and Social Change,* F. Geyer (ed.), Salinas, CA, Intersystems Publications, pp. 41-60.

Swenson, R., 1991b, Order, Evolution, and Natural Law: Fundamental Relations in Complex System Theory, in *Cybernetics and Applied Systems*, C. Negoita (ed.), New York, Marcel Dekker Inc, pp. 125-148.

Swenson, R., 1992, Autocatakinetics, Yes — Autopoiesis, No: Steps Towards a Unified Theory of Evolutionary Ordering, *International Journal of General Systems*, vol. 21, no. 2, pp. 207-228.

Swenson, R., 1997a, Autocatakinetics, evolution and the law of maximum entropy production: A principled foundation toward the study of human ecology, in *Advances in Human Ecology*, vol. 6, L. Freese (ed.), Greenwich, CT, JAI Press, Inc., pp. 1-46.

Swenson, R., 1997b, Thermodynamics and evolution, in *The Handbook of Comparative Psychology*, G. Greenberg and M. Haraway (eds.), New York, Garland Publishing.

Swenson, R., 1997c, Evolutionary theory developing: The problem(s) with 'Darwin's Dangerous Idea', *Ecological Psychology*, vol. 1, no. 9, pp. 47-96.

Swenson, R., 1997d (in press), *Spontaneous Order, Evolution, and Natural Law: An Introduction to the Physical Basis for an Ecological Psychology*, Hillsdale, NJ, Lawrence Erlbaum and Associates.

Swenson, R.; Turvey, M.T., 1991, Thermodynamic Reasons for Perception-Action Cycles, *Ecological Psychology*, vol. 3, no. 4, pp. 317-348.

Turner, J.; Maryanski, A., *Functionalism*, Menlo Park, CA, Benjamin/Cummings.

Turvey, M.T.; Shaw, R.E., 1995, Towards an ecological physics and a physical psychology, in *The Science of the Mind: 2001 and Beyond*, R. Solso and D. Massero (eds.), Oxford, Oxford University Press, pp. 144-169.

Von Bertalanffy, L., 1952, *Problems of life*, London, Watts.

Weiss, P., 1973, *The science of life*, Mt. Kisco, NY, Futura Publishing.

KLAUS KORNWACHS

PRAGMATIC INFORMATION AND THE EMERGENCE OF MEANING

INTRODUCTION

The battle over the meaning of meaning within the theory of evolutionary systems seems to cover, among others, the struggles about mind-matter problems, about the *res cogitans* and *res extensa* distinction, about constructivistic versus naturalistic views, holistic versus reductionistic models. Beyond these controversies it is agreed that we can observe meaningful "sentences" having effects in the real world. The term "sentences" has to be interpreted very generally, referring to strings in DNA as well as statements in communication processes, actions of system surfaces as well as distinguished behavior within a certain context. In this context we describe the emergence of new properties as meaningful, particularly for evolutionary systems, if we are able to connect these properties with actual given "sentences". The reason for doing so lies in the fact that we interpret "becoming" as more meaningful than "being". In order to explain the experience, or even to figure out preconditions for the possibility of such experience, of meaningful events or properties it would be necessary to answer the following questions: (1) By what means can one state that a given sentence (or event) in a given context is meaningful?; (2) By what means can one observe effects in the real world?; (3) By what means can one state relations between the meaningfulness of sentences and real effects in the real world?

To overcome the difficulty we firstly substitute the term "sentence" with "information". Indeed, sentences express, while events contain informational contents, even if we disregard intentional or illocutionary aspects of the act of "utterance". This step is not equivalent to decontextualization— each event (process, signal, string of symbols, data set or else) becomes meaningful if there is a model with which it can be interpreted, and if there is a context within which it can be interpreted. Secondly, we connect the term "information" with the term "system". What we know from systems can be expressed in terms of information. But systems usually react upon information after having received it. Thus, the interaction between system and information becomes important.

A receiving system is only an abstraction of a general system concept. A system can always be described as part of an overall system. The system in consideration interacts with the overall system via a surface (Kornwachs, 1996). So it is possible to define operatively what can be regarded as an input and as an output. The distinction between input and output is usually equivalent to a classification of causing and caused variables. From a descriptive point of view, systems *per se* do not exist, but it is possible to describe objects and processes and to separate them from the rest of the world (areas of things) *as* systems. From this point of view, the representation of information is performed on system surfaces, i.e. by the overall behavior.

How can meaning be measured? This basic problem of semantics has never been

G. Van de Vijver et al. (eds.), Evolutionary Systems, 181–196.

solved satisfactorily on a mere semantical level. Therefore the concept of pragmatic information put forward the idea that meaning can only be quantified by an effect of information. This effect can only be stated in relation to a system — the receiving system must become a sender at which the effects can be observed, and the sending system becomes a receiver.

A pre-theory of pragmatic information has been developed within the last twenty years and the concept is starting to be applied in several areas of cognitive and natural sciences (cf. Gernert, 1985, 1995; Kornwachs, 1987, 1988, 1990, 1992, 1993, 1996; von Lucadou, 1984, 1989; Weizsäcker, 1974 and for applications cf. Atmanspacher, 1989; Atmanspacher et al., 1990, 1992; Kornwachs & Jacobi, 1996). One of the main theses of this theory is that pragmatic information is able to build up new possibilities of informational exchange (in terms of Shannon's theory of communication these possibilities are channels). Another thesis is that pragmatic information generates new (pragmatic) information and that meaning can be generated within the context of a well defined system-system interaction. This meaning seems to "emerge" from the system, but it seems that this concept of emergence is a misleading one. This is one reason why this contribution has been written.

MEANING IN THE CONTEXT OF SYSTEM THEORY

Evolutionary systems are entities, described as systems, which have been generated within the framework of an evolutionary process. A lot of misunderstandings have been provoked by the fact that such an evolutionary process can be described also as an evolutionary system — here the distinction between inside view (endo-view) and outside view (exo-view) enters the discussion (cf. Kampis in this volume; cf. also Atmanspacher & Dalenoort, 1994). Evolutionary systems are products of evolutionary systems and they produce them. If one describes the evolutionary process as a system — taking the exo-view — then this evolving system is a self-organizing, autonomous[1], in the very beginning of its history an "autopoietic" super-system. It generates new sub-systems and it shows emergence as far as new properties appear during any chosen period of observation or operating interval. Within the context of this exo-view one hopes to develop a theory which is able to explain the creation of new sub-systems as part of the dynamics of the whole system. According to the definition of emergence, which states that new properties appear which cannot be ruled out by a predictive theory, such a future theory would make the term emergence superfluous. Nevertheless this theory is not yet available. If one takes the endo-view, the term emergence is still valid: an evolutionary system is suddenly confronted with a new environment, including new species with new properties with which it has to cope. Here the distinction between environment and system remains valid, too. Within the context of an exo-view the distinction between system and environment is substituted by the system-system-distinction. The sub-systems are interacting with each other and are the subject of mutual change (cf. also Csányi, this volume).

Meaning within the context of evolutionary system theory can be characterized by the following features:

1. If there are relations between structural and behavioral patterns and systems (or in the endo-view between systems and their environment), this fact is expressed in different terms: some people are talking about "Widerspiegelung" (reflection) of external states by internal states (there is a clear inside-outside distinction), others are talking about adaptation, fulfilling of goals, learning, autonomy, intentionality or purposeful behavior. All these concepts seem to be a *façon de parler* which hides the lack of a predicting theory about the dynamics of evolving systems. To observe patterns of behavior, on whatever level of description, presupposes a kind of filter that must be available by the observer. This filter is an observer's theory (at least a prescientific phenomenological approach or model) about the system.

2. A correlation may be made between the environment (in terms of a real world concept) and what a system is actually doing (as a behavior pattern, too), i.e. the system is doing something with the world and it is solving problems, definable in terms of system-environment-interaction. Intentionality is a metaphor for this kind of view, the circularity of the fitness of the fittest is on this level, too. "Meaning precedes finality" (Atlan, 1992), but finality is equally a concept only definable in terms of a semantic closure.

3. Environmental patterns and system patterns are interpreted in the same context by the describer of elements and systems. This is the epistemological rectification of a semantic closure (cf. Umerez, this volume). In the very beginning of any description there is a degree of freedom to choose what patterns should be important and relevant, but after having made a choice, the rectified semantic closure starts to play a role as constraints.

4. An input of a system is considered as meaningful when it triggers the system into another state, wanted or anticipated by an observer. This again presupposes a semantic closure. An output is considered as meaningful when the observer or another system has "understood" what it means: it can refer to inner states and their history or it can refer to a reaction to an input given earlier. State space dynamics is interpreted as meaningful when the system is driven from one important attractor to another, more important attractor (Haken, 1988). What is called important depends again on the semantic closure connecting observer and system. An outer force (interpreted as information) or any internal dynamics (autonomous creation of new attractors) may be considered as meaningful.

To establish the term "meaning" in the context of systems, science is faced with some severe problems. Everything and every process can be described as a system. Systems do not exist as an ontological but as an epistemological entity. Every system has an author and the author pursues his own interests with the help of a system. Thus the cut between system and environment, in the endo-view, is a mere question of adequacy[2], not of necessity. Systems science offers a lot of mathematical tools (dynamical system theory, control theory, state-space-approach, theory of decision and games, algorithmic systems, automata theory, statistical physics) that can be used for these kind of descriptions. But system theoretical description does not suspend the describer from having knowledge available about the area the description is coming from, i.e. system theory is no substitute for biology, physics or sociology.

The notion of meaning within the above context runs into difficulty if one tries to measure meaning in concrete cases. Looking at a pattern, there are scarcely criteria which are really observer-independent, i.e. which do not require a kind of pre-theory. Applying signal extraction and filter theory, it can be found out without any observer preknowledge *that* a pattern may be contained within the signal or behavior protocol, but it is not possible to deduce the meaning itself, or to what it may refer. Therefore the observer must know the context. The system behaves meaningfully, i.e. it is able to move itself in a world without loss of integrity, identity and existence.[3] Taking the system as an object this statement becomes trivial — it holds for each electron. Therefore without any given semantical relation it is not possible to make meaningful propositions about system behavior related to the external world. Moreover there is no observability independence on wishes, purposes, anticipations and selections made by the observer. Reference is like a genuine semantic concept which is not reducible to an observer-independent context. There is no objective theory with which the weights or degrees of importance for states or attractors could be deduced.

To summarize: it is not possible to state something about meaning independent of an observer. But one is often inclined to eliminate the observer. This is done by physics and by classical system theory. Therefore the concept of meaning is usually the first victim of Ockhams Razor. It is sufficient to keep the concept of "being meaningful", or "to carry meaning" instead of saying "such and such explicit meaning". In this way it is possible to operationalize the concept of "having meaning": if information has a meaning for a receiver, this must be observable by interpreting its behavior after receiving this information within a semantic closure. If the behavior is the same as before, there are no means to say something about information and its meaning. Only if the receiver becomes an emitter is there a basis to judge how a reaction may be interpreted. This is the basic idea of pragmatic information.

PRAGMATIC INFORMATION

There have been many attempts[4] to design a theory of pragmatic information. They all share a common motivation: they have tried to surmount the obvious limitations of the theory of communication by Shannon and Weaver (1948) which describes only the features of information transfer. The new approach is to conceptualize information as an interacting term between systems. Moreover, it is tried to come to a quantitative concept of meaningful information.

As could be shown in earlier contributions (cf. Kornwachs, 1988; Kornwachs & von Lucadou, 1989; Kornwachs, 1992, further literature is given there), pragmatic information, conceived as a meaningful interaction between systems, can only be treated within the framework of a classical theory as long as the interacting systems themselves are described in classical terms.[5] This may be helpful as a first step in examining the possibilities included in classical system theory and therefore this contribution will use classical arguments too. But pragmatic information has been conceived to be able to modify systems receiving it. It has been pointed out that systems which change structure and behavior cannot be described in terms of classical system theory (Kornwachs & von Lucadou, 1984, 1989).

As an *ideé directrice*, the notion of complementarity may be useful to look for examples of non-classical systems. With respect to the non-classical character of a theory of pragmatic information, pairs of conceptual entities have been introduced which are complementarily related to each other. If one could formulate them in terms of system-theoretical descriptions, all those pairs, like novelty and confirmation (the components of pragmatic information), behavior and structure, autonomy and reliability, and others, should form products which have the dimension of an action, i.e. [energy . time].[6] Some formal analogies to the structure of quantum theory could already be found. Therefore within the context of system theory, observables have been related to these in pairs-related terms. The meaning of such observables is the following: within a system theoretical description, the "values"[7] of an observable quantity can be ruled out from an explicit system description like the wave-function in quantum mechanics by an application of an operator belonging to such an observable quantity to a formal system-theoretical representation. Some of these operators have been discussed elsewhere.[8]

The usefulness of the concept of pragmatic information has induced some applications in physics, chemistry, even in astrophysics and in cognitive psychology as well (in physics cf. Atmanspacher, Kurths et al., 1992; Atmanspacher, 1989; Atmanspacher & Scheingraber; Gernert, 1994; cf. also further contributions in Kornwachs & Jacobi, 1996; Atmanspacher & Dalenoort, 1994, 1989; von Lucadou, 1994 in cognitive psychology). The main idea of this concept is that information is generated by a system[9] — the working strength of information can have the same result as the system itself. In other words: a system can be transformed into working information. On the other hand, pragmatic information is able to build up systems with certain properties. In terms of operating with systems, the effect of information can substitute the effect of a system and vice versa.[10] Therefore the very fundamental question arises, how can information be brought into a system in order to have an effect therein and how is information delivered by a system?[11]

Under the assumption that information, and also pragmatic information, is represented by signals which can be interpreted as signs under certain circumstances, the next question is how signs can interact with real or physical (i.e. material) systems. The problem of the "physics of reading and writing" has been discussed first by H. Pattee (1977), later by S. Rosen (1978), and some further discussions have been evoked about these issues (cf. the contribution of Kornwachs & Hinderer, 1991). Within this context the term of system surface may gain increasing importance. The assumption is that signs as time-spatial configurations, which serve as elements of informational representation, are only observable on the surface of a system — to speak metaphorically, they are "living" on those surfaces (Hinderer, 1990, oral communication).[12]

The approach by Shannon and Weaver (1948) can be compressed here into the idea that information is quantifiable by a measure of uncertainty with respect to a given situation not yet decided. But the given situation is not necessarily as objective as a map of a town with finite alternatives. It can also be defined by expectations, prejudices and pre-knowledge of the concrete receiver. The receiver may use given information in order to decide an undecided situation and to transform an uncertain situation into a certain one. This can be treated in terms of classical system theory.

The concept of pragmatic information starts with the basic assumption that the degree and the quality of change, caused by information within the receiver, is dependent on the structure already given and on the behavior of the receiver. The change in the receiver may be expressed by an altered behavior and/or an altered structure.[13] The change in behavior can be expressed by detecting a new pattern within the time dependent dynamics. The change in structure can be expressed by the creation and annihilation of elements, sub-systems and relations between them. As a measure for degree of change, Gernert (1985, 1995) has proposed to take the difference between the efficiency (η) of the receiving system before and after the impact of the information, i.e.

$$\text{"}\mathit{strength}\text{"}\ \mathit{of\ pragmatic\ information} = \frac{\eta_{after} - \eta_{before}}{\eta_{before}} \tag{1}$$

as a measure for cases where η is definable according to particular systems.[14]

In order to obtain a more predictive theory about pragmatic information it is necessary to regard the receiver as a system. Using the classical view, the influence of information is clearly separated in terms of time order (before and after the impact) and in terms of spatial order.[15] The latter can be defined by the borderline between system and non-system, analogous to a distinction between system and its environment. The uncertain situation is related to a measure, which is definable *inside* the system and this uncertainty is removed therein when information has been applied.

In order to gain information about a system it is necessary to observe its behavior. Or it is necessary to "open" it in order to learn something about its structure. One can try to analyze a system by calculating its overall behavior from the structure and behavior of its elements.[16] The operations "to observe", "to open", "to analyze" correspond to certain operators[17] which are applied to the system description. These operations are aimed at getting information from the inside of the system to the outside, i.e. for an outside observer. The object "receiver", described as a system, can be observed. Then it is conceived to be a source of information, or as a system which actually generates information. In both cases the information can be used to say something about the system.

Looking again at the two kinds of information, information *for* a receiver and information *from* an emitter, we may couple the receiving and the emitting system in a way which is more than a simple emitter-channel-receiver scheme: if no receiver reaction can be observed, nothing can be said about the influence of the received information. Operatively speaking, the receiver must therefore become a sender and the sender must become a receiver.[18]

NOVELTY AND CONFIRMATION

The basic postulates of pragmatic information theory have already been presented elsewhere in earlier contributions.[19] We only use those basic traits which are necessary for our arguments here.

Both components of pragmatic information, i.e. novelty and confirmation, have been introduced by the intuitive notion that complete confirmation or complete novelty do not provide any meaningful information to a receiver (cf. von Weizsäcker, 1974; Gernert, 1996). What novelty or confirmation is, depends on the pre-knowledge or on the pre-structure of the receiver. As von Weizsäcker (1974) has pointed out, a minimal pre-structure of the receiver must be available, otherwise pragmatic information cannot have an effect on this system.

If one tries to get information about a system, what the receiver does if it is receiving information from the environment or a neighboring system, one can use operators which apply to the system description. Such an operator model has been discussed in Kornwachs (1990, 1992). The basic point with this concept is that a difference between an expected and an observed behavior or structure is used to determine novelty or confirmation. Thus pragmatic information is strongly dependent upon a concrete context, given by a concrete system description.

The usual system-theoretical definitions of systems are based on set theoretical concepts (Klir, 1969; Mesarovic, 1972). They are designed as constructs (descriptions) whose elements are sub-systems which are not to be analyzed in more detail. The actual depth of analysis depends on the goals and interest of a system describer. The elements and sub-systems are connected by relations, called structure, which can either be defined by graphs (arrows and lines) or by relations between variables (Klir, 1985). Sub-systems or elements have a certain behavior, given by a local input-output mapping, also expressible by relations, analytical functions, differential equations or automata state transition functions.

Let a system be given by a quintuple $S = \{\alpha, \phi, \sigma, \pi, T\}$. The variables of the systems are given by $\alpha = \{\alpha_x, \alpha_y, \alpha_z, t\}$ with $\alpha_x \in \{X\}$ as variables of input, $\alpha_y \in \{Y\}$ as variables of output, $\alpha_z \in \{Z\}$ as variables of states and $t \in \{T\}$ as a time base. The behavior of the system is given by a state transition function δ and an output function λ, so that $\phi = \{\delta, \lambda\}$ and with

$$\delta: \alpha_x \otimes \alpha_z \rightarrow \alpha_z$$

and (2)

$$\lambda: \alpha_x \otimes \alpha_z \rightarrow \alpha_y.$$

With the decomposition of a system into sub-systems, these sub-systems (or elements) are connected by a structure π, expressible by a connectivity matrix $\pi = \{\kappa_{ij}\}$ with $\kappa_{ii} = 1$, i.e. the diagonal elements are defined as 1, the cells within the matrix contain the 0 and 1 representing the connection between the elements and/or sub-systems. The connections may represent time or causal relations, flows, impacts, influences and others. The sub-systems can include further sub-systems which may be defined analogously by $\sigma = \{S'_s\}$ with $S' = \{\alpha_{S'}, \phi_{S'}, \sigma_{S'}, \pi_{S'}, T\}$ and for a further level of hierarchy (second level) $\sigma_{S'} = \{S''_S\}$ with $S'', = \{\alpha_{S''}, \phi_{S''}, \sigma_{S''}, \pi_{S''}, T\}$ and so on (Mesarovic, 1972).

The overall behavior, expressed by δ and λ, can be calculated in the following way: let ϕ_i be the local behavior function of the sub-systems, indexed by i. This can be expressed by $y_i = \phi_i(x_i)$ with $x_i \in \{\alpha_x\}$. If the sub-systems are connected by $\pi = \{\kappa_{ij}\}$, more than just x_i may influence the i-th sub-system. If the ϕ_i are expressible as linear operators, the linear equation system

$$[\Phi X = Y - \Phi K Y]$$

with

$$\Phi = \|\phi_i \delta_{ik}\|, \; K = \|\kappa_{ik}\| \text{ and } X = \begin{pmatrix} \cdots \\ x_i \\ \cdots \end{pmatrix}, \; i = 1, \ldots, n \tag{3}$$

$$\mathbf{Y} = (\ldots y_j \ldots), \; j = 1, \ldots, m$$

can be solved. It should be noted that there is a certain kind of degeneration with respect to structure and behavior. A certain behavior can be put into practice with a wide class of possible structures, but if a structure is given, the behavior is determined, presupposing that all classical conditions will hold.

Whether the effect of pragmatic information can be described by a change of behavior, $\Delta\phi$, and by a change of structure, $\Delta\sigma$, depends on the way the system can be described. In a very aggregated representation (i.e. black box), only a change of behavior may provide knowledge about the effect of received information. In a rather disaggregated representation (i.e. the system is described by elements, sub-systems and relations between them), the effect of pragmatic information can be stated by changes in elementary behavior, by changes in the number of elements or sub-systems (creation and nihilation) or by changes in the relation between them.[20]

In the case of black-box representation, the amount of novelty can be estimated by looking for the maximum difference d (as a general distance function) between the observed change in behavior $\Delta\phi = d(\phi_{\text{before}}, \phi_{\text{after}})$ and the expected change in behavior $\Delta\phi_r$ (the index r runs over a number of distinguishable expectations of the receiver) as

$$\textit{Novelty}\ N = \max_r d(\Delta\phi, \Delta\phi_r) \tag{4}$$

and for the case of a disaggregated representation, we have to look for the change in the structure and the elementary behavior $\Delta\sigma = d(\sigma_{\text{before}}, \sigma_{\text{after}})$, compared with the expected change (like growing), i.e. $\Delta\sigma_r$, such that

$$\textit{Novelty}\ N = \max_r \langle d(\Delta\phi, \Delta\phi_r); d(\Delta\sigma, \Delta\sigma_r)\rangle \tag{5}$$

The semicolon means a generalized notion of *and/or*, depending on concrete system descriptions. It should be noted that $d(\Delta\phi,\Delta\phi_r)$ and $d(\Delta\sigma,\Delta\sigma_r)$ are not independent, if one is able to calculate the changed overall behavior ϕ due to the observed structural change and the changed elementary behavior. Therefore, novelty can only be estimated if a concrete system description is available. Depending upon this description, the distance function d may be definable.

In the case of black-box representation, the difference between a potential (normalized) "full" or complete information, denoted symbolically as **1**, and minimum difference between observed and expected behavioral change can be used to estimate

$$\textit{Confirmation}\ C = 1 - \min_r d(\Delta\phi, \Delta\phi_r) \tag{6}$$

If only structural issues are known, the difference $d(\sigma_{\text{before}}, \sigma_{\text{after}})$ can be useful in estimating

$$\textit{Confirmation}\ C = 1 - \min_r \langle d(\Delta\phi, \Delta\phi_r); d(\Delta\sigma, \Delta\sigma_r)\rangle. \tag{7}$$

If there is no possibility ruling out behavioral issues, the first term, i.e. $d(\Delta\phi,\Delta\phi_r)$ cannot be estimated. d and d are conceived as distance measures which must be specified according to the particular system.[21] From (4)-(7) the estimation strategy depends upon the representation type of system description — in most cases the system is not overly disaggregated such that (4) and (7) would be the most preferred estimation strategies.

Up to now we have only discussed the means of getting pragmatic information from a system. A further conjecture about pragmatic information has already been mentioned: if the receiver really understands information this can only be observed by its reaction to the information. Operatively speaking — the receiver must become an emitter for the observer. Pragmatic information is not only received by managing some signal transfer — otherwise the receiver is only a channel —, but by an alteration of its behavior or its structure. This point leads us to abandon the classical view of the problem.

OPERATORS

Analogous to the operators discussed in order to describe how to obtain information about a system (or environment), we can try to define operators which describe the effect of information within the receiver.

The basic assumption can be expressed as[22]

$$N \otimes C \approx \mathbf{f}(\Delta V \otimes \Delta S) \tag{8}$$

in words: the product of confirmation C and novelty N gives rise to a change of structure ($\sigma_{before} \rightarrow \sigma_{after}$) given by a structure operator ΔS and to a change of behavior ($\phi_{before} \rightarrow \phi_{after}$), given by a behavior operator ΔV whereas the function **f** has to be estimated more precisely for particular receivers. If the assumption is true that a system can be transformed into the information which is contained in it, the reverse formulation can be conjectured: information can build up a system, starting from a pre-system or a pre-structure. Every process to establish an organization shows this fact.

Thus, the inverse function (symbolically written as far as it can be defined)

$$\mathbf{f}^{-1}(N \otimes C) \approx \Delta V \otimes \Delta S \tag{9}$$

should be investigated. The operators within the arguments of **f**, namely the structure operator and the behavior operator, represent procedures: that which has to be done in order to gain insights into the system dynamics and the structure. It has been presumed that

$$N \otimes C \neq C \otimes N$$

and

$$\Delta V \otimes \Delta S \neq \Delta S \otimes \Delta V \tag{10}$$

i.e. changing of the order of application of those operators does not lead to the same results.[23] If one looks at $\mathbf{f}^{-1}$, one can ask what kind of structural change, for instance, is given by what function $\mathbf{f}^{-1}$. The most elementary action would be to establish a new relation between at least two elements or to sever a relation between them. By analogy it should also be possible to establish new elements. If one takes this elementary act into account, one can state with a given pre-structure κ_{ij}, that another κ_{nm} has been achieved there.[24]

It could now be argued that all those elements which are able to switch from "through" to "stop" or vice versa could serve as such structural building or changing elements. But this would require an already complete network of such elements and in general a growing process cannot be explained by this interpretation. Therefore a new relation is conceived to be generated by creating a new element with such switching properties. Consequently this would require that pragmatic information is able to "create" new elements within the structure of the system concerned (i.e. within the receiver). An element has at least a minimal behavior — this elementary behavior is given by the property to make a two-state transition, if an appropriate input is applied.

If one creates a new element this requires generating a two-state configuration, i.e. a meta-stable state as already proposed in the definition of a sign: signs require a meta-stable configuration "into" which can be written, deleted and read.

Therefore it seems better to conceptualize the effect of pragmatic information, i.e. the building up of new structures or creating new elements, by a generation of something that enables the system to bring different states into existence: pragmatic information creates new attractors. Thus, structure and behavior become dynamic, time-variant entities.

The main effect of pragmatic information can be represented as a change of already "existent" systems with respect to behavior and/or structure. A structure can be changed by introducing or deleting elements, by introducing not yet existent relations between elements or sub-systems or by severing them. By the means of changing structure, the behavior can be altered due to the classical formula (3). But it is also possible to reinterpret a changed behavior (altered pattern) by a changed system surface. One of these changes could be represented again by another choice of the cut between system and environment. Since elements are sub-systems which are not further analyzed by the system describer, the effect of pragmatic information can also have the effect of building up new systems or destroying existing ones.

As far as pragmatic information is provided by changing systems and change is induced by pragmatic information, a recursive self-organizing characteristic could also be induced by pragmatic information: let the starting system be described by behavior and structure

$$\Delta_0 V \otimes \Delta_0 S(t) \tag{11}$$

Assuming that the structure is the subject of change, $\Delta_0 S(t)$ may provide the pragmatic information $N(t+1) \otimes C(t+1) = PI(t+1)$. This pragmatic information PI may be applied by the system to alter its structure such that

$$PI\,(t+1) \rightarrow \Delta V(t+1) \otimes \Delta S(t+2) \tag{12}$$

If PI and $\Delta S(t)$ are equivalent for t and t + n for a given n, a certain kind of stabilization holds, the system shows invariances and it can be conceived as a "*Ding*".

This recursive formula is either leading to convergence, i.e. the system remains the system with "new" functionality (new, but not too far away from expectation, i.e. confirmed behavior and structure) or the system is "wandering" through behaviors and structures, losing its identity and its coherence with the system history — it is changed into a completely different system. This is equivalent to the destruction of the old and the generating of a new system. It is again the semantical closure that determines what kind of change (quality, magnitude) still affects the identity of a system.

EMERGENCE OF MEANING

Emergence is closely connected to novelty and surprise. Surprising behavior can be shown only by non-classical systems (per definition) where they are observed under an endo-view. The lack of information is principally comparable with that in complex systems or in quantum mechanical systems. It is only removable by changing the view — either going from endo to exo or changing the level of description. If we take an evolutionary system as an evolving system, the change in the behavior of a single sub-system is no longer surprising if one has a theory about the overall system. But within the endo-view, only regarding (sub)systems and their surface in contrast to their environment, this may provide a surprise since a behavioral change is not within the range of expectation.

The statistical description is only a means to operationalize what can be figured out as regularities within the observed behavior. In the case of a contingent lack of information, statistics can always (in principle) be removed by giving explicit ensemble equations. There are practical reasons which prevent an estimation of this equation such that one is forced to restrict oneself to counting frequencies. In the case of a non-contingent lack of information, there is a real random behavior (not chaotic) as in quantum mechanical or thermodynamical noise. A symptom for this is the well-known non-compressibility of a description (i.e. no shorter than a real-time behavior protocol) holds.

Emergence as the appearance of properties on a higher level, not explainable or predictable from a lower level, depends on what levels are selected by the system author. Synergetics, for instance, explains the generation of structure from chaos, and bifurcation theory is another way to explain that new structures come out. Here emergence is explained by the master-slave principle or by bifurcation. But an explained emergence is not an emergence anymore. Therefore emergence is a denotation not for an ontological but for an epistemological concept — constructed to design no-knowledge.

Meaning can be conceived as an non-extensive[25] entity. Non-extensive entities as variables are difficult to handle with respect to a measurement theory: the non-additivity of meaning shows that it is an undividable entity — it cannot be divided in parts. This holds, for instance, for an understanding of the proof of Pythagoras, of a joke or of a speech act. A context-free analysis, like the method of Kalmann-filtering or exponential smoothing, cannot superpose single meanings of single results according to a superposition rule.

It is always possible to analyze time series, x(t), according to entropy or other integrable measures. If there are non-vanishing results such that for any density function $\rho(x)$

$$\int \rho(x) \ln \rho(x) dx > 0 \tag{13}$$

one can state *that* there is information, but the measure is an extensive one. If $\Delta_0 V \otimes \Delta_0 S(t) = f(N \otimes C)$ and $\Delta_0 V \otimes \Delta_0 S(t) \geq 0$, pragmatic information is delivered by the system and the meaning can be operationalized by formula (1), interpreted with η as a performance measure over $\eta = \eta(V,S)$ and $\eta_1 - \eta_2 = \eta(\Delta V, \Delta S)$. If $(N \otimes C)$ is given, i.e. N, C can be defined in a context, the context refers to a system which has already been defined. If $N \otimes C \neq 0$, a system with V and S must exist. And vice versa: if V and S exist, there must be $N \neq 0$, and $C \neq 0$ or, if V and S are stable in time, $N = 0$ and $C = 1$ or $C = 0$ and $N > 0$.

SELF-ORGANIZATION AND EVOLVING DEVELOPMENT

The evolution of emergent properties is a tautology: evolution means a certain kind of development of system characterizations within an endo-view of the subject of evolution: the individuals or the species. Emergentism is a *façon de parler* to express this, but if one moves to the description of the evolutionary system as an embedding system, evolution and emergence of "new properties" can be substituted by system dynamics within a non-classical system theory which should explain the effect of cutting and inserting structures, changes of behaviors and so on. One of the pictures with which we can describe system-system interaction is that of pragmatic information. Due to (8) and (9) the pragmatic information can be substituted by the system emitting it and a system can be substituted by the emitted pragmatic information.

Developing systems (evolution is probably the endo-metaphor) are such that systems are self-interacting with the help of pragmatic information. A little change to the system provides pragmatic information which gives rise to a new change. It is irrelevant whether such a recursion process involves more than one system, because a cycle with changing systems is also expressible as an interaction between two sub-systems.

Self-organization can be stimulated by pragmatic information, but it can also be generated as a result of pragmatic information. On the other hand, self-organization can be characterized by the feature that this process is able to produce pragmatic information. The possibility of stating meaningfulness and to observe self-organization seem to be two sides of the same coin. Self-structuring is then the building up of new structures by internal sub-systems using the recursion with pragmatic information. The creating of new elements is an expression within the endo-view, the choice of activated elements from a set of present elements describes the same phenomena within the exo-view.

Development recurs to the inside view. If something is developing, it is always an evolution of something, but of what? Evolved systems (generated by evolution processes) and evolving systems are epistemologically different. But with respect to the exo-view they can be conceived as hierarchically nested: evolving systems contain what they are evolving.

The model of pragmatic information suggests that the whole system is evolving. A developing system is conceived as an accumulation of pragmatic information which is meaningful within it. Any accumulation of meaningfulness corresponds to a stabilization of the context and, with respect to the whole system, to a stabilization of the whole system.

It may be that evolutionary system theory presupposes what it wants to explain.

Brandenburgische Technische Universität Cottbus, University of Ulm
Ulm, Germany

NOTES

[1] With all the difficulties in circular definitions of this concept, cf. Küppers (1991).

[2] Adequacy means here: the chosen system description is adequate to the user of the description if it is more economic than others. Less variables or less complicated relations between the variables may serve the same purpose (for instance control, prediction or explanation). Therefore model simplification is an important issue in system theory.

[3] Diettrich calls an object an invariance property of its movement (private communication, Ulm, 1995 cf. Diettrich, 1989). One could argue that the invariances of a system operator (i.e. to describe something as a system) are: integrity, identity and continuity. All this is subsumed under the term "existence" of an object, described as a system.

[4] A survey is given by Gernert in Kornwachs & Jacobi (1996).

[5] The concept of "classical" has been discussed widely by three essentials: deterministic behavior, locality (of theory) and predictability. The non-classical concept is characterized by violations of at least one of these essentials. Non-predictability exists in chaotic systems for example, non-deterministic behavior exists in stochastic (random) and in quantum mechanical systems, non-local behavior (i.e. behavior which cannot be assigned to a finite chain of state-to-state transitions in a finite dimensional real state space) even exists in quantum mechanics (cf. Kornwachs & von Lucadou, 1989). Within a non-classical system theory, it is expected that complex systems show comparable violation of these three essentials.

[6] This is shown more explicitly in Kornwachs (1992).

[7] Which need not be quantitatively measured.

[8] As behavior, structure, novelty, confirmation (see Kornwachs, 1988), autonomy and reliability (see Kornwachs & von Lucadou, 1989).

[9] Conceptually there is a difference between real systems and formal systems. Formal systems (like mathematical tools or descriptions) can be used to describe objects or processes *as* systems. Therefore the notion of system is perspectivistic: there is no way to make any ontological statement beyond this descriptive level. As a consequence, information is a descriptive and perspectivistic concept, too.

[10] An effect of information can substitute an effect of other information — presupposing a defining system is available (for instance the effect of so-called macros in computer software).

[11] One should avoid posing essentialistic questions as to what information really *is*. One should rather try to describe how certain kinds of information cause effects from and within systems.

[12] Hinderer has modified this metaphor into "systems are living on surfaces of systems". Due to the mentioned substitution there is no contradiction with the first metaphor.

[13] Structure in a system theoretical sense is given by relations (i.e. connections) between elements or subsystems; a behavior is given by the time dependent dynamics of all relevant variables. By changing a structure, the behavior of a system can also be changed. By changing behavior, nothing is said about the structure, if only a black-box representation of the system is available.

[14] Marko (1996) has shown that conditioning actions on a controllable information source can change characteristic measures of signal and state transition probabilities like entropy. This can be appreciated as an initial good approach to make these ideas more quantifiable. The term conditioning action is closely related to the biological term of formative influence.

[15] Whereas the concept of space should be conceptualized in a very generalized sense: not only the three-dimensional ordinary space, but also (multidimensional) state-spaces (or phase-spaces) may be taken.

[16] This holds only for "simple" linear types of behavior in the elements or the subsystems.

[17] As proposed in Kornwachs (1990).

[18] Two cases can be discussed: 1. The information from a sender may substitute the sender itself for a receiver. This means that we substitute the presence of a system by information about it. 2. The information for a receiver can be used to (re-)construct a system by using the receiver as a pre-system. This inverse operation can be used to construct a system by applying information, i.e. the substitution of information by a system. To "erect" a system presupposes a pre-system, which is able to be a "condensation nucleus" for structural growth. E.U. von Weizsäcker (1974) talks about a presystem which represents the least amount of confirmation necessary for the pragmatic information to be effective. Pre-system and system to be erected can be considered as the same "thing": a receiver changes its own system structure and its behavior by reacting to the information received. Every organization can be taken as a model for this process.
[19] The basic lines of a theory of pragmatic information can be found in von Weizsäcker (1974); Kornwachs & von Lucadou (1984, 1989, see further references there).
[20] A mathematical model for changing relations has been given by Gernert's investigating graphs (1981).
[21] For special systems one can compare probability distributions, using Hamming distance measures (cf. Kornwachs, 1990).
[22] Here, ⊗ expresses the product of two operators.
[23] This non-commutativity of system operators has been discussed elsewhere in Kornwachs (1990).
[24] As an example for i=n and j=m, the transition may have happened

$$\kappa_{ij} = \begin{pmatrix} 1 & 0 & 0 \\ 0 & 1 & 0 \\ 0 & 0 & 1 \end{pmatrix} \Rightarrow \kappa_{ij}{}' = \begin{pmatrix} 1 & 1 & 0 \\ 0 & 1 & 1 \\ 1 & 0 & 1 \end{pmatrix}$$

i.e. a transitive closure has been introduced and the distance $\boldsymbol{d}$ (k_{ij}, k_{nm}) - presupposing a distance $\boldsymbol{d}$ is definable, for instance by graph grammars (cf. Gernert, 1981, 1996) — as ΔS is given by two possibilities: the structure of the three elements has been changed in such a way that a new relation (in the sense of a local relation by G. Klir, 1969) can be expressed, or a hidden element between element one and three has changed its behavior in such a way that it has lost its opaqueness and can serve as a perfect channel.
[25] System variables (input, state, output) are expressed as quantitative entities if possible — according to the leading concept of physics these are extensive entities. They can be characterized by congruence, order relations, positivity, Archimedic axiom, additivity and a homomorhism to the body of real numbers with respect to + and > (cf. Carnap, 1966).

REFERENCES

Atmanspacher, H., 1989, The Aspect of Information Production in the Process of Observation, *Foundation of Physics*, vol. 19, no. 5, pp. 553-577.

Atmanspacher, H.; Dalenoort, G. (eds.), 1994, *The Inside versus Outside. Endo- and Exo-Concepts of Observation and Knowledge in Physics, Philosophy and Cognitive Science*, Springer, Heidelberg.

Atmanspacher, H.; Kurths, J.; Scheingraber, H.; Wackerbauer, R.; Witt, A., 1992, Complexity and Meaning in Non-linear Dynamical Systems, *Open Systems and Information Dynamics*, vol. 1, no. 2, pp. 269-289.

Atmanspacher, H.; Scheingraber, H., 1990, Pragmatic Information and Dynamical Instabilities in a Multimode Continuous-Wave Dye Laser, *Canadian Journal of Physics*, vol. 68, pp. 728-737.

Atlan, H., 1992, Ends and Meaning in Machine-like Systems, in *New Perspectives on Cybernetics. Self Organization, Autonomy and Connectionism*, G. Van de Vijver (ed.), pp. 35-48.

Carnap, R., 1966, *Philosophical Foundation of Physics*, Basic Books, New York.

Dalenoort, G. (ed.), 1989, *The Paradigm of Self-Organization I*, Gordon and Breach Publ., New York.

Diettrich, O., 1989, *Kognitive, organische and gesellschaftliche Evolution*, Parey, Berlin, Hamburg.

Gernert, D., 1981, Distance or Similarity Measures which Respect the Internal Structure of Objects, *Methods of Operation Research*, vol. 43, pp. 329-335.

Gernert, D., 1985, Measurement of Pragmatic Information, *Cognitive Systems*, vol. 1, pp. 169-176.

Gernert, D., 1994, The Physical Basis of Self-organization, in *The Paradigm of Self-Organization II*, G. Dalenoort (ed.), Gordon and Breach, London, pp. 33-40.

Gernert, D., 1996, Pragmatic Information as a Unifying Concept, in *Information — New questions to a Multidisciplinary Concept*, K. Kornwachs and K. Jacobi (eds.), Akademie, Berlin, pp. 147-163.

Haken, H., 1988, *Information and Self-Organization. A Macroscopic Approach to Complex Systems*, Springer, Berlin, Heidelberg.

Klir, G.J., 1969, *An Approach to General System Theory*, Van Nostrand, New York.

Klir, G.J., 1985, *The Architecture of Problem Solving*, Plenum Press, New York.

Kornwachs, K., 1987, *Offene Systeme und die Frage nach der Information*, Habilitationsschrift, Universität Stuttgart 1987.

Kornwachs, K., 1988, Cognition and Complementarity, in *Nature, Cognition and Systems I*, M. Carvallo (ed.), Kluwer Akad. Publ., Amsterdam, pp. 95-127.

Kornwachs, K., 1990, Reconstructability Analysis and its Re-Interpretation in Terms of Pragmatic Information, in *Computer Aided Systems Theory — Eurocast '89*, F. Pichler and R. Moreno-Diaz (eds.), Springer, *Lecture Notes in Computer Science*, no. 140, Berlin, pp. 170-181

Kornwachs, K., 1992, Information and der Begriff der Wirkung, in *Physik and Informatik - Informatik and Physik*, D. Krönig & M. Lang (eds.), Springer, *Informatik-Fachberichte*, no. 306, Berlin, pp. 46-56.

Kornwachs, K., 1993, Information und Wechselwirkung, in *Mensch-Computer-Kommunikation*, H.D. Böcker, W. Glatthaar and Th. Strothotte (eds.), Springer, Berlin, pp. 263-173.

Kornwachs, K., 1996, Pragmatic Information and System Surface, in *Information — New Questions to a Multidisciplinary Concept*, K. Kornwachs and K. Jacobi (eds.), Akademie, Berlin, pp. 163-185.

Kornwachs, K.; Hinderer, W., 1990, Physik der Symbolverarbeitung, *Newsletter der Deutschen Gesellschaft für Systemforschung*, vol. 2, pp. 25-29.

Kornwachs, K.; Jacobi, K. (eds.), 1996, *Information — New Questions to a Multidisciplinary Concept*, Akademie, Berlin.

Kornwachs, K.; von Lucadou, W., 1984, Komplexe Systeme, in *Offenheit-Zeitlichkeit-Komplexität*, K. Kornwachs (ed.), Campus, Frankfurt a.M., New York, pp. 110-165.

Kornwachs, K.; von Lucadou, W., 1989, Open Systems and Complexity, in *The Paradigm of Self-Organization I*, G. Dalenoort (ed.), Gordon and Breach Publ., New York, pp. 123-145.

Küppers, B.O., 1991, Information and the Origin of Life, MIT Press, Cambridg (MA).

Marko, H., 1996, Causal Information Theory, in *Information — New Questions to a Multidisciplinary Concept*, K. Kornwachs and K. Jacobi (eds.), Akademie, Berlin, pp. 63-86.

Mesarovic, M.D., 1972, A Mathematical Theory of General Systems, in *Trends in General System Theory*, G.J. Klir (ed.), Wiley Interscience, New York, pp. 251-269.

Pattee, H.H., 1977, Dynamic and Linguistic Modes of Complex Systems, *International Journal for General Systems*, vol. 3, pp. 259-266.

Rosen, R., 1986, Causal Structures in Brain and Machines, *International Journal for General Systems*, vol. 12, pp. 107-116.

Shannon, C.; Weaver, W., 1949/1969, *The Mathematical Theory of Communication*, Urbana, Chicago, London, first published in *Bell System Techn. Journal*, 1948, vol. 27, pp. 379-423.

Van de Vijver, G. (ed.), 1992, *New Perspectives on Cybernetics. Self-Organization, Autonomy and Connectionism*, Synthese Library, vol. 220, Kluwer, Dordrecht.

Von Lucadou, W., 1994, Wigners's Friend Revitalized?, in *The Inside versus Outside. Endo- and Exo-Concepts of Observation and Knowledge in Physics, Philosophy and Cognitive Science*, H. Atmanspacher and G. Dalenoort (eds.), pp. 369-385.

Weizsäcker, E.U. von, 1974, Erstmaligkeit und Bestätigung als Komponenten der Pragmatischen Information, in *Offene Systeme I*, E.U. von Weizsäcker (ed.), Klett, Stuttgart, pp. 82-113.

DANIEL M. DUBOIS

EMERGENCE OF CHAOS IN EVOLVING VOLTERRA ECOSYSTEMS

INTRODUCTION

This paper is an attempt to give new avenues of research and development in the mathematical modelling of evolving systems. The best example of a natural evolution deals with the successive appearance, transformation and extinction of biological species on earth. The fundamental causes of such changes are not yet well established. Many theories were proposed to explain evolution. The first two were Lamarckism and Darwinism. Lamarck thought that the function creates the organ, but this cannot explain the memorization of the new characters in the genetic code. Darwin considered two complementary processes: on the one hand, an evolution of species by some random mutations and, on the other, a natural selection of the species which fit best their environment.

I propose to consider any explanation of evolving systems in this regard as dependent on the human interpretation of the meaning of events. Lamarck was speaking of "function" and Darwin of "fitting". Meaning involves final cause, as was also stated by F. Baillly (1991): "Le rôle de la signification comme principe organisateur est primordial. Or la signification n'a, elle, de pertinence que relativement à une représentation finalisée de l'action". Therefore, we propose to see the emergence of meaning as an evolving system.

Stanley Salthe (1993) states that "there cannot be evolution without a development to carry it". He defines development as "predictable irreversible change" and evolution as "the irreversible accumulation of historical information".

An evolving system is a system becoming more and more complex. This complexity emerges from the dynamics of the system. One of the main questions in this regard is whether the laws of evolution of artificial systems are similar to the natural ones. The main difference is that artificial systems are designed by men, that is to say by conscious minds with meaning and anticipatory possibilities (final causes). This was analyzed and described by Aristotle's four causes. Consider the building of a house. The material cause is the set of materials, the formal cause is the plan, the efficient cause is the making action, and the final cause is the purpose. Different interpretations of this causality were proposed (see for example, Rosen, 1986; Von Glasersfeld, 1991; Salthe, 1993). Salthe considers that Newtonian mechanics deals only with efficient cause and biology with material cause. The natural laws of evolution are also interpreted by men with their meaning and anticipation. The danger is to look at natural evolution as a man-made system.

This paper considers the mathematical modelling of ecological systems as a tool to define the meaning of mechanisms of change. I consider discrete dynamic systems described by finite difference equations. In this framework, change can be defined by two derivatives: a backward derivative related to the past and a forward derivative related to the future. I will consider very simple ecological systems represented

G. Van de Vijver et al. (eds.), Evolutionary Systems, 197–214.

by the growth of biological populations constrained by a quantity of nutrients. Indeed, the conservation law of the total materials in the ecosystem is a constraint, noted "c", given by the sum of the nutrients and the populations. General rules are obtained by numerical simulations. When the constraint c is low, no population can survive; when the constraint c increases populations emerge and when the constraint c becomes large, bifurcations and then chaos occur. In space-time models, patterns emerge from diffusive chaos. It is shown that chaos can be controlled by incursion. Finally a simple model of hyperincursion shows an ecosystem exhibiting multiple solutions. A survey of incursion and hyperincursion is given in Dubois (1996a-b).

MATHEMATICAL MODELS OF EVOLVING ECOSYSTEMS

How to detect a change of value of a variable x(t)? In measuring two successive values at time t and t-Δt, where Δt is the time interval:

$$(1a)\ \Delta_b x(t)/\Delta t = (x(t) - x(t-\Delta t))/\Delta t$$

which is a backward derivative, that I called a *recursive derivative* (Dubois, 1995). Indeed, it is impossible to measure the value of a variable at a future time t+Δt step at the current present time t step. This is true for any physical, biological or human system. If a system is evolving, it will change the value of any variable x(t) at the present time t step to another value at the future time t+Δt step by performing an action:

$$(1b)\ \Delta_f x(t)/\Delta t = (x(t+\Delta t) - x(t))/\Delta t$$

which is the classical definition of the forward derivative, which I called an *incursive derivative* (Dubois, 1995). Indeed, it is impossible to change the already realized past and present values of any variable, only the future value at time t+Δt step can be changed. If a change has occurred by a realized action during the time interval (t, t+Δt), the forward derivative at time t will be equal to the backward derivative at time t+Δt:

$$(1c)\ \Delta_b x(t+\Delta t)/\Delta t = (x(t+\Delta t) - x(t))/\Delta t = \Delta_f x(t)/\Delta t$$

which can be measured.

It must be pointed out that the mathematical analysis of derivable continuous systems defines the derivative for a time interval Δt tending to zero, when the backward and the forward derivatives are identical. But, for discrete or non derivable continuous systems, the backward and the forward derivatives have different values which must be taken into account. Numerical simulations by computer of mathematical models are only performed with discrete equations. Let us now give some examples in ecology for the growth of populations.

Environmental constraint for the emergence of a population

All individuals are built with the same fundamental chemical materials. For example, let us consider explicitly the nutrients N(t) for the growth of a population X(t) with a Malthusian growth, the discrete equations are given by

$$\text{(2a)}\ N(t+\Delta t) = N(t) - a.\Delta t.N(t).X(t) + b.\Delta t.X(t)$$
$$\text{(2b)}\ X(t+\Delta t) = X(t) + a.\Delta t.N(t).X(t) - b.\Delta t.X(t)$$

where a is the birth rate, b the death rate of the population X(t), and the growth rate is a function of the quantity of nutrients N(t). From the conservation law of materials, it is assumed that the dead population is recycled to nutrients so that the total quantity of materials is a constraint parameter:

$$\text{(2c)}\ N(t+\Delta t) + X(t+\Delta t) = N(t) + X(t) = c$$
$$\text{(2d)}\ N(t) = c - X(t)$$

The total quantity of materials is a global constraint which gives rise to a non-linear saturation in the growth rate of the population. Indeed, in replacing N(t) given by eq. 2d in eq. 2b, we obtain the following equation

$$\text{(3a)}\ X(t+\Delta t) = X(t) + a.c.\Delta t.X(t) - a.\Delta t.X(t)^2 - b.\Delta t.X(t)$$

which is a form of the Pearl-Verhulst map. For a constant constraint $c = 1$ and parameters $\Delta t = b = 1$, eq. 3a is

$$\text{(3b)}\ X(t+1) = a.X(t).(1 - X(t))$$

which is the well-known Pearl-Verhulst map which gives rise to chaos (May, 1976). Populations with a low birth rate $0 < a < 1$, disappear $X(t) = 0$, populations with $1 < a < 3$ have a stable stationary state $X(t) = 1 - 1/a$ and populations for which $3 < a < 4$, show bifurcations and then chaotic behavior. It is interesting to consider a population in function of different constraint conditions c in taking $\Delta t = a = b = 1$ in eq. 3a, which gives

$$\text{(3c)}\ X(t+1) = c.X(t).(1 - X(t)/c)$$

Figure 1 gives the numerical simulation for values of c varying from 0 to 4 with an initial condition $X(0) = 0.1$. For $c < 1$, there is no emergence of the population. For $1 < c < 3$, the population emerges and stabilizes at the stationary state $X(t) = c - 1$. For $c > 3$, the stationary state becomes unstable and the population shows bifurcations and then chaos. This means that under a certain threshold of nutrients, no population can emerge and above another threshold of nutrients, bifurcations and chaos occur. Between these two thresholds, the population emerges and stabilizes to a fixed point. In the chaotic zone, the determinism doesn't give a prediction of the future of the system from its initial conditions. In fact the chaotic system loses the memory of its past. But it is important to point out that an averaged value, or the bounds in which the variable can take its values can be known: it is only the precise values at the successive time steps which are not predictable. The local behavior is unpredictable while the global behavior is predictable. Bifurcation could be a mechanism for the emergence of new populations and chaos, a mechanism for a quick adaptation to variations of the environment. Let us show how a bifurcation of one population could be a mechanism for the emergence of two populations.

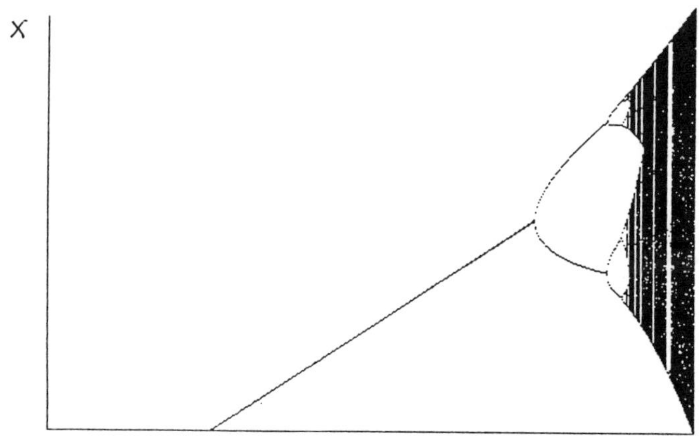

Two competing populations

For a system with two species X_1 and X_2, the equations with nutrients N are

$$(4a)\ N(t+\Delta t) = N(t) + \Delta t.(b_1.X(t) + b_2.Y(t) - a_1.N(t).X(t) - a_2.N(t).Y(t))$$
$$(4b)\ X_1(t+\Delta t) = X_1(t) + \Delta t.(a_1.N(t).X_1(t) - b_1.X_1(t))$$
$$(4c)\ X_2(t+\Delta t) = X_2(t) + \Delta t.(a_2.N(t).X_2(t) - b_2.X_2(t))$$

with a and b the birth and death rates respectively. The dead populations are recycled to nutrients. So, the total quantity of materials is a constraint parameter noted c

$$(4d)\ N(t+\Delta t) + X_1(t+\Delta t) + X_2(t+\Delta t) = N(t) + X_1(t) + X_2(t) = c$$
$$(4e)\ N(t) = c - X_1(t) - X_2(t)$$

In taking $a_1 = a_2 = a/2$ and $b_1 = b_2 = b/2$, the addition of the two populations $X(t) = X_1(t) + X_2(t)$ by eqs. 4b-c gives the previous eq. 2b for one population. It means that the population X(t) can be split into two populations $X_1(t)$ and $X_2(t)$ with the same birth and death rates. Let us see what happens when these rates change. In replacing eq. 4e of N(t) in eqs. 4b-c, we obtain

$$(5a)\ X_1(t+\Delta t) = X_1(t) + \Delta t.[a_1.c.X_1(t).(1 - X_1(t)/c) - a_1.X_1(t).X_2(t) - b_1.X_1(t)]$$
$$(5b)\ X_2(t+\Delta t) = X_2(t) + \Delta t.[a_2.c.X_2(t).(1 - X_2(t)/c) - a_2.X_1(t).X_2(t) - b_2.X_2(t)]$$

where the effective growth rate of the populations shows a Pearl-Verhulst saturation and an effective competition term between the two populations appears as $a.X_1(t).X_2(t)$ as in the Volterra model. It is important to notice that in fact the two populations do not really compete. It is the global constraint on the total quantity of materials which creates such an effect. Depending on the values of the parameters, different behaviors occur.

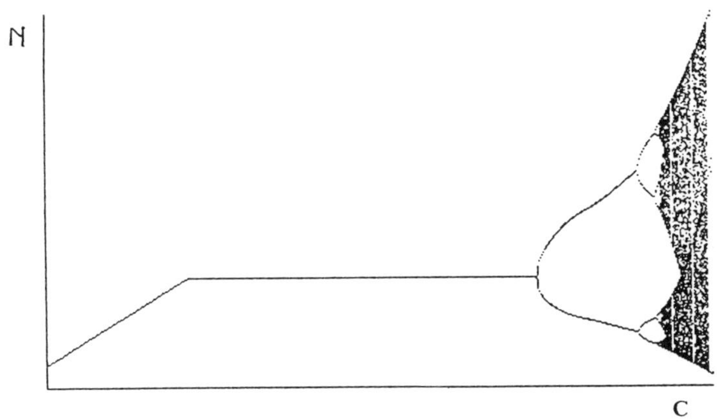

Figure 2a

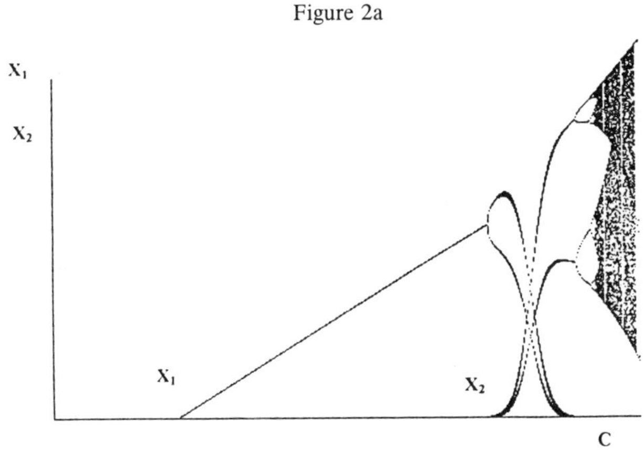

Figure 2b. Figures 2a-b show the numerical simulations of eqs. 4a-b-c, with $a_1 = 1$, $a_2 = 0.95$, $b_1=1$, $b_2 = 0.955$, $\Delta t = 1$ and the global constraint c varying from 0.2 to 4. Figure 2a is the bifurcation diagram of the nutrients. Figure 2b shows that the population X_1 emerges for c = 1, increases, bifurcates and becomes extinct; when the first population bifurcates, the second population X_2 emerges with a cascade of bifurcations and then becomes chaotic. There is a narrow zone for the global constraint in which the first population coexists with the second.

Prey predator populations

Now, let us show the influence of the nutrients in the dynamics of predator-prey populations and then two competing populations with the Volterra model. For a system with a prey X and a predator Y, the equations with nutrients N are

$$(6a)\ N(t+\Delta t) = N(t) + \Delta t.(d.Y(t) - a.N(t).X(t) + b_1\, X(t))$$
$$(6b)\ X(t+\Delta t) = X(t) + \Delta t.(a.N(t).X(t) - b.X(t).Y(t) - b_1\, X(t))$$
$$(6c)\ Y(t+\Delta t) = Y(t) + \Delta t.(b.X(t).Y(t) - d.Y(t))$$

where

$$(6d)\ N(t+\Delta t) + X(t+\Delta t) + Y(t+\Delta t) = N(t) + X(t) + Y(t) = c$$
$$(6e)\ N(t) = c - X(t) - Y(t)$$

The total quantity of materials is a constraint, that is to say the dead populations are recycled to nutrients. In replacing eq. 6e of N(t) in eqs. 6b-c, we obtain

$$(7a)\ X(t+\Delta t) = X(t) + \Delta t.[a.C.X(t).(1 - X(t)/C) - (a + b).X(t).Y(t) - b_1\, X(t)]$$
$$(7b)\ Y(t+\Delta t) = Y(t) + \Delta t.(b.X(t).Y(t) - d.Y(t))$$

where the effective growth rate of the prey shows a Pearl-Verhulst saturation and the effective predation in the prey is enhanced by a. Depending on the values of the parameters, different behaviors occur.

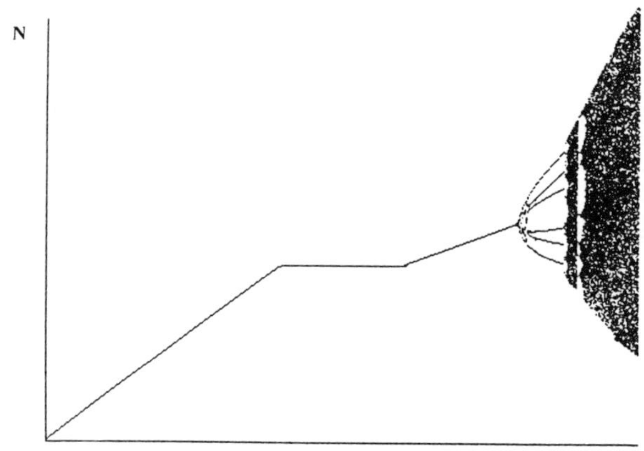

Figure 3a. Figures 3a-b-c show numerical simulations of eqs. 6a-b-c with $a = b = d = 1$, $b_1 = 2$, $\Delta t = 1$ for the global constraint c varying from 0.03 to 5. For low values of the constraint c, only the prey exists as shown in Figure 3b in taking all the nutrients, equal to zero as given in Figure 3a. Then for $c > 1$, the predator emerges as shown in Figure 3c and the quantity of nutrients is a non null constant until $c = 2$. For a greater constraint c, the nutrients and the two populations exhibit a chaotic behavior. Here also we can conclude to a general rule: two threshold effects appear at low and high values of the constraint with the emergence of a population and chaos.

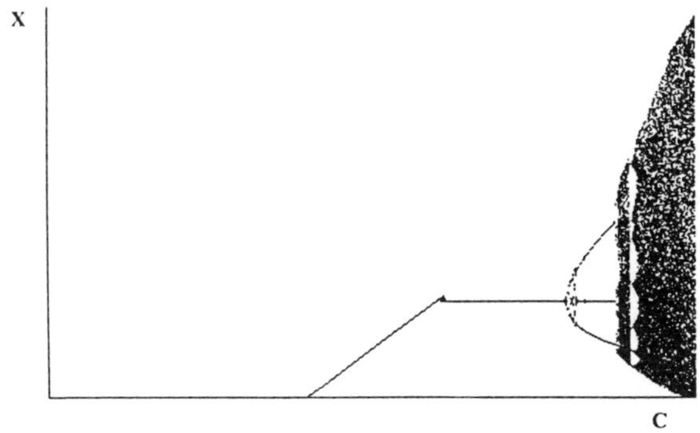

Figure 3b

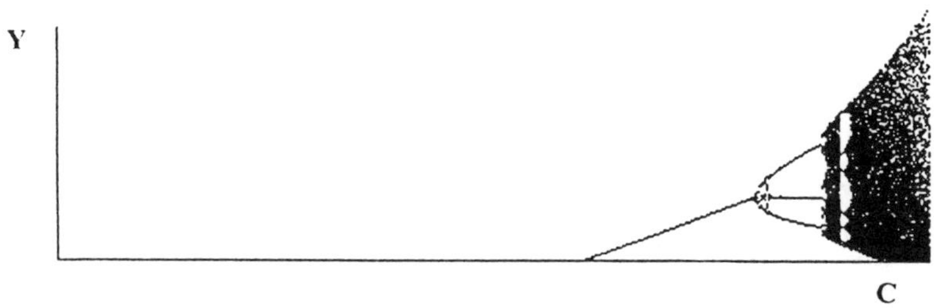

Figure 3c

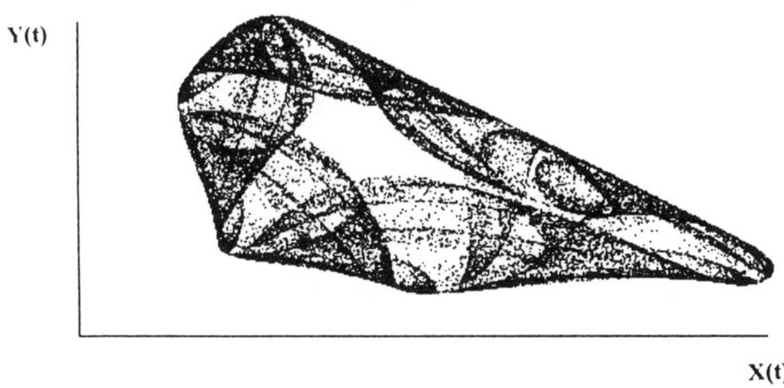

Figure 3d. Shows the prey-predator chaotic populations in the phase space (X(t), Y(t)) for c = 4.465.

DIFFUSIVE CHAOS IN SPACE-TIME ECOLOGICAL SYSTEMS

Until now, only the time dynamic of populations was considered. In adding a space diffusion in the populations, new effects appear. Turing (1952) proposed a morphogenesis theory from "diffusive instability". Dubois (1975, 1977, 1979, 1981) showed that space-time structures called patchiness appear in marine plankton populations (phytoplankton and zooplankton). The diffusion is due to water turbulence. Let us show a new mechanism of space-time patchiness by what I call "diffusive chaos" (Dubois, 1996c).

Diffusive chaos with two competing populations

With a space diffusion, eqs. 4a-b-c are written

(8a) $N(s,t+\Delta t) = N(s,t) + \Delta t.(b_1.X(s,t) + b_2.Y(s,t) - a_1.N(s,t).X(s,t) - a_2.N(s,t).Y(s,t)) + \Delta t.D.(N(s-\Delta s,t) - 2.N(s,t) + N(s+\Delta s,t))/\Delta s^2$

(8b) $X_1(s,t+\Delta t) = X_1(s,t) + \Delta t.(a_1.N(s,t).X_1(s,t) - b_1.X_1(s,t)) + \Delta t.D.(X_1(s-\Delta s) - 2.X_1(s,t) + X_1(s+\Delta s))/\Delta s^2$

(8c) $X_2\ (s,t+\Delta t) = X_2(s,t) + \Delta t.(a_2.N(s,t).X_2(s,t) - b_2.X_2(s,t)) + \Delta t.D.(X_2(s-\Delta s,t) - 2.X_2(s,t) + X_2(s+\Delta s,t))/\Delta s^2$

where the nutrients N(s,t), and the two populations $X_1(s,t)$, $X_2(s,t)$ are functions of the space s and the time t, and D is a diffusion parameter.

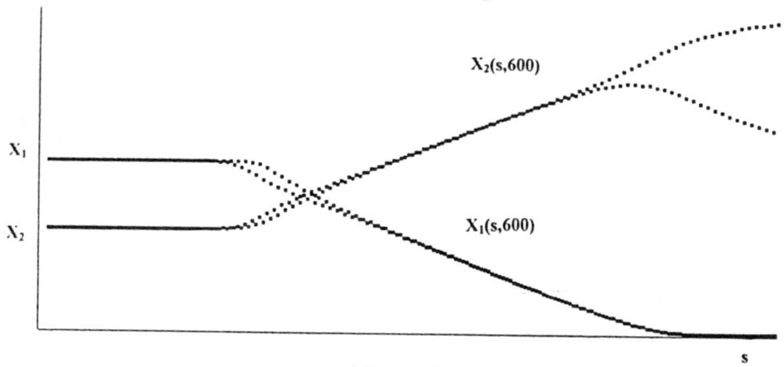

Figure 4a

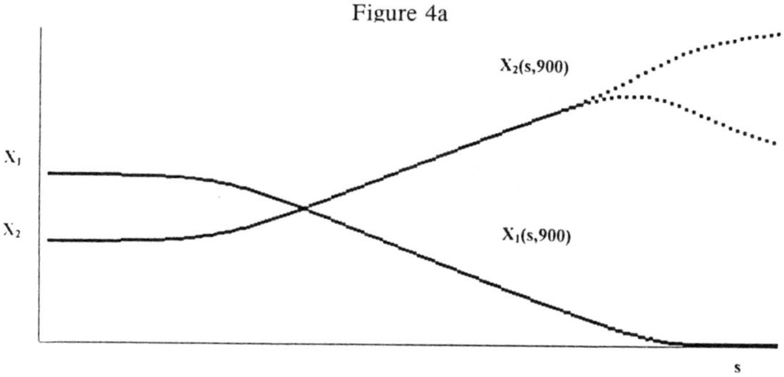

Figure 4b

Figures 4a-b show the numerical simulations of eqs. 8a-b-c with $a_1 = b_1 = 1$, $a_2 = b_2 = 1.5$, $\Delta t = \Delta s = 1$, the constraint c = 2.1 and D = 0.1 + 0.15.s/200, where s = 1 to 200 (this means that the diffusion increases along the space axis). Initial conditions are given by N(s,0) = 1.9, $X_1(s,0) = 0.1$, $X_2(s,0) = 0.1$ for all s = 1 to 200 except for s = 190: $X_1(190,0) = 0.05$ and $X_2(190,0) = 0.15$. This means that the initial conditions are spatially homogeneous except a small perturbation in s = 190. Without the perturbation, the system stabilizes to a fixed point where the two populations co-exist. With the perturbation, bifurcations appear and a space structure of the populations emerges.

Figure 4a shows the space-dependent populations at time t = 600: for low values of the diffusion, the two populations coexist; for intermediate values of the diffusion, bifurcations appear and the most important population decreases and the less important one increases in magnitude; for large diffusion, one population becomes extinct and the other shows a bifurcation (as with the time simulation). It is really important to notice that without space diffusion, the populations coexist and no bifurcation occurs because the constraint $c = 2.1$ is not great enough. Thus, diffu-sion is the cause of the emergence of bifurcations and the extinction of one popula-tion: I call this new phenomenon "diffusive chaos". In fact other simulations show chaotic behaviors by diffusion (Dubois, 1996c). I wanted to point out with this simulation another interesting phenomenon, given in Figure 4b, which is the stationary space-time populations along the space variable at time $t = 900$. It is well-shown that a bifurcation is present, only when one population becomes extinct. For low values of the diffusion the bifurcations vanish: these were transient effects leading the system to a space structure. The cause of this structure, the bifurcations, disappears when the system does not further evolve, but is in a developmental dynamic. So it can be concluded that bifurcations and chaos play a central role in evolving systems.

Diffusive chaos with prey-predator populations

With a space diffusion, eqs. 6a-b-c (with $b_1 = 0$) are written

$$\begin{aligned}
(9a)\ N(s,t+\Delta t) &= N(s,t) + \Delta t.(d.Y(s,t) - a.N(s,t).X(s,t)) \\
&\quad + \Delta t.D(s).(N(s-\Delta s,t) - 2.N(s,t) + N(s+\Delta s,t))/\Delta s^2 \\
(9b)\ X(s,t+\Delta t) &= X(s,t) + \Delta t.(a.N(s,t).X(s,t) - b.X(s,t).Y(s,t) \\
&\quad + \Delta t.D(s).(X(s-ds) - 2.X(s,t) + X(s+\Delta s))/\Delta s^2 \\
(9c)\ Y(s,t+\Delta t) &= Y(s,t) + \Delta t.(b.X(s,t).Y(s,t) - d.Y(s,t)) \\
&\quad + \Delta t.D(s).(Y(s-\Delta s,t) - 2.Y(s,t) + Y(s+\Delta s,t))/\Delta s^2
\end{aligned}$$

where the nutrients $N(s,t)$, and the two populations $X(s,t)$, $Y(s,t)$ are functions of the space s and the time t and D is a diffusion parameter. Figures 5a-b show the numerical simulations of eqs. 9a-b-c with $a = 0.2$, $b = d = 1$, $\Delta t = \Delta s = 1$, the constraint $c = 2.2$ and the diffusion is constant $D(s) = 0.35$. Initial conditions are given by $N(s,0) = 2$, $X_1(s,0) = 0.1$, $X_2(s,0) = 0.1$ for all $s = 1$ to 200 except for $s = 100$: $Y(100,0) = 0.05$. The boundary conditions are periodic. This means that the initial conditions are spatially homogeneous except a small perturbation in $s = 100$ for the predator. For small values of the diffusion, the system remains spatially homogeneous, but for a larger diffusion ($D = 0.35$) a space-time structure of the populations emerges as shown in Figure 5a at time $t = 300$. This space-time structure is called patchiness in plankton populations (Dubois, 1975). Figure 5b shows the patchiness along the space axis at time $t = 600$. The system never reaches a homogeneous space state. This patchiness is due to the diffusion, what I call "diffusive chaos" (See also Dubois, 1996c). Indeed there is a sensitivity to initial conditions as in deterministic chaos. This means that the deterministic patchiness is unpredictable. It is remarkable to see that the space heterogeneity is rather smooth and not chaotic. But the set of all the values presents a chaotic pattern.

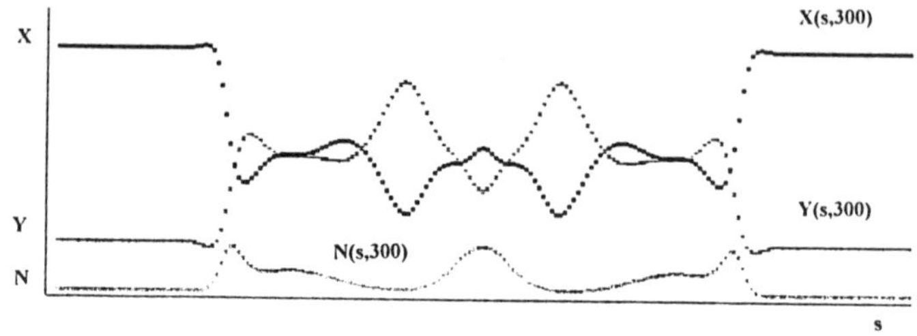

Figure 5a

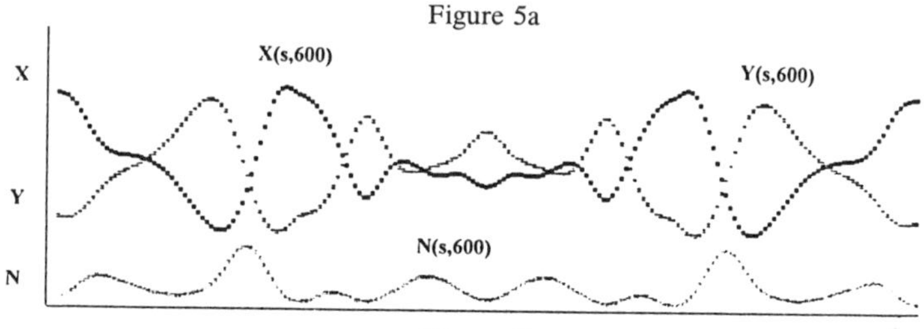

Figure 5b

INCURSIVE CONTROL OF CHAOS

If chaos is essential to explain evolving systems, it seems also necessary that these systems have some possibilities to control it. Let us give a simple example with one population. Figure 6a shows the simulation of chaos for eq. 3b with $\Delta t = b = 1$ in function of a from 0 to 4. In defining a backward derivative given by eq. 1a for eq. 2a of the nutrients, we obtain

(10a) $N(t) = N(t-\Delta t) - a.\Delta t.N(t).X(t) + b.\Delta t.X(t)$
(10b) $X(t+\Delta t) = X(t) + a.\Delta t.N(t).X(t) - b.\Delta t.X(t)$

and

(10c) $N(t) + X(t+\Delta t) = N(t-\Delta t) + X(t) = c$
(10d) $N(t) = c - N(t+\Delta t)$

the transformation of the nutrients to the population takes a time step Δt. An intermediate step must be taken into account to satisfy the conservation of materials: this is the case with enzymatic chemical reactions in the Michaelis-Menten model. In replacing N(t) in eq. 10b with $c = b = \Delta t = 1$

(11a) $X(t+1) = a.X(t).(1 - X(t+1))$

the incursive Pearl-Verhulst equation (Dubois, 1992) is obtained. This means that the future value of the population X(t+1) is a function of itself at the present time t and at its future value itself. It is impossible to control a system with future external inputs at the present time. The philosophical conclusion of this fact is that the final cause of Aristotle must be an actual cause which cannot be always transformed to another cause (Dubois, 1996a-b). But it is possible in considering a self-referential system. Indeed, it is possible to transform the incursive eq. 11a to the following recursive equation

$$(11b)\ X(t+1) = a.X(t)/(1 + a.X(t))$$

where the growth rate is now of the Michaelis-Menten-Monod type. In this case, populations with a birth rate $0 < a < 1$, disappear, and populations with $1 < a < 4$ show a stable stationary state $X(t) = 1 - 1/a$. No more bifurcations and chaos occur as shown in Figure 6b. At the stationary state given by $X(t+1) = X(t)$, this equation is similar to the Pearl-Verhulst equation. What is the difference? The incursion transforms the unstable state $X(t) = 1 - 1/a$ for $a > 3$ to a stable state. In replacing X(t+1) in the growth rate of eq. 11a by eq. 11b, we have

$$(11c)\ X(t+1) = a.X(t).(1 - a.X(t)/(1+a.X(t))$$

So, now the saturation term is recursive and represents an anticipation, a final cause. The evolution equation (11b) is well-seen as a part of itself in eq. 11c. This means that a part of a system can impress the behavior of whole the system to its image: this is actually the role of the genetic code DNA. In fact X(t+1) in the saturation rate of eq. 11a constitutes the program of the development of the population in controlling its growth rate.

The case presented here is a very elementary example. In fact, with incursion, it is also possible to generate bifurcations and chaos (Dubois, 1992, 1995).

If such incursive control of chaos exists in the neural network in the brain, an incremental anticipation is possible in considering a self-referential neural network. As stated by Bailly, the final cause represented by the meaning is the organizer of the mind.

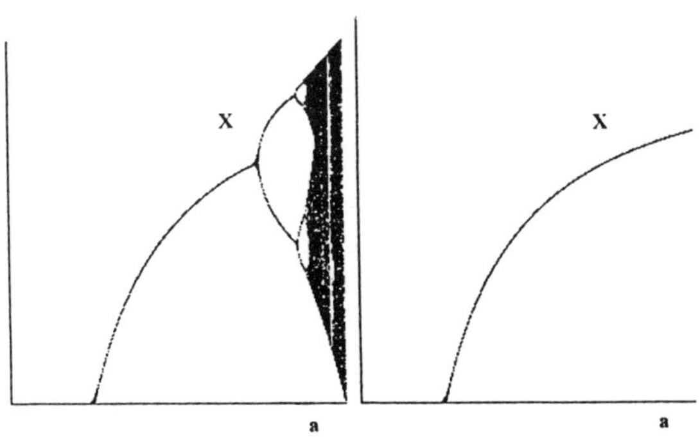

Figures 6a-b

A MODEL OF HYPERINCURSIVE SYSTEM FOR EXPLAINING EVOLUTION

We have shown that two derivatives can be defined, the time forward and the time backward derivatives. For a Malthusian growth

(12) $\Delta X(t)/\Delta t = r.X(t)$

the forward and backward discrete equations are

(13a) $X(t+\Delta t) = X(t)+\Delta t.r.X(t) = X(t).(1+r.\Delta t)$
(13b) $X(t) = X(t-\Delta t)+\Delta t.r.X(t)$ or $X(t+\Delta t) = X(t)+\Delta t.r.X(t+\Delta t)$ or
$X(t+\Delta t) = X(t)/(1-r.\Delta t)$

when $r.\Delta t$ is small, the two equations give the same solution. Lets us remark that if Δt is replaced by $-\Delta t$ in the first equation, the second one is obtained. There is time invariance and no arrow of time. The forward evolution is equal to the backward evolution. With a non-linear equation as in Pearl-Verhulst equation, that is to say a Malthusian growth with saturation (the term in X^2), the forward and the backward equations give different solutions. Indeed, in defining the Pearl-Verhulst model with a time backward derivative

(14) $X(t)-X(t-\Delta t) = \Delta t.r.X(t).(1-X(t))$

The recursive transformation of this equation has two solutions

(15) $X(t) = [1/2-1/(2.\Delta t.r)] \pm \sqrt{[(1/(2.\Delta t.r)-1/2)^2 + X(t-\Delta t)/(\Delta t.r)]}$

We have then a hyper-recursive relation: each iteration generates two successive iterates. So, if the laws of the Nature are fundamentally discrete (the quantum mechanics says this with the quantum of Planck), then a natural arrow of time emerges from the discrete aspect of the time. With the discrete backward equation there are many futures at each time step and only one past. In the forward equation, in replacing Δt by $-\Delta t$ in eq. 14

(14a) $X(t+\Delta t) = X(t) + \Delta t \,.\, r\, X(t)\, (1-X(t))$

which is the classical Pearl-Verhulst map exhibiting deterministic chaos and there is only one unpredictable future but many possible pasts. In looking at living systems, it is well shown that Nature found always many "parallel" solutions during the evolution of species coming from the first initial cell. The genetic code has memorized the main steps of each species. This is the reason why it is possible to reconstruct the genealogy of the different species. There is only one past but there are many futures. *I think that evolutionary systems models could be based on hyperincursive equations with multiple solutions at each step. Many possibilities emerge in nature and some are then selected.*

A GAME OF EVOLUTION

At the image of the game of life of Conway, I would like to propose a game of evolution. Let us start with a very elementary system with one nutrient N(t) and one species $X_1(t)$ evolving with time t. Let us reconsider the model of two competing species given by eqs. 4abc. I have shown that a route to chaos exists in increasing the total quantity of materials c. This is an externalist view of evolution but it does not explain how c can increase. In fact, it is also possible to go to chaos with a constant c in playing with the growth and decay rates a_1, a_2 and b_1, b_2. In an internalist view of evolution, I will consider a constant $c = X_1(t) + N(t) = 1$ (taking $c = 1 = 100\%$ means to consider proportions of nutrient and species), and create a game for one species $X_1(t)$ for which its growth and decay rates (related to birth and death rates) a_1 and b_1 are firstly well-defined:

$$(16)\ X_1(t+1) = X_1(t) + a_1X_1(t)(1 - X_1(t)) - b_1X_1(t)$$

In taking $a_1 = 3$ and $b_1 = 2$, for example, the population stabilizes to the value $X_1 = 1 - b_1/a_1 = 1/3$.

If a small part of the species $\Delta X_1 = X_{12}$ gives rise to random mutations at a time T which have the effect to change the growth and decay rates to the values $a_2 = a_1 + \Delta a_1$ and $b_2 = b_1 + \Delta b_1$ where Δa_1 and Δb_1 can be separately positive, null or positive values, this mutant species X_{12} will compete for the nutrient with the rest of the original species $X_{11} = X_1 - X_{12}$ so that $c = X_{11}(T+t) + X_{12}(T+t) + N(T+t) = 1$. The new system can be described by

$$(16a)\ X_{11}(T+t+1) = X_{11}(T+t) + a_1X_{11}(T+t)(1-X_{11}(T+t)-X_{12}(T+t)) - b_1X_{11}(T+t)$$
$$(16b)\ X_{12}(T+t+1) = X_{12}(T+t) + a_2X_{12}(T+t)(1-X_{11}(T+t)-X_{12}(T+t)) - b_2X_{12}(T+t)$$

The numerical simulation shows different solutions depending on the values of the mutations.

1. If $a_2 = a_1$ and $b_2 = b_1$, there is no evolution.
2. If $a_2/b_2 < a_1/b_1$, the mutant population X_{12} disappears.
3. If $a_2/b_2 > a_1/b_1$, the mutant population increases and the original population X_{11} disappears. This is a phenomenon of succession which is important during evolution (see Salthe, 1993). In this case, the original species shows three phases: growth, stabilization (maturity) and decay, and the mutant species increases and reaches a biomass value $X_{12} = 1 - b_2/a_2$ greater than the biomass of the original population. For example, $a_2 = 3$ and $b_2 = 1$, which means that the mutant population shows a smallest mortality and its biomass is given by $X_{12} = 1-1/3 = 2/3$. Now if a new

mutation occurs, a new species will succeed to the old one if $a_3/b_3 > a_2/b_2$. For example, $a_3 = 4$ and $b_3 = 1$, which means that the mutant population shows a greater growth rate. But here a new phenomenon occurs: the population shows chaos and its average biomass is again greater than the preceding one: $X_{13} = 1\text{-}1/4 = 3/4$. Let us remark that in reality, the number of mutations to reach the chaos must be more than two, but the final state is always the same. The general law is that the species evolves so that its biomass increases: evolution tends to maximize the biomass and at the same time, goes to chaos where the system becomes then locally unpredictable. But the things are a little more complicated. Indeed, from chaos theory, there is no solution for a new mutation $a_4/b_4 > 4$ with $a_4>4$ and $b_4<1$. Happily, there are two ways for evolution.

4a. The first solution is that the mutations dininish the turnover of the dynamics of populations. So, a mutant species $_{Ex}$ with $a_4 = 2$ and $b_4 = 0.4$ so that $a_4/b_4 = 5$, will replace the chaotic species and its biomass will be higher $_{Ex} = 1 - 0.4/2 = 0.8$. The evolution goes thus to the continuation of the maximization of the biomass but with a slower dynamics and chaos disappears. At the limit case, a mutant species with $a_5 = 2$ and $b_5 = 0$ will give the maximum biomass $X_{15} = 1$. Figure 7a gives the numerical simulation of this game of evolution.

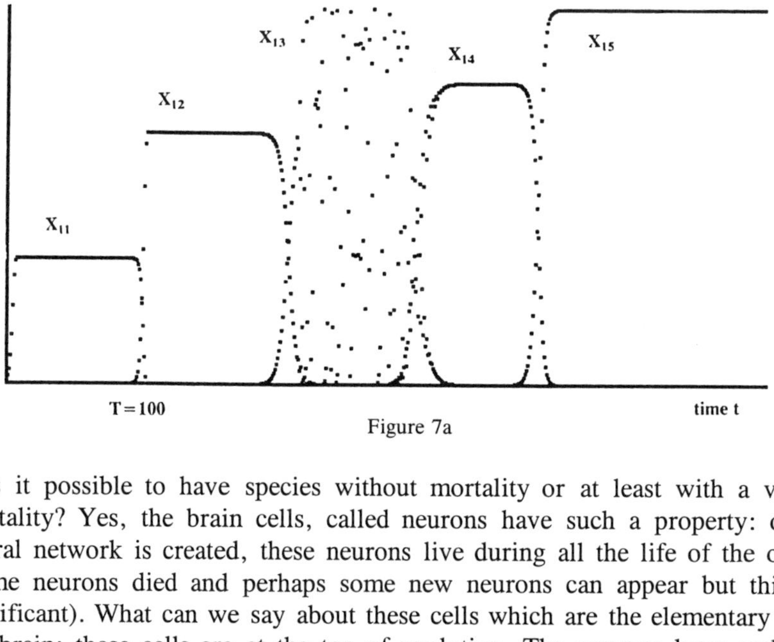

Figure 7a

Is it possible to have species without mortality or at least with a very low mortality? Yes, the brain cells, called neurons have such a property: once the neural network is created, these neurons live during all the life of the organism (some neurons died and perhaps some new neurons can appear but this is not significant). What can we say about these cells which are the elementary parts of the brain: these cells are at the top of evolution. The neurons have realized the dream of evolution: maximize the biomass and the immortality for only one category of cells (no diversity).

Remark 1: With a weak or null mortality, the growth parameter is limited by the value $b = 2$, at the edge of chaos, because there is no more nutrient.

Remark 2: for some values of the parameters it is possible to obtain the coexistence of a species and its mutant species (see section on competitive populations for such a case).

4b. There is another possible solution for $a_4/b_4 > 4$ for which the dynamics is faster than in the preceding case (diminution of the turnover). If the mutation is incursive, there will be a change in the structure of the equation of the species instead of a simple change in the values of its growth and decay values.

Figure 7b gives the numerical simulation of the competition between a recursive species X_{13} and the mutant incursive species X_{14} given by the equation system:

(17a) $X_{13}(t+1) = X_{13}(t) + a_3N(t)X_{13}(t) - b_3X_{13}(t)$

(17b) $X_{14}(t+1) = {}_{Ex}(t) + a_4N(t)X_{14}(t) - b_4X_{14}(t)$

(17c) $N(t+1) = [N(t) - a_3N(t)X_{13}(t) + b_3X_{13}(t) - b_4X_{14}(t+1)]/[1 + a_4X_{14}(t+1)]$

This system obeys the conservation of materials given by

(17d) $c = N(t) + X_{13}(t) + X_{14}(t+1) = 1$

For example, the chaotic species X_{13} is computed with $a_3 = 4$ and $b_3 = 1$ and the mutant incursive species is computed with $a_4 = 4.1$ and $b_4 = 1$. The mutant species replaces the other.

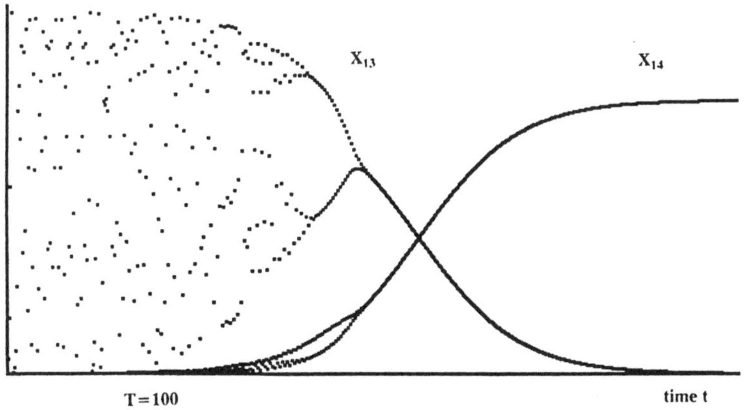

Figure 7b

So, the mutant incursive population X_{14} obeys the incursive equation:

(17e) $X_{14}(t+1) = X_{14}(t) + a_4X_{14}(t)(1-{}_{Ex}(t+1)) - b_4X_{14}(t)$

which can be transformed to the recursive equation (see the section on the incursive control of chaos):

(17f) $X_{14}(t+1) = [X_{14}(t) - b_4X_{14}(t) + a_4X_{14}(t)\,]/[\,1 + a_4X_{14}(t)]$

which is an equation of the Michaelis-Menten-Monod type (see Odum, 1983, for such models). Let us notice that J. Monod shows experimentally that the growth of bacteria follows such a law.

In conclusion, I have demonstrated with a simple model that some natural effects of evolution can be modelled and simulated. Are the dynamics shown here Darwinian or Lamarckian processes?

When random changes in the parameters give rise to new species by competition, the process is similar to the Darwinian natural selection of the fittest. The fittest means in this game of evolution: maximization of the biomass which leads to chaos and then a diminution of the dynamics of the population with the disappearance of chaos and immortality. Neurons seem to be such evolutive cells with a slow dynamics.

When a structural change in the equation of evolution occurs, the incursive solution is rather similar to a catastrophic process. E.C. Zeeman (1972-1977) proposes a Lamarckian evolution complementary to a Darwinian one.

In my game of evolution, the incursive solution suppresses chaos and at the same time, the dynamics of the population can increase beyond the possibilities of chaos in maximizing also the biomass (see also Odum, 1983, for the maximum biomass related to Lotka maximum power principle). Growth in a Michaelis-Menten-Monod way plays a central role in ecological models (Odum, 1983).

This game of evolution is evidently too simple to be realistic and must evolve, but its results are rather surprising and not obvious at all.

CONCLUSION

This paper deals with mathematical models and simulations of evolving ecosystems represented by finite difference equations. In discrete systems two derivatives can define a change: the forward and the backward derivatives related to measure and action.

First time dynamic of populations in relation with the nutrients are considered. The law of conservation of materials in nutrients and populations is used as an environmental constraint parameter for evolving systems. Malthusian growth is then transformed to a Pearl-Verhulst map for which the population emerges and stabilizes beyond a first constraint threshold, while beyond a second threshold, bifurcations and then chaotic behavior of the population occurs. With such a constraint, Volterra models of competing and prey-predator populations show the emergence of chaos. With two populations competing for the nutrients, one or two populations emerge. In varying the value of the constraint, a drastic change can occur in the evolution of the populations: one disappears and the other one shows a growth with bifurcation. For populations in a prey-predator relationship, the prey emerges first and secondly the predator in function of the constraint. Chaos appears for large values of the constraint.

Secondly, space-time dynamics of populations are considered in taking into account diffusion. Turing stated a theory of morphogenesis based on "diffusive instability". I propose in this paper a new principle for the emergence of space-time structures based on what I call a "diffusive chaos". The space diffusion of populations by water turbulence, for example, is another important parameter which gives rise to the emergence of space-time structures and patchiness.

Thirdly, it is proposed to control chaos in self-referential evolving systems by incursion, an inclusive recursion. This is realized by a backward derivative for the nutrients equation. Incursion is an incremental anticipation as a final cause. Malthusian growth is then transformed to a Michaelis-Menten-Monod system.

Finally the Pearl-Verhulst is considered with a backward derivative which defines a hyperincursive system where multiple solutions emerge at each time step. Such a system shows only one past and many futures. The genetic code memorizes this past. Evolving systems must generate many solutions during their evolution and only a part is selected.

ACKNOWLEDGEMENT

I would like to thank Gertrudis Van de Vijver for many discussions and for the critical reading of the preceding draft papers.

Institut de Mathématique, Université de Liège
Liège, Belgium

REFERENCES

Bailly, F., 1991, Niveaux d'organisation, changements de niveaux, finalité, in *Self-organization and Teleology*, Gertrudis Van de Vijver (ed.), vol. II, *Philosophica*, no. 47, pp. 31-47.

Dubois, D.M., 1975, A model of patchiness for prey-predator plankton population, *Ecological Modelling*, vol. 1, pp. 67-80.

Dubois, D.M., 1977, On mathematical modelling of complex ecosystems: Applications to marine planktonic patchiness, Invited paper, *Helgolander Wiss. Meerekunders*, vol. 30, pp. 76-82.

Dubois, D.M., 1979, State-of-the-Art of Predator-Prey Systems Modelling, Invited paper, in *State-of-the-Art in Ecological Modelling*, Jorgensen (ed.), Pergamon Press, pp. 163-217.

Dubois, D.M. (ed.), 1981, Progress in ecological engineering and management by mathematical modelling, *Proceedings of the International Congress of ISEM* (International Society for Ecological Modelling), à Liège, CEBEDOC, 1014 pp.

Dubois, D.M., 1990, *Le labyrinthe de l'intelligence: de l'intelligence naturelle à l'"intelligence fractale*, InterEditions/Paris-Academia/Louvain-la-Neuve.

Dubois, D.M., 1992, Hyperincursivity: Inclusive recursivity without time arrow, in *Proceedings of the 13th International Congress on Cybernetics*, Namur, pp. 152-156.

Dubois, D.M., 1992, *The Fractal Machine*, Presses Universitaires de Liège.

Dubois, D.M., 1992, The hyperincursive fractal machine as a quantum holographic brain in *Communication & Cognition - Artificial Intelligence*, vol. 9, no. 4, pp. 335-372.

Dubois, D.M., 1996a, Introduction of the Arsitotle's Final Causation, in CAST: Concept and Method of Incursion and Hyperincursion, Computer Aided Systems Theory - Eurocast '95, in *Lecture Notes in Computer Science*, F. Pichler, R. Moreno Diaz & R. Albrecht (eds.), Springer, vol. 1030, pp. 477-493.

Dubois, D.M., 1996b, A Semantic Logic for CAST related to Zuse, Deutsch and McCulloch and Pitts Computing Principles, Computer Aided Systems Theory - Eurocast '95, F. Pichler, R. Moreno Diaz, R. Albrecht (eds.), *Lecture Notes in Computer Science*, vol. 1030, Springer, pp. 494-510.

Dubois, D.M., 1996c, Emergence of Space-time Structures from Diffusive Chaos in Pearl-Verhulst and Lotka-Volterra Cellular Automata, *Cybernetics and Systems '96*, vol. 1, R. Trappl (ed.), published by Austrian Society for Cybernetic Studies, pp. 100-105.

Dubois, D.M.; Resconi, G., 1992, *Hyperincursivity: A new mathematical theory*, Presses Universitaires de Liège.

Dubois, D.M.; Resconi, G., 1993, Introduction to hyperincursion with applications to computer science, quantum mechanics and fractal processes, in *Designing New Intelligent Machines (European COMETT programme), Communication & Cognition - Artificial Intelligence*, vol. 10, nos. 1-2, pp. 109-148.

Dubois, D.M.; Resconi, G., 1994, Holistic Control by Incursion of Feedback Systems, Fractal Chaos and Numerical Instabilities, in *Cybernetics and Systems '94*, R. Trappl (ed.), World Scientific, pp. 71-78.

Dubois, D.M., 1995, Total Incursive Control of Linear, Non-linear and Chaotic systems, in *Advances in Computer Cybernetics*, G. Lasker (ed.), Int. Inst. for Adv. Studies in Syst. Res. and Cybernetics, vol. II, pp. 167-171.

Dubois, D.M.; Resconi, G., 1995, *Advanced Research in Incursion Theory applied to Ecology, Physics and Engineering*, COMETT European Lecture Notes in Incursion, Edited by A.I.Lg., Association des Ingénieurs de l'Université de Liège, D/1995/3603/01.

Lotka, A.J., 1925, *Elements of Physical Biology*, William and Wilkins, Baltimore.

Mandelbrot, B., 1983, *The Fractal Geometry of Nature*, Freeman, San Francisco.

May, R.M., 1976, Simple mathematical models with very complicated dynamics, *Nature*, vol. 261, pp. 459-467.

Michaelis, L.; Menten, M.L., 1913, Die kinetik der invertinwirkung, *Biochemische Zietschrift*, vol. 49, pp. 333-369.

Monod, J., 1942, *Recherches sur la croissance des cultures bactériennes*, Hermann et Cie, Paris.

Monod, J., 1972, *Chance and Necessity*, Harper Collins, London.

Odum, H.T., 1983, *Systems Ecology, An Introduction*, J. Wiley & Sons.

Pearl, R., 1924, *Studies in human biology*, William and Wilkins, Baltimore.

Rosen, R., 1986, Causal Structures in Brains and Machines, *Int. J. General Systems*, vol. 12, pp. 107-126.

Rosenblueth, A.; Wiener, N.; Bigelow, J., 1943, Behavior, purpose and teleology, *Philosophy of Science*, vol. 10, pp. 18-24.

Salthe, S.N., 1993, *Development and Evolution: Complexity and Change in Biology*, The MIT Press.

Scheid, F., 1986, *Theory and Problems of Numerical Analysis*, McGraw-Hill Inc.

Schoffeniels, E., 1976, *Anti-Chance, A reply to Monod's Chance and Necessity*, Pergamon Press.

Turing, A., 1952, The Chemical Basis of Morphogenesis, *Proc. Roy. Soc. Lond.*, no. B237, pp. 37-72.

Van de Vijver, G. (ed.), 1992, *New Perspectives on Cybernetics*, Synthese Library, vol. 220, Kluwer Academic Publishers, Dordrecht.

Verhulst, P.F., 1847, Nuov. Mem. Acad. Royale, Bruxelles, vol. 18, no. 1, 1845, vol. 20, no. 1.

Volterra, V., 1931, *Leçon sur la théorie mathématique de la lutte pour la vie*, Gauthier-Villars.

Von Glasersfeld, E., 1991, Teleology and the Concepts of Causation, in *Self-organizing and Complex Systems I, Philosophica*, G. Van de Vijver (ed.), no. 46, pp. 17-43.

Weisbuch, G., 1989, *Dynamique des systèmes complexes: une introduction aux réseaux d'automates*, InterEditions/Editions du CNRS.

Zeeman, E.C., 1972-1977, Appendix on Lamarkian Evolution, in *Catastrophe Theory*, Addison-Wesley Company, Inc, pp. 136-139.

HENRI ATLAN

IMMANENT CAUSALITY: A SPINOZIST VIEWPOINT ON EVOLUTION AND THEORY OF ACTION

INTRODUCTION

Spinoza's monism is seldom taken into consideration when the mind-body problem is stated and its traditionally proposed solutions are discussed. Sometimes, Spinoza's philosophy is referred to as a "parallelism" which is tantamount to a revised Cartesian dualism. Sometimes, it is considered as a materialist monism, thereby setting Spinoza as a precursor of the 18th materialist philosophers, like Diderot and d'Holbach. In fact, Spinoza's theory of psychophysical identity is neither of these. I will argue that it is particularly well-adapted to a discussion of the mind-body problem in the framework of present day natural sciences. In fact, Spinoza's philosophy can only be understood if one takes into consideration his notion of immanent causality. The cause of itself, causa sui, which pertains to the Substance, is distributed in the modes through their essences or conatus, although the modes in their existence are produced by one another, come to existence and are destroyed in their infinite chain of efficient causes. With such a notion of immanent causality, Evolution can be seen as the unfolding of a dynamic system, or a process of complexification and self-organization of matter, produced as a necessary outcome of the laws of physics and chemistry. In this process, new species come to existence one after the other as effects of mutations and stabilizing conditions working as their efficient causes, whereas their particular organizations are particular instances of the whole process. This view of Evolution is compatible with the idea of a dynamic evolutionary landscape with peaks of local stability. The whole dynamics created by the physical constraints of composite organized bodies is an *intemporal* theoretical description of possible organisms. The actual peaks of stability are populated one after the other in a historical, partially contingent fashion, which constitutes the *temporal* evolutionary processes. The latter may be oriented by adaptive natural selection, but that is not always necessarily the case. The temporal evolution is also a self-organizing process, driven by a random walk, whereby more "sophistication" (Koppel & Atlan, 1991; Atlan, 1995) or "logical depth" (Bennett, 1988), in the sense of functional meaningful complexity, can be memorized and accumulated.

The so-called parallelism is largely based on the famous prop. 7 of the 2nd part of *The Ethics*, where Spinoza deals with the relationship between matter and mind: "The order and connection of ideas is the same as the order and connection of things". However, Spinoza himself never used the word "parallelism" and explicitly states his stance as an ontological monism underlying conceptual dualism: "(...) thinking substance and extended substance are one and the same thing, which is now comprehended through this and now through that attribute. Thus also a mode of extension and the idea of that mode are one and the same thing, but expressed in two manners (...)" (*The Ethics*, II, 7, scholium).

G. Van de Vijver et al. (eds.), Evolutionary Systems, 215–231.

On the other hand, the so-called materialism of Spinoza is based on statements like: "(...) No one will be able to understand this [the union of the mind and the body] adequately and distinctly unless, at first, he is sufficiently acquainted with the nature of our body" (*ibid.,* II, 13, scholium). "Each one judges concerning things according to the disposition of his own brain or rather takes for things that which is really the modifications of his imagination" (*ibid.,* I, Appendix). However, the attribute of Thought seems to prevail, in a way, over all other attributes, including Extension, in that it produces not only "ideas", but "ideas of ideas". Ideas are not only modes of the Substance as it is expressed by the attribute of Thought, in the same way as bodies are modes of the Substance as it is expressed by the attribute of Extension. Ideas are also ideas of all other modes of the Substance as it is expressed by all other attributes. As such, they contribute to the activity of the "infinite understanding", which is itself the infinite mode of the attribute of Thought. In other words, Thought entails an epistemological superiority over matter, in that matter is understood and "intellected" by means of ideas with a reflective nature, i.e. "ideas of ideas". If an idea is the mode in Thought associated to a body in Extension, the idea of this idea expresses the reflective nature of Thought through the capacity of "intellection", i.e. knowing and understanding. In addition, *The Ethics* ends with several propositions concerning "those points which appertain to the duration of the mind without relation to the body" (*ibid.*, V, 20, end of the scholium), such as "eternity" (*ibid.,* V, 23). This part of the book seems so strange and foreign to a materialist stance that many philosophers view it as a kind of enigma in the work of Spinoza who, all of a sudden, appears to be a mystic. In fact, close to the end, Spinoza warns the reader against an idealist misinterpretation by recalling that "He who has a body capable of many things, has a mind of which the greater part is eternal" (*ibid.*, V, 39). However, seeing this statement as an expression of a materialist ontology would also be a misinterpretation.

In fact, Spinoza's monism is an original one, as it is neither materialist nor idealist. It is difficult to grasp within the framework of existing ontologies, both monist and dualist. That is probably the main reason why Spinozism is seldom considered as a serious alternative to the conflicting solutions to the mind-body problem.

The purpose of this paper is to suggest a way of understanding Spinoza's original stance on psychophysical identity, starting from what is called his "physics", that is, the small set of axioms, lemmas and postulates between propositions 13 and 14 in the second part of *The Ethics*. Firstly, we shall see that these few statements are more relevant to what we would today call a biophysical theory of the organism than to physics *per se*. Then, based on a remark of Hilary Putnam (1981), I shall refer to the idea of the *synthetic identity of properties* to show that a similar way of thinking, which is different from the usual philosophical analytical conceptualization, is at the basis of both the psychophysical identity as it is conceived by Spinoza, and the identity of microphysical and macrophysical quantities as it is conceived in contemporary physics. This will help us solve a puzzle encountered by Donald Davidson (forthcoming) when he explicitly referred the "anomalous monism" at the basis of his theory of action to Spinoza's psychological monism.

SPINOZA'S PHYSICS

In Spinoza's own terms, his physics is "poor" since it is limited to "a few (*pauca*) premises on the nature of the bodies" (*The Ethics*, II, 13, end of the scholium). His purpose was not "to lecture on the body", but only to mention what he needed in order to deduce more easily what he wanted to demonstrate, regarding the "Nature and Origin of the Mind", to which the second part of *The Ethics* is devoted.

In the eyes of a modern Newtonian and post-Newtonian physicist, these few statements are not only irrelevant, but uninteresting, even from a historical point of view, and therefore contrast with the physics of Pascal, Descartes, and more particularly with that of Huygens and Leibniz. However, this opinion is based on a misunderstanding, since Spinoza's concern in his "physics" is not mechanics, i.e. a physical theory of motion which would be considered to be the precursor of today's physics. Rather, these few statements in *The Ethics* outline a theory on "the nature of the simple and compound bodies". These premises of a theory are more relevant to what we would consider today as chemistry or biophysics, that is, a physical theory of composite individuals with no basic difference between living and non-living bodies. For example, as H. Jonas (1965) already pointed out, one can easily recognize metabolism, growth, internal activity and locomotion in lemmas IV, V, VI and VII respectively. These are various ways in which "an individual can be affected and, despite this, preserve its nature". The "nature" of an individual is defined by its "form", i.e. by the fact that its constitutive parts "remain united with one another" in spite of the changes that they can undergo, provided that "a certain ratio of movement and rest" is retained. This invariant ratio of movement and rest is the "form" or "nature" of the individual, and it works like a law of organization which defines a given individual. However, this form is not imposed on the body by a soul, as it is, for example, in Leibniz's conception. Nor does it work like a machine as it does in Descartes *animal-machine* based on the model of an hydraulic or mechanical device, where an individualized source of energy, similar to an internal fire, is necessary to activate the parts. The law of organization results only from the way in which the parts are in contact with one another, and that what is necessary to maintain it invariant, is produced by the metabolism and activity of the parts themselves. According to the scholastic tradition in which movement meant not only displacement in space, but also change of any sort — like, e.g. a chemical transformation —, this "certain ratio of movement and rest" defining a given individual must be understood as a certain ratio of change and invariance. This is reminiscent of what we call today dynamic systems, such as those made up of several chemicals which are able to react with one another, where a global law of organization is defined by local kinetics of chemical reactions. It is easier for us to define precisely such laws of organization because we can now make use of differential calculus which was inexistent at that time. From this point of view, there is no doubt that we are indebted to Leibniz and Newton who discovered it a little later, although their respective ontologies and metaphysics are hardly acceptable to us nowadays.

At the end of this brief summary of Spinoza's physics, let us remind ourselves that it applies equally to all individual bodies, living or not. As a result, one should not be misled by Spinoza's statement on the generality of the definition of the human mind, as pertaining to all things and "no more to man than to the other individual things which are all, though in various grades, animate" (*The Ethics,* II, prop 13 and its scholium). This statement should not be confounded with universal animism or panvitalism. It is the very opposite, as is clear from the context. The last quotation follows the remark that "in truth, no one will be able to understand [the union of the mind and body] adequately and distinctly unless, at first, he is sufficiently acquainted with the nature of our body" (*ibid.*). The whole scholium is a commentary on the definition of the human mind as the idea of the body which actually exists (*ibid.,* II, 13). Being applied to any individual, living or non-living, this definition therefore implies that a stone has a mind. But the mind of a stone must not be confounded with a soul which would keep the stone alive. This mind, like the human mind, is merely the idea, as a mode of Thought, which is always united in Nature with an existing material object, as a mode of Extension, of which it is the idea. Therefore, the "mind" of a stone must not be confounded with the soul of a human being or another "animate" being. It is just the opposite. The mind of a human being is the idea of his body, just like that is the case for the idea of a stone or any other thing. The differences between minds reflect differences between the bodies of which they are the ideas. Our distinction between animate and inanimate beings results only from our insufficient and inadequate knowledge of the nature of the composite individual bodies.

Of course, the more complex the bodies, made of various parts which in turn, are composite bodies, the more complex their idea is (*ibid.*, II, 14 and 15). The ideas of human bodies, assumed to be the most complex, exhibit a greater capacity for understanding and reasoning, i.e. for producing ideas with the reflexive nature of "ideas of ideas". In other words, consciousness and reason are properties of the idea of the human body which accompany the complex nature of that body, i.e. the large number of ways in which the human body can be affected and can do things without losing its "form" or nature. Contrary to the idea of a stone whose object is relatively less complex than a human body, "the human mind is apt to perceive many things, and more so accordingly as its body can be disposed in more ways" (*ibid.*, II, 14).

Thus, under certain conditions related to the complexity of the body, i.e. the many ways in which it can be disposed, the mind may not only be an idea, but may also produce and have ideas, like the infinite Understanding of which it is a part.[1]

In this way, Spinoza showed that the "soul", or more correctly the "mind" (*mens*), should not be confused with the classical soul as the origin of movement and "fire" for the living body. Moreover, and contrary to what our inadequate immediate knowledge usually leads us to believe, even in living and moving bodies "*the mind cannot determine the body to remain in motion, or at rest, or in any other state (if there be any)*", and similarly, "*the body cannot determine the mind to think*" (*ibid.,* III, 2, my emphasis).

This proposition is at the core of the Spinozist psychophysical theory and it is certainly one of the most difficult to understand. Donald Davidson has written an important article (forthcoming) in which he tries to reconcile this statement with his own theory of action and perception. This endeavor is significant because Davidson's theory is based on his "anomalous monism" which, according to him, is closely related to Spinoza's monism and "parallelism".

Before turning to this thesis, I summarize Putnam's remark on the synthetic identity of properties in physics. I will refer to it later in my discussion of Davidson's theory of action in the light of Spinoza's theory of psychophysical identity.

THE SYNTHETIC IDENTITY OF PROPERTIES

According to Putnam (1981), the notion of physical magnitudes which we employ in physics, implies the existence of a "synthetic identity of properties", which is to be distinguished from an analytic identity or equivalence. For example, the physical magnitude "temperature" is identical to the "mean molecular kinetic energy". However, this identity is not analytic because the two sentences "a gas has a temperature T" and "its molecules have a mean kinetic energy equal to 3/2 kT" are not synonymous, even though we learn from the kinetic theory of gases and statistical thermodynamics that they represent two different ways of expressing the same property. If two identical physical properties were linguistically definable as predicates, then we should be able to describe them by means of synonymous enunciations, i.e. sentences with identical meaning or, in other words, sentences which are analytically equivalent. The example taken from the kinetic theory of gases shows that this is not the case, since two physical properties (having a particular temperature, and having a certain molecular mean kinetic energy) are identical despite the fact that their descriptions are not synonymous. Thus, we must admit that physical properties can exist which are "synthetically identical" without being analytically identical, i.e. without being conceptually identical, in the usual sense of a concept expressed by a meaningful unequivocal enunciation.

Therefore, we can safely say that such a synthetic identity may exist between mental states and brain states, although we may not be able to describe it in a unique, non-equivocal enunciation. A brain state may be identical to a mental state and also to the sensation of a given qualitative state, even though we need non-synonymous expressions to describe each of these. In other words, according to Putnam, "what the physicist means by 'physical magnitude' is something different from what philosophers call a 'predicate' or a 'concept' (...) Properties, as opposed to predicates, can be 'synthetically identical'". Therefore, "if there is such a thing as *synthetic identity of properties*, then why shouldn't it be the case that the property of being in a certain brain-state is *the same property* as the property of having sensation of a certain qualitative character (very much in line with Spinoza's thinking) — even though it is not a conceptual truth that it is, even though, in fact, it seems to many to be *a priori* false?" (Putnam, 1981).

Spinoza himself often takes another example from geometry. The definition of a triangle entails that the sum of its angles equals the sum of two right angles. This property is a consequence or an effect of what Spinoza calls the generic definition of a triangle, which is its reason or its cause. However, this property is identical with what is expressed by "the sum of the angles of a triangle equals 180 degrees", although the two expressions are not synonymous.

For the purpose of my discussion of Davidson's interpretation of Spinoza, it is important to notice that these examples raise the problem of a possible causal relationship between the two properties, which are synthetically identical but analytically different. This problem is solved simply by noting that such a causal relationship between identical properties would be meaningless (a cause and its effect cannot be identical), even though it may seem meaningful linguistically. Although described in two different ways, the two properties cannot be cause and effect since they are identical: temperature is neither the cause nor the effect of the molecular mean kinetic energy because it is this energy. Similarly, "the sum of the angles in a triangle is equal to two right angles" is neither the cause nor the effect of "the sum of the angles in a triangle is equal to 180°" because they are two different enunciations of the same property. (This geometrical example may seem less obvious than the previous, physical, one. It seems that the sum of the angles equals 180° *because* the sum of two right angles equals 180°. In fact, the two properties can be demonstrated independently by means of two right angles or one single angle of 180°. In any case, the example of temperature is more instructive because it can be easily applied to all thermodynamical quantities such as pressure, volume, free energy, entropy, which can be defined at two different levels, macroscopically and microscopically). In the same way, a mental state is not the cause or the effect of a given brain-state since it *is* this brain-state, even though we cannot describe the mental state and the brain-state by synonymous expressions.

This view is exactly the same as that of Spinoza, who states explicitly not only that body and mind are two different aspects of the same, but also that no causal relationship can exist between them, which is obvious in view of the aforementioned discussion.

ACTION AND PERCEPTION. DAVIDSON'S "ANOMALOUS MONISM"

However, this lack of causal relationship is hard to reconcile with our immediate experience of perception and action: we feel spontaneously that our voluntary movements are caused by some mental state which corresponds to our experience of making a decision. Conversely, it seems that our perceptions are effects on our mind of modifications of our bodily sense organs. Davidson is certainly one of the contemporary philosophers who has devoted most of his work to elaborating a naturalist theory of action and perception which does not assume a kind of dualist interactionism imbued with the same mystery as Descartes' "solution" of the pineal gland. In this, Davidson invokes Kant and the necessity of establishing the reality of human actions caused by voluntary decisions in order to save the reality of ethics based on free choice and responsibility (Davidson, 1980). On the other hand, in a kind of summary of his theory, Davidson presents it as a modern version of Spinoza's "parallelism". The association of Kant with Spinoza here is strange since it is well known that Spinoza's *Ethics* is a search for freedom which denies the reality of free will. Therefore, in a Spinozist, non-Kantian universe, there is no need to establish the reality of bodily movements which are caused by decisions of our mind, in order to build a "model of human nature" which would lead us to freedom and happiness (*The Ethics*, pref. to part IV).

In any case, Davidson's argument goes as follows:

Let us consider two individuals, or two consecutive states of one individual, I and II, such that I causes II. Each of them is a union of the mental and the physical. Let A and B respectively be the mental and the physical in I, and C and D the mental and the physical in II, as is represented in the following diagram.

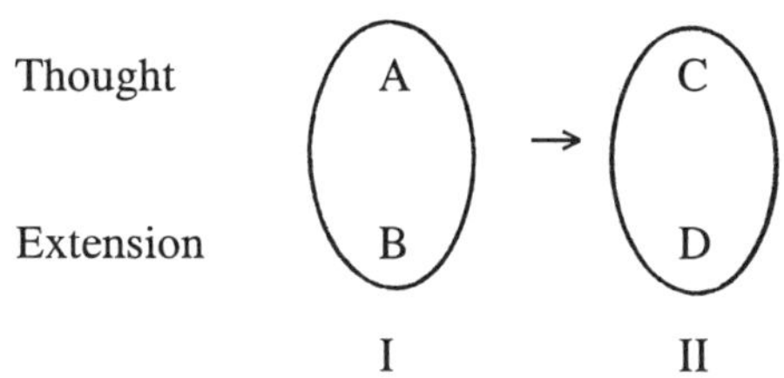

In Spinozist terms, A and C arc respectively modes I and II of the Substance under the attribute of Thought, and B and D are modes I and II of the Substance under the attribute of Extension. Prop. 7 in the second part of *The Ethics* states: "The order and connection of ideas is the same as the order and connection of things", which means that the causal relation between A and C, or rather the production of C by A, is the same as the causal relation between B and D, through which D is produced by B. (This becomes obvious in the reformulation of prop. 7 when it is used to demonstrating prop. 9: "the order and connection of ideas is the same as the order and connection of *causes*" (my emphasis)). In other words, the causal connection between A and C, through which an idea A is the cause and the reason of an idea C, is the same as the causal connection between B and D, united with A and C respectively.

However, the two attributes of Thought and Extension are only two different expressions of the same Substance which is only one. Contrary to Descartes, Spinoza holds that the "thinking substance and extended substance are one and the same thing, which is now comprehended through this attribute, and now through that attribute. Thus, also a mode of Extension and the idea of that mode are one and the same thing, but expressed in two manners (...)" (Scholium on prop. 7, part II). This means that A is identical to B, and C is identical to D. Incidentally, it is for this reason that A as a mental event cannot be the cause of B as a physical thing: it is not because A and B are two different substances with no interaction between them, but because A and B are one and the same. The one cannot be the cause of the other, since cause and effect must be different (except for God, i.e. Nature as a whole, which is *causa sui*, the cause of itself). This is stated explicitly by Spinoza in part III, prop. 2: "The body cannot determine the mind to thinking, and the mind cannot determine the body to be in motion, or at rest, or in any other state (if there be any other)". And the scholium of this proposition contains an explicit reference to the above mentioned prop. 7 of part II.

However, Davidson questions the meaning of this proposition (*ibid.*, III, 2) when it is applied to our scheme. Whereas it is clear to him that A cannot be the cause of B, because they are identical, A, nevertheless, must be the cause of D because C and D are identical and A is the cause of C. The same holds for the physical event B, which must be the cause of the mental event C because it is the cause of the physical event D, and D and C are identical. Thus, it is clear that the body B determines the mind C to thinking by means of this cross-causality or transitive causality: "B causes D, and since C is identical to D, B causes C". The same also holds for the mental event A which causes the physical event D. Therefore, how can Spinoza say that the body cannot determine the mind to thinking and the mind cannot determine the body to motion?

As Davidson (forthcoming) puts it, "The difficulty, described in this way, is so apparent that one must assume that we have misinterpreted Spinoza on some essential point, or have failed to make a distinction that was crucial to his position".

After having quoted conflicting interpretations of Spinoza which obviously miss the point, he suggests his own solution which fits his theory of action and perception. A usual misinterpretation is to refer to Spinoza as a materialist monist. Davidson only mentions Curley, but several commentators adopt the same line of thought. Their reasoning goes as follows: as a materialist monist, Spinoza denies the reality of the mental and therefore for him the interaction between mind and body does not exist and does not pose any problem. However, it can easily be shown that Spinoza's denial of interaction does not stem from any denial of the reality of the mental and, hence, that he cannot be seen as a materialist in the modern sense of the word. Another misinterpretation is the distinction between the two attributes and their so-called parallelism. This amounts to denying that Spinoza is a monist, referring to the fact that the mind and body cannot be causally related to one another because the attributes of Thought and Extension are separate. Davidson attributes this interpretation to Alan Donagan, but it can be traced back to Leibniz, who invented the term "parallelism" to describe the relationship between mind and body described in Spinoza's *Ethics* (prop. 7 of part II). As Davidson rightfully comments on Donagan's interpretation, "If Spinoza is not a genuine monist, mind and body are not really identical, and the apparent contradiction [that he] pointed disappears. But then, so would the apparent contrast between Spinoza's and Descartes's metaphysics also disappear, and Spinoza would need some substitute for the pineal gland" (Davidson, forthcoming).

As for Leibniz himself, his substitute for the pineal gland was his principle of universal harmony, according to which the lack of interaction between mind and body was compensated for by the mysterious action of God, who maintains harmony and adequacy between the two parallel realms of the mental and the physical by ways unknown to men. At this point, Spinoza departs from both Descartes and Leibniz because they resort to the mystery of an unintelligible God, thereby entering into what he calls the "asylum of ignorance".

What Spinoza said about Descartes' theory of the pineal gland and animal spirits is well known: "I cannot sufficiently wonder that a philosophic man, who clearly stated that he would deduce nothing save from self-evident bases of argument, and that he would assert nothing save what he perceived clearly and distinctly — one, moreover, who so many times reproved the scholastics for wishing to explain obscure things by means of occult qualities, should take an hypothesis far more occult than all the occult qualities" (*The Ethics*, Pref. to part V). Of course, Spinoza could not say anything about Leibniz since he died first, but we can easily imagine that he would have reacted in a similar way to the principle of universal harmony.

Having discarded these misinterpretations of Spinoza, Davidson addresses the important question of how we are to understand the concept of causation? More specifically: "What did Spinoza mean when he wrote 'The body cannot determine the mind to thinking and the mind cannot determine the body to motion (...)'? Did he mean by 'determine' what we mean by 'cause'?" (Davidson, forthcoming).

Davidson's answer is based on the idea that a causal explanation of something is not always the same as the actual cause of that thing. In other words, he states that a logical reason (*ratio*) for something may be different from its physical cause (*causa*). This distinction is of course not an ontological one, since this would lead us again to an ontological dualism. It is merely the result of our frequent inability to have an *adequate* explanation of something, or some event, that is, one which fully "deduces the occurrence (or probability of occurrence) of the event to be explained from the laws of nature and a statement of the prior conditions" (Davidson, forthcoming). Although we may not have such a full explanation and therefore know only partial causes (which Spinoza himself calls "inadequate"), the "inadequacy of our knowledge of the cause and the effect does not throw in doubt the causal connection". The mind-body problem is an example of inadequate knowledge, since we do not have a vocabulary allowing us to describe mental events in terms of physical ones and vice versa, and we do not know psycho-physical laws which would allow us to reduce mental events to physical ones. Such a situation is not inconsistent with ontological monism. This position is Davidson's own "anomalous monism", i.e. an ontological monism of mind and body "compatible with the failure of nomological reduction, that is, with the absence of strict psychophysical laws (...) Monism, coupled with the failure of nomological connections, implies that a complete or adequate explanation of a mental event cannot be given in physical terms, and a complete and adequate explanation of a physical event cannot be given in mental terms" (*ibid.*). For Davidson, this anomalous monism is akin to Spinoza's ontological monism coupled with a dualistic (or multiple) explanatory apparatus.

The question I want to address now is whether Davidson's reading of Spinoza is a correct interpretation of Spinoza's concept of causation. His reading allows him to save the reality of causal relations between the mental and the physical, in spite of the assertion that the body cannot determine the mind to thinking and vice versa. The latter assertion is understood as belonging to the realm of causal (inadequate) descriptions or explanations and not to the realm of causes. "Causal relations in nature are indifferent to how we describe them (...) Causal *relations* as [Davidson] conceives them are between events however described; causal *explanations*, on the other hand, depend on the vocabulary or concepts used to describe events and to formulate laws" (*ibid.*).

ACTION AND PERCEPTION IN THE LIGHT OF SPINOZA'S MONISM

In the following passage, I want to question Davidson's reading of Spinoza and to suggest a related but different way to solve the above-mentioned apparent contradiction. As mentioned by Davidson, this contradiction decidedly stems from a misinterpretation of Spinoza. However, this misinterpretation does not concern the question of determination by causes versus the explanation by reasons. I will argue first that the contradiction disappears within the system of *The Ethics* if we keep in mind the immanent and non-transitive nature of the causation of all things produced as modes of the Substance by the Substance itself (see further, *The Ethics*, I, 18). Then, in the following section, I will make use of the idea of the synthetic identity of properties and draw an analogy with today's physics, which can help a modern reader to better conceive of the kind of union between mind and body implied by Spinoza's doctrine. In effect, it seems that for Spinoza, even if the *existence* of this union was asserted and demonstrated by deduction from the unity of the unique Substance, its *nature* would still be an object of investigation that one would be "able to understand adequately and distinctly" only "by being sufficiently acquainted with the nature of our body" (see above, *ibid.*, II, 13, scholium), i.e. through further physical research and knowledge.

As far as the question about the meaning of "determination" in prop. 2 of part III is concerned, it is difficult to accept Davidson's answer that it belongs to the realm of inadequate knowledge of causes, since the demonstration of this proposition makes use of the basic idea in *The Ethics* that ideas are produced and can be known (adequately) as modes and effects of Nature only under the attribute of Thought, and that physical events are produced and can be known (adequately) as modes and effects of Nature only under the attribute of Extension. Moreover, the distinction between *causa* and *ratio*, or between causal relation and causal explanation is contradicted by Spinoza's insistence that they are the same, albeit viewed through different aspects, as noted in his prop. 7 of part II and his own commentary. Davidson himself acknowledges that there is some difficulty here which leaves him with a "remaining doubt" about the correctness of his interpretation. Nowhere does Spinoza seem to imply that the absence of a causal relation between the physical and the mental is due to our lack of knowledge. Indeed, the opposite is true. Our usual experience that a decree of our will can cause a bodily movement is due to our inadequate knowledge of the real causes of both our bodily movements and our ideas, thoughts, emotions, unconscious and conscious feelings, which, *being identical to them*, always accompany them. Thus, if it is true that "a complete or adequate explanation of a mental event cannot be given in physical terms, and a complete and adequate explanation of a physical event cannot be given in mental terms", this does not mean that we must accept the necessity of causal relations between the mental and the physical which would be real, even though we can only describe them in partial, confused and inadequate ways. It is clear that, for Spinoza, even if we had a perfect infinite knowledge of Nature ("God's infinite understanding"), we would still maintain that there cannot be causal relations between the mental and the physical. Then how can we solve the apparent contradiction that was analyzed schematically above: if A causes C, and C is identical to D, how can we say that A does not cause D? (The same applies for B, D, and C). What is at stake here is not the nature of the causal explanation which, when adequate, cannot be distinguished from the causal relation. What is at stake, rather, is the nature of the identity between C and D, and the transitivity of the causal relation between A and D that we deduce from this identity. A transitive cause is such that its effects are produced beyond itself. The effects do not occur in the same entity as the cause itself. That is why, for Spinoza, "God is the immanent and not the transitive cause of all things" (*ibid.*, prop. 18, part I).

The effects of Nature (i.e. *natura naturata*) are produced within the same entity (namely "God, i.e. Nature") as that of the causes (i.e. *natura naturans*). According to modern logics and mathematics, a relation r is transitive, if from A r C and C = D, we can deduce A r D. This itself results from our axiomatic agreement that equality is a transitive relation: if X = Y and Y = Z, then X = Z. Thus, our reasoning in the above problem implies that the causal relation between A and C is transitive. This however means that we consider D as being not *really* identical to C, but merely related to C by a relation of equality or equivalence. D and C are therefore two different entities. Again, we come back to a dualist position which is certainly not consistent with the Spinozist context of our analysis.

The only way one can circumvent this is to postulate a real identity between C and D which are united in II, and the same for A and B united in I, as required by Spinoza's monism, and to consider the causal relation between A and C to be non transitive. Thus, we should not say: A causes D because A causes C and C = D, i.e. because of the transitivity of the relation between A and C. Rather, apparently, we should say: A causes C or D indifferently (similarly B causes C or D indifferently) because C and D are one and the same. But can we say so? In fact, this would suggest that the *descriptions* of C and D are synonymous, and that they can be replaced by one another in the description of their causal relation with A or with B. In other words, to go back to Putnam's distinction, this would be possible only if the identity between the mental C and the physical D (or likewise, the mental A and the physical B) was analytical. Or, to use Davidson's terms, it would imply a possible reduction of the mental to the physical by means of a common vocabulary and strict psychophysical laws. Since that is not the case, and since we must consider the identity between C and D as synthetical, we cannot replace C by D in the relation A causes C (the same holds for replacing D by C in the relation B causes D).

It seems that we are arriving at the same conclusion as Davidson, since our reasoning apparently implies that a causal relation still exists between A and D, although we do not have the means to describe it as a causal explanation. This is not quite the case, however. The causal relation existing in nature that we are unable to describe is not between A and C, or between B and D, but between I (i.e. A and B together) and II (i.e. C and D together). It is simultaneously a causal relation and an explanation, i.e. *causa* and *ratio*. The relation between A and C, and that between B and D are *both* different ways of conceiving and describing the same relation between I and II. Thus, both the relation between the mental events A and C, and the relation between physical events B and D can describe the only causal relation which exists, namely between I and II. The fact that this causal relation can only be described through its expressions in the two attributes, has nothing to do with inadequate knowledge or hidden reality. This situation results from the fact that we can only have access to the essence of the Substance and to its causal production of its modes through the attributes. This does not preclude the possibility of adequate knowledge in either one of the attributes. In addition, even if we knew strict psychophysical laws allowing to equate mental and physical states, this would not necessarily imply analytical identities between them. The same applies for microscopic and macroscopic descriptions of physical quantities, which are synthetically but not analytically identical. The causal descriptions in the mental and the physical domains can be adequate without there being any other cross-causal relations. Replacing the mental state C by the physical state D would result in mere confusion and meaninglessness, since the identity between C and D is synthetical and not analytical.

The difference with Davidson's position is that we introduce the dualism of descriptions, not within the causal relation which remains as one and which includes both the causal relation and the explanation, assumed to be adequate, but within the identity of the events, where the mental C and the physical D, although identical, need different descriptions which cannot replace one another when related to A and B respectively.

THE ANALOGY WITH PHYSICAL MAGNITUDES

A comparison with synthetically identical physical magnitudes can help one to visualize the situation better. Let A be the temperature of a gas, and B the mean kinetic energy of molecular motion of that gas. As mentioned above, A and B are synthetically but not analytically identical. Similarly, the pressure of a gas is synthetically identical with the exchange of momentum of the molecular collisions of that gas with the walls containing its volume. Let C be the pressure and D the force of the molecular collisions corresponding to that pressure.

As is well known, an increase in the temperature A of the gas causes an increase in the pressure C. Similarly, an increase in the molecular motion B causes an increase in the molecular collisions. However, there is no sense in talking about the cross-causal relations between A and D and between B and C. The only thing we can say is that an increase in temperature causes an increase in pressure, and that this is accompanied by an increase in molecular collisions. This is so because pressure and molecular collisions are identical, even if their definitions are not exchangeable in the description of what is going on, since the latter refers to molecular properties and the former to thermodynamic properties of a macroscopic sample of matter. In other words, the causal relation between A and C acts *as if* it were transitive and passed on to D because of the identity between C and D, but *it is not* transitive because the identity between C and D cannot be described as an epistemic (analytical) identity.

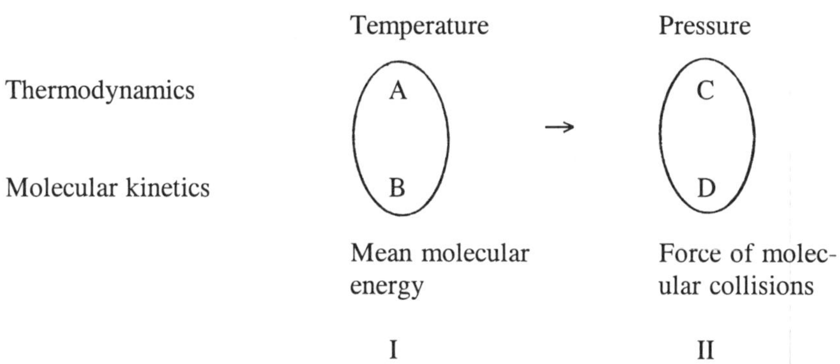

Our solution to the puzzle presented by Davidson is closer to the letter of Spinoza's *Ethics*. It is also closer to its spirit, in that it does not dissociate explanation (*ratio*) from cause (*causa*) and does not assume that the apparently contradictory propositions (*ibid.*, III, 2 and II, 7) concern only inadequate causes, which is obviously not the case as can be judged by their contexts and demonstrations.

Thus, Davidson was forced to depart from Spinoza in order to save his causal theory of action and perception, although he is obviously very sympathetic to Spinoza's unique association of ontological monism and epistemic dualism.

Of course, nothing forces us to be orthodox Spinozists. We must try to weigh the advantages and disadvantages of this radical denial of a causal relation between the mental and the physical, which is associated with a synthetic identity which needs different vocabularies with different analytical meanings.

An obvious disadvantage of this position is that it contradicts our common sense experience which tells us that a decision of our will may be the cause of our bodily movement, and that a physical affection of our body may be the cause of our perception and thinking.

However, in order to save our common sense experience — and also maybe the Kantian moral *a priori* of the existence of free will — we must pay the price and assume the reality of relations depicted by crossing arrows in our schema. Is this price too high? As mentioned above, these arrows represent causal relations which are supposed to be real but which are hidden and can only be described by means of inadequate causal explanations. How could inadequate explanations then refer to something real? If there is no adequate causal explanation between the mental and the physical, then what is the point in talking about a hidden causal relation that we can only conceive of confusedly, simply because it is suggested by our common sense experience? Do we need to save a causal theory of action and perception at any price?

I will not go into the classical moral argument that states that if we did not assume a free will, there would be no responsibility and no ethics. This argument has been dealt with by all proponents of absolute determinism such as the Stoics and Spinoza himself, who have clearly dissociated the possibility of moral conduct and responsibility from the necessity of free will. After all, Spinozist ethics is a path to freedom and joy based, among other things, on the knowledge that free will is illusory.

Instead, I will make use of present trends in cognitive sciences which support a non-causal theory of action and perception compatible with ontological monism. In this context, the Spinozist stance allows us to make better sense of what is implied by these trends, and to overcome the difficulties which stem from our common sense experience. Contradictions between theoretical physics — quantum theory and relativity — and common ideas about space and time have already taught us to be cautious about common sense as a source of insights into reality. Pragmatic theories of cognition have shown how acquisition of knowledge and action on the world are mutually dependent as far as their attribution of meaning is concerned (see Shanon, 1993, for a review and discussion).

FUNCTIONAL SELF-ORGANIZATION

With the help of models of functional self-organization (Atlan et al., 1986; Atlan, 1987), it is possible to understand, at least in principle, how meaning and intentionality can be created by a more or less frequent repetition of events leading to self-observed behaviors of dynamical systems. This model is made by associating memory devices with self-organizing networks of neuron-like automata. The memorization of self-organizing behaviors is what generates meaning and intentionality in a process that we call intentional self-organization (Atlan, 1992, 1995). In the light of this model, we can understand that pattern recognition and decision-making are basically the same process, even if the former seems to be a recollection of the past and the latter a projection into the future. The differences between the two are only a matter of emphasis in a unified process of creation of meaning by self-observed, self-organization of behavior. If, for instance, we are more sensitive to the aspect of observation, we experience what we call perception or pattern recognition. By accentuating the behavioral aspect on the other hand, we experience what we call will and decision-making. However, as Spinoza said in his concise way: "Will and Understanding are one and the same" (*The Ethics*, part II, prop. 49, corollary.). In other words, our usual distinction between perception and knowledge on the one hand, and conscious decision-making of voluntary action on the other, is due to our "imaginary knowledge", i.e. to our confusion between an "idea, or a conception of the mind, and the images of things which we imagine (...) which are found in us by the concourse of bodies" (*ibid.*, part II, prop. 49, scholium). More generally, this confusion is due to our limited knowledge of the true nature of the union of our mind and body. This is what makes "men think themselves free on account of this alone, that they are conscious of their actions and ignorant of the causes of them; and, moreover, that the decisions of the mind are nothing save their desires, which are accordingly various according to various dispositions [of their and other interacting bodies] " (*ibid.*, part III, prop. 2, scholium).

This quotation is taken from Spinoza's comment on his famous prop. 2 in part III, which denies the determination of the body by the mind and vice versa. Here, he makes his point even clearer: "Now all these things clearly show that the decision of the mind and the desire and decision of the body are simultaneous in nature, or rather one and the same thing, which when considered under the attribute of Thought and explained through the same we call a decision, and when considered under the attribute of Extension and deduced from the laws of motion and rest we call determination." (*ibid.*) The last sentence shows that Davidson's suggestion that, for Spinoza, "determination" means explanation and not causal relation is hard to accept, since what he calls determination here is both a causal *production* of a bodily movement, i.e. something "considered under the attribute of extension", and its causal *explanation* by laws of physics, i.e. something considered under the attribute of Thought.

The scholium continues with the development of "another point (...) namely, that we can do nothing by a decision of the mind unless we recollect having done so before (...)" (*ibid.*). This additional point exemplifies the role of memory in decision-making. Its further development will lead to the above-mentioned identity between deciding and knowing, or "will and understanding" (*ibid.*, II, 49).

From all this, we can see how consistent and radical Spinoza's stance on psychophysical monism and causalism is. As a matter of fact, his philosophy inspired a "model of man" in the form of what he called a "spiritual automaton", which is more adapted to the kind of mechanistic models we can build today than to most ideas of his time (Spinoza, 1954, p. 133).

CONCLUSION: A SUPPORT FROM NEUROPHYSIOLOGICAL DATA?

Although such a stand may be difficult to reconcile with some of our usual experiences, it provides the best theoretical framework for understanding spectacular and apparently paradoxical neurophysiological data on the chronological order of decision-making and action. In a set of experiments involving subjects under brain surgery who had to remain conscious, Libet (1985) consistently found that a conscious decision to act corresponds to an electrical brain event which occurs 200 to 300 milliseconds *after* the beginning of action. This finding triggered a lot of controversy (Libet, 1992) because it does not fit in with the theory of action where the conscious decision to act is an efficient cause of the action. However, this observation can easily be understood in the context of our model. It means that the action is triggered by some neuronal unconscious stimuli, and thus supports the Spinozist statement that the cause of a bodily movement can only be a physical event in the body. A conscious observation with an understanding of our action accompanies that action, but is not its cause. We can interpret it as a decision of our will which determines the action, because we do not know the unconscious events in our body which are its real causes. Moreover, there is a slight delay between the triggering of the action and our being conscious of it, because consciousness and understanding take time. They need to be retrieved from memory. As Jeannerod (1992) so nicely put it in his comments on the neurophysiology of consciousness, "the where in the brain determines the when in the mind".

Thus, we can make sense of Spinoza's radical denial of the causal relation between mind and body entailed by their identity, since we can conceive of how "understanding and will are one and the same". This can be achieved if we have a better understanding of what is meant by causality and identity in the framework of Spinoza's *Ethics*.

Causality must not be understood as an abstract relation between events, but as a production process, which is non-transitive from the mental to the physical (and vice versa) because they are synthetically but not analytically one and the same. Following Putnam's suggestion, an examination of the kind of synthetic identity of properties exemplified today by physical quantities apprehended at two different levels, allows us to grasp what kind of mind-body identity is at the core of the Spinozist system.

ACKNOWLEDGEMENT

This work was supported by a Ishaiah Horowitz Scholarship in residence in Philosophy and Ethics of Biology, at the Hadassah University Hospital Human Biology Research Centre, Jerusalem.

Hadassah Medical Organization, University of Jerusalem
Jerusalem, Israel
CHU Broussais - Hôtel Dieu and EHESS
Paris, France

NOTES

[1] The "idea of God" is one of the ideas of the human mind. It is the "infinite Understanding of God" (in both senses of understanding God and being understood by God) that is the same idea in the human mind — which, being finite, is a part of this idea — as it is in itself as the whole of this idea. This conception, which lies at the basis of Spinoza's theory of knowledge, is, however, outside the scope of this article.

REFERENCES

Atlan, H.; Ben Ezra, E.; Fogelman-Soulié, F.; Pellegrin, D.; Weisbuch, G., 1986, Emergence of Classification Procedures in Automata Networks as a Model for Functional Self-Organization, *J. Theoret. Biol.*, vol. 120, pp. 371-380.

Atlan, H., 1987, Self-Creation of Meaning, *Physica Scripta*, vol. 36, pp. 563-576.

Atlan, H., 1992, Self-organizing networks: weak, strong and intentional. The role of their underdetermination, *La Nuova Critica*, Rome, vol. I-II, Quaderno 19-20, pp. 51-70.

Atlan, H., 1995, Projet et signification dans des réseaux d'automates: le rôle de la sophistication, in *L'intentionnalité en question*, D. Janicaud (ed.), Paris, Vrin, pp. 261-288.

Bennett, C.H., 1988, Logical Depth and Physical Complexity, in *The Universal Turing Machine. A Half-Century Survey*, R. Herken (ed.), Oxford University Press, pp. 227-258.

Davidson, D., 1980, Mental Events, in *Essays on Actions and Events*, Oxford University Press.

Davidson, D., forthcoming, Spinoza's Causal Theory of the Affects, *Jerusalem Spinoza Lecture*, under the auspices of the Spinoza Institute.

Jeannerod, M., 1992, The where in the brain determines the when in the mind, (Commentary/Dennett & Kinsbourne: Time and the observer), *Behavioral and Brain Sciences*, vol. 15, pp. 212-213.

Jonas, H., 1965, Spinoza and the Theory of Organism, *J. of the Hist. of Phil.*, vol. 3, pp. 43-57.

Koppel, M., 1988, Structure, in *The Universal Turing Machine. A Half-Century Survey*, R. Herken (ed.), Oxford University Press, pp. 435-452.

Koppel, M.; Atlan, H., 1991, An Almost Machine-Independent Theory of Program Length Complexity, Sophistication and Induction, *Information Sciences*, vol. 56, pp. 23-33.

Libet, B., 1985, Unconscious cerebral initiative and the role of conscious will in voluntary action, *Behavioral and Brain Sciences*, vol. 8, pp. 529-566.

Libet, B., 1992, Models of conscious timing and the experimental evidence (Commentary/Dennett & Kinsbourne: Time and the observer), *Behavioral and Brain Sciences*, vol. 15, pp. 213-215.

Putnam, H., 1981, *Reason, Truth and History*, Cambridge University Press, pp. 84-85.

Shanon, B., 1993, *The Representational and the Presentational*, Harv. Wheatsheaf, Simon & Schuster.

Spinoza, B. de, (naturae humanae exemplar) *The Ethics*, Preface to part IV.

Spinoza, B. de, 1954, *Tractatus Intellectus Emendatione*, par. 85 (French transl. R. Caillois), *Traité de la Réforme de l'Entendement*, Paris, Gallimard, La Pleiade, p. 133.

ALICIA JUARRERO

CAUSALITY AS CONSTRAINT

INTRODUCTION

Suppose you are trying to arrange furniture you already own in your new house. "Each piece of furniture has some weak constraints associated with it. For example, the bed must go against a wall because the headboard is rickety; two small tables go on either side of the bed; the couch is missing its back legs and so sits on books, which means it must go against a wall so that the back is not visible, and so on. The constraints associated with each piece of furniture are weak because each one can be satisfied in many ways; the bed, for example, can be put against any of several walls. However, once the bed is placed, the entire arrangement may be determined: there may be only one other wall large enough for the couch, the tables must go next to the bed, and so forth" (Kosslyn & Koenig, 1992, p. 111).

How are we to understand the relationship between the constraints and the particular furniture arrangement that results? Despite deliberately sidestepping causal language in formulating that question, it is difficult to avoid the sense that "simultaneous constraint satisfaction" as occurs in Kosslyn and Koenig's example is somehow responsible for the way the furniture gets arranged. This example suggests that constraints satisfy at least one oft-touted requirement of causal relationships: constraints support what philosophers call "counterfactual conditionals", that is, if it were not for the constraints the resulting furniture arrangement would not be what it is. And yet it is clear that the constraints don't cause the furniture arrangement the same way cue sticks cause cue balls to move. As Salthe says, although constraints carry information that has meaning for the entities involved in natural processes, the information is not dynamically involved in the processes (Salthe, 1985, p. 71).

The concept of constraint was first used formally in the Physical Mechanics. Although never actually defining it, Lindsay uses the term in his introductory textbook first to refer to the way in which the motion of a simple pendulum or a particle on an inclined plane (Lindsay, 1961, p. 35) is "*compelled by the geometry of its environment* to move on some specified curve or surface" (*ibid.*, p. 239, my emphasis). Later on, in his explanation of oscillations, Lindsay states that "some of the most important cases of constrained motion are those in which particles are *connected by rods and strings*", and cannot, therefore, move any which way. In his explanation of D'Alembert's Principle, Lindsay states that "If the masses were subject to no constraints (i.e. if they were not connected in any way or forced to move along certain curves or surfaces) ..." (*ibid.*, p. 251); and in his discussion of Gauss' Principle of Least Constraint Lindsay states, "The system being subjected to certain constraints (i.e. the masses being perhaps connected to each other by rods or cords, or constrained to move along certain curves or surfaces) ..." (*ibid.*, p. 254). I assume that in the last clause the word "constrained" means, as in the earlier quotes, either

G. Van de Vijver et al. (eds.), Evolutionary Systems, 233–242.

"forced" or "compelled". Those are the only four references to the term in the book!

In mechanics, then, constraints are features either of the object's connections with the environment or of the environment in which the object is embedded. Constraints are, therefore, emphatically not examples of what used to be called "primary qualities", intrinsic features of a system which it exhibits regardless of the circumstances in which it is placed. Constraints are relational properties. But they are not simply relationships among components within a system, which is why Pattee refuses to label a chemical bond a constraint. Constraints are relational properties components acquire in virtue of being embedded in a higher level system.

The connection of the tibia and the peronei to the knee joint *constrains* the movement of the lower leg in such a way that it makes no sense to examine the tibia's physiology, for example, independently of the knee. The tibia's connection to the knee gives the former characteristics which it wouldn't have otherwise: it can move in some ways but not others. The constraints which the connections subject the lower leg to reduce the number of ways in which the leg can move: it can bend backwards but not forwards, for example. In this example a constraint is a reduction of the leg's state space. This is the most common understanding of the term "constraint".

In the rest of the paper I wish to show that constraints not only reduce the alternatives — they also create alternatives. Constraints, that is, can also *create* properties which a component exhibits in virtue of its embeddedness in a system, properties it would otherwise not have. We need, therefore, to understand how constraints can both close up and open up options in this way.

CONSTRAINTS IN INFORMATION THEORY

In Information Theory the concept of constraint is used to refer, not to physical features, but to rules for reducing randomness in order to minimize the transmission error rate. Random, equiprobable signals are "noise", they cannot communicate. In order to be able to transmit or decipher a message sent by radar or telegraph, there must be a clear demarcation between message and background noise. The transmitter as well as the receiver, therefore, must reduce the randomness in the sequence of signals: the transmitter must encode the message according to certain rules; the receiver, in turn, will also apply the rules to decipher the message.

When signals in a sequence are random and equiprobable (the information counterpart to equilibrium and maximum entropy), what Lila Gatlin calls *potential* information or message variety is at its maximum (Gatlin, 1972). However, although in those circumstances you *could* say much, you in fact *do* say nothing, an important point since information simpliciter is often erroneously equated (by Hayles, 1990, for example) with equilibrium, suggesting that random signals can transmit actual information. Not so; when signals are equiprobable message variety is a great but idle potential. No contrasts = no message, pure noise. "Capacity is of no value if it cannot be utilized" (Gatlin, 1972, p. 99). *Actual information content* requires an ordering process that takes the system away from total randomness. Altering the probability distribution of signals in order to turn potential information into actual information is called "redundancy", and it is "one of the most important concepts in information theory. Redundancy is essentially a constraint [which reduces] the number of ways in which the various parts of a system can be arranged" (Campbell, 1982, p. 28).

Shannon & Weaver's theorems *are* two such ordering processes.

(D1): *Context-free redundancy*

The "most random state is the maximum entropy state and this is characterized by events which are both independent and equiprobable" (Gatlin, 1972, p. 87). Below I will examine the role of constraints in thermodynamic systems; let us look at their role in communications first.

Since random, equiprobable signals can carry no information, to convey a message the random distribution of signals must be constrained such that it *diverges from chance, randomness and equiprobability*. One way to do so is to narrow the number of alternatives. At the limit (one alternative with probability 1), there would be one signal repeated over and over. Regularly pulsing flashes from a lighthouse carry information precisely because their regularity is far from randomness and equiprobability, and is differentiable from the background noise. The probability distribution of letters in a particular language is another form of this D1-type of "context-free" constraint. Some letters in the English language appear more often than others. Given that x's and z's appear infrequently in English, finding an overwhelming number of x's and z's in a text and relatively few a's and e's sends the message that the text may by Hunglish, but is definitely not English. D1-type constraints on signal transmission increase improbability by creating order, pattern, which then has the capacity to carry information. Even a Martian who sees the regularly pulsing flashes of light can tell that there is a message. Robert Artigiani reminded me of Marshall McLuhan's comment that a lightbulb is pure information (personal communication). Thus far the information-theoretic concept of constraint parallels that of mechanics.

The price paid for depending solely on D1-type constraints, however, is high because reliability of transmission is inversely related to message variety. At first glance, therefore, there is an inverse correlation between imposing constraints so as to be able to send some message and the ability to say a lot in that message. At a certain point the amount of redundancy stifles the variety of what can be said. No pattern whatsoever conveys no information, but the same pattern repeated over and over again conveys no new information. The drawback with D1-type constraints, in other words, is that they are "expensive": "if increased too much, they place severe limits on the variety of messages which can be sent" (Pattee, 1973, p. 119).

Is there a thermodynamic version of all this? A container filled with evenly diffused molecules of gas is in equilibrium. A system such as this with no temperature gradient can perform no work. Moving a piston such that the gas molecules are constrained to one side of the container takes the system far from equilibrium in the sense of inhomogeneity, the thermodynamic counterpart to the regularly pulsing flashes of light. The system can now do work, the thermodynamic counterpart to "carrying information".

If nature relied solely on D1-type thermodynamic constraints, however, matter would clump but there would be no complexity, the thermodynamic analog to increased message variety. Gatlin quotes Weaver as saying that "this word 'information' in communication theory relates not so much to what you do say, as to what you could say". What is needed, therefore, is a type of constraint which limits the amount of randomness without eliminating disorder altogether; sufficient leeway must remain for new messages to be expressable. This is precisely what Shannon proved: that the need to reduce error by constraining the number of alternatives need not restrict the ability to transmit messages, for there exists a code that will provide the desired degree of accuracy without cramping the ability to transmit messages. Although, as noted earlier, one ordinarily thinks of a constraint as limiting freedom, "control constraints must also create freedom in some sense" (*ibid.*, p. 85). Actual information, that is, treads a fine line between total randomness and pure disorder. How to decrease entropy while simultaneously increasing the number of possible messages?

(D2): *Context-sensitive constraints*

In every language some letters or words are more likely to occur not just because of the overall probability distribution of letters, but depending on what letter or letters, *preceded* them, i.e. depending on what happened around them. In English, the occurrence of the letter "q" dramatically raises the probability that the next letter will be a "u". D2-type constraints are such that the *components* of a system no longer are independent of each other. This kind of constraint imposes conditional probabilities on the relationships among the letters on top of the context-free constraints mentioned earlier (which make the letters of the language in any alphabet no longer equiprobable). Gatlin calls this type of constraint "context-sensitive" constraints. There can be a series of such constraints. Syntactic requirements for a well-formed formula make nouns, verbs, adjectives and adverbs in a sentence no longer independent of each other. On top of the conditional probabilities that increase the probability (in English) that a "u" will appear *given a "q"*, syntax subjects word order to even higher level contextual constraints: for example, given the word "the" it is highly unlikely that an adverb will follow. And so on up a hierarchy of such constraints to the sophisticated poetic level of the requirements for writing sonnets or haiku.

The advantage of context-sensitive constraints is that they permit unlimited possibilities in message variety despite limited channel capacity. A great variety of messages can be sent without being hampered by the error-rate: context sensitive constraints are as efficient but not as expensive as context-free redundancy. "It can be increased by a reasonable amount without cramping the message source too severely (...)" (*ibid.,* p. 119). The fact that spoken Mandarin limits words to one or two syllables requires adding the context-sensitive constraint of inflection to permit potential message variety.

Imposing D1-type constraints to an extreme would result in the repetition of only one signal, the regularly pulsing flash of light from the lighthouse, say. The information transmitted using this D1-type of constraints is self-referential in the sense that it identifies the regularly pulsating flash of light as information. As Bob Artigiani puts it, for the Martian the D1 constraint on a sequence of signals amounts to transmitting "Message", "Message", "Message". It is the regularity's divergence from equiprobability that conveys information, but the only information it conveys is *that* there is information. D1 constraints cannot carry semantic meaning, nor can they create information. Since D2-type constraints, on the other hand, embody conditional probabilities, that is, of the probability of x *given* the context in which it is embedded, i.e. given the presence of y and z, signals or processes constrained by conditional probabilities refer not just *to* themselves; they embody and transmit information about what happened before and what is going on around them. That is, signals subject to D2 type constraints refer to the contextual web (temporal and spatial) in which that particular signal or event is embedded. The information such signals or processes carry is *of* the organization of the network; the actual information content is *about* the overall *web*, not just of its components. The potential information contextual constraints create is thus *relational* information. Relational constraints on *sequences* of letters allow words with the ability to refer to be expressed. Progressively higher level constraints allow sentences expressing even more complex semantic information to be articulated. As Bob Artigiani puts it, "meaning is the experience by parts of their interaction with the environment". It is because of his awareness of a network of relationships that the sailor, unlike the Martian, can understand that the flashes of light he sees mean "Land!". The sailor sees the flash of light *as* a node in a complex network of relationships, in particular Morse code, plus the convention that the sequence three short, three long and three short MEANS help!, etc. Indeed, the sailor is a *sailor* (and not just a man adrift on a boat) only because he himself is embedded in a network of D2-type relationships (such as navigation, tacking into the wind, etc).

Constraints, therefore, do not function just by closing off possibilities; contextual constraints are also the mechanism whereby the creation of a new level of organization with greater degrees of freedom, takes place. The amount of actual information, Gatlin says, "is a measure of all the constraints placed upon a sequence of symbols" (Gatlin, 1972, p. 94). Altering the contextual constraints creates the possibility of new meaning. The creative possibilities of metaphor are a function of the capacity of sentences (not letters) to convey semantic information, a capacity created by the very contextual constraints operating first at the level of letters, then at the level of words, and so on.

Can we find any analogs to this kind of process in dissipative and autocatalytic structures?

The emergence of the macroscopic structure of Bénard cells coincides with the appearance of contextual constraints as molecules become correlated (i.e. when the behavior of each molecule depends on what is happening around it). To borrow an example from Waldrop illustrating autocatalysis, assume that you start out with just many small molecules floating around randomly in a primordial soup. No constraints = no form or structure. Assume next that some molecules (catalysts) are able to bind two other molecules together into a large molecule. Suppose molecule A catalyzes the formation of molecule B, which in turn is able to catalyze the formation of a third molecule C; C catalyzes D, and so on, so that "somewhere down the line you might very well have found a molecule Z that closed the loop and catalyzes the creation of molecule A" (Waldrop, 1993, p. 123).

The "organizational closure" autocatalysis effects is such that a boundary between the autocatalytic web and the background soup from which it emerged has been formed. The autocatalytic web is distinguishable from its environment; it is a structure of processes (Earley, 1981). Interactions among molecules created the web; but once the molecules are captured by the web in which they are embedded (a web which, from a different point of view, the very connections among the molecules produced) means that they are no longer independent of each other. *Given* the presence of molecule C, for example, the appearance of molecule D becomes more likely. Just as u follows q in English, the probability that molecule C or D will occur is no longer a matter of simple deviation from equiprobability (whether any such probability distribution occurs, see Ulanowicz, 1994): it has become a matter of conditional probability.

CONSTRAINTS AND MESSAGE VARIETY

Just as words and sentences can be meaningful in a way that phonemes alone cannot, the emergent higher level of organization (the autocatalytic web as a whole) can access more and different states than the isolated components from which it was formed. Proteins, which in one sense are nothing other than a folded up chain of amino acids, can perform enzymatic or catalytic functions that a linear chain of amino acids cannot. Chemical phenomena can access states that physical phenomena cannot; biological phenomena can access states that chemical processes cannot. The same applies *mutatis mutandis* to the emergence of e.g. the level of a cell from the level of organelles, tissues from cells, etc. Greater number of degrees of freedom is one way increased complexity is identified. The higher level of organization is more than the sum of its parts in that sense. The explosion of potential message variety characterizing the appearance of each new level of organization (the fact that a higher level can do qualitatively different things than the earlier one) correlates with the greater number of degrees of freedom that each new dimension provides. Gatlin argues that the explosion of phenotypes that took place with the appearance of the vertebrates occurred because vertebrates managed to maintain D1 constant while allowing D2 to expand. Perhaps this combination accessed what Lewin identifies as the area in which complex structures emerge, "the site of maximum computational capability", Langton's "edge of chaos", or Ulanowicz's "window of vitality" (see Lewin,

1993; Langton, 1995; Ulanowicz, 1994).

Contextual constraints thus perform double duty. From the combined effects of contextual constraints operating on matter and energy flows, structures and patterns at increasingly higher levels of organization emerge. Furthermore, the orderly context of the system in which the parts are now embedded alters and redefines their behavior. The dynamical framework of the whole constrains the behavior of the components, a form of interlevel (in this case top-down) causality. The dynamics of the system as a whole thus provides "the framework for the behavioral characteristics and activities of the parts (Zeleny, 1980, p. 20). Molecules C and D are now different than they were before — they are *components* of a whole, and their behavior is therefore constrained (in the sense of having a reduced number of alternative ways they can behave) by the higher level in which they are now embedded. C exhibits certain traits only because of the presence of D, E and F, that is, only because it is captured in an autocatalytic web. In contrast to grains of sand, which acquire no new properties in a sand dune, the relationship of C to D is now unlike the merely external, spatial relationship among grains of sand in a sand dune: C and D are internally related in the sense that, unlike grains of sand, molecules C and D have characteristics they would not have outside the network of relationships. The constraints which the whole imposes on the parts, while restrictive in the sense that they reduce the number of ways in which the parts can be arranged, are also creative in a different sense. Once captured into a hive, bees become "drones", "workers", "queen bees", etc. A "component" or a "node" is such only because of its role in a web of relationships.

D2 constraints, that is, create new possibilities while at the same time increasing order. How is this possible? The answer is a version of Russell's type-token distinction[1]: the *potential* message variety in the sense of new possibilities which the D2 ordering process creates is at a level other than that on which the constraints operate. *The local order that contextual constraints effect at the component level is more than offset by the increased potential message variety occasioned by the new level of organization which the contextual constraints enable.* D2 type constraints create hierarchies. Once the hierarchy is established, the "bits" that created them acquire — at the cost of a reduced number of ways they can be arranged — an identity they previously did not have: they are now "components" or "nodes". of a higher level whole. In organizing the higher level whole by correlating the parts, contextual constraints increased the number of states the newly created *system as a whole* can access. Contextual constraints are thus nature's own answer to the problem posed by Maxwell's demon. Contextual constraints are a demon with a Janus nature: acting as Dr. Jekyll they impose order by limiting alternatives; as Mr. Hyde they create potential message variety through reintroducing disorder at a higher level. And it is as Mr. Hyde that they are responsible for the creative evolutionary progress of nature. The increasing complexity of evolution is a function of the operation of contextual constraints. Parts no longer independent of each other constitute the self-organization of a higher level; as such contextual constraints are the "agents" of interlevel, bottom-up causality. Acting top-down they simultaneously create new roles for those parts as they correlate them.

Because the word "constraints" suggests a reduction in the number of options a system has open to it, not the possibility of opening up a new repertoire of alternatives through self-organization, some authors (Popper, 1990; Depew & Weber, 1994; Ulanowicz, 1994) have called the creativity of enabling constraints the system's "propensities". Insofar as the term "propensities" carries with it the connotation of "predictability", however, it too is problematic: emergent properties are unpredictable. Salthe calls enabling constraints "initiating conditions". Regardless of the terminology agreed upon, for Depew and Weber the "propensities" of organisms (Depew & Weber, 1994, p. 487), because they are real and objective, are responsible for self-organization and can "provide an objective foundation for probability theory itself" (*ibid.*, p. 488). The problem with phrasing it this way, however, is that since the term "dispositional" can only be cashed out in terms of probability, explaining propensities in terms of an object's "dispositional properties" already presupposes the notion of probability. Ulanowicz analyzes propensities in terms of the reduction of probabilities, which is fine for D1-type constraints. But as we have just seen, D2 context-sensitive constraints not only reduce the number of ways the parts of a system can be arranged; they simultaneously increase the things that the system can do. While altering the behavior of the components (in the sense of reducing the number of alternative ways they can behave), contextual constraints simultaneously create a level of organization with a renewed capacity to surprise (Casti, 1994)!

Depew and Weber claim that the probability distribution of propensities is not just a matter of epistemology, that is, of the subjective probability involved in the reduction of ignorance or the renewed capacity to surprise. They claim that propensities are more than an epistemological description; propensities identify ontological processes that can serve as the basis for assigning probabilities. If this paper is correct, the ontology can be located in the way contextual constraints (on a linear sequence of amino acids, for example) self-organize the related components into a higher level (a protein) which in turn can access more and different states (enzymatic function). It is the emergence of a new function, enzymatic activity, that surprises. Contextual constraints operating "bottom up" (as "enabling constraints") are responsible for the formation of higher level wholes with emergent properties. In fact, as recent newspaper reports on the fact that scientists continue to discover that proteins can do more and different things that were ever thought possible, all the possibilities which contextual constraints create are not yet known. The potential which enabling constraints create is real.

CONSTRAINTS AND STATE SPACE

What are constraints, then? Constraints are alterations in the probability distribution of a system's state space. As enabling constraints operating bottom up, contextual constraints free up a set of states which the higher level system of relationships they create can now access. Systems of relationships themselves can in turn become related (the earlier relationships now becoming the relata in a new relationship), thus evolving into systems of ever higher level relationships with creative new properties of their own. Once the higher systemic level is in place, it acts as a top-down selective constraint on the (now) lower level components from which it organized, altering the number of ways they can be arranged.

D2-type contextual constraints partially decouple the whole from its environment and from its components by creating and regulating a whole/part hierarchical system. The dynamics of the system function so as to renew the system as a whole, not its individual components. As structures of relationships Bénard cells are Bénard cells whether they consist of water or other viscous materials. An autocatalytic structure orders the processes whereby molecules are taken in as food and those dumped into the environment as waste. Furthermore, autocatalytic structures streamline their processes (Ulanowicz, 1994) so as to increase the energy throughput (the autocatalytic version of "efficiency") by preferring certain pathways and discarding others. To identify a set of processes as a "Bénard cell" is thus to identify an emergent, supervenient property which contextual constraints have created, that is, it is to identify a macrostate which can be accessed by different microstates (Depew & Weber, 1994, p. 471).

Likewise, the dynamics of the system as a whole determines which molecules are "fit" to survive or to be imported into the system, and which are not. The property "fitness" thus refers to D2 constraints operating top-down: the dynamical framework in which the components are embedded (for example, the ecological niche in which a species is found) rules on the adequacy of the components by selecting among those components. "Selecting" here, once again, means "reducing alternatives". But as in the opening example (the furniture arrangement), such top-down selective constraints are often weak constraints: they can be satisfied in many ways. As Depew and Weber note, "fitness" is a supervenient property. The point of the above discussion, of course, is that the creation of a new level of organization effected by D2 contextual constraints *is* the emergence of a supervenient property which, as such, is partially independent of its specific components.

When several weak constraints must be satisfied simultaneously, however, the situation can begin to look deterministic. In the furniture example the many different weak constraints that needed to be simultaneously satisfied makes the system converge on only one possible arrangement. I suspect that it might be possible to reconceptualize those forces referred to as "causes" within the mechanistic framework as the limit of constraint satisfaction: those cases in which the probability of the occurrence of a particular situation given a certain set of constraints approaches unity. But that is another paper.

CONCLUSION

I close with the following observations pertinent to philosophy of mind: for centuries the problem of intentional action has been a philosophical headache. How does an intention "cause" action? How do a bunch of neurons "become" a mind capable of intentionality? Certainly while causality was explained mechanistically, as the collision-like relationship of billiard balls, intentional behavior was inexplicable. And if a whole could have no properties over and above those of the aggregation of its components, then the emergence of mind was indeed a mysterious process. If causality is reconceptualized in terms of the operation of constraints, intentional behavior can be rethought as an example of constraints operating top-down. The emergence of mental properties like intentionality can in turn be understood in terms of the workings of contextual (enabling) constraints operating bottom up. Likewise, while causality was understood mechanistically, formal and final causes, and in particular their role in self-organization (in the sense which Kant uses it in the *Critique of Teleological Judgment* — leaves produce a tree which in turn is produced by the leaves) were inexplicable because they involved "a kind of causality unknown to us", as Kant phrases it. Appreciating the Janus nature of contextual constraints makes this kind of whole/part and part/whole interactions understandable.

Department of Philosophy, Prince George's Community College
Largo, Maryland

NOTES

[1] I Recall Russell's problem: A barber in a village shaves all the townspeople who do not shave themselves. Who shaves the barber ? Russell's solution was to argue that classes belong to a higher logical type than the elements that compose that class.

REFERENCES

Campbell, J., 1982, *Grammatical Man*, New York, Simon and Schuster.

Casti, J.L., 1994, *Complexification*, New York, Harper Collins.

Depew, D.J.; Weber, B.H., 1995, *Darwinism Evolving: Systems Dynamics and the Genealogy of Natural Selection*, Cambridge, Mass., MIT Press.

Earley, J., 1981, Self-Organization and Agency: in Chemistry and Process Philosophy, *Process Studies*, vol. 11, pp. 242-258.

Gatlin, L., 1972, *Information and the Living System*, New York, Columbia University Press.

Hayles, N.K., 1990, *Chaos Bound: Orderly Disorder in Contemprorary Literature and Science*, Ithaca-NY, Cornell University Press.

Holyoak, K.J.; Thagard, P., 1995, *Mental Leaps*, Cambridge MA, MIT Press.

Kosslyn, S.M.; Koenig, O., 1992, *Wet Mind*, Free Press.

Langton, C., 1995, *Artificial Life: An Overview*, Cambridge MA, MIT Press.

Lewin, R., 1993, *Complexity: Life on the Edge of Chaos*, New York, Macmillan.

Lindsay, R.B., 1961, *Physical Mechanics*, Van Nostrand.

Pattee, H., 1973, *Hierarchy Theory*, New York, George Braziller.

Popper, K., 1990, *A world of Propensities*, Thoemmes.

Salthe, S., 1985, *Evolving Hierarchical Systems*, New York, Columbia university Press.

Ulanowicz, R., 1994, The Propensities of Evolving Systems, in *Social and Natural Systems*, E.L. & K. Boulding (eds.), Cambridge.

Waldrop, M., 1993, *Complexity*, Touchstone.

Zeleny, M., 1980, *Autopoiesis, Dissipative Structures, and Spontaneous Social Orders*, Westview.

GERTRUDIS VAN DE VIJVER

EVOLUTIONARY SYSTEMS AND THE FOUR CAUSES: A REAL ARISTOTELIAN STORY?

INTRODUCTION

Let us grant, with Couloubaritsis, that philosophy is by definition a historical enterprise. If we accept that the birth of Greek philosophy took place as a consequence of a new, more radical form of astonishment, and if we accept that this astonishment leads philosophers to more radically question their experience of being in the world, that is, to drastically shuffle the dogmatic, imposed or naive interpretations of it, then it is quite natural to see philosophy as a matter of interpreting the world in an ever new, different and critical way. In other words, it is natural to define it as a construction of historical meanings (cf. Couloubaritsis, 1994, pp. 11-21).

Does this imply that philosophy is 'just' a matter of adding one meaningful story after the other? Certainly not. If at each moment one looks for the *particular* conditions that lead to certain historical meanings, if one attempts to analyze particular solutions as answers to particular questions, it becomes difficult to sustain a 'just so' viewpoint on stories. In this regard, basic questions are: What story do we prefer? How and why do we give preference to a particular story? What principles guide our choices?

I suggest taking this 'particularizing' attitude as a guide for analyzing the way in which Aristotle's theory of causality is currently used in evolutionary systems theory (cf. Alvarez de Lorenzana, this volume; Minch, this volume; Salthe, 1993; Swenson, 1990, 1992; Ulanowicz, 1990). The aim is: 1. to elucidate the way in which particular answers were connected with particular problems at the time of Aristotle, 2. to clarify what is at stake today in particular forms of actualization of this philosophy, in other words, to elaborate here also the relation between particular problems and particular answers, 3. to make explicit the relevance and appropriateness of using old problems and solutions for tackling new ones.

Aristotle is a very special philosopher to do this kind of exercise with. Indeed, in the context of evolutionary theory it appears that his thinking is quite close to the developmental thread that has remained largely implicit since Darwinian and neo-Darwinian formulations of evolution (see a.o. Depew & Weber, 1995; Amundson, 1994). Besides this, however, his philosophy of nature inspires many scientists and philosophers who feel uncomfortable with some of the philosophical and scientific implications of Modern Science. As a consequence, there is frequently a double-layered story in 'neo'-Aristotelian evolutionary thinking: on the one hand there are the more or less concrete arguments in favor of a developmentalist stance, and on the other we have the much more complex reasons, philosophical and/or epistemological, to try to arrive at an alternative conception of what science is or ought to be. I will attempt to show on the basis of the causality debate that the major issue will be whether and under what conditions these two aspects are to be

G. Van de Vijver et al. (eds.), Evolutionary Systems, 243–254.

connected or disconnected. In other words: in what way do questions about evolution and development oblige us to adopt a radically different viewpoint on scientific knowledge?

Let me begin by sketching the Aristotelian view on causality.

ARISTOTLE ON CAUSALITY[1]

It is impossible to understand the place causality occupies in Aristotle's philosophy without presenting his theory of nature, and within it, the place of development.

Developmental thinking

Many fragments of Aristotle's texts on nature indicate that development and change, in living and non-living systems, in human and non-human systems, have a central place in his thinking. In saying, for instance, that the behavior of nature towards itself is such that, "what grows *qua* growing grows from something into something" (*Phys. II*, 1, 193b, 16-17)", Aristotle clearly rehabilitates the possibility of development in nature (cf. Couloubaritsis, 1991, pp. 7-8).

Faithful to the idea of development, Aristotle considers that a *theory of nature* cannot but deal with the relation between being (*to on*) and motion: "each [natural being] has within itself a principle of motion and of stationariness (...) which seems to indicate that nature is a principle or cause of being moved and of being at rest in that to which it belongs primarily, in virtue of itself and not accidentally" (*Physics*, Book II, 1, 192b, 14-23). As a consequence, a theory of nature has to deal with a particular sort of substance[2], namely that "which has the principle of its movement and rest present in itself" (*Metaphysics*, Book E, 1, 20-21), and which develops, inseparably from matter, towards its 'raison d'être', its reason of being.

The latter viewpoint on substance implies an important shift with regard to the Platonic conception of nature, in which development, inherently lacking stability and scientificity, had to be discarded in favor of the eternal Ideas. To Plato, substance is a transcendent essence inaccessible to the senses. To Aristotle, however, it is set apart from the other categories (such as quality, quantity, relation, ...) and put forward as reality *par excellence*. Substance is a 'distinct' reality and ultimate subject of attribution. Substance is the thing itself in its reality. As we will see further on, the unity of form and matter is for Aristotle what constitutes a thing in its reality.

This attention for substance shows that Aristotle not only considers things in their being present, he also considers them as things that are according to the categories (essential modalities of being). To him, the first of these categories is substance, and the most important development is development of substance.

A theory of development in general

In undertaking a more detailed description of all development of things in general, as described in Book I of *Physics*, Aristotle arrives at three principles that account for development: specificity, privation and matter. There clearly is a difference in things that are (beings) and things that become: beings are one and identical to themselves[3], and things that become are numerically one but multiple as to the specificity (*eidos*). As a matter of fact, all becoming is in the meantime the initial being and also the privation of the being that is to come. In that sense, in becoming, the initial being submits its own specificity (*eidos*) to the specificity of a new being. It is, as a consequence, a substratum or a subject of becoming (i.e. matter, *hypokeimenon*), which conceals the specificity of that which will come, but only in a negative sense, as a privation, as an absent *eidos*.[4] In summary, *specificity* is transmitted on the basis of movement and realizes the phenomenal form of a thing; *matter* is the most immediate and closest substrate (the subject) of it; *privation* is specificity in a negative sense.

A physical theory of development: the four causes

If a physical science is to be established, and not just a science of development in general, Aristotle has to undertake to describe the specificity of natural things in their development. This is the issue in Book II of *Physics*. Here, Aristotle describes development in terms of the causes of natural beings. In other words, the question here is: what are the causes of development in nature? An additional question is: what makes natural things different from other things, such as those produced in art? Underlying these questions is Aristotle's basic attitude with regard to physics: its aim is to precisely describe the objective conditions of development. That is what he endeavors with his theory of the four causes, situating it thereby clearly in a perspective of knowledge.[5]

In this regard, it is interesting to note that to Aristotle, no being, human or non-human, living or non-living, produces anything by chance. His viewpoint on chance and spontaneous movement indicates why he attaches so much importance on final causes. Indeed, Aristotle refuses to see chance and spontaneous movement as the causes of the universe, because otherwise they would be anterior to intelligence in the case of human activity, and anterior to nature if natural development is understood to be a process that happens always or frequently. It is important to stress this viewpoint, because, if accidental causes are to be posterior to 'essential' causes, it implies a search for the ultimate causes of things in terms of their essential and final causes. This clarifies the fact that Aristotle is looking to ground finality in nature, much more than to ground nature in mechanistic terms. Any apprehension in mechanistic terms, reduced to material necessities, is thus submitted to form, and hence, to the grip of final causes. So, it is only because things happen that are different from those that happen always and frequently, that chance and spontaneous movement exist. This is what constitutes the metaphysical background of Aristotle's theory of nature.[6]

How then does the theory of the four causes fit in the task of describing the particular manifestations of the various causes in nature? How does it, moreover, fit in the general picture of Aristotle's physics? The four causes can be defined as follows:
— The *formal cause* is the specificity (*eidos*, structure) that is transmitted to matter on the basis of movement.
— The *material cause* is the material, bodily reality that functions as the substrate of change. In this, matter and form are inextricably linked: the unity of form and matter is what constitutes a thing in its reality (hylemorphism).
— However, without the intervention of movement, development is excluded. The *efficient cause* is the movement that transmits the specificity to matter and that, as such, realizes proximately the development of beings.
— And finally, there is the goal at which change is directed, the goal of relative perfection in which a being is considered to have optimally accomplished its nature or to have realized its form: the *final cause*. Form is to be situated at the side of final cause, specificity at the side of formal cause. The specificity (*eidos*) realizes the phenomenal form of the thing.[7]

Causality at a junction

It are the relations between the various causes that give us the clue to understanding the coherence of Aristotle's physics. They will also lead to a better understanding of the Aristotelian conception of causality in evolutionary systems theory. What do we know about these relations?
— We know that there is an inherent link between formal and final causes: the specificity (*eidos*), that is in accordance with the reason of being of the entity in question, is realized through the form, i.e. the final cause. In other words, form is only manifest through a particular organization of matter (structure).
— What makes this "formation of matter" possible is the efficient cause, which allows the specificity of a system to be transmitted. In other words, Aristotle's conception of development depends upon efficient cause. Developmental systems confront us with the problem of the interaction between the internal determination of systems and the transmission and development of their specificity. In this sense, Aristotle states that the formal, final and efficient causes often coincide, for "the what and for the sake of which are one, while the primary source of motion is the same in species as these. For man generates man — and so too, in general, with all things which cause movement by being themselves moved; and such as are not of this kind are no longer inside the province of natural science, for they cause motion not by possessing motion or a source of motion in themselves, but being themselves incapable of motion" (*Physics*, Book II, 7, 198a, 25-30).
— The relation between these three causes (i.e. the formal, the final and the

efficient) and the material cause is one of necessity. "The necessity in nature, then, is plainly what we call by the name of matter, and the changes in it" (*Physics*, Book II, 9, 200a, 30-35). In this regard, Aristotle rejoins his initial viewpoint about the relation between form and matter, and indicates that the unity of the three causes as described above, is included in a more fundamental unity, namely that of the four causes. "Perhaps the necessary is also present in the definition. For if one defines the operation of sawing as being a certain kind of dividing, then this cannot come about unless the saw has teeth of a certain kind; and these cannot be unless it is of iron. For in the definition too there are some parts that stand as matter" (*Physics*, Book II, 9, 200b, 4-8). Clearly, at the end of Aristotle's second Book of *Physics*, we arrive at the quite surprising conclusion that, for natural beings, the material cause is in a way included in its reason of being.

This short presentation of Aristotle's physics brings me now to the following questions:
— Does evolutionary systems thinking promote an Aristotelian viewpoint on causality because it stresses the structural aspects of developing systems? Is the attention to structure to be equated with Aristotle's 'substantialism'? What about the actual interpretation of the structural and morphological aspects of the Aristotelian project?
— Is it the case today, in the Aristotelian 'wing' of evolutionary systems thinking, that the relation between, on the one hand, the formal/final/and efficient causes, grounding a 'substantialist' viewpoint, and on the other, the material cause, similar to Aristotle's viewpoint? Is the material cause here also in some sense included in the 'specificity' of natural beings?
— Is it today the case that the efficient cause is what makes self-organizational dynamics, i.e. the 'formation of matter', possible? Is the current conception of the efficient cause the same as with Aristotle, i.e. 'the same in species', or in other words, is it intricately linked with the definition or specificity of the natural being?
— What can be concluded out of this for the modern scientific causality picture? Is it indeed so that in modern science the focus is on efficient cause, to the detriment of the three other causes? Or is the latter a biased physical viewpoint inadequate, a.o., for the domain of biology?
— And finally, to come back to the question I addressed in the introduction: Do we have here, as with Aristotle, a similar link between the attention to development and the metaphysical background and methodological options? Or in other words, do evolutionary systems thinkers believe that chance and spontaneous movement are posterior to final causes, and that natural systems, as a consequence, *cannot but be* apprehended on the basis of final causes?

It would certainly be worthwhile to address all these questions in detail, in the context of the various uses made of Aristotle today. However, in the rest of this paper, I can only focus on one position in EST, namely the interpretation of Salthe and Matsuno (in particular, Matsuno & Salthe, 1995; Salthe, 1993, this volume). In this regard, my aim is to show 1. that the current attention to Aristotelian causality is grounded in the belief that natural systems are continuously evolving, 2. that structure occupies a central, though problematic place in the new Aristotelian picture 3. that the basic metaphysical and methodological assumptions are not quite comparable to Aristotle's: in Aristotle we have what I would call a *"developmentalism of one"*, leading to clear-cut classifications and a quite static overall understanding of nature close to linguistic categorizations[8], expressing what is nowadays called an externalist viewpoint, whereas in EST, we have a *"developmentalism of many"*, in which the development of structure is understood in emergentist and self-organizational terms, expressing radically different metaphysical and methodological options, sometimes identified as internalism, sometimes considered as a combination of internalism and externalism. I will attempt to delineate the major reasons for the divergences between both approaches, and formulate some suggestions that can serve a further elaboration of a modern developmentalist position, and more generally, of a genuine historical conception of science.

EVOLUTIONARY SYSTEMS AND ARISTOTLE

Structure: the core of current Aristotelianism?

I believe it is the attention to structure that drives most evolutionary system thinkers in criticizing the neo-Darwinian approaches to evolution. Whereas the latter consider natural systems as black-boxes, the evolution of which is to be described in terms of blind variation and natural selection, evolutionary systems theorists consider that the boxes do have insides, which need to be concretely, materially studied when dealing with their evolution.

However, it appears that a particular kind of structuralism is at work here: the attention to structure is of a dynamical, radically historical kind.[9] Indeed, this type of dynamic structuralism considers the realization of structure and generic form as the result of dynamical processes. Therefore, it aims at getting at the "momentary, local causal capacity of contingency" related to the local dynamics that are basic in this development (Matsuno & Salthe, 1995, p. 320): "Materialism is the preferred term for the fact that nothing can be realized in our world without the opposition of friction. This includes the accession of deep structures in surface realizations. In other words, the difference between *deep structure (reality)* and *surface structure (actuality)* is the scarring of history made necessary by contingency (caused by the fact that there is not only a single thing going on in the world). Structures are realized only with the elaborations necessitated by history" (cf. Salthe, 1996, personal communication, my emphasis). Hence, evolutionary systems is a "*relatively materialist enterprise*", that stresses the details of the actual generation of systems and their variants (Salthe, this volume; see also Amundson, 1994; Depew, this volume; Kampis, this volume).

The focus on structuralism and materialism at first sight is the most straightforward Aristotelian point in current EST. Already in the formulation, it appears that deep structure is close to the Aristotelian idea of specificity and substance, and surface structure to the idea of form. As in this context Aristotle's causal categories frequently serve to articulate precisely the idea of structure, as well as the emergence of structure starting from the material basis (Salthe, 1993, p. 42; Matsuno & Salthe, 1995), it is appropriate to have a closer look at the way in which the four causes are used in this context.

A structuralist view on causes

We saw that evolutionary systems theory aims at getting beyond the neo-Darwinian approach, that is, endeavors to describe the emergence of structure starting from the "local capacity of contingency". In this, it is believed that efficient causes alone cannot be sufficient to explain the evolution of complex systems. A complex causality is required: "Material, formal and even final causality must be involved at every local coordinate" (Matsuno & Salthe, 1995, p. 321).

Here, the *material cause* refers to the uniqueness of local dynamics, i.e. energy flowing through associated systems at contiguous levels, which proceeds sequentially. *Formal causes* are identified with boundary conditions from the environment — environmental parameters, and *efficient causes* with episodic, external perturbations. *Final causes* have to do with the securing of whatever consistency (conservation rules, and so on) of local dynamics: "(...) local operations are by no means merely the efficient causes of simultaneous operations globally. Local dynamics must be finalistic as well in necessarily securing whatever consistency (...) that must be identified in subsequent records" (Matsuno & Salthe, 1995, p. 316).

A comparison with Aristotle's categorization cannot but raise many questions. Apparently, we have gone a long way, and this appears most clearly in the formal and the final causes. Indeed, what is the relation between a formal cause defined in terms of boundary conditions from the environment, and a formal cause defined as the specificity of natural systems? What is the relation between the final cause as the perfect form to which any natural system develops, and the (descriptional) aim of securing consistency?

We know that for Salthe the formal and final causes are inherently linked, although they are not identical, but it remains difficult to conceive of a linkage between the two when they are not related, in one way or another, from within the natural system. It is difficult because organization or structure is precisely what gives a kind of unity to the system. Also, the efficient cause is not included in some way in the specificity of the natural system, as with Aristotle, because it refers to episodic perturbations only, of internal of external origin. So, at first sight, it looks problematic to define a 'unity' between the efficient, final and formal causes on the one hand, and the material cause on the other hand. Clearly, this is not an 'identitarian' conception of the four causes.

Let us grant, however, that it is the notion of structure that comes to secure the linkage between the four causes. As Aristotle believed, natural beings do have a structural integrity, natural beings do develop towards structural integrity, always or frequently. Structure expresses this organizational, systemic aspect of natural beings. In the current context, structure could be taken to refer to that 'organization from within', even if the linkage between formal, final and efficient causes is not as straightforwardly 'internal' as with Aristotle. But at present we can leave the question open as to whether it is to be situated at various levels, or whether there is a basic level of identification of structure, and in what way it acquires a unity through a particular history.

My main point is the following: if the notion of deep structure is what comes to secure the linkage between the four causes, it remains to be seen, firstly, in what precise sense structure relates to the idea of emergentist local dynamics. How can an emergentist self-organizational development explain the passage of structure from one generation to another? — and secondly, how does a structure relate to its expression(s), its actuality(ies), i.e. its form(s). Whereas with Aristotle the structure (specificity) realizes the phenomenal form, in this case it is doubtful whether the form (i.e. the final cause, that is, a description) ever fully represents the structure of the system. One of the reasons for this is that the final cause is not defined from within the system; it is part of the *description* of the system, in a very particular way.

In order to try to grasp the latter point, it is necessary to have a look at the broader philosophical and metaphysical frame, that is, I believe, in major part responsible for the divergences between Aristotle and evolutionary systems thinking.

About some metaphysical and methodological options

Here we arrive at one of the most radical aspects of Matsuno's and Salthe's proposal, namely the intricate relation between local and global, i.e. local and global dynamics and local and global descriptions. It is radical because it indicates a radical choice for the logical priority of particularity, locality and history. Priority is given to local descriptions of dynamics: "Local descriptions of dynamics put priority on concrete particulars, while global consistency becomes secondary and derived — a consequence in a subsequent record of an integration of preceding, uncertain local dynamics and individuations" (Matsuno & Salthe, 1995, p. 312). Hence, local dynamical activities have temporal priority over global dynamical descriptions, leading, as we saw above, to a materialist reductionist assumption: "From the point of view of any single local situation, the global is almost (but of course not fully) epiphenomenal" (Matsuno & Salthe, 1995, p. 315).[10]

The basic reason for giving priority over local perspectives, is the conviction that observers only have limited access to the systems they are studying, i.c. complex systems. Complex, that is, continuously evolving systems, can only be grasped locally, hence partially and inadequately. "Our recognition of causal relations between local dynamics and description forces us to transcend mechanistic limitations" (Matsuno & Salthe, 1995, p. 318).

The fact that observers have only limited access, indicates the impossibility of a global viewpoint, and hence the necessity of taking part in the local dynamics. Therefore, description and observation are, as essentially material processes, related to the local dynamics: "The relationship between dynamics and its description must (...) be considered to be mutually causative within any schema of local dynamics (...) historical description (and the concept of self-organization as well) is made possible only when description is local" (Matsuno & Salthe, 1995, pp. 316-317). Therefore, in a local process of observation and description, the final cause cannot be taken to 'fully' express the realization of relative perfection or the attainment of perfect form. The final cause is expressing something quite different. It indicates the fact that natural systems are driven by local dynamics and cannot but write history on this basis, being meanwhile driven by what can be seen as the treasure of past global records. The latter constitute a source of future dynamics, but not in the form of a program that guides the future actions. As a matter of fact, as in each new dynamics the past frozen records are being localized, they are inevitably rewritten and they force the system to continue its semi-anticipating, hasty, almost Beckettian way to be in the world.

What is then the basic reason for the ineluctable loss of the 'beautiful' unity of the Aristotelian causal schema ? To me it is the idea that contingency and locality are basic and temporally prior.[11] Whereas with Aristotle we have a refusal to see chance and spontaneous movement as the causes of the universe and of natural beings, in this case we have a prominent role played by contingency, which leads to the impossibility to consider that which appears as an expression of structure (form, similarity, ...) as a full, undistorted and timeless expression of it.

This is what constitutes the metaphysical background of Matsuno's and Salthe's viewpoint, and what leads them, methodologically, to go far beyond the mechanistic, a-temporal, basically a-historical viewpoint of modern science.[12]

CONCLUSION: WAYS OF KILLING TIME?

We saw that for Aristotle development is of primary importance, and in the case of natural beings the essential development is the development of substance. Aristotle finds it unacceptable that natural beings be governed by chance or spontaneous movement, and opts for a description in terms of final causes, the latter being inherently linked to the specificity of the system in question. This leads to a developmentalism of one, as I called it. It clearly has the advantage of considering form as the expression of specificity, thereby bridging the gaps between specificity (structure), phenomenal reality (form) and language. But it has the disadvantage of stopping time when it comes to describing natural systems. Symptomatic in this regard is the difficulty of combining the autonomous, self-organizational side of natural systems with the fact that they are influenced by the environment. Aristotle solved it on the basis of a technical model, in which the idea of contact (movement) was essential. Aristotle's conception of causality is not a self-organizational, emergentist one.

For Salthe and Matsuno, development is certainly of primary importance, and in the case of natural beings, the essential development is the developmental realization of structures. However, natural beings are clearly governed by inherently contingent processes — this is the basic characteristic of local dynamics. In the description of these local processes, various causes interfere and so a complexification of causality is deemed necessary because efficient causes alone, leading to a basically mechanistic apprehension, cannot account for the development of structure.

Evolutionary systems do not encounter Aristotle's difficulty in describing the evolution of natural systems: chance, spontaneous movement, contingency are the basic sources of any material evolution. At the other end, however, it is necessary to describe the repeated appearance of generic form, that transcends local dynamics, as something timeless emerging out of those dynamics. There *is* something robust in this world. And here arises a basic difficulty, to which I have no answer at the moment: how to deal, theoretically, with structure, if it is not clearly linked to form, and if it does not lead in this way to determined descriptional categories, as with Aristotle?

Another sense of understanding and explanation comes to mind, however. This is a sense in which no concession whatsoever is made as to the historical way of being in the world. It is indeed fundamentally different to choose contingency and particularity first, and robustness second, than the other way round. In the frame of a locally multiplied, fractionated vision of dynamics and description, according to which no description whatsoever is faithful to the underlying local dynamics because of the fact that the observer is implicated in the process, the most basic questions to be asked are perhaps ethical questions, related to local choices. For science, this opens up some perspectives that probably are for many scientists too risky to even think of, but that would allow one to genuinely address the question Evelyn Fox Keller (1995) formulated, namely: "What do you want science for?".

ACKNOWLEDGEMENTS

My sincere thanks to Stan Salthe and Rod Swenson for their extensive and illuminating remarks on this and a previous draft of this paper.

Department of Philosophy and Moral Science, University of Ghent
Senior Research Associate FWO
Ghent, Belgium

NOTES

[1]For more detailed descriptions of Aristotle's physics, see Couloubaritis, 1991, 1994. For this section on Aristotle, and in particular for the subdivision in three parts, I follow the introduction of Couloubaritsis

to Aristotle's Book II of *Physics* (1991, pp. 5-39).

[2] Usually, the term 'ousia' is translated as substance. This translation is clearly problematic, for substance was the medieval translation of the term, and supposes a reductionist interpretation of the aporia's of Book Z. of *Metaphysics*. Translating the term by 'essence' is still more problematic, as its Platonic characteristic obliterates the originality of Aristotle on the point of development. In French, the term has been translated by 'étance' (Fédier, Couloubaritsis), which is a substantive made of the present tense of the verb to be ('être'). Moreover, the etymology of the term 'étance' refers to *stare* (to stand up), which is equally the case for the term substance (cf. Couloubaritsis, 1991, p. 152). Here, I will use the traditional translation, i.e. substance.

[3] Couloubaritsis stresses that with Aristotle the essence of natural beings would be without any subsistence if there were not something that ascertains their unity. The issue of Being is inextricably linked to the issue of One. Questions in this regard could well be of importance for the metaphysics and the methodology of Aristotle — so he says — but remain largely unaddressed today (see in particular, Couloubaritsis, 1991, pp. 10-13). We will see further that it is exactly this aspect which is central in evolutionary systems theory, generally addressed in terms of structure.

[4] In this way, Aristotle can consider privation as a certain specificity. This reasoning in terms of specificity and privation paves the way for Aristotle's philosophy of actuality and potentiality.

[5] A theory of causality has to provide explicative principles of things; it has to be pertinent — the reality in which the causes find their origin has to be the criterion of assessment —; it has to be productive — on the one hand, it has to have in view a generic description that clarifies the foundation of things, and on the other it has to account for the non-scientific domain of the accidental where spontaneous movement and chance reign.

[6] For more details, see Couloubaritsis (1991, pp. 31 sq). It is good to have this in mind in reading the current propositions of evolutionary systems thinking. It gives rise to the following question: does the attention to Aristotle's view on causality imply equally the anteriority, or priority of the final causes over 'chance and spontaneous movement' ?

[7] I believe that in some proposals in developmental biology, the concern for structure is mistakenly considered as the concern for form. This becomes particularly clear in Amundson's comparison between Darwinism and developmentalism on the one hand, and behaviorism and cognitive or neurological models in psychology on the other hand (see Amundson, 1994, p. 575-576).

[8] This is sometimes called Aristotle's 'constitutive essentialism' (cf. Depew & Weber, 1995, p. 36 ff.).

[9] That the interpretation of dynamic structuralism varies, can become clear from Petitot's theory. In his morphodynamical approach (cf. Petitot, 1993) the emergence of form is stressed in both biology and cognition, the various constraints at work in these processes are analyzed, i.e. those that are independent of the various levels, and those which are to be considered as mere contingencies. Petitot will say, with regard to cognition in particular, that structure *expresses* the internal dynamics, and that we have to attempt to reconstruct the internal dynamics starting from the external expression. This is clearly different from Salthe's viewpoint, in which, as we will see further, form is distinguished from structure, in that the latter is in the first place, a 'deep structure'.

[10] It remains vague in what sense it is 'not fully' epiphenomenal, but it would lead too far to disentangle here the issue of reductionism (cf. Van de Vijver, 1995).

[11] Salthe states in this context that deep structures, as structural attractors, preexist local dynamics. Structuralism to him refers to the fact that there exist universal, unchanging structural attractors that are influential in all local dynamics. Salthe would call his interpretation of structures more indebted to Plato than to Aristotle. I presume this is not the viewpoint of Matsuno though.

[12] It also leads them, quite logically, to dialectics, idealism and semiotics, namely to questions related to the description of evolving systems from a global viewpoint. It would, however, require another article to deal with this aspect.

REFERENCES

Amundson, R., 1994, Two concepts of constraint: adaptationism and the challenge from developmental biology, *Philosophy of Science*, vol. 61, pp. 556-578.

Aristotle, *Physics*, in *The Complete Works of Aristotle*, 1984, J. Barnes (ed.), Princeton, Bollingen Series LXXI, vol. 1, pp. 315-447.

Couloubaritsis, L., 1991, *Aristote: Sur la Nature (Physique II)*, Introduction, Librairie Philosopique J. Vrin, pp. 5-39.

Couloubaritsis, L. 1994, *Aux origines de la pensée européenne*, De Boeck, Bruxelles.

Depew, D.; Weber, B., 1995, *Darwinism Evolving. Systems Dynamics and the Genealogy of Natural Selection*, MIT Press, A Bradford Book.

Fox Keller, E., 1996, Just a phrase they're going through, interview in *The Times Higher Education Supplement*, Feb. 9, 1996, pp. 16-17.

Matsuno, K.; Salthe, S.N., 1995, Global Idealism/Local Materialism, *Biology and Philosophy*, vol. 10, no. 3, pp. 309-337.

Petitot, J., 1993, Phénoménologie naturalisée et morphodynamique: La fonction cognitive du synthétique a priori, *Intellectica*, vol. 17, no. 2, pp. 79-127.

Rosen, R., 1986, Causal Structures in Brains and Machines, *International Journal of General Systems*, vol. 12, pp. 107-126.

Rosen, 1985, *Anticipatory Systems. Philosophical, Mathematical and Methodological Foundations*, Oxford, Pergamon Press.

Roth, G.; Schwegler, H., Self-organization, Emergent Properties and the Unity of the World, *Philosophica*, vol. 46, no. 2: Self-organizing and complex systems I, G. Van de Vijver (ed.), pp. 45-65.

Salmon, W., 1984, *Scientific Explanation and the Causal Structure of the World*, Princeton, Princeton University Press.

Salmon, W., 1993, The Value of Scientific Understanding, *Philosophica*, vol. 51, no. 1, pp. 9-21.

Salthe, S.N., 1985, *Evolving Hierarchical Systems*, New York, Columbia University Press

Salthe, S.N., 1993, *Development and Evolution: Complexity and Change in Biology*, Cambridge, MA, MIT Press.

Swenson, R., 1990, Evolutionary Systems and Society, *World Futures*, no. 30, pp. 11-16.

Swenson, R., 1992, Autocatakinetics, Yes — Autopoiesis, No: Steps Towards a Unified Theory of Evolutionary Ordering, *International Journal of General Systems*, vol. 21, no. 2, pp. 207-228.

Ulanowicz, R.E., 1990, Aristotelean causalities in ecosystem development, *Oikois*, vol. 57, pp. 42-48

Van de Vijver, G., 1995, The relation between causality and explanation in emergentist naturalistic theories of cognition, *Behavioural Processes*, vol. 35, Elsevier Publishing Company, Amsterdam, pp. 287-297.

Van de Vijver, G., 1996 (in Japanese), Internalism versus externalism: a matter of choice?, *Contemporary Philosophy*, sept. 1996, vol. 24, no. 11, pp. 93-101.

GEORGE KAMPIS

EVOLUTION AS ITS OWN CAUSE AND EFFECT

INTRODUCTION

Darwinism is fundamentally incomplete (or, if we are less permissive, fundamentally wrong).

This statement needs some clarification. It would be difficult to deny the existence of differential reproduction, to reject the notion of Malthusian growth, or to challenge the relevance of gradual changes. It is these that bring forth the amazing adaptations we admire when studying evolution. In this sense, Darwinism rests on solid grounds.

In other words, adaptation and selection can certainly be taken as granted. If we are only interested in adaptation and selection, then Darwinism is, in one or another of its modern forms, the very theory we need.

However, I believe that is exactly one half of the story. Evolution theory based on adaptation and selection alone is fundamentally one-sided. It is not just organismic constraints, the Bauplan idea and its relatives, that can modify the adaptationist picture[1]; the problem has a subtler, and at the same time more central element as well.

Relying on variation and selection in the first place, is like having a theory of classical mechanics based on purely kinetic considerations. Kinetics deals with the speed and the direction of the motion, *when the forces have already perfected the shaping of the path*. In kinetics, you have trajectories, and you can compute everything about them, but you give no account of the forces, the factors that produce the trajectories. Just as the science of mechanics needed kinematics to explain how forces yield motion, Darwinism in order to become an evolution theory needs a *Doppelgänger* to explain the nature and origin of selection, responsible for the commencement and sustenance of evolution.

To express the same idea differently: current evolution theory is preoccupied with the "how" of the evolution process (where evolution itself — the very occurrence of the process — is taken for granted); we need now a complementary theory that deals with the "why" part as well, explaining the *causes* of evolution. We shall see that this *generativist* program has suggestive consequences, as well as desirable side effects, such as the reduction of omniscience and anticipation built into the theory.

DARWINISM AS A SCIENTIFIC THEORY

Do we need an entirely new theory? It would be tempting to turn and twist Darwinism, so as to incorporate whatever is missing. Yet I think this is impossible on two grounds. The first is a complex issue, the subject of our paper, whereas the second is a much simpler semantic matter that can be sorted out right

G. Van de Vijver et al. (eds.), Evolutionary Systems, 255–265.

now.

The scientific status of evolutionary theory is associated with well-known difficulties. Take the tautology debate of the late seventies.[2] The issue revolved around the circularity of fitness: survival depends on how fit you are, but what determines when you are fit and when you are not? It seems survival is the only proof of this pudding. (Whether there are other ways of pudding-eating we put aside now.) This logic leads to an *a posteriori* definition of the concept of fitness; we get the tautology "the survivor survives".

Based on the tautology argument, evolution theory was deemed empty by the critics. As a response, defenders pointed out that the theory successfully predicted the existence of survivors and non-survivors, hence it can't be tautologous. Of course, the solution of the paradox is that a theory can still have a certain explanatory value when saying it is a game in which there will be winners and losers, even if it is not able to foretell who wins. Yes, but is this what the theory wanted to be about?

At this point we face another, more serious problem. Darwinism as a theory has never been clearly and unambiguously defined. A good part of the otherwise (from the point of view of this paper) unimportant tautology debate was concerned with various formulations of the theory. To put it boldly, nobody knew exactly what Darwinism was. The epigrammatic, general statements offered by J. Maynard Smith or Mary Williams were of little help when it came to the mud-wrestling; it was an open game.

It is widely known that Karl Popper was a passionate opponent of Darwinism exactly on grounds of its notorious flexibility and its openness to endless reinterpretations. He depicted Darwinism as an irrefutable, metaphysical research program. Popper drew a forceful parallel with Marxism, another semantic chameleon of the time, making his point an easy grasp. Popper later withdrew his critique, but it remains an interesting fact that many, sometimes fundamental amendments and rejoinders have been accommodated in Darwinism without any major change in the general rhetoric or the name of the theory.[3]

Yet we do not have to share Popper's radical taste to recognize that there are limits to obey. It is perhaps reasonable to say that versions of Darwinism (just as of any other theory) possess some core concepts that lends them a common identity. For our purpose, we will say that if this core changes, then we deal with a different theory (research program, paradigm, etc.). And it is in this sense, not for the sake of being provocative, that the present paper suggests the necessity of a new theory of evolution.[4]

THE PAN-COMPETITIVE VIEW

Let me start the body of the paper with an hypothetical, strawman version of the Darwinian theory. This caricature of Darwinism will be claimed to hold that the various evolutionary events (including the "small business" of gene kinetics and the "big business" of speciation) occur as responses to one and the same type of pressure — the one exerted by differential survival. If evolution proceeds by differential survival, *and that is all there is*, then every species must necessarily compete against every other. The reason why there will be new species is that they do the same job better and better.

Hidden in this seductively simple picture there are several problems, such as those of micro- versus macro-evolution, or the evolution versus ecology controversy. Lack of clarity notwithstanding, it is interesting to look at the consequences. The competitive view leads to an essentially "technological" misrepresentation of evolution: if the needs of different species were brought to a common denominator, the latter could consist of nothing but inorganic factors, such as the ones related to efficiency, energy turnover, the mechanics of locomotion, and the like. This picture offers an image of life as an enterprise run in an unfriendly and non-supportive environment, at a place where you have to fight against cold, the wind, scarcity of food, the lack of water, or its opposite, the abundance of water (in the form of rain, floods, etc.) — a game against Nature that those will win who have a better heat isolation, a bigger ear to hear the prey, or just an umbrella to stay dry.

Needless to say, this bizarre picture has a clear orthogenetic flavor to it, with evolution shown as essentially a functional optimization process. This is the story of how we get from those poor cold-blooded reptiles to the victorious mammals with their controlled body temperature, from gymnosperms to angiosperms, from external to internal fertilization systems, from eggs to wombs, from skin to fur, from r to K strategy, and so on.

Few people would be ready to defend such an oversimplified view, so manifold are its problems. Nevertheless (see note 4), this is the skeleton of every Darwinian argument.

We said there are many problems swept under the carpet: one of them is that the story says nothing about why technically inferior solutions don't die out. In reality, they tend to coexist with the "modern" organisms or may (as in the popular case of Gyngko Biloba) spread anew where the technologically more advanced species die out. Also, the story has nothing to say about the origin of such complex adaptations as behavior. Behaviors have little to do with environmental needs whatsoever. They arise "out of nothing", and have a completely internal evolutionary origin. The behavior of the rabbit is a response to that of the eagle, and the behavior of the eagle reflects that of the rabbit. These behaviors (just as many others) can't exist without each other, and they are not necessitated by any global condition or selection pressure.

Furthermore, the technical story is silent about why there is diversity and progression at all, why evolution does not come to a halt, to an endstate characterized by a single best species, or maybe (if we permit competition to be confined to populations that share common needs) one best species for every lifestyle. Relevant here is the fact that no truly major technical invention took place in the last 70 million years (when we already had those "best" solutions at hand), and yet this was the perhaps most intensive evolutionary period we know of in life's history.

I believe the point where the selective-competitive picture is essentially misleading is that it suggests that resources are independently given and therefore the job is to adapt to them. However, resources cannot only be consumed but can also be produced; this is true for ultimately energetic terms (e.g. plants are ecological producers and can also multiply the net energy influx by changing the color of a territory), but also in terms of subtler resources, such as the providing of shelter, or (if you are a carnivore) the formation of large centralized food deposits called big animals. All these goodies depend on entirely biological factors — nothing outside the very process of evolution is responsible for them.

It is other species above all, and not the physical surrounding that constitutes a given species' environment. Evolution takes place in the web of interactions among the species — therefore, natural selection is not a *cause* but a *consequence* of the change of this web.

THE TRAGICOMEDY OF DARWINIAN CO-EVOLUTION

One of the main points of the paper is this: I propose to rediscover and rethink the notion of co-evolution. I say "rediscover", since co-evolution is not a new idea. Yet with perhaps L. Van Valen's *Red Queen Hypothesis* (1973) as the only exception, the notion of co-evolution (or mutual evolution) has never been seriously accepted in evolutionary theory. Even the Red Queen is a far too careful (and, therefore, controversial) formulation of a deep fundamental idea. The Red Queen keeps selectionist notions such as fitness and adaptation, and, more importantly, it continues to convey the false impression, amplified by the competitive view, that evolution is a game against something "out there".

What the Red Queen Hypothesis states is that successful adaptation in one species is perceived as the worsening of the environment by others. (Note the term "environment", as if there were any environment independently from evolution; and if my success makes your harm then we are still in competition, aren't we?). As an implication, so continues the argument, other species are challenged to undergo new adaptations to cope with the new conditions. On a mutual basis, this process extends along a progressive braid. You have to enter the rat race in order to stay where you are, or after a few turns you are out.

Expressed properly, I believe this idea has far-reaching consequences. But terms such as "worsening" and the like may not befit situations other than the proverbial rabbit-and-the-fox, where a "better" rabbit means a hungrier predator. How the evolution of species that open various new niches for themselves and for others to interact with can be incorporated in this picture is an open problem, and in fact it is equally unclear how to talk with this theory about anything that does not conform to the logic of arms race where, just as in the Tom and Jerry cartoons, all you will ever get is simply more of the same.

What about the betterment of the environment? Or what about qualitative changes, allowing not for better-or-worse but altogether different, new types of outcomes?[5]

The later fate of the Red Queen speaks for itself. It was incorporated as part of Darwinian evolutionary ecology and population dynamics[6], and today it plays the role of a somewhat perplexed, naive forerunner of Maynard Smith's game theoretic models which do allow for a rational study of the plasticity of evolutionary forces in a multiple actors framework, something the Red Queen has failed to offer.

The game theoretic ESS is a good tool for dealing with genetic variability in populations, and with behavioral feedback. Still, ESS does not apply to co-evolution. You can't consider new genes because they don't have well-defined fitness functions (or payouts), central to the conception. In this interpretation, "co-evolution" is an endless, monotonous wandering in one and the same space of solutions (or worse still, an approach to a fixed point). Furthermore, no word is said about the context-dependence of fitness, nor about how fitness is related to the presence or absence of other species, or about how changing aspects of the phenotype enter "fitness" determination (Levins, 1968).

The situation is much worse if we turn to the standard, more limited interpretations within evolutionary theory of the term "co-evolution". Used for such a narrow purpose, co-evolution is often just another name for Darwinian evolution. For the entomologist, co-evolution may amount to changes in the vegetation, to be included as a parameter on the bug's parallel evolutionary time scale. Whether this sort of biotic simultaneity has to do with the definition of the basic concepts of evolutionary theory, or with the condition that plants and insects can evolve at all, is none of the entomologist's concern; not his department.

EMERGENTISM AND THE OPEN UNIVERSE

Where do the evolutionary forces come from, in the first place? Or where do parameters of the ecological coupling have their origin? Are they present, from time zero on, in an encoded form in the primordial soup? Or is there some causal mechanism that produces them de novo?

Many theorists have articulated the view that evolution must be capable of supplying or generating its own causes en route; the co-evolution argument may be the most biological way of coming to a similar position. Other ways are purely philosophical. The emergence of new selective forces as a biological problem will now be considered a subcase of this broader philosophical issue.

The emergentists H. Bergson, Ch.L. Morgan, or J.S. Mill have spoken for a dynamic or "creative" view of life that supports causal forms of evolution. Interestingly, the thread begins with Darwin already. He understood that his picture of evolution implies an image of the Universe where things are not created all at once by some omnipotent and omniscient God, but are being created as time goes by. Darwin realized that no essences, no archetypes, no anticipated forms can exist for such transitory entities of biology. It is a grin of history, that he did not seek open confrontation with Christianity, and he did not amplify these aspects of the theory.[7]

Bergson was concerned with general conceptual problems arising with the metaphors of determinism. He characterized evolution as a productive process where they don't apply. Of the several facets of his criticism of contemporary science, the most pervasive were the ones about time. He spoke of the possibility of motions that do not occupy successions of welldefined states but are open to shape themselves in an interval of time (which he called duration, or *durée*).[8]

The emergentist view received its perhaps clearest and certainly most advanced formulation in Popper's "*Open Universe*" conception. Popper speaks of an unfinished Universe whose future outcomes are not fully contained in its present — but hints (what he calls *propensities*) exist. Propensity, or physical possibility, is interpreted by Popper as the occurrence of an individual probabilistic event. Applied to biological evolution, the Popperian view offers the vision of a process with irreducible internal degrees of freedom; in Popper's conception, evolutionary development is seen as an inherent trait of matter, not a response to something else, selection or otherwise.

The Popperian conception is bound to problems and inconsistencies. It denies determinism under all circumstances; with propensities taken seriously, even clocks may not work properly. (That they sometimes don't is a score for Popper, at least within his own system of thinking.) Then, there is the awkward issue of *causality versus determinism*, where Popper never keeps a clear distinction, obfuscating the question. The line could be continued. Yet it is difficult not to acknowledge that Popper's concept is a truly modern and properly materialistic version of the same theme that bothered many earlier causal evolutionists.

TROUBLES WITH THE CAUSAL VIEW

The causal view is not easy to develop, even assuming the truth of one or another of the above constructions. Putting evolution's driving forces inside the living system leads to notorious intellectual puzzles. Let us unveil one of best known specimens of this blend.

The competitive view had an easy job, as its ground terms were well defined at the outset. We have the population, on the one hand, and its environment, on the other. How these terms can be operationalized, and whether they together suffice to explain evolution is another issue, but as far as the logic goes, it is easy to understand on its own. The logic conforms to a scheme familiar from a number of sciences, starting with the example of classical mechanics: there is A (the environment) that causes B (genetic change in the population).

At first glance, mutualism may seem to do no harm to this. A makes B evolve and B makes A evolve, no problem as long as we know what we mean by A and B — that is, as long as we assume that *they are a priori there*, having a separate origin, independent from the evolutionary subprocess we are just considering.

This was the case of the bug and the plant; presumably, a bug has evolved under bug-making selection forces, and a plant has evolved due to plant-making ones. Now they meet and start evolving together. So be it; all we have to do is to single out those factors that alter the one and alter the other, and we get two nice evolutionary histories that run in parallel.

But already at the ecosystem level things start to bend dangerously backwards, leaning towards themselves. Species of the ecosystem do not have an independent existence any more, not if decoupled from the kinds of evolutionary events they themselves cause when responding to exactly these kinds of events. If we ask what makes the whole ecosystem evolve, and what made it to become what it is now, we are studying a system that has virtually no causal powers outside itself. Studying one part (A) at the expense of another (B) imports the danger that when reversing things (B against A) we end up in a circularity.

In a closed system, how is it possible to render a mechanism to its origin? If change in A causes B to appear, and change in B causes A to appear, then what makes A+B appear? Where do we start? The Archimedean point is lost.

Incidentally, it was probably this sort of argument that led Maturana and Varela to adopt a neutralist perspective in the theory of *autopoiesis*. Autopoiesis is a theory often misunderstood and cited on the wrong occasion. It is a conceptual account of biological existence, with implications for evolution. Maturana and Varela recognized (and doubtlessly they are right) that terms of selection and adaptation do not apply to the systemic level of life. They also realized that it is the systemic level that determines evolution as such (which they express by saying that the evolving system has a certain "autonomy"). At the same time, not being able to associate a governing principle to the closed network of A-to-B-and-B-to-A-type evolutionary interactions, they advanced a neutralist ("drift"-based) theory.

But this thinking is just as erroneous as was neo-Darwinism's optimism that things at large are the same as things in the small. The neutralist or probabilistic perspective (something Popper would be happy to hear about) is biologically empty. It does not conform to the facts. There are just too many adaptations to throw them out.

CIRCULARITY ELIMINATED

"Self-evoked causes", circularities, or bootstrapping mechanisms like Baron Münchhausen's infamous self-elevating trick always import a suspicion of ill-definedness. Circularity is difficult to digest: in fact, autopoiesis may be the only scientific theory that supports vicious circles, *Teufelskreise*, and the like.

I am going to argue that the circularity of co-evolution is a pseudo-problem. The self-referential paradox of evolution is an illusory by-product, yielded by a static re-expression of temporally defined dynamic evolutionary forces for which there is no constant "self", apart from a linguistic artifact, to which we could refer.

What I mean to say is this: "A causes A" does not necessarily imply that A begets itself (in which case A would have to be truly circular: empty, undefined or non-existent). There is another possibility. As a starting point, we should clearly understand that statements of the type "A causes A" are made possible by the rules of the scientific language. These rules require that the symbol A stands for the object it stands for: in other words, it assumes that there is such and such an object, conceptualized as an identity translated into propositional content. Then, it is also clear that the problem of self-reference arises at the level of the propositional content and not at that of the causal process.

Paradox can be avoided in more than one way. Of importance for our subsequent discussion is the possibility that *A has never been quite A* — that there is no static object underlying evolution. That is, the paradox is bypassed if we think of the objects of evolution as arbitrary labels associated with essentially unknown, complex and flexible things that cannot be exhausted by describing them as A (or B or anything else).

This idea is so central to our discussion that I take the liberty to re-formulate it now, first the paradox, then its solution. Take interactions in A, an ecosystem, and take an algebraic model of this system. Interactions of A can never produce A by a transformation over the connectivity matrix that determines these interactions. The matrix should be given or it will never come about — there is no third possibility. No system has access to its own definition, to its own primitives. Now, how to solve the paradox: *what I offer is a rejection of the notion of primitives*. The suggestion is that primitives are different factors in a system's material composition than the ones represented at any time in its mirror image, A, that propel A towards new elementary evolutionary events. In this way A can cause its own evolution (or, in the long run, its own existence made possible by evolution), by making use of properties and interactions not yet represented in the connectivity matrix, that is, by something in fact not in (what we come to call) A.

PROCESS PHILOSOPHY AND RELATIONAL BIOLOGY

Here we come to the topic of process philosophy: we deal with imperfections of existence and lack of constant identity, as consequences of dormant or previously untested modes of interaction getting activated while others get cancelled.

The content of process philosophy is often identified with Heraclitus' *panta rhei* or some other idea of irreversibility. We have commonplaces like "everything moves" or its twin "you can never step twice into the same river".

Process philosophy's modern expression is in Whitehead. He draws attention to the fact that identity (of any A together with its propositional content or description) is always a fiction. Whitehead uses trivial examples. We may think of the pendulum that gets rusty, breaks into parts, and stops swinging, to point out the transitory nature of every property — with the permissive remark that science is normally interested in those periods when identity is a good approximation; these are the periods in which pendulums swing and do nothing or little else.

The escape route offered for many sciences where identity change may be interesting but not important does not apply to evolution. Not only does evolution indeed deal with the longest time intervals possible, and does so almost by definition, but also, co-evolutionary scenarios may require a direct use of changing or imperfect identities.

One concept particularly interesting to reconsider is that of *relationalism*. Relational uses of structures are quite common in evolution: F. Jacob's notion of "evolutionary tinkering" (1982) or Rosen's "principle of function change" (1985) provide examples where they made it to theory. Jacob formulates the thesis that evolution proceeds by re-using old structures in new tasks. In discussing the phenomenon, his concern is with the sub-optimality of evolutionary design due to the use of improper materials, but there is another message as well, that of the *non-committed* nature of structure. The same organ can function in several ways, depending on the circumstances. Likewise, the same organism (that is, the same body) can show different traits, depending on the context. Rosen's "function change" is a concurrent formulation of the same idea.

In the relational view, there is no such thing as the organism. There are several organisms in one, depending on the interaction modality. If that changes, the same entity presents a different aspect of its fluid phenotype, redefining the evolutionary situation. This suggests we put an emphasis on the flux of spatio-temporal and other factors of interaction instead of any kind of material structure, when looking for the determiners of evolutionary properties and evolution itself.

RECURSIVENESS IN EVOLUTION

Applying process thinking, the units of evolution present themselves as relative to the temporary organizations that encompass them, and lend them a character of transitory identity. Elementary evolutionary change means adaptation to the co-extant "properties", and a simultaneous change of these "properties", without structural transformation.

It is instructive to play with the thought that evolution proceeds by a recursion of such elementary steps. This offers an hypothesis for a causal mechanism that permits evolution, and at the same time explains what sustains it.

That there is a certain recursiveness in evolution was already noticed by several authors, most explicitly by M. Conrad (1979) and O.E. Rössler (1979). One particular instance where recursion is seen at work is the origin of evolutionary capability: life is evolving, but its evolvability should be a result of previous evolution. Sexuality conforms to the same pattern of thought, one of its assumed roles being the maintenance of high genetic variability in populations (i.e. giving food for evolution), with the simultaneous possibility of producing viable offsprings (i.e. reducing the number of selective deaths).

With the incorporation of shifting relational properties, recursiveness can obtain a new dimension in describing co-evolution. It may be a basic mechanism for evolutionary change — an idea deserving more discussion than permitted by the limited length of this paper.

ACKNOWLEDGMENT

This paper was written with the support of a research grant obtained from the Hungarian National Research Fund (OTKA T6428). The first version of the paper was completed during the author's stay at the Department of Logic and Philosophy of Science, University of Basque Country, San Sebastian, Spain. Both sources of support are gratefully acknowledged.

Department of History and Philosophy of Science, Eötvös University
Budapest, Hungary

NOTES

[1] Here I refer to the developments of dynamic structuralism. Its representatives put an emphasis on the origin of variations in organisms, a point largely neglected in the now-classical theory of the modern synthesis. For most recent publications, see Goodwin (1994) and Meinhardt (1994).

[2] The debate started with the paper of Peters (1976) and the reply of Stebbins (1977).

[3] It is not the purpose of this paper to use sociology of knowledge or philosophy of science techniques to discuss the cognitive mechanisms underlying this practice, nor is it our aim to look for justifications or counterarguments. A liberal Lakatosian would speak of the normal maturation process of a research program. At the other extreme, an imaginary Vienna Circle neo-positivist would insist that, in lack of a canonical set of sentences uttered in some well-defined, fixed language, Darwinism has never achieved the status of a theory.

[4] For the pedantic, by (neo-)Darwinism hereinafter we mean any explanation that is based on (i) heritable variability and (ii) differential survival; see e.g. Maynard Smith (1993).

[5] The point is that ecological evolution or co-evolution is not limited to parallel adaptation but includes something that can be called "innovation", meaning the biological production of new evolutionary forces — in Darwinian terms, new selective forces. This is not restricted to ecology: even a single species is subjected to conditions selected by its own phenotype. For instance, whether its habitat is water, soil, or the air — and so on, up to the minute details — helps determining the physical parameters to which it has to "adapt".

[6] See in particular Stenseth and Maynard Smith (1984).

[7] D. Dennett's most recent book, *Darwin's Dangerous Idea* (1995) brings a fair review of Darwin's own concepts; another source is Wuketits (1987).

[8] A good summary of Bergson and other radical emergentist ideas is found in Blitz (1992); Nagel (1961) provides famous arguments for why (he thinks) all these ideas are wrong.

REFERENCES

Bergson, H., 1984, *Creative Evolution*, New York, University Press of America (repr. of 1944 English transl.).

Blitz, D., 1992, *Emergent Evolution: Qualitative Novelty and the Levels of Reality*, New York, Kluwer.

Conrad, M., 1979, Bootstrapping on the Adaptive Landscape, *BioSystems*, vol. 11, pp. 167-182.

Dennett, D., 1995, *Darwin's Dangerous Idea*, New York, Simon and Schuster.

Goodwin, B., 1994, *How the Leopard Changed its Spots: The Evolution of Complexity*, New York, Scribners.

Jacob, F., 1982, *The Possible and the Actual*, New York, Random House.

Levins, R., 1968, *Evolution in Changing Environments*, Princeton, NJ, Princeton University Press.

Maturana, H.R.; Varela, F.J., 1980, *Autopoiesis and Cognition*, Dordrecht, Reidel.

Maynard Smith, J., 1982, *Evolution and the Theory of Games*, Cambridge, Cambridge University Press.

Maynard Smith, J., 1993, *The Theory of Evolution*, Cambridge, Cambridge University Press (3rd edition of the 1969 original).

Meinhardt, H., 1994, *The Algorithmic Beauty of Sea Shells: Shell Patterns as Dynamic Systems*, Berlin, Springer.

Mill, J.S., 1980, *Collected Works*, London, Routledge.

Morgan, C.L., 1977, *Emergent Evolution*, New York, AMS Press (2nd reprint edition of the 1923 original).

Nagel, E., 1961, *The Structure of Science*, New York, Harcourt, Brace and World.

Peters, R.H., 1976, Tautology in Evolution and Ecology, *American Naturalist*, vol. 110, pp. 1-12.

Popper, K.R., 1992, *The Open Universe: an Argument for Indeterminism*, London, Routledge.

Rosen, R., 1985, *Anticipatory Systems*, New York, Pergamon.

Rössler, O.E., 1979, Recursive Evolution, *BioSystems*, vol. 11, pp. 193-199.

Stebbins, G.L., 1977, In Defense of Evolution: Tautology or Theory?, *American Naturalist*, vol. 111, pp. 386-390.

Stenseth, N.C.; Maynard Smith, J., 1984, Co-evolution in Ecosystems: Red Queen Evolution or Stasis? *Evolution*, vol. 38, pp. 870-880.

Van Valen, L., 1973, A New Evolutionary Law, *Evolutionary Theory*, vol. 1, pp. 1-30.

Varela, F.J., 1979, *Principles of Biological Autonomy*, New York, North-Holland.

Whitehead, A.N., 1979, *Process and Reality*, New York, Free Press (corrected edition of the 1926 original).

Wuketits, F.M., 1987, *Charles Darwin*, München, Piper.

RUPERT RIEDL

DEALING WITH COMPLEX SYSTEMS OR HOW TO DECIPHER LANGUAGE AND ORGANISMS

INTRODUCTION

We use the word 'complexity' to refer to systems that are composed of different subordinate and superimposed components. And we call 'system' a complex whole with functional interconnections between its parts and/or its origin, including specific structural and functional boundaries.

A HIERARCHIC COMPOSITION

If one considers the sub- and superimposed components of a complex system, a stratified structure is the consequence. But since there are in general several to many subimposed parts composing and being held together by a superimposed one, the composition is of a hierarchical type. If — for brevity's sake — we speak about 'layers', layers of a hierarchical composition must be taken into account.

The number of layers in complex systems can be large, but it is always functionally limited. The same applies to the constituents of every layer, where the former also prevail. For composed units, where the number of components can be almost unlimited and has little influence on the whole are not systems in the former sense. Molecules of a liquid or grains of a sandbank may be of this kind.

Matter itself is hierarchically layered: quarks compose elementary particles which compose atomic nuclei which compose, together with electrons, atoms which eventually compose molecules. This may be considered trivial. It is not trivial, however, that this principle extends in many ways from macromolecules to organisms and societies and from simple forms of communication (such as 'sign-' or 'body language') to language and great poetry.

In the case of language we distinguish strata such as letters, words, sentences and context; but each context is also embedded within additional frameworks such as authors, traditions and 'Zeitgeist'. In the case of organismic organization up to twelve and eighteen hierarchical levels may be expected, reaching from biomolecules to organisms, such as human beings. Cells, tissues and organs are the generally better known intermediate levels (references in Riedl, 1977, 1985).

INSERTIONS BETWEEN THE LOWEST CONSTITUENTS AND THE WHOLE

With regard to the origin and development of such systems, it has to be taken into

G. Van de Vijver et al. (eds.), Evolutionary Systems, 267–280.

account that all the hierarchical levels have developed after the instalment of the lowest constituents as well as of the whole (wholeness) (Riedl, 1978-79; cf. Salthe, 1985). They all have been inserted in between them; they all are in interpositions.

In the beginning this is the case with all cosmic structures. The basic constituents were hydrogen atoms, the whole were the fields of gravitation as the consequence of the patchy expansion of this matter. Within these cosmic clouds originated the galaxies, within them the protosun-systems, and only in the marginal whirls (*Randwirbel*) of the latter did the planets condense to their bodily existence. This may be considered as self-evident. But the same principle is at work in systems of higher complexity, such as our examples: language and organisms.

In speaking or writing, the phonemes, the single sounds or the single letters are the basic constituents, and the whole is the intention to transmit specific information: from a single shout to the recitation of the Odyssey, and from the warning sign 'attention' to the *Encyclopedia Britannica*. And this must have been so from the beginning of human vocal communication. Only then single sounds could have developed to special meanings, grouped to semantic significance, such as words. And only between words and an intended context could sentences and syntactic structures have been developed, as useful intermediate layers.

The same can be reconstructed with regard to the intermediate layers in organismic organization. For simplicity's sake let us take the cell as the basic constituent, and the organism as the whole. Then it is clear that tissues could have developed only between the two, and organs between tissues and the organism.

At this point the reader may be aware of a common fallacy. The obvious fact that first cosmic evolution and then the evolution of organisms and communication have stepwise developed systems of higher complexity can easily lead to a wrong conception. One could envision every new layer of complexity being built on top of the former. For there seems to be no doubt that the simple communications of our human ancestors have grown to enormous achievements and single animals gave rise to colonies or man to tribes, societies and cultures.

It can be easily overlooked that there was always a superimposed principle under the guidance of which such a development occurred. Concerning inorganic evolution we speak of maintenance conditions of matter, and in organic evolution of the fitness conditions of species. With respect to artifacts such as speech and literature, we speak of overall intention. The environment, determining the required fitness of a species, exists always before the species itself. And the concept for an encyclopedia existed before the items in it became arranged in alphabetical order.

PRE- AND POST-SELECTIVITY

When thinking about selectivity one is accustomed to envision an environment determining the conditions under which one of the systems existing in it maintains its stability. If, however, one considers the emergence of new complex systems, the mode of its origin requires a wider perspective.

Before an emerging system can be tested by its environment, the availability of its necessary constituents must be taken as a first step of selection. Many different systems could be imagined with good stability-conditions for whatever environment. But this kind of 'pre-selection' allows only a few or none of them to develop.

The same holds for more suitable components. Evidently flying mammals such as bats would do better with feathers than with delicate skin on their wings. But the feather is an invention of ancestral birds. It was not available when bats 'decided' to fly. And marine mammals such as dolphins would do better with gills, but although they pass a gill-stage in their embryology, this organ became substituted by lungs and cannot be restored any more.

It is only after this 'pre-selection' that the environment decides 'post-selectively' under which conditions and circumstances a mammal can fly or live in the sea.

The same holds for communication. Existence of a larynx is among the preconditions of verbal speech. Environment decides for French or English.

It is useful to distinguish between these two kinds of selective processes to structure a complex system, because both provide necessary, but not sufficient explanation, and they also act differently.

Take, e.g. the task to bridge a canyon. The material at hand: ropes or steel wires, or, on the other hand, bricks or building blocks, will influence the resulting structure as well as its possible later function whether it is to serve pedestrians or as a railroad. These constituents, the pre-selected material's function, condition the resulting form in a different way than does the post-selective test of the overall function: whether the bridge can carry a train or pedestrians only. According to Aristotle this corresponds to the difference between *causa materialis* (material cause) and *causa formalis* (cause of form).

In Aristotle's contributions, collected in what has been labelled "Oldest Metaphysics" (Aristoteles A1-3, p. 980a, line 11 to p. 983b, line 3) he states: "Only then can we say that we understand something, if we assume to know the cause at the outset. Of causes, however, we speak in a fourfold way". And later in "Metaphysics" book 5 (p. 1013b, line 17-29) he gives definitions. *Causa efficiens* he defines as "from where motion or rest originates", the *causa materialis* (hypokeimenon) as "the substrate, out of which something develops, for example the components", the *causa formalis* (tó ti en einai) "out of which results the wholeness and the composition of the form", and *causa finalis* "wherefore or to which objective something is happening".

TWO COGNITIVE SYMMETRIES OF CAUSATION

Since the 'Galilean revolution', with its intention to safely confine scientific observations to the measurement of quantities, and boosted by the time of the Enlightenment, the paradigm of (inorganic) sciences today takes 'power', in the sense of the fundamental physical forces of matter as the major source of a sufficient explanation for what causes the phenomena of this world.

Selection and elimination, though they can also be measured in quantities, are obviously of another kind or quality. And even goal-seeking functions and purposefulness have not remained within the paradigm of what is considered to be hard or exact science. However, if we take our example of bridging a canyon, power in the sense of energy, labor or money, is doubtlessly required to produce the materials, to bring them together, and to assemble the bridge. But power alone has never accomplished a bridge. Selecting suitable materials and selecting their appropriate position while assembling them, are also unavoidable: as is someone setting an aim, or whatever intentional process to bridge the canyon.

Beyond the Galilean revolution

This was known already to Aristotle. As reported, he divided our potentials to sufficiently understand or to describe the causes for the establishment of a complex system by four modalities. Two of them have already been mentioned: *causa materialis* and *causa formalis*, corresponding to what I called 'pre-' and 'post-selective' procedures.

His *causa efficiens* (often translated as 'efficient cause') corresponds to nothing but power and not at all what the term 'efficient' may suggest, i.e. a cause, "that which makes a thing what it is" (Oxford Dictionary), or "the potentiality which brings the thing into being" (Encyclopedia). This misleading interpretation from Greek to Latin and from Latin to English might have been one of the sources to put causation on the wrong track.

His *causa finalis* (the 'final cause') is concerned with the purpose or end aimed at, in general with the goal-seeking modality, 'at the end' or 'coming last'.

Besides this division of modalities to perceive causality, Aristotle discovered a symmetry. Putting his insight in modern terms, he found *causa materialis* and *formalis* acting inside as system (in mind), where, in contrast, *causa efficiens* and *finalis* act from outside on the system. We acknowledge this observation in the sense of a cognitive dualism (Riedl, 1979, 1984, 1985).

Actually the results of the two selective processes look different in every hierarchic layer under observation. This is the case for materials available. Atoms, cells, organs, populations are different qualities as are letters, sentences and poems. This is also true for selection and elimination, for the conditions they act on look different to us on every level. One may compare the 'Pauli prohibition' in excluding types of electrons in an orbit of an atom to the selection of organs in an organism, and the selection of a person in a society. And one may also compare the selective processes in spelling, in syntax or in the metrics of poem.

Causa efficiens and *finalis* on the other hand look the same on every level. Energy of gun powder, of labor or capital can be equally converted into Watt or Joule, as a goal-seeking process obviously follows the same objective, whether a cell fulfills its function for a tissue, an organ for an organism, a letter for a word or a sentence for a poem. As a consequence they may actually be thought to reach beyond a system in mind. Since the roots of power propelling a system can be traced back to the elementary forces of nature, and the aim of a sentence, beyond its author to a *Zeitgeist*.

In addition to this Aristotelian symmetry we observe a second one. As soon as one considers the layered pattern of complex systems, two of the causes are acting from the bottom, namely *causa efficiens* and *materialis*, the two others *causa formalis* and *finalis* from the top.

Bottom-up and top-down causation

Observations of this kind have been published by several authors using the terms 'bottom-up' and 'top-down causation' (e.g. Campbell, 1974; Salthe, 1985). Yet a deeper differentiation with respect to the four causes has to be added. This is particularly so because the first symmetry, bottom-up as well as top-down causation contain one type of causality which, for human insight, is passing unchanged through the layers of complexity, whereas the other is changing its appearance stepwise.

For top-down causation regarding an organism or a poem they are determined by goals such as the survival of the species or the aim of the author (*causa finalis*), as well as the stepwise series of selecting processes for organs, tissues and cells, or for metrics, syntax and semantics (*causa formalis*). Whereas bottom-up-causation is composed of the energy the organism or the author employs (*causa efficiens*), as well as, in a stepwise manner, the material which may be at hand, for an animal including, e.g. the abilities to feed on seeds, to fly and to communicate vocally, for an artist access to marble, to hammer and chisel and the command of Greek for producing a commemorative plate.

If we now keep the former analyses in mind, the hierarchy of complex systems, the way they are assembled, the kinds of selective processes and the double symmetry of our apprehension for causality, we may proceed to decipher our way to analyze (and synthesize) the selected examples.

OPTIMIZATION BY AN ITERATIVE PROCESS

We define optimization as a process which leads to the best or most favorable solution, even if not always the final one. Next we deal with the term iteration. This term is also commonly used for the repeated use of an algorithm.

Evolutionary Epistemology, on which this paper is based, has instructed us about the origin and the preconditions of human reason (Lorenz, 1978; Vollmer 1990; Riedl, 1984, 1985, 1987, 1992). We derive the *apriori* of our rational thinking *a posteriori* from genetic adaptation in the phylogeny of our ancestors. And we call this ability a 'ratiomorphic apparatus', or 'inherited hypotheses', guiding us to reasonably adapted interpretations of our environment, but having nothing to do with conscious reasoning. It is related to what one may call 'unconscious action by common sense'.

A paper of this kind cannot go deeper into this extensive matter. For our immediate purpose, we explore just two of the consequences: optimization by iterative processes, and the underestimated accomplishments of our unconscious thinking.

The ground for the process of optimizing a solution by iteration lies in the adaptive process of the genome. Great numbers of reproducing wild forms as well as mutants are tested by the environment, selected or eliminated and in a stepwise way adapted to new conditions. In associative learning, numerous comparable perceptions constitute an expectation, starting from the conditioned response to unconscious thinking, which gets support or is weakened by new instances of observation. However, in conscious thinking we create a hypothesis or theory by induction, starting from observed cases, to submit them logically by deduction to verification or refutation in the face of new data.

The principle of optimization by iteration remains the same, corresponding to the highly redundant phenomena of this world, which does not, however, repeat itself identically. A spiral process, circle after circle never returning to its origin. The rise of the circles corresponds to the gain of adaptation, predictability or knowledge.

As concerns the performance of our unconscious thinking, the reader may be reminded that he perfectly deciphers uncommon handwriting, without knowing the process. And, regarding our second example: comparative anatomists and systematists have classified the representatives of two million species in an order of natural relationship, without rationalizing the cognitive procedure.

A PROCESS OF MUTUAL ENLIGHTENMENT

In this chapter unusual terms, such as enlightenment are unavoidable. One may think of the period of the Enlightenment, or of instructing or informing a person. But 'to shed light' on something sounds already rather poetical. It is not used in exact sciences. And more: what is (then) mutual enlightenment?

This term recognizes the fact that in complex systems, both discovery and explanation of their subsystems must be based on a mutual operation; none can be understood without the other. Today one may better speak of a recursive or feedback operation. But they are not defined clearly enough. And since much thought has been devoted to the concept of 'mutual enlightenment', particularly in the late 18th and in the early 19th centuries, this insight shall be acknowledged.

The operation itself is not avoidable, the simple reason being that those layers of subsystems gave rise or originated from each other as their pre- and post-selective mutual agents.

Actually, this process leading to discovery and explanation was in the first half of the 19th century called 'hermeneutics', but idealistic philosophy has entangled the matter, until bewildered scientists and contemporary epistemologists have renounced the term.

How to decipher a letter

Let us start with deciphering a letter, something everybody may have experienced. It means dealing with a structural hierarchy. Let us assume the writing is very strange, almost illegible, but the letter is important. How do we proceed?

First we adopt three hypotheses from what we know from written texts. First: the lowest constituents will be letters of the alphabet, second: they will be grouped in layers, such as words and sentences, and third: the whole will be a context in a language familiar to us and it will make sense. Actually our ratiomorphic thinking starts with hypotheses on all levels at the same time and is using all 'background knowledge' simultaneously. But to rationalize the process, we must advance stepwise.

For didactic reasons we may first regard the letters. There are signs with resemblance to m, n or u, another group like a, o or e, finally signs like g, y, or j. How does one differentiate? Cases of words will tell. If we find words e.g. composed of three letters, with g, y or j? in the first, and aoe? in the second position, they, most likely, will represent y and o (for you). Epistemologically, cases on the level of words have allowed us to form hypotheses for the level of letters.

Now we look the other way. How do we understand words? Cases of letters will tell. Assuming the y and o are deciphered, cases of combinations of y-o and finally of 'y-o-u' will appear. This may go forth and back many times, until the predictability of letters and words seems to be mutually optimized. A mutual enlightenment, still waiting for further confirmation.

For this purpose we next investigate the level of words. We assume that we have deciphered a word like 'bow'. Can we be sure of what it may mean? A curve, the fore-end of a ship, bending of head and body, or to bow a violin? No doubt, the sentences will tell. And vice versa, the sense of a sentence becomes clear by the cases of combined words. And this feeds back confirmation to words and letters.

But how do we find out, now on the level of sentences, whether a sentence is meant ironically or not? The sentence itself cannot tell. It is the context. Cases of sentences will allow us to form a hypothesis about this point. And no doubt the context becomes understandable by nothing but the cases of sentences. And only until everything makes sense in the letter may we trust the deciphering.

With regard to the causes involved, *causa efficiens* corresponds to the effort of the process of writing. Regardless whether the text is carved in marble or put on paper with a pen, it is effort, reaching from each single letter up to the whole text.

Also *causa finalis* is running equally through the layers; now from the opposite side, starting from the aim of the writer to communicate something, down to each sentence, word and letter, corresponding to the purpose of the whole.

Causa materialis runs again bottom-up, providing the materials, such as letters. And one may know the situation when the text requires an o (o crosse) or c (c subscribed by i) (as for German, Swedish or French) and the typewriter does not provide them. What would happen, on the level of words, if the key for intervals would not work? And even a level higher; punctuation marks, to be pre-selected, could fail.

Causa formalis goes also stepwise, now top-down. Agreement between sender and receiver selects the language, the structure of the context selects the sentences, syntax conditions the post-selection of the position of words, semantic rules determine for the spelling.

How to decipher a new species

One may decipher a new letter from time to time, but not so commonly a new species. I had the privilege to discover a few dozen new species of tiny marine invertebrates, living under seaweed or between sandgrains (for example Riedl, 1969). And since the process of deciphering old inscriptions and consequently of bad handwriting was known to me, I became alert also for such occasions. The reader who has not yet made such experiences may follow trustfully.

Let us assume we have a slender, moving object under the microscope (and under coverslip), one or two millimeters long. Immediately the same pattern of theories comes to one's mind, as in the case of a handwriting. First: there will be lowest constituents e.g. cells, second: they will be grouped in layers, such as tissues and organs, and third: the whole will be an organism not completely unfamiliar to us, and it will make sense (confirming the 'natural system' of the living world).

But compared to the letter, we must take another step back: we must first ask whether or not it is a complete organism. The experienced biologist faces the possibility that it could be a part of a tentacle of e.g. a sedentary polychete or cnidarian, which may also creep around for days. A task, from which the envelope, the beginning and the end of the letter has saved us.

Yet, no site of fracture is discovered, and what we find by magnifying does not yet tell us much (a in Figure 1). And, as in the case of the letter, ratiomorphic thinking starts with hypotheses on all levels and by using all background knowledge simultaneously. But to rationalize the process we must again advance stepwise.

For didactic reasons we may first regard the whole: where is the anterior end? Normally the direction of its motion would tell easily. But some of these creatures move forward and backward likewise.

Since the whole, by itself, does not tell, a first theory is adopted, expecting that cases of organs will clarify the orientation. Higher magnification (Figure 1b) uncovers organs, containing stiff, presumably cuticularized parts, rodlike on one end, paired on the other. Parasitic rotifers (*Seisonidae*) have a rodlike, cuticularized prick in their pharynx. Pricklike penises, on the other hand, are found scattered in other groups of lower worms. Pincerslike parts are known from the pharynx of rotifers and from the proboscis of a group of flatworms (*Kalyp torhynchia*), but rarely in connection with a genital apparatus.

What have we learned? We trust that cases of organs will allow to form a theory for the orientation of a worm-shaped animal. As, vice versa, many cases of front and rear ends of such types of animals are supporting the theory that the ends of the body hold characteristic organs (such as pharynx- and genital-apparatus). The 'type' may compare to the language of a letter and the orientation to the expectation that the context will make sense. Now the remaining uncertainty compares to the specific meaning of a sentence.

A further magnification brings us to the next lower level of mutual theory-forming: that between organs and tissues. We trust that cases of tissues will allow for the formation of a theory of the specific kinds of organs. Since we foresee, turning the view around, that many cases of organs allow to form expectations about the tissues they will be made of.

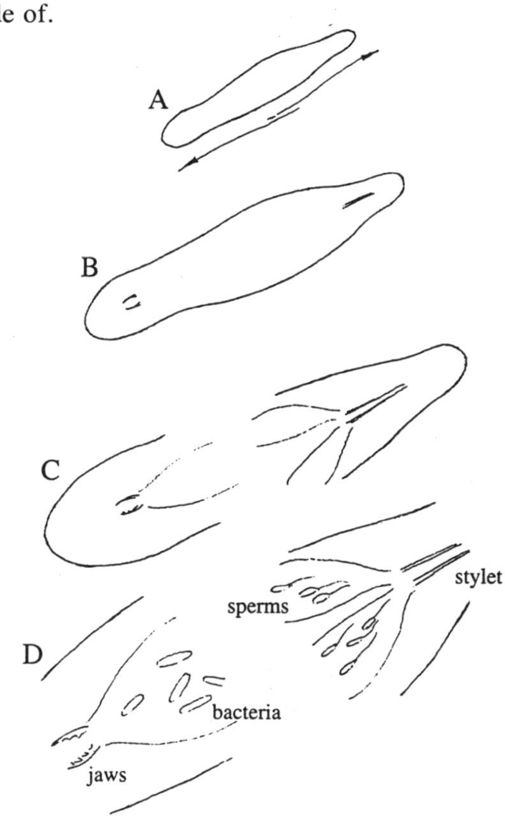

Figure 1a,b,c,d. How to decipher a new species: later recognized as the representation of the new phyla, Gnathostomolida. Positions a to d correspond to stepwise increased magnification (in the microscope), and to the development of mutually interconnected theories or expectations (see text).

Our observation is consequently led to the expected tissues, and we learn (Figure 1c) that the organ with the paired hard-parts is connected with an unpaired tissue, while the other, with the rodlike part, with paired tissue-sacks. Paired sacks, among many things, can be salivary glands as well as testes; an unpaired sack an ovary or the gut as well. Similar uncertainties we found in the meaning of words.

Higher magnification (oil-immersion) is now — and only now — necessary, because the following mutual theory-formation must lead between tissues and cells. We trust that cases of cells will tell us the kind of the tissue, and similarly we expect that many cases of tissues allow us to predict the kinds of cells they contain.

The observation (Figure 1d) reveals the unpaired sack containing food-like particles, the paired sacks containing sperms. One must confess, however, that food particles are often not clearly definable, and sperms can be found in testes, as sperms from another specimen in a compartment of the female genital apparatus (*Receptaculum or Bursa seminalis*).

Yet the whole organization starts to make sense. Although the remaining insecurity of the food particles and the origin of the sperms compares in the letter to the 'a-o-e' and the 'g-y-j'-uncertainty. But, as in the letter, mutual confirmation now combines all the four levels of organization. Sperms, paired testes and the penis-rod confirms the rear-end, as do food particles, gut, and a pharynx with a pair of jaws the front, and both together the orientation of the whole.

Clearly enough I could have described the investigation starting from the pincers, from the cell or the organ level as well. As in the letter, the deciphering could have started from a well readable word or from the greeting formula at the end. And as in the letter where we relied on all the words and sentences, in our creature the process relies on the confirmation by all its tissues and organs. The principle of mutual enlightenment is the same.

From structural to class hierarchies

What we have treated until now were structural hierarchies. Focusing on complex systems, particularly on their hierarchic structure, this may have been justifiable. Yet it was a simplification, for complex systems usually have a history and are rarely isolated. Usually there are many of a kind and we find ourselves challenged in all such cases to compare them and to be supported by their comparison. What we treat in such cases is a class of comparable objects, that is to say: complex systems. And in cases such as languages and organisms, we find groups of classes again arranged in a hierarchical order, as will be substantiated later on.

In the case of the letter, I postponed handling the matter because the letters we normally receive are expected to be written in a familiar language. This is not to be expected in professionally deciphering documents of perished languages. In all such cases, the scholar fully relies on his knowledge of comparable languages and symbols and depends on his correct estimation of their relationship.

In a writing as in our example, we could have referred to our knowledge that, e.g. the word 'bow' has several, but a limited number of defined meanings. This would not have been so, if the letter would have been in Papuán or in Yanomami.

In the case of the organism (actually a representative of one of the new phyla of the animal kingdom: *Gnathostomulida*) the logic had to include comparison already. My arguments relied on the knowledge of comparative anatomy and systematics. If I had to write to learned helminthologists only, suppositions could have allowed a comparable kind of presupposing argumentation.

Consequently, without presuppositions, comparison must be considered. And, as mentio ned before, these classes of comparable things are again in hierarchical order. This needs proof and explanation.

One may recall that both languages and organisms gained their differentiation by insertions between the basic constituents and the whole. These are in our examples e.g. cells or sounds on one end, and fitness of the species or effectivity (communicability) on the other. They remain unchanged in principle. The inserted differentiation however is subject to permanent changes due to advancing adaptation and organization. In addition, these accomplishments are either inherited or handed down by tradition. In short: they have a history.

But they do not only have a history, their paths through history experienced bifurcations. So, what is inherited or handed down throughout time includes selected alternatives; the more, the more time has passed. In other words: the older the accomplishment, the more general, broader as well as comprehensive, is its representation. This leads to class hierarchies of the structurally hierarchic representatives of complex systems with history.

And consequently the same patterns of four causes, with two cognitive symmetries, as I described these for structural hierarchies, are to be applied to sufficient understanding and explanation of class hierarchies as well. For they are composed by nothing else but by structurally hierarchic representatives.

Simply enough, one can justify the concept of 'Indo-Germanic languages' only if one knows that first: many languages are of a different kind, and second: that the difference, e.g. between English and French can be included; that one can talk of vertebrates only if none of the many other animals contain their characteristics, but the difference between fish and mammals can be incorporated. And this holds for every level of class hierarchies.

THE SUBSUMPTION SCHEME

What has been described under the title 'mutual enlightenment', such as deciphering structural hierarchies and arranging them into class hierarchies of natural relationship of high reliability, was worked out until now by common sense only. Guided by what I called a "ratiomorphic apparatus", it worked unconsciously. The reader had no need to disentangle and rationalize the procedure he applies when effectively deciphering strange handwritings. And systematists have put hundreds of languages and two million species into natural order, without ever having disclosed the method of their most remarkable accomplishments. This fact might be embarrassing for many. The little that was known from the procedure of mutual enlightenment, was, as mentioned, soon tangled by idealistic philosophy, then refuted by scientists and contemporary philosophy of science. And to justify this refutation, attempts have recently been made to demonstrate the logical inadmissibility of its basic assumption. Stegmüller (1979) has brought this to the point by stating: one cannot explain A out of B, when B has to be explained out of A. Putting it this way, the accusation sounds very reasonable, a logical circle.

Where is the mistake?

Two amendments have to be added. The first argumentation faces a fact that we have been using already. A and B have alternate functions. In one direction A is providing the cases and B assembles the expectation, hypothesis or theory. In the other direction, it is the other way round. So A and B do not remain the same. But the circuit still looks symmetrical like a logical circle.

The second argument holds the point: the two circuits, although they supplement or complement each other, deliver their messages independently from each other. Each is bound within its own hierarchy of cases and theories, in the sense that theories on one level are forming the cases on the next level. This has been described by Hempel and Oppenheim (1948) as the subsumption scheme.

Assume the letters "bow" are clearly recognizable. Then the sentence: "The bow splintered" gives a first hint of what could be meant. But there are many sentences which have to do with problems either in violin playing or in navigation (e.g.: "Yet, the wren did not stop singing"). Only if the context clearly has to do with a wrecker, "bow" must definitely mean the front of a ship and not a fiddlestick (and "wren" not a bird, but a ship-maiden).

If, on the other hand, the sentence may read: "The bow splintered", "x" standing for a misprinted letter, other words may clarify the matter (such as "the wren did nxt stxp singing"). Then on the level of letters we may simply find out that the "o" on the typewriter was demolished.

Consequently sentences (A) and words (B) supplement or complement each other in enlightenment. But they are linked with their super- and subimposed theories, namely of the context and the letters. They are part of their own system; and no logical circle exists.

The same applies to the case of clarifying the organization of an organism. The clarifica tion of an organ (A) and a tissue (B) do of course complement each other. But there are always several organs contributing to the understanding of the whole, and many tissues linked with the level of cells.

What was overlooked in the "subsumption scheme" is the fact that in complex systems it must be used in a symmetrical or mirror-image way. This is for the simple reason that complex systems gained their differentiationsymmetrically. And the failure of the acceptance of this fact by reductionistic sciences represents one of the major sources for not sufficiently understanding complex systems.

THE PATHS OF DISCOVERY AND EXPLANATION

The two processes run in opposite directions. We normally grasp a phenomenon from within, and open it to our sensory perception; and the path to disclose the causes leads us into two fields of more and more abstract connotations.

Take the case of organismic organization. Sensory observation tells us about specimens. Species are already abstractions and remained a controversial concept till the end of the 18th century (e.g. Bonet, 1769). Meanwhile, the grouping of taxa developed so convincingly by comparative anatomists and systematists that a reason for this orderly pattern had to be considered: a theory of evolution (e.g. Lamarck, 1909).

And there was still a long way to disclose the most basic constituents, the strings of DNA. And what concerns the uppermost regime responsible for differentiation and these orderly patterns, such as adaptation and constraints by functional burden, is still being discussed (Riedl, 1977).

Clearly enough, the double path of explanation runs in the opposite direction. At the lower end the explanation must start with the conditions of reproducible memory which directs the pre-selective installment of disposive differentiation (*causa materialis*) and the basic mechanism to assemble energy for growth and maintenance, far from physical equilibrium (*causa efficiens*). At the upper end the principles of post-selectivity must be understood, such as selection by the environment and by system conditions to explain adaptation and organization (*causa formalis*). And overall fitness must be understood, as well as the drive for species-preservation (*causa finalis*) in order to comprehend why there are organisms.

We find the same situation as regards languages. What catches our interest is what people say: statements, offers or warnings. The reasons underlying them, which have led to communication, and in particular phonetical communication, was beyond reach, as, from the top, why we are bound to a most specific spelling and grammar.

And again the double path of explanation is starting from both ends: from the basic disposition of phonemes (*causa materialis*) and the efforts to communicate (*causa efficiens*); from the top the cultural history of our language (*causa formalis*) and our purpose of communication (*causa finalis*).

THE PATHS OF EXPLANATION AND ORIGINATION

The path of explanation recapitulates the path of origination. The reader may recall the condition from which science has learned to explain the existence of, e.g. the living world; in such a case the origin of physically reproducible memory on one hand, and post-selective processes to steer adaptation and organization on the other, are taken as the primary causes. All the rest, leading up to two million species, their differentiation and order, are consequences. And obviously enough we also reconstruct their origin in the same way.

In the same way the basic explanations for the existence of linguistic communication correspond to how we think about the causes of its origin. The physical preconditions to produce and moderate sounds on one hand, and the aim to communicate, are what we take as the basis of our explanation, and those must also have been the real sources of its materialization. And all human languages and poetry starts from there, differentiating the included possibilities.

So far this may sound trivial. It seems less so, the further the phenomena and the causal models are removed from mesocosmical dimensions to which our perceiving abilities have been adapted, such as in cosmology. There we have been led to the expectation that even the elementary forces have originated from each other: two of them, governing the meso- and megacosmic world, the other reigning over subatomic dimensions.

I end this contribution on the expectation that whatever complex system will be investigated epistemologically in the future, it will reveal the same pattern of correlation between our cognitive abilities for sufficient understanding and the outer world within which we have developed.

Konrad Lorenz Institute for Evolution and Cognition
Altenberg/Donau, Austria

REFERENCES

Aristoteles, *Hauptwerke*, 1977, Ausgewaehlt, uebersetzt und eingeleitet von W. Nestle, Stuttgart, Kruener.

Bonet, C., 1769, La palingénésis philosophique, 2 vols., Genève, D. Philibert & B. Chitol.

Campbell, D., 1974, Downward causation in hierarchically organized biological systems, in *Studies in the philosophy of biology. Reduction and related problems*, F. Ayala & Th. Dobzhansky (eds.), New York, Macmillan.

Hempel, C.; Oppenheim, P., 1965, Studies in the logic of explanation, *Philosophy of Science*, vol. 15, pp. 135-175.

Lamarck, J., 1909 (first French edition), *De: Zoologische Philosophie*, Deutsch von H. Schmidt, Leipzig, Kruener.

Lorenz, K., 1978, *Behind the Mirror*, Methuen, London and Harcourt-Brace-Jovanovic, New York. Original German edition: Lorenz, K., 1973, *Die Rückseite des Spiegels. Versuch einer Naturgeschichte menschlichen Erkennens*, München, Zürich, Piper.

Riedl, R., 1969, Gnathostomulida from America. First record of the new phylum from North American shores, *Science*, vol. 163, pp. 445-452.

Riedl, R., 1977, *Order in living organisms*, Wiley, London. Original German edition: Riedl, R., 1975, *Die Ordnung des Lebendigen*, Hamburg, Berlin, Paul Parey.

Riedl, R., 1978-79, Über die Biologie des Ursachen-Denkens-Ein evolutionistischer, systemtheoretischer Versuch, *Mannheimer Forum*, no. 7, pp. 8-70.

Riedl, R., 1984, *Biology of Knowledge. The Evolutionary Basis of Reason*, Chichester, New York, Brisbane, Toronto, singapore, J. Wiley & Sons. Original German edition: Riedl, R., 1980, *Biologie der Erkenntnis. Die stammesgeschichtlichen Grundlagen der Vernunft*, Hamburg, Berlin, Parey.

Riedl, R., 1985, *Die Spaltung des Weltbildes. Biologische Grundlagen des Erklaerens und Verstehens*, Hamburg, Berlin, Paul Parey.

Riedl, R., 1987, *Begriff und Welt. Biologische Grundlagen des Erkennens und Begreifens*, Hamburg, Berlin, Paul Parey.

Riedl, R., 1992, *Wahrheit und Wahrscheinlichkeit. Biologische Grundlagen des Für-Wahr-Nehmens*, Hamburg, Berlin, Paul Parey.

Salthe, S., 1985, *Evolving Hierarchical Systems*, New York, Columbia University Press.

Stegmüller, W., 1979, *Rationale Rekonstruktion von Wissenschaft und ihrem Wandel*, 1979 Stuttgart, Reclam.

Vollmer, G., 1990, *Evolutionäre Erkenntnistheorie*, Stuttgart, Hirzel.

JESPER HOFFMEYER

THE UNFOLDING SEMIOSPHERE

NATURE'S TENDENCY TO TAKE HABITS

In 1891 the American chemist and philosopher Charles Sanders Peirce published five articles in *The Monist* which in many ways anticipates present day paradigmatic discussions of science. They also present an evolutionary cosmology which, as I shall try to show, fits well into modern scientific conceptions. Thus in 'The Architecture of Theories' we can find the following: "Uniformities are precisely the kind of fact that need to be accounted for. That a pitched coin should sometimes turn up heads and sometimes tails calls for no particular explanation; but if it shows heads every time, we wish to know how this result has been brought about. Law is *par excellence* the thing which wants a reason. Now the only possible way of accounting for the laws of nature and the uniformity in general is to suppose them results of evolution. This supposes them not to be absolute, not to be obeyed precisely. It makes an element of indeterminacy, spontaneity, or absolute chance in nature" (Peirce, 1955 [1891], p. 318). Peirce in this section goes on to launch an attack on Herbert Spencer's ideas of evolution based on mechanical principles. Among other things he points out, that "exact law obviously never can produce heterogeneity out of homogeneity" and that "the law of conservation of energy is equivalent to the proposition that all operations governed by mechanical laws are reversible; so that an immediate corollary from it is that growth is not explicable by those laws, even if they be not violated in the process of growth" (*ibid.*).

Peirce held a philosophical position which may nowadays appear strange: he combined idealism and conceptual realism. His realism of course was not of the naive sort which nourishes the anti-philosophical arrogance of the scientific establishment. Peirce's realism is based on a classical argument: had we not been of this world we should never be able to grasp it: "Our minds having been formed under the influence of phenomena governed by the laws of mechanics, certain conceptions entering into those laws become implanted in our brains, so that we readily guess at what the laws are. Without such a natural prompting, having to search blindfold for a law which would suit the phenomena, our chance of finding it would be as one to infinity" (Peirce, 1955 [1891], p. 317). Peirce's realism does not claim any simple isomorphy between our scientific theories and the part of reality they concern. He defines the real as that "which is not whatever we happen to think it is, but is unaffected by what we may think of it". And "truth is that concordance of an abstract statement with the ideal limit towards which endless investigation would tend to bring scientific belief, which concordance the abstract statement may possess by virtue of the confession of its inaccuracy and one-sidedness, and this confession is an essential ingredient of truth" (Peirce, 1931-1958, 5.565). Idealism on the other hand is a philosophical

G. Van de Vijver et al. (eds.), Evolutionary Systems, 281–293.

choice which Peirce takes to follow more or less directly from the commitment to realism. For if feelings are 'real' materialism must fail: Materialism "requires us to suppose that a certain kind of mechanism will feel, which would be a hypothesis absolutely irreducible to reason — an ultimate, inexplicable regularity; while the only possible justification of any theory is that it should make things clear and reasonable" (Peirce, 1955 [1891], p. 322). Instead Peirce goes on to state his own opinion, that: "The one intelligible theory of the universe is that matter is effete mind, inveterate habits becoming physical law" (Peirce, 1955 [1891], p. 322).[1]

This leads us to Peirce's metaphysics of First, Second and Third: "First is the conception of being or existing independent of anything else. Second is the conception of being relative to, the conception of reaction with, something else. Third is the conception of mediation whereby a first and a second are brought into relation" (...) "Chance is First, Law is Second, the tendency to take habits is Third. Mind is First, Matter is Second, Evolution is Third" (Peirce, 1955 [1891], p. 323). Peirce calls this a 'Cosmogonic Philosophy' and expounds it in the following words: "It would suppose that in the beginning — infinitely remote — there was a chaos of unpersonalized feeling, which being without connection or regularity would properly be without existence. This feeling, sporting here and there in pure arbitrariness, would have started the germ of a generalizing tendency. Its other sportings would be evanescent, but this would have a growing virtue. Thus, the tendency to habit would be started; and from this, with the other principles of evolution, all the regularities of the universe would be evolved. At any time, however, an element of pure chance survives and will remain until the world becomes an absolutely perfect, rational, symmetrical system, in which mind is at last crystallized in the infinite future" (Peirce, 1955 [1891], p. 323). Thus according to Peirce necessity does not rule the universe. Quite to the contrary: the universe is free to create necessity. This is the tendency of nature to take habits, to evolve into new regularities, new constraints (Hoffmeyer, 1993a).

HISTORY IN NATURE

The idea that habits perpetually grow up in our universe makes *history* central to our understanding of nature. History however means different things depending on the level at which the term is applied. Thus human history is generally taken to work in ways very different from the workings of organic evolution, and most biologist see organic evolution as very different from thermodynamic evolution.

Obviously, in the Peircean vision the same fundamental 'law' is operative at all levels. But instead of the reductionist idea of unifying our world through natural laws working at the most general level, he considers these laws themselves a result of an even more fundamental law, i.e. the law of mind.

Let us now try to consider these ideas in the light of modern conceptions of cosmic evolution. The Danish physicist Peder Voetmann Christiansen has given the following example: "I share (...) the attraction to the idea that physical laws are special cases of the law of mind, and that habit formation goes on in the Quasi-mind of nature. One example of such habit formation is the way of peaceful coexistence of planets in the solar system due to their nearly circular orbits. The initial conditions for such a dynamical state are just too improbable to be accidental, they can only be considered as results of a selection principle acting over a long period: slight perturbations from other bodies disturb eccentric orbits more than circular orbits, so near circularity of the orbit makes a body fitter for survival. The perturbations are random and rare like the mutations in biology, but of course they do not contradict Newton's laws, just like a move in a game of chess may be unpredictable but still has to obey the rules of the game" (Christiansen, 1994). Here the term selection might be the focus for further discussion. As biologists we use this term much too easily, I think. In everyday language selection presupposes an intentional person, say a teacher selecting the 'best' boys for a football game. A selection presupposes a selector. Now, as is well known, Darwin was careful to underline that natural selection was a process very different from artificial selection in that no intention or purpose lay behind it. Natural selection was a selection without a selector (or even a selection principle since organic evolution had no privileged direction). I think it is fair to say that the semantic ambiguity thus implied by the term natural selection is the main reason why the Darwinian theory of evolution has caused so much controversy inside of, and misunderstanding outside of, biology. After all, if it is not humbug to ascribe selective power to nature, then nature should be described as capable of exercising such power.

The Peircean answer to this challenge would be, I guess, that *the principle of natural selection is itself a product of evolution*. The capacity for selective processes to unfold in the world has itself evolved (related views have been expressed by several of the participants in this volume (cf. Brooks, Depew, Kampis, Swenson and Weber). Selection is not an either/or principle separating the human sphere (a sphere of selective processes) from the pre-human sphere as seems to be presupposed by the humanities, neither is selection in the human sense of that word present in pre-human nature as biologists often seem to think. Selection is a concept of more-and-less, in the sense that there may be different degrees of intentionality involved in selective processes, i.e. selection. It has its own natural history, and historicity may well be the essence of evolution. Or to state it differently: even historicity has a history.

This idea immediately confronts us with the problem of discontinuity, the discontinuity between life and non-life, or between nature and culture. More-or-less is a concept of continuity and yet we should certainly acknowledge the emergence through time of complex systems exhibiting behaviors which cannot be explained solely by reference to initial conditions, i.e. to the behavior of the simpler systems from which they evolved. New boundary conditions are formed which must be taken into account in any description of such systems (Polanyi, 1968; Eldredge & Salthe, 1984; Salthe, 1985) .

In the macro-evolutionary perspective we can distinguish at least three dominating instances of emergence: the emergence of galaxies, the emergence of life, and the emergence of linguistic culture. Each of these changed the rules of the evolutionary game to such an extent that no evolutionary theory can be taken as satisfactory without dealing with them. More specifically, in each case the historical nature of our universe acquired qualitatively more powerful ways of expressing itself. Let us consider the three events of emergence successively.

The formation of galaxies

The formation of galaxies began shortly after the big bang (i.e. at some 700.000 years). At this time the cooling of the universe reached a level at which radiation and matter became uncoupled. Stanley Salthe puts it this way: "As the universe expands, it does so at a rate too fast for thermodynamic equilibrium to be achieved anywhere. The result of this is that (in Peircean fashion) matter is precipitated from energy, and clumps of matter build up frictionally as bits of it get in one another's way as they randomly seek equilibrium configurations, leading to the emergence of gravity, which then pulls together larger clumps" (Salthe, 1993, p. 120). Now, the arrival of this primary symmetry break in our universe is of course a major mystery. Having accepted the idea that our universe was actually born at a big bang, it still remains a mystery why this universe should not just stay uniform. Knowing the laws of physics we can of course retrospectively deduce that a separation of the uniform plasma into matter and radiation had to occur at a certain temperature, but this is not an explanation, only a description. Why were the laws not different, why didn't the separation occur at another temperature, or why did it occur at all?

The point is that the constraint imposed on the universe at some 700.000 years after the big bang most probably revealed an inherent capacity of that specific universe of ours which just happened to be present there all the time (reflecting maybe "the generally associative properties of our world" (Wicken, 1987). Its actual realization, however, at this specific time in cosmic history resulted in an immense change of the rules governing the fate of the universe. It allowed an enormous increase in entropy production (cf. Swenson, 1989) through a diversity of proto-selective processes choosing among potential channels for energy conversion, thus creating among so many other entities a planet like the Earth placed in an irreversible stream of energy well fitted for the evolution of carbon-based life.

In the semiotic view presented in this paper the essence of this event was the emergence of difference, i.e. the creation of lumps in the middle of nothing. Semiotics is about distinctions, and going from difference to distinction is a crucial jump. This jump is what the formation of life is about.

The formation of life

Living systems are not only 'autopoietic' systems (Maturana & Varela, 1980) although autopoiesis is certainly a prerequisite for life. For the dynamics of life to be set off, one further characteristic had to be present, the capacity which I have termed *code-duality* (Hoffmeyer, 1987; Hoffmeyer & Emmeche, 1991).[2] All living systems contain a self-description in a digital code, i.e. the code of DNA (or RNA if certain virses are to be included). Contrary to the general image created by textbooks, there is no simple relation between the DNA coded messages and the construction of the organism, whether single-celled or multi-cellular. What is described in the DNA-text mostly concerns the amino acid sequence of the backbones of proteins. How those backbones are actually folded into three-dimensional protein molecules is not itself directly specified. Furthermore, as discussed by Luis Rocha (this volume), so-called RNA-editing processes may even introduce context dependent modification of the DNA text at the level of transcription. Neither is it fully specified how the virgin proteins should be put into the right place in the nearly unbelievably complex architecture of the cell, or how and when, in multi-cellular organisms, cells divide, differentiate or migrate in the embryonic tissue. As Richard Lewontin has said: "First, DNA is not self-reproducing, second, it makes nothing and third, organisms are not determined by it" (Lewontin, 1992). What all this amounts to is a simple but crucial fact: DNA does not contain the key to its own *interpretation*. In a way the molecule is hermetic. In the prototype case of sexually reproducing organisms only the fertilized egg 'knows' how to interpret it, i.e. to use its text for the construction of the organism. The interpretant of the DNA message is buried in the cytoskeleton of the fertilized egg (and the growing embryo), which again is the product of history, of the billions of molecular habits having been acquired through the evolution of the eukaryotic cell (Margulis, 1981) in general and in the successive phylogenetic history of the species in particular.[3]

Life, then, exhibits a non-trivial, that is semiotic, interaction between two states, the analog coded state of the organism itself and its redescription in the digital code of DNA. And this principle of code-duality in fact can be taken as a definition of life. As analog codes the organisms recognize and interact with each other in the ecological space giving rise to a horizontal semiotic system (the ecological hierarchy of Eldredge and Salthe, 1984), while as digital codes (after eventual recombination through meiosis and fertilization in sexually reproducing species) they are passively carried forward in time between generations. This of course is the process responsible for nature's vertical semiotic system, the genealogical hierarchy (Eldredge & Salthe, 1984). Thus, heredity should be understood as 'semiotic survival' (Hoffmeyer, 1993a).

The emergence of code-duality takes us from the sphere of difference to the sphere of distinction, i.e. Gregory Bateson's "a difference which makes a difference" (Bateson, 1979), which is in fact the essence of a sign in the Peircean sense. Peirce has defined a sign as 'something which stands to somebody for something in some respect or capacity.' (Peirce, 1955 [1897], p. 99). The sign is a triadic unity of the sign vehicle (e.g. the smell of seaweed), the object to which the sign vehicle refers (e.g. the nearby presence of the sea), and the interpretant, i.e. the key (or context) relating the sign vehicle to the object (e.g. mind processes in the observer). At another level: the sign vehicle might, for instance, be the genome, the object might be the ontogenetic trajectory whereas the interpretant would be hidden in the cytoskeletal architecture of the fertilized egg (or the growing embryo). Students of human semiotics (anthroposemioticians) often, especially if trained in the Saussurian tradition of structural linguistics, tend to think the existence of real sign processes is a property exclusively of the human sphere. But from an evolutionary point of view this appears as an unnecessary limitation on the Peircean sign concept, which was indeed very broadly intended by Peirce himself. In my view the distinctive mark of linguistic culture is not just the presence of interpretative processes. The distinctive mark is the appearance on the planet of self-interpretation. This leads us to the emergence of linguistic culture.

The formation of linguistic culture

I will say only a few words on this 'final' emergence, which I have dealt with elsewhere (Hoffmeyer, 1993a, 1993b). The difference between human life and other kinds of life is dramatic and can be shortly stated as follows: living beings are selves who don't know themselves. Human beings are selves who know they are selves. Being self-conscious selves they are the result of the evolutionary creation of a whole new kind of code-duality, a 'meta'-code-duality so to say, a duality of reality as analog coded experience perpetually interacting with its digital linguistic redescription in an unending chain of translations back and forth. The dynamic properties and creativity of this code-duality, which is the core of cultural evolution, were responsible for a dramatic turn in the evolutionary process, speeding it up to the unequalled pace of human history. In its formal aspects, however, this meta or secondary code-duality resembles the primary code-duality from which it has evolved (Hoffmeyer & Emmeche, 1991). Because of this specifically human code-duality (wo)men are existentially alienated from themselves and therefore they can know themselves. For knowledge presupposes a distance or a non-coincidence, i.e. the self cannot know itself without being able to step back from itself as it necessarily does when talking about itself. The creation of the specifically human digital code, language, holds the key to self-consciousness.

WHAT IS IT TO BE A SUBJECT?

The moment we accept the idea of 'a difference, that makes a difference' we also introduce the problem of the subject. What does it take to be someone (to whom it makes a difference) and not just something? Gregory Bateson struggled with this problem when he tried to establish his six criteria for being a mind (Bateson, 1979). He used terms such as mapping and coding, but clearly what he had in mind was a self-descriptional semiotic system. Bateson also advocated the very Peircean idea of evolution as a kind of mind process.

The appearance of life, and thus self-reference based on code-duality, added a new dimension to historicity: experience. Life's strange way of persistence is based on the fact that the mortality of code-dual systems not necessarily implies the death of the digital descriptions which often enough survive through reproduction (or, in the case of the human meta code-duality, as texts or ideas). Fertilized eggs assure the reinterpretation of the surviving mix of digital descriptions which are thus given a new try as analog coded messages, organisms. This mechanism provides for the capacity of living systems to carry out a selective incorporation of the pasts into their future, i.e. for experience (Hoffmeyer, 1992). Systems based on code-duality form branching lineages so that cohorts of systems are created, populations, and this is the level at which selection takes place. While selection in the case of pre-biotic processes (as when we talk about selection of circular planetary orbits) can only be said to be true selection if we accept the idea of a quasi-mind of nature as a selector, selection in the biotic case does have a more particular selector: the lineage. The lineage stores a selection of its past, in the form of the genetic texts carried by its individual members (the 'kimflok' in Hoffmeyer & Emmeche, 1991), in order to cope with the future. Or, in other words, the present is measured towards an internal standard reflecting the past, and the logical structure of this process very much resembles the principle of selection as the term is used in the human context. As has been argued in more detail elsewhere (Hoffmeyer, 1992), the lineage has the character of a true subject: it is a temporal being captured in the strange loops of digital-analog self-referential translations. Although the lineage has no consciousness or free will and certainly no moral responsibility, I nevertheless propose that we extract the essence of the concept of subjectivity as being exactly the two characteristic properties of lineages (as well as all other code-dual systems): temporality and non-coincidence (or alienation) which were in fact the very terms used to characterize 'subjectivity' in the phenomenology of Maurice Merleau-Ponty (Langer, 1989).[4]

This leads to the view of subjectness as a more-or-less concept thereby allowing for an evolutionary explanation of the specifically human way of being a subject: self-consciousness. As I said, only human beings know that they are selves. The evolution of the human skeleton is not just by chance accompanied by the evolution of the human spirit. In stead, the human skeleton is an epiphenomenon created as an accompanying consequence of the general unfolding on the planet of the principle of subjectness or, in general, of 'semiotic freedom' (Hoffmeyer, 1992).

SEMETIC INTERACTIONS

With the creation on planet Earth of living systems, 'nature's tendency to take habits' became very much accelerated in these quarters of the universe. Living systems exhibit extreme semiogenic behavior through the process I have termed 'semetic interaction' (of Greek semeion = sign, and etos = habit; Hoffmeyer, 1994a, 1994b). Semetic interactions refer to interactions in which regularities (habits) developed by one species (or individual) successively become used (interpreted) as signs by the individuals of the same or another species, thereby eliciting new habits in this species eventually to become — sooner or later — signs for other individuals, and so on in a branching and unending web integrating the ecosystems of the planet into a global *semiosphere* (Hoffmeyer, 1993a).[5] The semiosphere is a sphere like the atmosphere, the hydrosphere or the biosphere. It penetrates these spheres and consists in signification and communication: sounds, odours, movements, colors, electric fields, waves of any kind, chemical signals, touch, etc.

A prototype example of semetic interaction is the bird who lures the predator away from the nest by pretending to have a broken wing. This example actually gives a chain of at least two semetic interactions. First, the predator takes the awkward movement of the bird as a habitual sign that the bird is an easy catch. Next, the bird having interpreted the behavior of predators takes advantage of the predators habit of going for an easy catch. Notice that it makes no difference whether this interaction is built on 'knowledge' written in the genes or 'knowledge' written in brains. The principle works in both cases although the operational time scales are of course very different.

Semetic interactions were central to evolution from the very beginning of life. An interesting example is the development of light perception in heterotrophs discussed by Swenson and Turvey (1991, p. 340): "Photo pigments were first used in photosynthesis, and in locating or moving toward or away from places where the wavelength of light was suitable or not suitable for photo chemistry (...) At some point, when [photosynthesizing] cyanobacteria are presumed to have constituted a major portion of the biomass on earth, they themselves represented a field potential on which heterotrophs (...) began to feed. The heterotrophs used the same photopigments for detecting light, but not to photosynthesize; instead the pigments were used to detect light that was specific to where the autotrophs (photosynthesizing cyanobacteria) were feeding (on the light). Light distributions specifying not light as food itself, but *information about* the location of food, was evolutionarily instantiated in its modern sense". Thus, to the heterotrophs certain light distributions came to signify the habitual presence of prey.

The principle of semetic interaction is not exclusive to a single organizational level. The semiosphere penetrates into the interior of organisms, or rather it emanates from within organisms. An amusing example of semetic interactions 'across the skin' is exhibited by the female ring dove (*Streptopelia risoria*). Before a female ring dove lays her eggs, she and her mate go through a series of courtship displays. As courtship proceeds, hormonal changes in the female trigger the growth of follicles in her ovaries, each of which eventually bursts to release an egg. Now, recent experiments have shown that if a female dove is operationally hindered from making the so-called 'nest-coo', she will not be able to ovulate, despite enthusiastic courting by males. In control experiments tape recordings of nest coos were played to females with no males present. Now follicles began to grow. The conclusion is simple: female doves that coo during courtship are not cooing at the male — they are cooing at their own ovaries, to trigger the release of eggs.

One can only guess of course how such a strange mechanism has actually evolved. But at least it seems to indicate that the set of habits involved in courtship and thus the cooing was in fact more safely correlated with the timing of mating, than any simple endogenous release system would have been. The obvious, although speculative explanation, is that the cooing behavior is a measure of the state of the relation between two birds, which is superior to a simple measure of the state of the female organism itself (Hoffmeyer, 1994a).

The recognition of semetic interaction patterns implies a radical extension not only of the range but of the dynamics of co-evolution. If the semiogenic dimension of evolutionary innovations is to be integrated into the theory of evolution, we shall have to reconsider the organism-environment interface, which can no longer be treated in terms of the simplistic concept of the ecological niche. Rather, the communicative creativity taking place at this interface must be placed at the center of theory.

SEMIOTIC FREEDOM

At all levels of animate nature, from the cell to the ecosystem, biochemical processes are interconnected to webs of 'habits' on which are built other 'habits' in an unending variety. As I noted above, evolution spent two billion years to create that enormously complicated web of chemical habits which we call the eukaryotic cell, and the genome would not work if it were not put into the context of the historically appropriated competence of that cell. Ecosystems would not be stable were it not for the millions of semiotic processes built on habits which themselves were formerly built on other habits. And the human brain would not function without the historically developed patterns of communication between some hundred billion highly organized nerve cells.

Sitting here in front of my Mac, and having just scrutinised my 'mail-box' to find 14 postings coming from three continents, I have no difficulty in interpreting the pattern disclosed by the reflections offered in this paper. Clearly, my mailbox is a continuation of the trend I have been describing, the unfolding of the semiosphere: the perpetual growth of *semiotic freedom* (Hoffmeyer, 1992). As far as cultural evolution is concerned, there can be no doubt that the single most impressive and, and in fact decisive thing, to have grown throughout human history is the capacity for processing and communicating knowledge (not information). Democracy may or may not have facilitated justice or equality, but there can be no doubt it has facilitated freedom of knowledge, communication and thought, i.e. semiotic freedom. Today we can — in principle — think everything which has ever been thought. But thousands of years ago much of what we think today would have been simply impossible to think.

Freedom is an ambiguous word however and the concept 'semiotic freedom' is not easy to capture in a more exact definition. Freedom of speech, for instance, is useless if you have nothing important to say. Lack of emotional depth is a kind of enslavement. Semiotic freedom does not simply refer to the wealth or quantity of semiosic processes around, but rather to the quality of such processes: the depth of meaning a culture, an individual or a species is capable of communicating.

That some concept like 'depth' is necessary in communication science as a supplement to the concept of information has been increasingly clear in recent years. And 'depth of meaning' obviously should be seen as related to Charles Bennet's concept of 'logical depth' which measures not how much information is contained in a given message but rather the amount of calculation time logically needed to arrive at this information (Bennet, 1988). Tacitly implied by this measure is of course the world of human purpose. The reason why anybody would calculate a given piece of information is not itself part of logical depth, but there has to be some such reason. Turning to the concept of semiotic freedom, we must refrain from defining a fixed point outside of its domain of operation from where to make quantitative measurements. As scientists we are not in the favorable position of baron von Münchhausen, who could lift himself by the hair. As human subjects we are part of the dynamic system which we try to understand. We must admit that the universe has produced us, and that this fact cannot possibly be eliminated from our understanding of that universe. Therefore the dream of making a quantifiable measure of evolutionary progression is futile. As a parameter semiotic freedom is free to be negotiated.

This does not mean, of course, that the concept cannot be given a more precise content. Semiotic freedom is an emergent property and should always be analyzed in relation to the proper level. Thus, the semiotic freedom of the cell probably decreased to a certain extent when the unicellular organism became a multi-cellular organism. The 'somatic ecology' of the body (Buss, 1987) constrained the freedom of single cells, but at the level of the organism an enormous gain in semiotic freedom was thus attained. Through the differentiation of its tissues the multi-cellular organism has acquired a much larger capacity for processing and communicating knowledge in the sense that it can deal with larger parts of its environment in space as well as in time. Or, to use the expression of Jakob von Uexküll, its *umwelt* has increased (Uexküll, 1982 [1940]).

Ultimately, the growth of semiotic freedom in our universe would seem to follow from the principle of maximum entropy production. In the words of Swenson and Turvey: "Progressive evolutionary ordering entails the production of increasingly higher ordered states — higher order symmetries of the world itself in its own becoming — and *perception-action is the physics at these levels*. There is no other way to co-ordinate the dynamics at these levels" (Swenson & Turvey, 1991, p. 344). Obviously, the idea of fitness is only understandable in this context. Genetic fitness may be a useful term in genetics, but if evolution is concerned, what matters is not genetic fitness but *semiotic fitness*. After all, fitness depends on a relation: something can be fit only in a given context. Genes may be fit only under certain environmental conditions, or environments might perhaps be said to be fit in the sense that their self-sustaining dynamic capacity has been adapted to the actual genotype resources offered to them. But if genotypes and envirotypes (Odling-Smee

& Patten, 1994) reciprocally constitute the context on which fitness should be measured, it seems we should rather talk about the fit in its relational entirety, that is as a semiotic capacity. The evolutionarily relevant fitness concept, semiotic fitness, should ideally measure the semiotic competence or success of natural systems in managing the genotype-envirotype translation processes. The optimization of semiotic fitness results in the continuing growth in the depth of interpretative patterns accessible to life. And this is the nearest thing we can come, I guess, to defining what should be meant by an increasing 'depth of meaning' at the level of biology.[6]

Returning to the macro-evolutionary perspective, we can now describe the workings of the Peircean 'law of the mind' in more explicit terms. Peirce himself explained the law in the following way: "The one primary and fundamental law of mental action consists in a tendency to generalization. Feeling tends to spread; connections between feelings awaken feelings; neighboring feelings become assimilated; ideas are apt to reproduce themselves. These are so many formulations of the law of the growth of mind" (Peirce, 1955 [1891] p. 321). Looking back into the cosmic origins, what we see is at first just the cosmic ocean of vibrations, which then suddenly at a point crystallizes into galaxies suspended in near empty space, i.e. differences appear in the world. Reacting upon the primary difference, a multitude of regularities, habits, take form reflecting the selective workings of the quasi-mind of nature. From there on selective processes arise in ever more variant and specific form, resulting finally in the appearance of self-referring entities capable of digital redescription i.e.living entities. Selective processes now become organized around cohorts of such entities reflecting the kind of mind-process (modus Bateson) we call organic evolution. Differences began to make a difference to 'someones'. But the quasi-mind of nature was not content to let itself represent only in the very general form of organic evolution. It slowly reappeared as a more individualized mind-process connecting members of mammalian (and perhaps avian) species into social structures based on sophisticated and shared umwelts. And finally the concretization of nature's quasi-mind in the form of single personalized and self-conscious minds such as yours and mine had to happen. From there on the cosmic process took on a whole new dynamics, albeit of course only in this very, very limited area of the universe we call Earth — or who knows, maybe also in some other local regions. All this seems to be just another way of deriving the specification hierarchy (Salthe, 1993). If we dare think that semiosis, and thus in a sense humanness, was in fact present, at least in a proto-form, in the universe from the very beginning, then we can link ourselves into the universe. If we dare not, only miracle can explain our existence. Ironically, the traditional scientific understanding rests its existence (as human endeavor) on the truth of exactly that miraculous intervention which it was its whole ambition to explain away. While the semiotization of nature, often looked upon with suspicion by scientists, makes it possible to think of the world as one unified process, requiring only that one miracle of existence as such, which no thinking can surpass.

Institute of Molecular Biology, University of Copenhagen
Copenhagen, Denmark

NOTES

[1] A note of caution might be in order here. Ontological quarrels have a tendency to evoke strong emotional stands. Allow me therefore to make a personal statement. I always thought of myself as a decided materialist. But much to my surprise I one day suddenly found myself an admirer of an idealist philosopher, C.S. Peirce. It then occurred to me that my resistance towards idealism was always based on a general disliking for easy ways to escape our human duty to try understanding the world as far as possible in terms of 'the knowable'. And I still in no way think we should abandon that strategy. Also, idealism all too often has been associated to right-wing ideas tending to disregard the material needs governing poor people's lives. This whole question would require a special treatment, and here I just want to state that I see no reason why Peirce's kind of idealism should lead us to such ideas. And in fact, the core of Peirce's argument for ontological idealism is parsimony: That mind can crystallize into mechanical laws, habits, seems to be less challenging to human intuition than is the materialist idea of a mechanism by itself evolving into mind. How could a mechanism evolve at all ?

[2] Back in the seventies Howard Pattee (1977) formulated a related view where he distinguished between the dynamic and the linguistic mode of complex systems.

[3] It took evolution more than two billion years to produce this marvelous entity, the eukaryotic cell. Having accomplished this deed evolution spent only one and a half billion years on producing all the rest.

[4] The idea of the self as relational and self-referring is central also in the existentialist philosophy of Søren Kierkegaard (Kierkegaard, 1944 [1848]).

[5] The concept 'semiosphere' was originally introduced by the Russian Estonian semiotician Yuri Lotman although he used it in a somewhat different way (Lotman, 1990).

[6] I am indebted to Bernard Patten for starting this train of thought.

REFERENCES

Bateson, G., 1979, *Mind and Nature. A Necessary Unity*, New York, Bentam Books.

Bennet, C., 1988, Logical Depth and Physical Complexity, in *The Universal Turing Machine*, R. Herken (ed.), A Half-Century Survey, Oxford, Oxford University Press, pp. 227-257.

Buss, L., 1987, *The Evolution of Individuality*, Princeton, Princeton University Press.

Christiansen, P.V., 1994, Personal communication.

Eldredge, N.; Salthe, S.N., 1984, Hierarchy and Evolution, in *Oxford Surveys in Evolutionary Biology*, R. Dawkins & M. Ridley (eds.), Oxford, Oxford University Press, pp. 184-208.

Hoffmeyer, J., 1987, The Constraints of Nature on Free Will, in *Free Will and Determinism*, V. Mortensen & R. C. Sorensen (eds.), Århus, Aarhus University Press, pp. 188-200.

Hoffmeyer, J., 1992, Some Semiotic Aspects of the Psycho-Physical Relation: The Endo-Exosemiotic Boundary, in *Biosemiotics: The Semiotic Web 1991*, T.A. Sebeok & J. Umiker-Sebeok (eds.), Berlin, Mouton de Gruyter, pp. 101-123.

Hoffmeyer, J., 1993a, *En snegl på vejen. Om betydningens naturhistorie*, København, Rosinante/ Munksgaard. (English: *Signs of Meaning in the Universe*, Bloomington IN, Indiana UP, 1997).

Hoffmeyer, J., 1993b, Biosemiotics and Ethics, in *Culture and Environment. Interdisciplinary Approaches*, N. Witoszek & E. Gulbrandsen (eds.), Oslo, Centre for Development and the Environment, pp. 152-176.

Hoffmeyer, J., 1994a, The Global Semiosphere, in I. Rauch & G.F. Carr, 1997, pp. 933-936.

Hoffmeyer, J., 1994b, The Swarming Body, in I. Rauch & G.F. Carr, 1997, pp. 937-940.

Hoffmeyer, J.; Emmeche, C., 1991, Code-Duality and the Semiotics of Nature, in *On Semiotic Modelling*, M. Anderson & F. Merrell (eds.), New York/Berlin Mouton de Gruyter, pp. 117-166.

Kierkegaard, S., 1944, *The Sickness unto Death*, London, Oxford University Press. First published in 1849 (in Danish).

Langer, M.M., 1989, *Merleau-Ponty's Phenomenology of Perception. A Guide and Commentary*, Tallahassee, Florida State University Press.

Lewontin, R.C., 1992, The Dream of the Human Genome, *The New York Review*, May 28, pp. 31-40.

Lotman, Y.M., 1990, *Universe of the Mind. A Semiotic Theory of Culture*, London, I.B. Taurus & Co.

Margulis, L., 1981, *Symbiosis in Cell Evolution: Life and Its Environment on Earth*, San Francisco, Freeman.

Maturana, H.; Varela, F., 1980, *Autopoiesis and Cognition. The Realization of the Living*, Dordrecht, Reidel.

Odling-Smee, F.J.; Patten, B., 1994, *The Genotype-Phenotype-Envirotype Complex: Ecological and Genetic Inheritance in Evolution*, Manuscript.

Pattee, H., 1977, Dynamic and Linguistic Modes of Complex Systems, *International Journal for General Systems*, vol. 3, pp. 259-266.

Peirce, C.S., 1931-58, Collected papers of Charles Sanders Peirce, vols. 1-6, C. Hartshorne & P. Weiss (eds.), Cambridge, Mass., Cambridge University Press.

Peirce, C.S., 1955, The Architecture of Theories, in *The Philosophical Writings of Peirce*, Buchler, Justus (ed.), New York, Dover. First printed in The Monist, 1891.

Polanyi, M., 1968, Life's Irreducible Structure, *Science*, vol. 160, pp. 1308-1312.

Rauch, I.; Carr, G.F. (eds), 1997, *Semiotics around the World: Synthesis in Diversity*, Proceedings of the Fifth Congress of the International Association for Semiotic Studies, Berkeley 1994, Berlin/New York, Mouton de Gruyter.

Salthe, S.N., 1985, *Evolving Hierarchical Systems. Their Structure and Representation*, New York, Columbia University Press.

Salthe, S.N., 1993, *Development and Evolution. Complexity and Change in Biology*, Cambridge, Mass./London, MIT Press.

Swenson, R., 1989, Emergent Attractors and the Law of Maximum Entropy Production, *Systems Research*, vol. 6, pp. 187-197.

Swenson, R.; Turvey, M.T., 1991, Thermodynamic Reasons for Perception-Action Cycles, *Ecological Psychology*, vol. 3, no. 4, pp. 317-348.

Üxküll, J. von, 1982, (1940), The Theory of Meaning, *Semiotica*, vol. 42, no. 1, pp. 25-87.

Wicken, J.S., 1987, *Evolution, Thermodynamics, and Information. Extending the Darwinian program*, Oxford, Oxford University press.

KOICHIRO MATSUNO

COMPETENCE OF NATURAL LANGUAGES FOR DESCRIBING THE PHYSICAL ORIGIN OF LIFE

INTRODUCTION

Difficulties in tracing the physical origin of life on the primitive earth are both empirical and theoretical. Isotope dating of carbon and other atoms has pinned the beginning of life on earth down into the narrow window region of far less than 10 million years centered around 3.80 billion years ago (Moorbath, 1994; Schidlowski, 1988). However, empirical details on how life would have started on earth have been extremely scanty and have hardly been preserved. The oldest fossil record of micro-organisms available is at most 3.45 billion years old (Schopf, 1993). Tracing the course of evolution further back toward its origin can not be facilitated by empirical studies alone. The endeavor would have to be supplemented by experimental efforts.

Nonetheless, experimental investigation of the origin of life has its own limitation ascribed to the underlying queer methodology (Salthe, 1993). Controllability and reproducibility required of any experiments simply contradict the uncontrollability and historical uniqueness of the phenomenon called life on earth (Hahlweg, 1991; Küppers, 1992). The historical uniqueness and coherent integration of the sequence of events over 3.80 billion years since the origin of life set forth a formidable problem to experimental studies (Atlan et al., 1986). It is simply inconceivable to contrive an experiment on the origin of life that could guarantee a succeeding evolution for another 3.80 billion years. What could be possible instead is to design likely experiments at most piecemeal (Gatens-Robinson, 1993).

One principal objective of such piecemeal experiments for the origin of life is to design the onset of replicating molecules without assuming any biological organisms. In particular, RNA molecules having catalytic capabilities that have been demonstrated to be able to exhibit their replication if the constituent monomers and appropriate energy sources are available (Cech, 1987; Guerrier-Takeda et al., 1983). The present likelihood of RNA replication, though impressive (Noller et al., 1992; Picchirilli et al., 1992; Mohr et al., 1994), does not however answer the question of how molecular replication as a harbinger of the phenomenon called life could have been settled in the first place. The original RNA templates prepared in the experiments are controlled and imposed externally (Shapiro, 1986). The issue is how to get molecular replication started out of the material and energy resources that could presumably have populated the primitive earth about 3.8 billion years ago and to sustain the evolvable capability indefinitely since then.

One of the driving forces toward the possibility of more primitive replicating molecules resides in the theoretical idea asserting that molecular replications could exhibit evolvability if supplemented by errors in the processes as epitomized in the phrase of molecular Darwinian evolution (Sober, 1989). If making errors is equated

G. Van de Vijver et al. (eds.), Evolutionary Systems, 295–305.

to making slight differences in the heritability of reproducing organisms, Darwinian evolution originally conceived in the biological realm could readily be extended to the pre-biotic regime so long as the presence of molecular replication could be guaranteed. The problem of the origin then reduces to the nature of the errors to be made (Grene, 1990).

Making errors, however, sounds extremely anthropomorphic. It presumes what is correct and what is in error. Although it may be possible to conceive of the dichotomy between a molecular replication proceeding with perfect fidelity and exogenous disturbances perturbing the fidelity, the dichotomy would simply be inaccessible in any experiment. Whatever experiments on the onset of molecular replication may be intended, the complete separation between literal replication and making errors could be unattainable since there is no non-anthropomorphic standard to facilitate literal distinction between the two. A case in point is the possibility that molecular replication could not proceed without making errors. In this regard, admitting the inseparability between molecular replication and making errors may claim some methodological legitimacy, while noting an occurrence of separability at the hypothetically weakening limit of the inseparability.

One of the aids in employing the methodological inseparability between molecular replication and making errors can be found in the relationship between an arbitrary production process and its products (Matsuno, 1984). Insofar as an indefinite continuation of production is guaranteed out of component materials while the products would eventually disintegrate into their components, the likelihood of actualizing the production process producing products of like kind could be enhanced with the elapse of time. Those production processes that would fail in acquiring necessary material resources would be wiped out in due course of time. Molecular replication and making errors are just the functional attributes of a production process that can hold on indefinitely (Rieppel, 1990). Our approach to examining the origin of life both experimentally and theoretically now reduces to how production processes of indefinite continuation could be made possible in the first place.

LETTING PRODUCTION LAST

The relationship between production and products is subtle because of the contrast between the duration of time in production and a static configuration within products (Cheetham, 1993). Nonetheless, the contrast is not clearly demarcated (Leydesdorff, 1993). Preceding products constitute boundary conditions for subsequent production so long as production keeps itself surviving. The notion of boundary conditions now introduces a new dynamic attribute pointing to a process of detection that can serve as a means of implementing and identifying boundary conditions. Boundary conditions necessitate processes of implementation and detection proceeding exclusively on material grounds. At the least, detection of material origin, that is to say, internal measurement (Matsuno, 1985, 1989) as opposed to measurement by an external observer comes to take part in the interplay between production and products (Rössler, 1987; Kampis & Rössler, 1990).

Internal measurement, as compared to external measurement by an outside observer is local, and boundary conditions thus identified remain local (Weber et al., 1989). Internal or introspective boundary conditions imputed to internal measurement serve as a dynamic attribute of the products resulting as a consequence of the preceding production (Rychlak, 1991). On the other hand, external or extrospective boundary conditions perceived by an outside observer as being frozen in the record are global in the sense that the observer assumes the perspective making the measurement of the object external and global. These two different types of boundary conditions, introspective and extrospective, are however related to each other because introspective local boundary conditions, when transferred and integrated into the record, have to coincide with their extrospective global counterpart. Otherwise, the integrity of the notion of boundary conditions would collapse (Ulanowicz, 1990).

The reconciliation between introspective and extrospective boundary conditions is owing to a unique nature of measurement. Measurement is a kind of interaction taking place between two arbitrary bodies supplemented by the irreversible characteristic that measurement cannot identify what will be identified before the actual measurement. The anthropomorphic separation between an object to be measured and the measuring apparatus is not the rule, though it holds for external measurement on the part of an outside observer. Either party of a pair of arbitrary interacting bodies can serve as a measuring agent internally. This amounts to saying no more than that the two bodies interact with each other. Still, the present classification turns out to be good enough to distinguish between internal and external measurement. Although external measurement is unilateral in the direction of influence from an object to be measured to the measuring apparatus, internal measurement is bilateral in direction between an arbitrary pair of interacting bodies.

Internal measurement detecting introspective boundary conditions locally in turn serves as another introspective boundary condition to be measured internally in a mutually interchangeable manner. Internal measurement not only detects introspective boundary conditions but also implements them. The extent to which external measurement can access internal measurement is only to detect boundary conditions.

Implementing boundary conditions through internal measurement remains inaccessible to external measurement because of its unilaterality. It is by implementing and producing introspective boundary conditions that internal measurement can be more than what external measurement is all about. Of course, external measurement requires implementation of the measuring apparatus for the purpose. However, it is taken for granted that external measurement is methodologically immune to any procedures required for implementing the measuring apparatus. Internal measurement does not assume such methodological immunization against the implementation, and has to squarely face how introspective boundary conditions serving as both a means and an end of internal measurement can be implemented.

External measurement associated with extrospective boundary conditions, though lacking the bilaterality between an object to be measured and the measuring apparatus, has one great advantage. It can be consistent with the notion of state defined as a static representation of a moving body in its every detail. Extrospective boundary conditions take it for granted that there is something out there to be detected, and external measurement is assumed to detect it. Definite identifiability of state, when combined with dynamics, necessarily yields a one-to-one temporal mapping of the state, otherwise the identifiability would fail. Such a deterministic development of state is certainly the case with both classical and quantum mechanics, which are consistent with external measurement. Even the wave-function or a vector in the Hilbert space conceived as the quantum state can externally be measurable by applying various quantum interferometry techniques. Nonetheless, state dynamics of a one-to-one temporal mapping suffers a methodological shortcoming. Unless exerted upon externally, the state dynamics does no more than preserving the invariant state, once given, through its equivalent transformation. There is no room for generating de novo production and products within the framework of state dynamics.

In contrast, internal measurement associated with introspective boundary conditions dispenses with the notion of state. Instead, events come to assume the key role there because introspective boundary conditions are detected by local events that take time for execution. More than that, internal measurement is always right in the middle of completion but not completed in the sense that any internally formed measuring apparatus at one moment constantly makes itself serve as an object to be measured subsequently. What is now measuring internally cannot be measured simultaneously. It will be measured only subsequently. The successive alternation between measuring and being measured renders internal measurement to constantly carry a leftover that defies identification by external measurement, but causes subsequent internal measurement, indefinitely.

The leftover that internal measurement carries with itself is the on-going variation induced at the internally formed measuring apparatus while performing measurement, since measurement is taken to be an instance of making variations at the measurement apparatus. It is the leftover of the self-induced on-going variation that drives internal measurement, while external measurement owes its driving factor exclusively to the outside observer. Internal measurement thus incorporates into itself the capacity of generating and producing variations indefinitely. Internal measurement is endogenously generative, while the external counterpart is generative only exogenously. This distinction between internal and external measurement also induces a difference of the descriptive stance between the two.

INTERNAL AND EXTERNAL DESCRIPTIONS

If one intends to describe a consequence of external measurement, the outcome would be a description of the finished events frozen in the external record (Depew, 1986). Each event in external description has already been accomplished and remains invariant in the record (Hinde, 1991). The invariant nature of the finished events in the record is consistent with the invariant stance that external description requires. A clear-cut separation between a measured object and an outside observer certainly matches the distinction between subject and object. Nonetheless, external description remains incompetent in coping with the indefinite leftover of on-going variations induced in the process of internal measurement because of the interchangeability between measuring and being measured in the latter (Sattler, 1990). In order to describe the indefinite leftover with internal measurement, a descriptive stance other than the external is required (Emmeche, 1991; Muller & Wagner, 1991). Internal description only can follow how internal measurement would proceed.

Internal description, however, imposes upon itself a formidable burden in that the internal descriptive agents are widely distributed and there can be no monopoly of descriptive agents, while the consequence of internal description, as frozen in a record, has to be approached and identified as an external description by a single external descriptive agent. One of the constraints to which any internal description succumbs is to observe the principle of the excluded middle in the resulting external description if the integrity of external description is to be taken for granted (Ereshefsky, 1991). Since each internal descriptive agent can recognize what others have described only afterward, fulfillment of the principle of the excluded middle found in the consequences of internal measurement cannot pre-exist in an atemporal manner, but has to be accomplished at cost so as not to cause violation of this principle. Internal descriptive agents must drive themselves toward fulfilling the principle of the excluded middle at the least while being involved in description. This means that if internal descriptive agents stop acting, a violation of the principle of the excluded middle would come to the surface and the whole descriptive enterprise would collapse.

Contrast between internal and external description is inescapable in regard to the principle of the excluded middle. Needless to say, there is no room for violating the principle within the realm of external description that presumes the integrity of the descriptive subject. This is, however, simply a consequence of monopolizing descriptive agents by a single external agent, which is inevitable in any discourse (including the present article). At issue is how an external descriptive agent can properly appreciate the involvement of many internal descriptive agents. The stance that an external descriptive agent assumes while facing many internal descriptive agents within the descriptive object is ambivalent in observing the principle of the excluded middle on the one hand and action for the principle in the other. But this ambivalence is a necessary price to pay for adopting the distinction between internal and external description.

In spite of its methodological clumsiness, however, a most significant advantage for accepting the ambivalence latent in the dichotomy between internal and external description is that it enables one to perceive the generation of variations endogenously (Matsuno & Salthe, 1995). This sharply contrasts with external description seeking the source of de novo variations always in the outside regardless of their real origins unless supplemented by internal description.

An internal description to be integrated into an external description while fulfilling the principle of the excluded middle as a consequence would be a description of the relationship between internal and external measurement. The principle of the excluded middle is to external description an incontrovertible empirical principle like the principle of the conservation of energy. External measurement is by definition free from any conflicts interfering with incontrovertible empirical principles. By the same token, internal measuring agents involved in internal measurement act so as to eliminate any conflicts with the invariant empirical principle exemplified in a record (Gayon, 1990). Constant removal of conflicts by internal measuring agents inevitably induces further conflicts that may also be eliminated without being entangled in the record because of the absence of any central agency coordinating all the local actions globally in a synchronous manner. Internal measurement constantly generates and passes forward internal conflicts without leaving any of them behind in the record.

HANDLING INTERNAL CONFLICTS

Internal measurement requires internal measuring agents just as internal description requires internal descriptive agents. However, any discourse assumes only one agent as the author. This single authorship brings about a difficult challenge to both internal measurement and description because of the involvement of multiple agents in the latter. In this regard, external measurement and description are consistent with the scheme of discourse accepting only one agent that serves as the subject facing an object.

Internal description participated in and sustained by internal descriptive agents is like a conference attended by many conferees. Although the reporter as an outside observer can report on what the conference has come up with, it is not the reporter but the conferees as internal descriptive agents that determine the descriptive content of the outcome of the conference. This does not necessarily imply that descriptive integrity on the part of each conferee could hold indefinitely because the principle of the excluded middle must be observed in what the conference has brought about. Only those conferees that could conform with the conference can maintain themselves. Practicing the principle of the excluded middle on the part of each conferee is to constantly adjust each internal description to the context the conference has come to provide. Internal description must be written in a natural language because it admits context dependence in the fullest sense.

Nonetheless, observing the principle of the excluded middle is not directly relevant to internal descriptive agents because the principle applies only to external description pointing to a definite static nature of an invariant descriptive object. It comes about simply a consequence of making and adjusting to the context among the participating internal descriptive agents. Context searching and making capability is primary to internal descriptive agents.

What is unique to internal descriptive agents is an internal generative capability. This exhibits a marked contrast to external description by an external descriptive agent that perceives an object prepared by whatever means just as it is. Those internal causes and conflicts that drive internal descriptive agents remain unfathomable to external description. An irony about the unfathomable nature of internal causes can easily be seen if one considers, for instance, the origin of a human language. If external description is employed, the origin would have to be totally of an exogenous nature because of its methodological stipulation. On the other hand, if internal description is attempted, internal descriptive agents would now have no alternative other than to live with on-going internal descriptions. That means that internal conflicts driving internal descriptive agents should be focused even though their definite identification would remain constantly out of reach. The language that each internal descriptive agent lives with is a natural language.

The internal generative capability that drives each internal descriptive agent, however, takes the contrast between the inside and the outside for granted. In particular, activity from the inside toward the outside is focused. The internal generative capability is incompatible with external description which is what a formal language can cope with, using a set of irreducible invariant predicates. The prohibition of causes from the inside, of course, does not apply to internal description performed within the framework of a natural language, which would remain immune to any forcible intervention imposing invariant fundamental predicates of whatever sort. Natural languages provide with an opportunity to see how evolutionary processes would proceed in the eyes of the participants, even though they have had only several hundred thousand years' history behind themselves.

The origin of life on earth was about 3.80 billion years ago. In contrast, human language employed as a means of describing the origin is at most several hundred thousand years old (Bloom, 1994; Corvalis, 1994). The intriguing conversion of letting such a newcomer evolutionarily encompass antecedent processes could not be the rule. This problem could be handled only methodologically, if at all (Paton, 1992). The situation is quite different from the case of external description attempted with a formal language. Here anything can happen depending upon how the invincible outsider would intervene. If internal descriptive agents must find their precursors before the evolutionary appearance of a human language, one could legitimately conceive a double image of those precursors of natural languages.

One kind of the evolutionary precursors to internal descriptive agents that we have recognized is internal measuring agents. Both of these agents have the capacity of influencing and being influenced by others of the similar nature, while observing the constraints that the external agents would recognize in the record, such as the principle of the excluded middle or the empirical principle of the conservation of energy. Internal measuring agents constantly generate internal conflicts just so as to remove any of those conflicts that may violate the empirical principle.

In fact, it is no more than a historical contingency that terms such as measurement and interaction have been invented and accepted in natural languages. Could one extend the notion of internal descriptive agents processing natural languages to that of internal measuring agents including the former as a special case? This perspective reduces the problem of the physical origin of life to figuring out the physical characteristic of internal measuring agents available on the primitive earth at about 3.80 billion years ago.

PHYSICAL ORIGIN OF LIFE

The ubiquity of internal measuring agents is unquestionable. Atoms and molecules are undoubtedly legitimate candidates for agents. However, new internal measuring agents appeared at the onset of life, and afterward assumed a unique quality of continuity, even though they are alternated frequently. Internal measuring agents are intrinsically coherent in generating and passing forward internal conflicts without leaving any trace of them behind. Such coherent alternation of internal measuring agents may sometimes effectively be halted either by stopping the alternation or by having too frequent alternations involving too many agents. A crystal as a measuring agent would stay as it is indefinitely unless acted upon externally. On the contrary, in gas molecules, each molecule serving as an internal measuring agent would be alternated by too many other molecules too frequently, with the result that the consequence of internal measurement would reduce to an aggregation of incoherent molecular motions. Except these two extreme cases, though quite ubiquitous in the realm of physics, coherent alternation of internal measuring agents remains the rule.

What signifies coherent alternation of internal measuring agents is the generating and passing forward of internal conflicts without leaving any behind. In particular, in view of the fact that internal measurement acts at least so as to fulfill the empirical principle of the conservation of energy retrospectively, each internal measuring agent comes to exercise the capacity of wanting energetic resources from the inside (Matsuno, 1992). This agent is no more than a form of consumer. The onset of coherent alternation of internal measuring agents can thus be equated with the appearance of consumers to be alternated successively (Matsuno, 1995). If one takes behaviors as consumers grounded upon the capacity of wanting from within to be a major characteristic of the phenomenon of life (Rosen, 1991), those evolutionarily significant events that occurred on the primitive earth at about 3.80 billion years ago would certainly have included the emergence of consumers wanting carbon dioxide, water and light energy from the sun. Of course, a fine tuning of the coding mechanism between protein and nucleic acid molecules would have had to be installed almost at the same time, otherwise those primitive consumers, even if they got started, could not have sustained themselves afterward.

The present perspective on the physical origin of life can also suggest a new possibility for designing a successful experiment on the origin. The emphasis should be on the significance of internal measurement and on the emergence of sustainable consumers grounded upon the capacity from the inside. Although experiments deal with external measurement and are methodologically incompetent for internal measurement, the products that internal measurement has come up with are certainly visible to external measurement. If one can set up an experiment that could enhance energy concentration locally in microscopic regions of protocellular size, it may be seen as a consequence of energy intake on the part of those microscopic bodies behaving as consumers.

The possibility of approaching the physical origin of life on an experimental basis shed a new light on the relationship between experimental research as a methodology and scientific language as a means of description (Farre, 1994). Needless to say, description of an invariant configuration of experimental or empirical findings requires a formal language that can preserve the uncovered invariance while being supplemented by the principle of the excluded middle. However, the externality of a formal language, which is sometimes also referred to as a scientific language, is destined to be intrinsically incompetent in coping with variations and changes because of the imposed externality (Gunji, 1995). Although one can circumvent the invariant constraint latent in a formal language by supplementing it with a source of variations in the outside, this sort of modification would not work if the problem of the origin is our matter of concern. To say that the matter of the origin goes beyond any scientific discourse because of its externality may be one possibility, though smacking of a defeatism. Another possibility would be to switch the descriptive means from a formal or scientific to a natural language.

Appreciating the role of natural languages in experimental and empirical researches is at least twofold. One is to provide an internal generator of variations. The other, and more significant, is to open a perspective enabling one to see the problem of the origin as internally caused. The physical origin of life is certainly an internally caused material phenomenon that could perhaps be deciphered exclusively in natural languages.

CONCLUDING REMARKS

The competence of natural languages for describing evolutionary processes in general and the physical origin of life in particular resides within their capacity of admitting internal descriptive agents, since natural languages are grounded upon dialogues instead of monologues. Internal measuring agents serving as the dynamic of evolutionary systems could be the evolutionary precursors of such internal descriptive agents. Nonetheless, external description as a monologue, like the present article, is also inevitable. What is specific to external description is that the descriptive object, whatever it may be, is taken to be out there, and remains invariant. The external descriptive agent as the author of the description who is informed about the object neither increases nor decreases the information about it during the process of description. This implies that information is conserved in the eye of the external descriptive agent irrespective of how it could be identified externally (Küppers, 1995). In contrast, internal measuring agents are constantly gaining and losing information about the others through their internal measurement. Gain and loss of information to internal measuring agents is certainly consistent with the conservation of information to the external descriptive agent, since the difference is merely methodological. The only constraint on internal measuring agents is that they act so as to come to terms with the conservation of information in the eye of the external descriptive agent.

One particular instance of the conservation of information to external description and its gain and loss to internal measuring agents is seen in the evolutionary emergence of energy consumers. The conservation of energy is incontrovertible in the empirical material world, while energy consumers act towards energy resources so as to make energy intake and loss meet. Associating the physical origin of life with the emergence of durable, though alternated energy consumers, facing the conservation of energy lets us see the appearance of information consumers in the face of the conservation of information as a principal characteristic of evolutionary systems.

Department of BioEngineering, Nagaoka University of Technology
Nagaoka, Japan

REFERENCES

Atlan, H.; Ben-Ezra, E.; Fogelman-Soulie, F.; Pellegrin, D.; Weisbuch, G., 1986, Emergence of classification procedures in automata networks as a model of functional organization, *J. Theor. Biol.*, vol. 120, pp. 271-280.

Bloom, P., 1994, Generativity within language and other cognitive domains, *Cognition*, vol. 51, pp. 177-189.

Cech, T.R., 1987, The chemistry of self-splicing RNA and RNA enzymes, *Science*, vol. 236, pp. 1532-1539.

Cheetham, T., 1993, The forms of life: complexity, history and actuality, *Env. Ethics*, vol. 15, pp. 293-311.

Corbalis, M.C., 1994, The generation of generativity: a response to Bloom, *Cognition*, vol. 51, pp. 191-198.

Depew, D.J., 1986, Nonequilibrium thermodynamics and a philosophical perspective, *Philosophica*, vol. 37, pp. 27-58.

Emmeche, C., 1991, A theoretical reflection on biology, living signs and artificial life, *Biol. & Philos.*, vol. 6, pp. 325-340.

Farre, G.L., 1994, Philosophical reflections on the question of emergence, in G.E. Lasker & G.L. Farre (eds.), *Advances in Synergetics*, vol. I: Systems Research on Emergence, IIAS, Windsor Ontario, pp. 63-67.

Gatens-Robinson, E., 1993, Why falsification is the wrong paradigm for evolutionary epistemology: an analysis of Hull's selection theory, *Philos. Sci.*, vol. 60, pp. 535-547.

Gayon, J., 1990, Critics and criticisms of the modern synthesis of the viewpoint of a philosopher, *Evol. Biol.*, vol. 24, pp. 1-49.

Grene, M., 1990, Evolution, "typology" and "population thinking", *Amer. Philos. Quart.*, vol. 27, pp. 237-244.

Gunji, Y.-P., 1995, Global logic resulting from desequilibration process, *Biosyst.*, vol. 35, pp. 33-62.

Hahlweg, K., 1991, On the notion of evolutionary process, *Philos. Sci.*, vol. 58, pp. 436-451.

Hinde, R.A., 1991, A biologist looks at anthropology, *J. Roy. Anthropol. Inst.*, vol. 26, pp. 583-608.

Kampis, G.; Rössler, O.E., 1990, How many gods do we need?: endophysical self-creation of material structures and the exophysical mastery of the universal libraries, in *Cybernetics and Systems*, E. Trappl (ed.), World Scientific, Singapore.

Küppers, B.-O., 1992, Understanding complexity, in *Emergence or Reduction? Essays on the Prospects of Nonreductive Physicalism*, A. Beckermann, H.H. Flohr & J. Kim (eds.), Walter de Gruyter, Berlin, pp. 241-256.

Küppers, B.-O., 1995 (in press), The context-dependence of biological information, in *Physik, Philosophie und die Einheit der Wissenschaften*, L. Kruger & B. Falkenburg (eds.), Grundlagender exakten Naturwiss., Bd 10, Spektrum Akademischer Verlag, Heidelberg.

Leydesdorff, L., 1993, Is society a self-organizing system?, *J. Soc. Evol. Syst.*, vol. 16, pp. 331-349.

Matsuno, K., 1984, Is matter inanimate?: protobiological information from within, *Origins Life*, vol. 14, pp. 489-496.

Matsuno, K., 1985, How can quantum mechanics of material evolution be possible?: symmetry and symmetry-breaking in protobiological evolution, *BioSystems*, vol. 17, pp. 179-192.

Matsuno, K., 1989, *Protobiology: Physical Basis of Biology*, CRCPress, Boca Raton Florida.

Matsuno, K., 1992, Cohesive interaction for the emergence of eukaryotic cells, in *The Origin and Evolution of the Cell*, H. Hartman & K. Matsuno (eds.), World Scientific, Singapore.

Matsuno, K., 1995, Consumer power as the major evolutionary force, *J. Theor. Biol.*, vol. 173, pp. 137-145.

Matsuno, K.; Salthe, S.N., 1995, Global idealism/local materialism, *Biol. & Philos.*, vol. 10, pp. 309-337.

Mohr., G.; Caprqara, M.G.; Guo, Q.; Lambowitz, 1994, A tyrosyl-tRNA synthetase can function similarly to an RNA structure in the Tetrahymena ribozyme, *Nature*, vol. 370, pp. 147-150.

Moorbath, S., 1994, Evidence of the geological age of the oldest terrestrial rocks with the biogenic components, presented at the Conf. on the Structure and Model of the First Cell, Trieste.

Muller, G.B.; Wagner, G.P., 1991, Novelty in evolution: restructuring the concept, *Ann. Rev. Ecol. Syst.*, vol. 22, pp. 229-256.

Noller, H.F.; Hoffarth, V.; Zimniak, L., 1992. Unusual resistance to protein procedures, *Science*, vol. 256, pp. 1416-1419.

Paton, R.C., 1992, Towards a methaphorical biology, *Biol. & Philos.*, vol. 7, pp. 279-294.

Picchirilli, J.A.; McConnell, T.S.; Zaug, A.J.; Noller, H.F.; Cech, T.R., 1992, Aminoacyl esterase activity of Tetrahymena ribozyme, *Science*, vol. 256, pp. 1420-1424.

Rieppel, O., 1990, Structuralism, functionalism, and the four Aristotelian causes, *J. Hist. Biol.*, vol. 23, pp. 291-320.

Rosen, R., 1991, *Life Itself: a Comprehensive Inquiry into the Nature, Origin and Fabrication of Life*, Columbia Univ. Press, New York.

Rössler, O.E., 1987, Endophysics, in Real Brains, Artificial Minds, J.L. Casti & A. Karlquist (eds.), North Holland, New York, pp. 25-46.

Rychlak, J.F., 1991, *Artificial Intelligence and Human Reason: A Teleological Critique*, Columbia UP, NY.

Salthe, S.N., 1993, *Development and Evolution: Complexity and Change in Biology*, MIT Press, Cambridge Mass.

Sattler, R., 1990, Towards a more dynamic plant morphology, *ActaBiotheor.*, vol. 38, pp. 303-315.

Schidlowski, M., 1988, A 3,800 million-year isotopic record of life from carbon in sedimentary rocks, *Nature*, vol. 333, pp. 313-318.

Schopf, J.W., 1993, Microfossils of the early archean apex chert: new evidence of the antiquity of life, *Science*, vol. 260, pp. 640-646.

Sober, E., 1989, Is the theory of natural selection unprincipled? A reply to Shimony, *Biol. & Philos.*, vol. 4, pp. 275-279.

Ulanowicz, R.E., 1990, Aristotelian causalities in ecosystem development, *Oikos*, vol. 57, pp. 42-48.

Weber, H.B.; Depew, D.J.; Dyke, C.; Salthe, S.N.; Schneider, E.D.; Ulanowicz, R.E.; Wicken, J.S., 1989, Evolution in thermodynamic perspective: an ecological approach, *Biol. & Philos.*, vol. 4, pp. 373-405.

SUSANTHA GOONATILAKE

TOWARDS A "META-ETHIC" DERIVED FROM EVOLUTIONARY LINEAGES

INTRODUCTION

This paper addresses a central moral issue of the coming century brought about by changes in evolutionary systems, exemplified by the internal life of the cyborg. The purpose is to demonstrate that a meta-ethic can be arrived at through the dynamics of evolutionary lineages, which has repercussions in approaching evolution as an internal process.

My incursion into the set of problems is by considering the coming merger of three evolutionary lineages whose commonalities I had sketched in earlier discussions (Goonatilake, 1993). These three evolutionary lineages are those of the genetic-biological, the cultural, and an incipient one in the case of adaptive computer systems. All three lineages deal with information, though not in the Shannon sense. It is information in the sense of the ability of each lineage (or sub-branch of a lineage) to interact and deal with its environment.

EVOLUTIONARY LINEAGES

The evolutionary characteristics of the biological lineage are well known. It bifurcates into new species, sometimes there are sudden disjunctures due to punctuated equilibria, there is a phylogenetic ascendancy and the information is self-organizing. The encoded information is also a window to the external world, having its own 'subjectivity' on the world, which changes as the lineages course through time.

The second information lineage with evolutionary characteristics is that of human culture, which encodes another set of reactions to the external environment. It is also a means of responding and adapting to the environment. Like genetic information, it is passed down diachronically from generation to generation. The characteristics of these lineages have been described in detail by social scientists as to their adaptation, construction through historical encounters with the environment, transmissions down a lineage and collective subjectivities in the form of, say, class consciousness or orientation of a profession.

The most rigorous work in the cultural lineages has been in the social studies of one social group, scientists and their culture. Over the last two decades the mechanics, by which scientific knowledge is constructed through the operation of social factors, has been described through what amounts to an evolutionary epistemology of knowledge. In this formulation, the production of scientific knowledge is influenced by factors in the social environment as well by social factors within the social group doing science. These influences help push science in

G. Van de Vijver et al. (eds.), Evolutionary Systems, 307–318.

particular directions, changing its content as well as the trajectory of the lineage. The lineage that results is a tree of knowledge which is parallel to the tree of genetic information and having many characteristics common with the latter.

These characteristics include speciation (in this case to disciplines and sub-disciplines), continuation of a core memory (in the form of the discipline's core knowledge), a process of self-construction (through the continuous social interaction between scientists), "subjectivities" attached to each lineage (in the form say of a profession's or class' particular view on the world), and sudden disjunctures in cognition in the lineage in the form of changes in paradigms. To these two lineages is now being added an incipient lineage of information associated with computers (for a detailed treatment of these lineages see Goonatilake, 1991). This lineage is only a few decades old but is growing rapidly as computer use grows exponentially. Through this process a lineage of 'artefactual' information that extends backwards and forwards through time is created.

This lineage also interacts with its external world, its environment, through its input and output devices. The lineage is tightly controlled at the moment, without much internal flexibility apart from that given by its human mentors, just as, say, insects are tightly controlled by their genetic programming. But new developments including those that use learning systems, genetic algorithms and neural networks allow the lineage to become more adaptive to the environment, and so change its internal states in non-trivial ways corresponding to changes in the environment.

This lineage grows initially according to the disciplinary boundaries and classification criteria set by humans, but as the information stores within it are increasingly exchanged and operated on by internally generated rules, specially those set by learning systems, these boundaries begin to change according to new internal criteria. And so, new boundaries emerge in a process analogous to the formation of species.

This artefactual lineage is also self-constructed through interactions with the environment. The lineage only samples its environment through the particular information fed to it. Hence its 'view' on the external world is a particular view, a particular 'subjectivity' that changes as the lineage changes its internal information system. It has been also observed that the lineage sometimes changes abruptly, apparently exhibiting evolutionary characteristics similar to punctuated equilibrium in biology (Huberman & Hogg, 1988). So, the general evolutionary characteristics that we observed in the case of the other two lineages also appear in the artefactual lineage too.

MERGERS

The three lineages are increasingly interacting with each other, merging directly or indirectly their information streams. This is happening largely through biotechnology and information technology. One can give a brief overview of these mergings (for details see Goonatilake, 1993).

Thus, genetic and machine information is operated upon by human actions through cultural information. Drawing from the sociological literature of the social construction of science one can easily demonstrate that cultural information in the form of scientific knowledge is mapped directly in machine information as programs and data, and in genes through genetic engineering (*ibid.*, pp. 78-82). On the other hand, machine information as computer output is mapped within cultural information as it is internalized by computer users. Genetic information is mapped on culture through information processing techniques and by the limits of perception that biology gives to culture. For example, we can subjectively sample our environment only through particular genetic windows, e.g. we hear a different sound spectrum than would a dog or a bat. And, once these are changed through interventions in biology, our perceptions, and so some aspects of our culture change.

Current developments in the analysis, storage and eventual manipulation of gene sequences by computers, as in the work associated with the Human Genome Project also results in direct transfers between the genetic and machine information realms. Nascent developments in biochips and biocomputers promise interactions from genes to machines.

The three lineages therefore are increasingly beginning to interact with each other and to merge directly or indirectly their information streams. This process merges the information content as well as the modes of processing, resulting in changes in all three streams, the genetic, the artefactual and the cultural. The continuity of each lineage solely with its past is now broken as new 'memories' are added from outside a lineage. New mutant combinations arise that have different interactions with the environment. There are resultant changes in the evolutionary characteristics of the hybrid-whole that are interesting, and which I have sketched elsewhere (Goonatilake, 1993).

But this hybridization of the lineages creates two sets of value, two kinds of ethical questions. One relates to the identity created by the new interventions. The other to what 'value' orientation each of these hybrids should have. At the moment, in say genetic engineering's infancy, one sees only the beginnings of such vexed questions. Yet, a precursor to the nature of the identity question is seen in the non-genetic interventions that have already been made in reproductive technology and tissue transplants.

Such techniques as flushing of embryos, in-vitro fertilization, sex preselection, surrogate motherhood, surrogate embryo transfer and cloning have raised a hornet's nest of issues. Some of these interventions have seen confusion and dramatic reversal of biological and social kinships. For example, when a woman carries, in an act of surrogate motherhood, her own daughter's child, as has already happened; the grandmother and mother become the same person, and the child's identity is multiple — at the same time son, grand son, and stepfather (Gruson, 1993). In a reverse direction the identity of the parent can also be thrown into question. Thus who, for example is the parent of a child brought about by in-vitro fertilization from an egg donated by Mrs. A., combined with a sperm from Mr. B., implanted in Mrs. C's uterus and given for adoption to Mr. and Mrs. D.?

These very complicated social and ethical issues will intensify when genetic characteristics themselves can be excised-in or -out of chromosomes in the future. In such cases, possible parents could be spread over a large number of desirable gene donors, or for that matter, a computerized gene bank. In the extreme hypothetical example, there could be a different donor for each of a variety of genes standing respectively, for example, for a nose, for an eye, for a set of teeth, for a particular type of intelligence, for avoiding a particular disease and so on. If one transfers genes from other organisms (the gene for preventing a disease could even be from another species — a transgenic source — as has happened already in the plant field), "parentage" is a very complicated affair. In the ultimate — meaning in the next few decades — it becomes an unsolvable affair within current discourse.

There is another intrusion to human identity occurring through information technology. The senses, and so parts of the "mind", are through electronic means now being extended beyond one's person. By pressing a lever, clicking on an icon or speaking to a device I can effect an event remotely, and that effect can also be fed back to me so that I internalize this feed back experience. This, and the extensive cloning of human capacities through a computer tend to spread the human identity widely over artifacts.

With advances in biotechnology and information technology therefore, questions of identity, and of associated philosophical, social and ethical modes become complicated. A further important question also arises: what is the ethical framework from which to view the environment from one of these hybrid lineages, say hybrids of the cultural-artefact, artifact-genetic or the genetic-cultural streams?

The fact that the three streams can hybridize is further proof of their basic equivalence, made of the same basic 'stuff' with similar properties. So what is true of one lineage, say the mind lineage, the transmission mode for culture, must be true in some degree of the others. Has this class of philosophical, and ethical problems been met with, and discussed before, and so provided us with some hints to navigate such a future world? There have been attempts in the field of cultural information to tackle these hybrids, to which I will now turn.

One attempt is by Derek Parfit (1984) in a book *Reasons and Persons* which aroused considerable interest. *The Times Literary Supplement* reviewer considered it "by any standard the most notable contribution to moral philosophy in the last hundred years" (Scheffler, 1984). Among the issues raised by the book is the nature of the person. Parfit argues what would happen in three hypothetical instances of changes in the internal make up of individuals such as, firstly, in an Enterprise type scan (of the TV film series Star Trek) and recreation of a human at a remote end, and secondly of hypothetical surgical interventions in the brain with implantation of memories, and thirdly of brain transplants. All these raise deep questions regarding personal identity. But what can be believed with reason is only the physical and psychological connectedness from one instant to another of a person. That is all one can truly assert. From this description of personal identity and continuity Parfit argues against self-interest theory, the dominant moral theory in philosophical circles.

Parfit's argument is from the point of view of a 'person' whose interiority had become changed either through biotechnology or information technology and becomes a hybrid. He does not spread this hybridization problem more generally to that of hybrids in the other two lineages, or hybrids of all three lineages taken together. Especially the questions of morality and ethics, of looking at the world from one subjective framework of a given lineage or a combination thereof, is not examined.

There is another approach to such problems that uses some of the general properties considered in evolution. But before I discuss this, let me digress to some related issues which have been referred to in this volume.

THE VIEW FROM WITHIN

These issues relate to a view from within an evolutionary lineage, from its semiotics, from its 'subjectivity'. Many in this volume have evoked a set of classical writers that have dealt with some aspects of this problem. Among the names evoked here, by for instance Salthe, Kampis and Hoffmeyer, are Hume, Huxley, Mach, Peirce and Whitehead. They have dealt with problems of identity, continuity and process; issues central to the evolutionary perspective.

In this way, Salthe has suggested that evolution is semiotic rather than material. He proposes that such a perspective is necessary because the systems we are concerned with in evolution envelop us within them. This internalist viewpoint must take into account the observer, that is ourselves. As in Peirce, such a perspective would mean that semiotics becomes a kind of scientific epistemology (Salthe, this volume).

Jesper Hoffmeyer also takes a Peircian point of view, and has shown how species form interactions with other species whereby the 'habits' (a Peircian term) learnt by one become signs to be interpreted by another. This results in the interconnected total of species becoming a global web, a global semiosphere (Hoffmeyer, this volume).

Kampis points out that evolution takes place within an interconnected web of species. Natural selection is a result of the evolution of this web, and not vice versa. Moreover, Kampis finds that process philosophy, such as that of Whitehead, which rejects a notion of permanent identity is very useful to picture evolution. The units that undergo change in evolution are seen as subject to temporary organization with only a transitory identity. Kampis finds that such notions are necessary for co-evolutionary scenarios (Kampis, this volume).

But many of the classical writers, Hume, Huxley, Mach, Peirce and Whitehead, that have been evoked have been directly or indirectly compared with, or have drawn sustenance from, one of the largest philosophical and observational literature on process phenomena available, namely that of Buddhist philosophy. And the issues raised by Salthe, Hoffmeyer, Kampis and several others find deep cords of resonance in that literature.

Thus, there has been a large set of writings on the surprising parallels between Buddhism and Hume, one commentator even suggesting that Hume's central ideas, which parallel those of Buddhism, were the result of a direct transfer of ideas (Jacobson, 1969). Hume, as also Buddhism, found the intellect rooted in feeling, which combined the realm of the body with that of the mind (Jacobson, 1988, p. 19). Huxley's attempt to link evolution and ethics in the late 1880's also saw an excursion into Buddhism. His was a pioneering attempt to link some Buddhist positions with Western philosophical ones (Rajapakse, 1985).

Mach was sympathetic to Buddhism, and there are many parallels between his position and that of Buddhism (Blackmore, 1972; Baatz, 1992). There have been a number of articles that referred to Mach's restoration of the Buddhist analyses on the "self" and "ego". One commentator observed: "Mach's thought shows a remarkable agreement in its main characteristics with those of Buddha in the exclusion of metaphysics and the concept of substance" (Blackmore, 1972, p. 288).

But the philosophers that are more quoted in the current discussions of evolution are Whitehead and Peirce. Until these two, Buddhism was virtually the only serious process philosophy around (Jacobson, 1988, p. 67).

Peirce "related his convictions more specifically to Buddhism than anyone had previously done" (Inada & Jacobson, 1984, p. xii). Peirce's law of love, "agapism", also had similarities with Buddhist categories. And he, like Buddhists, equated truth with righteousness (Bishop, 1982, p. 265-271). A Buddhist text on logic of circa 600 AD, *Nyayapravesa* ("Introduction to Logical Methods"), has basic syllogisms that are like the retroductive-deductive schemes found in Peirce. In fact one syllogism, Lance Factor points out, is virtually identical with the retroductive form suggested by Peirce (Factor, 1983).

Ames shows that Whitehead, Peirce and Buddhism also have several other common themes. These include the immediacy of present experience, a naturalistic framework, testing of ideas on the basis of their results, the rejection of dualisms, and the idea of individual freedom (Ames, 1962). To both Peirce and Buddhists, the immediacy of experience is most important, and there is no notion of a self. In Peirce's words: "One is immediately conscious of his or her feelings but not that they are feelings of an ego — the self is only inferred; there is no time in the present for any inference at all" (quoted in Jacobson, 1988, p. 71).

Whitehead and Buddhists both accept the instant flow of phenomena and the interconnectedness of all phenomena (Inada & Jacobson, 1984, p. 133). And the idea of the non-self in Buddhism, as Kenneth Inada points out, is consistent with Whiteheadian process philosophy (Inada, 1979). Whitehead's work, it has therefore been noted, became the first fully developed metaphysical system with which Buddhism could find general sympathy. The contemporary process philosopher Charles Hartshorne finding the many commonalities between Whitehead and Buddhism, as well as with Pierce, says: "Ancient Buddhism and Whitehead agree, against Aristotle, and also against Bergson, but in nearer agreement with Plato's Timaeus, that concrete reality consists of momentary actualities that successively become, this succession being what we call change".

If we are looking for concrete definite unitary wholes of reality, we should recognize that the individual now is always a new such whole. The Buddhists, whom Peirce admired, saw this (Jacobson, 1988, p. 23).

Having overviewed some of these similarities, Robert Neville (1984) suggests that Buddhism, as the process philosophy with the longest unbroken tradition in the world, has much to offer modern process philosophers. Buddhism has concepts that bring together in a process philosophy the internal world of the flow phenomena and the external environment in which it operates, as well as the interconnectedness of all phenomena, elements common to the internalist view of evolutionary theory.

However, before discussing these aspects I should reiterate some other legitimizing factors for entering the evolutionary field through Buddhist insights. Varela has pointed out that the Buddhist experiences of observing the mind are in the tradition of scientific observation (Varela et al., 1993, p. 23). The validity of the Buddhist conclusions on mental processes, including those of the many practitioners who came after the Buddha, have been increasingly corroborated within the last twenty years, as western psychologists and physicians have taken these observations seriously and attempted to validate them (representative of this literature are: Kabat-Zinn et al., 1992; Walsh, 1988; Sweet & Johnson, 1990; De Silva, 1984; Goleman, 1981; Bograt, 1991; Donaldson, 1992).

As for uses in the evolution field itself, I should mention, that Varela's and other's concepts of *Embodied Mind* is basically Buddhist observational practice and theory transferred to the realms of evolutionary theory and cognitive science (Varela et al., 1993 p. xviii). There, they discuss aspects of what I have considered flows in the three lineages, namely the internal flow of our thoughts (that is of culture within our minds); the flow of genes (that is of evolutionary biology); and the flow of "artificial thoughts" (that is of Artificial Intelligence) (Varela et al., p. xiv).

Buddhist theory and practice have a large collection of technical matter that could be useful to our project. The observation and analysis there, in many ways parallel — albeit perhaps notably only in narrow realms such as this — some of the rigor that one observes in modern science. I will explore part of the Buddhist observational and theoretical approaches through one of our lineages, namely the flow of culture, where, in the minds of humans, culture acts out its role, taking the form of mental activity, and a stream of thoughts.

STREAMS OF "INFORMATION" IN BUDDHISM

Thoughts as they arise and flow internally in our minds have been subject to careful observation and analysis in Buddhism. In this literature a class of questions is raised that directly parallel in many ways some of the questions that occur today in the era of the cyborg and the gene transplanted hybrid.

In Buddhism, there is nothing durable or of static being. Physical elements change, as do mental phenomena, all are in a state of perpetual becoming. In the Buddhist analysis of identity, there is no individual, only a stream unraveling within the mind, fleeting strings and chains of events ("Anatta", *The Encyclopedia of Buddhism*, vol. 1, 1984, p. 567).

This analysis is partly arrived at from observing the innermost subjectivity of a person. In fact, one of the objectives of Buddhist mental exercises, 'meditation', is to observe, experience and describe for oneself this lack of self and of permanence from within one's own streams of thoughts and mental phenomena. From within our innermost subjectivity, the problem of identity and of an abiding "I" can be shown to be a false one.

If this were the real state of "I from both an external material point of view, and internally from a subjective point of view, then what would this entail for our own streams of information, the lineages referred to above? All three lineages have the characteristics described here. They are all ever changing, hybridizing streams, whose identity is not in a snapshot existence of being, but in a process of becoming, an unraveling.

From such a perspective then, the questions on identity raised by the two technologies are seen differently. The existential angst of being a hybrid, of having genes of plants and animals inside one is seen differently. The problem of one's self being spread over several artifacts now loses its potential terror. The threat of being a cyborg, of Frankenstein's creature — the concerns of a Jeremy Rifkin, the fundamentalist critic of biotechnology, is seen differently.

Living things, complained Jeremy Rifkin, "are no longer perceived as carrots and peas, foxes and hens, but as bundles of information. All living things are drained of their aliveness and turned into abstract messages. Life becomes a code to be deciphered. There is no longer any question of sacredness or inviolability. How could there be when there are no longer any recognizable boundaries to respect?" Further, he continued "as bioengineering technology winds its way through the many passageways of life, stripping one living thing after another of its identity, replacing the original creations with technologically designed replicas, the world gradually becomes a lonelier place" (Rifkin, 1987, p. 29). Buddhist philosophy stripped this seeming sacredness and identity more than two and some half millennia ago.

A gene does not make a sentient being. Only the stream of a being's existence, of an onwards flowing history, constitutes the human or the cyborg. A person does not exist as a unique individual but as a constructed, ever changing flow, an onwardly moving lineage. If to this lineage are added new elements, new parts, it is but in the very 'normal' nature of such constructed streams. The artificial introduction of elements say to the cultural flow from genes or artifacts is but another manifestation of the normal construction of such flows. From a Buddhist's perspective, there is no difference.

But experiencing the intrusion of the new technologies that remake us biologically and culturally is, in an internal sense, disturbing. It challenges our sense of self "This idea that I may not be, I may not have, is frightening to the uninstructed" as the Buddhist texts put it. And, as the belief in an abiding self is deeply rooted in humans, this position is against the current as the texts say on one other occasion (Rahula, 1978, p. 56).

In such circumstances one may desperately try to cling to the idea of the mind being unchanging. But in such instances, the Buddha himself was very firm, rejecting the views of persons who take the thing called the 'mind' or 'consciousness' to be an unchanging substance. In that case it was better, he argued, for a person to take the physical body as an unchanging 'self'; rather than thought, mind or consciousness, because the body at least appeared more solid than the mind (Rahula, p. 65).

Interiority and consciousness are demystified into mundane components. As the 5th Century Sri Lankan classic of higher Buddhist psychology Vissudhi Magga (Gunaratne, 1982, p. 12) put it: "There is no doer but the deed. There is no experiencer but the experience. Constituent parts roll on, this is the true and correct view".

When it is inevitable that we are constructed and reconstructed, from biology, culture and artefact, this should be our epistemological, philosophical, ethical and subjectively felt guiding principle. This would be an unafraid view on the interiority of robots and of constructed hybrids, as they navigate reality, and tunnel through time in our lineages. This is the view to be internalized in the inevitable day of the cyborg.

This Buddhist process philosophy therefore challenges some of the basic assumptions of the self as an ethical agent. Its doctrines of no-self, momentariness and conditioned origination raise, Peerenboom notes, questions of 'who' is the moral agent. Buddhism requires that the agent is not an assumed atomistic individual but a person within both a social and natural context. Ethical responsibility is now spread over the whole community. It also goes against reading an either-or, dualistic reading of freedom and determinism (Peerenboom, 1989).

From this perspective, one analyses oneself, knows oneself, only to realize that there is no self in the first place. This is not an intellectual knowledge but an internally observed, felt knowledge. This elimination of this sense of self "sets one free" in Buddhism. When the realization dawns that I am not a thing but a process, then the future becomes open ended. Buddhist theory and practice are self-referential, to know oneself is to make oneself, to guide the self that is not there (Kolm, 1986). In the Buddhist analysis, unsatisfactoriness and anxiety becomes essential to the 'I' because these are the 'I's response to its own groundlessness (Loy, 1992).

This internal experiencing of the non-self, that is, of the view from the interiority of the lineages, does not lead to a loss in integration, awareness or vitality of the mind. On the contrary, perception unclouded by what Buddhists call false perceptions leads to perceptual clarity. Perceptions of others are enlarged because there is an empathic openness based on a non-judgmental awareness (Page & Berkow, 1991). The fully mentally healthy person, is expected to have a state of continuous inner delight, attends keenly to all the circumstances of a situation and can respond with skill to every situation. Some of these observations have been validated in some recent Western psychological research.

This perspective, based on observation, has given rise to a profound altruistic moral code. And it is not entirely far fetched to think that it could also do so in this case of merged information streams discussed above. This phenomenology of flow of human thought could be extrapolated to hybridization in the other two lineages. These observations could be some pointers for a moral compass on the inevitable future mergings of the three streams.

Do we have any further hints of the ethical problems to be encountered in the merging from the only process philosophy to have developed elaborate ethical schemes based on process analyses? In Buddhism, the ability to get a "correct" understanding, or insight itself, on the experiential reality is considered to have ethical value. The person who is awake to the central tenets of existence, namely its process nature, therefore indulges in right action. In Buddhism, the process of development and growth, is also one of becoming, of evolution. In developing mental culture, the disciple does not obtain virtue, a final state; he becomes virtue, a process, an evolutionary state ("Bhava", *The Encyclopedia of Buddhism*, vol. 111, Fasc. 1, p. 10). From the Buddhist ethical point of view, thinking itself is of two kinds, "right thinking" and "wrong thinking". The former looks realistically at the world and its phenomena, and leads to freedom. The latter is speculative, unrealistic thinking ("Cinta", *the Encyclopedia of Buddhism*, p. 167). The Eightfold Path, which is the essence of Buddhist ethics, reiterates among others, these central positions.

Given these, how would one internally react to the massive inflow, into one's hybrid biological and mental interiority, that is in store in the next century? If it is not 'mine', and if in fact 'I' do not exist, how does one make sense of this alien intrusion? Let me recourse to a standard exercise in dealing with that interiority, to Buddhist observational practice, mindful 'meditation'. Here, one trains oneself to observe one's interiority to realize for oneself its constructed nature, its lack of an essential being. Secondly in this process of observing one self, one dispassionately notes also the coming and going away of one's thoughts.

To paraphrase a summary by Varela, when one is in this mindful state, one has an embodied, open-ended reflection. This reflection is not just on experience. It becomes a type of experience itself cutting through habitual thought processes and preconceptions. Reflection then becomes open ended, mindful and embodied. The Cartesian dichotomy of mind and body, is now found to be only a limited point of view when the mind and body are separated, an outcome of a specific experience of unmindful, disembodied reflection. This would be an example of the "right thinking", that is, the desired internal orientation from any lineage's interiority.

How would such a perspective appear as, say, one whizzes down a lineage? Experiences fly through as in a tunnel. In one's physical-mental evolution and becoming ("one" meaning a biological entity, a cultural being, or a computer with intelligent characteristics) different historical lenses, different evolving realities present themselves as an unreeling film. These snapshots come from the future, but one lives only in the present. Examining them into different components, one does not find anything 'there'. That realization makes one view the world afresh, with a noetic vision. Peirce's agapism is a partial approximation.

Given the above perspectives, can we now see whether some of the concerns expressed by our colleagues have been met from this Buddhist theoretical and observational platform? There are many resonant echoes.

Approaching evolution through the process philosophy of a Whitehead, Kampis reiterated the call for an 'endophysics', an internal physics, as suggested by O.E. Rössler, a science that takes into account observations from within (Kampis, this volume). Such an endophysics of processes from within is seen in the flow of thoughts in a person, which have been described and analyzed in great detail in Buddhism. Salthe states that the correct perspective for us to take for the internalist point of view would be to become disinterested observers at one time, and at another time interested actors (Salthe, this volume). To use a Buddhist expression, the observer should be like the lotus flower both within the waters (of subjectivity) and above it (in the realm of the objectivity), a metaphor used to describe the ideal observational platform in Buddhism. Eric Minch on the other hand, evoked the idea of a morphogenetic movie in evolution, a succession of structural snapshots (Minch, this volume). This parallels directly a Buddhist metaphor used to describe the endophysics of the thought process, the illusion of a moving line created by a rapidly spinning fireball. The interdependent web of Hoffmeyer, the global semiosphere, has also its parallel in the Buddhist interrelated web of dependent origination. And Peirce's 'habits' have their parallels in the Buddhist descriptions of 'the ruts and grooves of the mind' formed by the flow (Hoffmeyer, 1997).

It appears that an interaction between the semiotics and processes of evolution and of Buddhist observations and theoretical orientations could indeed be fruitful. If one wants to explore evolution and processes as an internal process then there is much in the vast storehouse of Buddhist technical literature. I have only indulged in a preliminary foray in this paper, but there are discussions, for example, on the perceptual grids that structure perceptions, which could be very useful for those working on the semiotics of evolution. Evolution viewed as a process driven from within and without would be enriched by interactions with the largest technical literature in process philosophy.

New School for Social Research
New York, USA

REFERENCES

Ames; Van Meter, 1962, *Zen and American Thought*, Honolulu, Univ. of Hawaii Press.

Baatz, U., 1992, Ernst Mach: The Scientist as Buddhist?, in *Ernst Mach - A Deeper Look*, Kluwer Academic Publishers, pp. 183-199.

Bishop, D.H., 1982, Peirce and Eastern Thought, in *Proceedings of the C.S. Peirce Bicentennial International Congress*, K.L. Ketner, J.M. Ransdell, C. Eisele, M.H. Fisch & C.S. Hardwick (eds.), Graduate Studies, Texas Tech University, pp. 265-271.

Blackmore, J.T., 1972, Mach and Buddhism, in *Ernst Mach, His Work Life and Influence*, J.T. Blackmore (ed.), University of California Press, pp. 286-355.

Bograt, G., 1991, The Use of Meditation in Psychotherapy: A Review of the Literature, *American Journal of Psychotherapy*, vol. 45, no. 3, pp. 383-412.

De Silva, P., 1984, Buddhism and behavior modification, *Behavior Research and Therapy*, vol. 22, no.

6, pp. 661-678.
Donaldson, M., 1992, *Human Minds An Exploration*, Penguin Books, London, p. 227.
The Encyclopedia of Buddhism, 1961-90, Government of Sri Lanka/UNESCO, Government of Sri Lanka Press, Colombo.
Factor, R.L., 1983, What Is the 'Logic' in Buddhist Logic?, *Philosophy East West*, vol. 33, pp. 183-188.
Goleman, D., 1982, Buddhist and Western psychology: Some commonalities and differences, *Journal of Transpersonal Psychology*, vol. 13, no. 2, pp. 125-136.
Goonatilake, S., 1991, *Evolution of Information: Lineages in Gene, Culture and Artefact*, Pinter Publishers, London.
Goonatilake, S., 1993, The New Technologies and the 'End of History', *Futures Research Quarterly*, vol. 9, no. 2, pp. 71-93.
Gruson, L., 1993, When 'Mom' and 'Grandma' are One and the Same, *New York Times*, Feb. 16.
Gunaratne, V.F., 1982, *Buddhist Reflections on Death*, Buddhist Publication Society, Kandy, p. 12.
Hoffmeyer, J., 1997, *Signs of Meaning in the Universe*, Bloomington IN, Indiana UP.
Huberman, B.A.; Hogg, T., 1988, The Behavior of Computational Ecologies, in *The Ecology of Computation*, B.A. Huberman (ed.), Amsterdam, North-Holland Publishing.
Inada, K.K., 1979, Problematics of the Buddhist Nature of Self, *Philosophy East West*, vol. 29, pp. 141-158.
Inada, K.K.; Jacobson, N.P. (eds.), 1984, *Buddhism and American Thinkers*, SUNY Press Albany.
Jacobson, N.P., 1969, The Philosophy of Oriental Influence in Hume's Philosophy, *Philosophy East and West*, vol. 19, no. 1, University of Hawaii Press, Honolulu, Hawaii.
Jacobson, N.P., 1988, *The Heart of Buddhist Philosophy*, Southern Illinois University Press, Carbondale and Edwardsville.
Kabat-Zinn, J. et al., 1992, Effectiveness of a Meditation-Based Stress Reduction Program in the Treatment of Anxiety Disorders, *American Journal of Psychiatry*, vol. 149, no. 7, pp. 936-943.
Kolm, S.-C., 1986, The Buddhist theory of 'no-self', in *The Multiple Self*, J. Elster (ed.), Cambridge University Press Cambridge, pp. 233-265.
Loy, D., 1992, Avoiding the Void: The Lack of Self in Psychotherapy and Buddhism, *The Journal of Transpersonal Psychology*, vol. 24, no. 2, pp. 151-179.
Neville, R.C., 1984, Buddhism and Process Philosophy, in *Buddhism and American Thinkers*, K.K. Inada & N.P. Jacobson (eds.), State University of New York Press, Albany.
Page, R.C.; Berkow, D.N., 1991, Concepts of the Self: Western and Eastern Perspectives, *Journal of Multi-cultural Counseling and Development*, vol. 19, pp. 83-93.
Parfit, D., 1984, *Reasons and Persons*, Oxford University Press.
Peerenboom, R.P., 1989, Buddhist Process Ethics: Dissolving the Dilemmas of Substantialist Metaphysics, *Indian Philosophical Quarterly*, vol. 16, pp. 247-268.
Rahula, W., 1978, *What the Buddha Taught*, Gordon Fraser, London.
Rajapakse, V., 1985, Buddhism in Huxley's Evolution and Ethics, *Philosophy East and West*, vol. 35, pp. 295-304.
Rifkin, J., 1987, Genetic Engineering May Threaten Humanity, Biomedical Ethics: Opposing Viewpoints, in *Biomedical ethics: Opposing viewpoints*, J.S. Bach (ed.), Greenhaven Press, St. Paul, Minnesota, p. 29.
Scheffler, S., 1984, Ergo: less Ego, *The Times Literary Supplement*, May 4th, no. 4231.
Sweet, M.J.; Johnson, C.G., 1990, Enhancing empathy: The interpersonal implications of a Buddhist meditation technique, Special Issue: Psychotherapy and religion, *Psychotherapy*, vol. 27, no. 1, pp. 19-29.
Varela, F.F.; Thompson, E.; Rosch, E., 1993, *The Embodied Mind: Cognitive Science and Human Experience*, The MIT Press Cambridge, Massachussets, p. xiv.
Walsh, R., 1988, Two Asian Psychologies and their Implications for Western Psychotherapists, *American Journal of Psychotherapy*, vol. 42, no. 4.

OLAF DIETTRICH

ON SOME RELATIONS BETWEEN COGNITIVE AND ORGANIC EVOLUTION

INTRODUCTION

Largely uncontested is that the evolution of cognitive capabilities has to be regarded as a continuation of organic evolution by different means. For example, Evolutionary Epistemology (EE, Campbell, 1973, Lorenz 1966, Riedl 1980, Volmer 1980, Wuketits 1984) is based on the assumption that our cognitive instruments and the categories of our thinking have evolved in ways similar to the evolution of organic tools for life management, tools such as metabolic and homeostatic instruments or limbs as instruments for locomotion and action.

Within EE, however, the question remains unanswered as to what cognitive instruments have evolved and how. The argument is sometimes used that our cognitive instruments must be outcomes of adaptation to the environment and to the problems presented there, otherwise man could not have survived, and that, therefore, our cognitive phenotype must contain some information about the character of our world. Here Campbell (1973) speaks explicitly of *"natural selection epistemology"*. Most problems, however, can be solved in various ways, and in ways that are often not related to one another. Horses and snakes, for example, although they may have developed in a similar physical environment, have entirely different organs of locomotion and these have no structural element in common.

So the criterion of survivability cannot explain why we have acquired the actual categories of human thinking and perceiving and not others. We cannot say, for example, that our visual perception of space is 3-dimensional because the world itself is 3-dimensional in character, and pre-human simians that did not see the world in 3 dimensions would have been unable to spring from tree to tree. Easy to show is that appropriate and successful survival strategies could well be based on 2- or 4-dimensional space-perceptions, and independent of the number of degrees of freedom actually available. Kinaesthetic animals, as described by Lorenz (1983), show us that even a 1-dimensional space-perception referring to nothing other than the linear context of their 3-dimensional pathways can be a helpful cognitive tool. An ape jumping from tree to tree is by no means obliged to interpret his visual perception in terms of a 3-dimensional perspective (geometrical approach). It could be quite sufficient if he saw his world in a 2-dimensional manner — as 2-dimensional as the picture at his retina is. In this case, however, the ape is obliged to attribute to his legs a capacity not only for propelling him through the air, but also the capacity, associated with jumping, for effecting changes in the sizes of objects, branches in this case, causing them to become larger, actually changing physically in a characteristic manner (physical approach), knowing that he has to grasp the branch envisaged, which he is seeing,

G. Van de Vijver et al. (eds.), Evolutionary Systems, 319–340.

just when its size and position achieve certain typical values. If he has learned to do this, an external observer would find no difference between the ape's movement strategies and those initiated on the basis of 3-dimensional perception. (It is evident that physical theories based on an inborn meta-theory saying that objects can be "deformed" not only by means of our hands, but also by means of walking or jumping, would have no similarities with the theories we are used to).[1,2]

Concluding that cognitive structures and instruments are unconditional or arbitrary because they are not, and cannot be derived from external boundary conditions, is mistaken, since internal boundary conditions must also be taken into account. Firstly, there are the developmental constraints of cognitive evolution itself; cognitive as well as organic evolution is subject to what has been evolved before. Cognitive evolution in our time, therefore, would find rather limited degrees of freedom. Further, cognitive instruments exert themselves in continuous co-evolution with organic instruments for meeting organically defined needs and requirements. This means that cognitive systems cannot be explained by reference to what is called their object, but only through their organic genesis. This justifies efforts made to look for a closer relationship between cognitive and organic evolution.

CONSTRUCTIVIST EVOLUTIONARY EPISTEMOLOGY (CEE)

To understand cognitive evolution from an organic point of view, we start from a constructivist extension (CEE, Diettrich, 1993) of classical evolutionary epistemology (EE). The particularity of the CEE is based on a methodological element used mainly in physics, the so-called operational definition of physical terms.

What does that mean? As is well known, classical physics failed to accommodate the phenomena of quantum mechanics and special relativity primarily because it got involved with a non verifiable syntax brought about by the use of terms that had not been checked as to whether they could be defined by means of physical processes.

In our day to day life this epistemological refinement is not necessary. We have a clear understanding of what the length or the weight of a body means, and we do not need confirmation from a tape measure or a scale for carrying on. The situation, however, is different with microscopic distances that are smaller than the atoms of the tape measure. Here, first of all, we have to decide what kind of experimental facility we will apply in order to define length or momentum. Physicists say that properties are defined as invariants of measurement devices. This even applies to the order in time of events which, under normal conditions, can easily be defined and detected. With very high speeds, however, the topology of events may depend on relative motions.

As this kind of experience may be repeated again and again, it suggests a generalization that can be summarized as follows: properties of whatever kind and of whatever subject have no ontological quality. Instead they are defined by the fact that they are the invariants of certain measurement operators. This contrasts with classical thinking in which properties are used for the objective characterization of objects. One of the most important properties we usually attribute to properties, namely, independent existence, is based simply on the assumption of their independent ontological quality. In day to day life this is incontestable. The length of a body and its color exist independently of each other and can be measured separately. This does not necessarily apply in subatomic regions, as we know. The position and momentum of microscopic particles cannot be measured independently of one another. Physicists learned from this that theoretical terms have to be defined operationally, i.e. they have to describe nature by means of theories in which only those terms which are defined by certain experimental facilities are acceptable, rather than by categories which are defined by protophysical common sense.

The crucial step of CEE is to suggest that not only theoretical terms have to be defined operationally, but also observational terms as well as mathematical and logical terms.

Theoretical terms are defined as invariants of operations represented by physical measurement devices.

Observational terms, comprising both the visually perceived regularities (patterns) and those we condense into theories and into what we call laws of nature, are considered to be invariants of phylogenetically evolved mental cognitive operators. These operators are physiologically implemented somewhere in our brain and can be considered as a kind of cognitive measurement device: measurement objects are the sensorial inputs and measurement results are perceptions, i.e. pictures, and, within these pictures, certain regularities, structures or patterns, rather than numbers or pointer positions. Therefore, the entire system of laws of nature we have derived from these regularities cannot be objective entities but only mental constructs.

In this context the often discussed dichotomy of observational and theoretical terms is reduced to a rather secondary difference: observational terms have developed phylogenetically in the unconscious parts of the human brain, whereas theoretical terms are the outcome of conscious and rational efforts. Nevertheless, observational terms remain privileged as the basic elements of any higher theories; we can modify theories according to observational data, but we can not modify the genetically fixed mental operators and their invariants according to the requirements of special situations.

THE OPERATIONAL DEFINITION OF SPACE, TIME AND CAUSALITY

The most crucial consequence of what has been said above, is that space, time and causality, which according to Kant are the necessary categories on which all external appearance is based, are not the only possible (and therefore necessary) categories. They are rather the phylogenetically evolved features of human perception and interpretation, defined operationally as invariants of certain actions and transformations. Let us look at this in more detail.

Following Piaget, *the spatial metric* of our perceptional space (and therefore the topology comprised) is operationally defined by means of motion. The identity of extended subjects, therefore, is defined as an invariant of locomotion (Üxküll: "a body is what moves together as a unit"). This definition is probably the main reason for the major difference between of what we call space and what we call time. Time is said to flow in an irreversible way; no one can retrieve any part of the past. We cannot move back and forth between two points in time. But we can do so quite well between two points in space. If we say we travel from point A to point B and than back again to A, we mean that the A where we started before arriving at B, and the A to which we arrived after leaving B, are not only equal but identical. To say so, however, is possible only if we can distinguish between "equal" and "identical" and if what we call identical is not influenced by our travel. This means that Identity is defined as the invariant of motion. And exactly this is the point. Only on grounds of such a definition can we call a change in spatial positions reversible, or more precisely: only on the basis of such a definition can we distinguish between the repeated return to the same A and travel along a sequence of equal As, i.e. between periodicity in time and space.

In a similar way, locomotion can change the visually perceived environment. We can transform the perception we call "forest" by means of walking, appropriately, into the perception we call "city". But this is not what we are accustomed to saying. More common is to speak in terms of an environment which, apriori, is multidimensional in character, i.e. comprising at the same time several structures which differ, first of all, in what we call their spatial positions. What we achieve, then, by means of our legs, is not a modification of the environment. We just "go" to places consisting of different structures and therefore experience different perceptions. What we call the multiplicity of the world, thus, is defined as invariant to changing our positions in that world. From the functional (and CEE) point of view, mentally generated spatial pictures belong to the most elementary theories we have at our disposal, by means of which we can forecast perceptions when walking — in the same way as the temporal structures stored in our memory inform us on what we can expect when repeating certain actions. So, both the formation of pictures and the formation of memory are first of all modi of extending life competence.

The arrow of time determines the difference between past and future. All efforts to define that arrow by means of an operator failed since every operator already comprises inherently the arrow of time: an operator transforms the status of its object before its application into the status after its application. This means that without a preceding definition of the arrow of time it cannot be said what an operator is doing. A consequence is that in all the endless cases where it is said that the arrow of time has been defined operationally it can be shown that the direction of time was already comprised implicitly in the preconditions of the experiment. A typical example is the following: shaking a box containing black and white balls that have been separated according to color will always lead to disorder and never again to the original order. In physical terms: entropy will increase in time, never decrease. Entropy therefore seems to operationalise the arrow of time. But in this case, the result will depend on what is done first, separating the balls or shaking them. First separating and then shaking will lead to disorder. First shaking and then separating will lead to order. So beforehand we have to know what the terms 'before' and 'after' mean before we can perform the experiment which is to tell us what before and after mean. Also, the often mentioned thermodynamic concepts cannot help, as one can discover in the specialized literature concerned (see also Diettrich, 1989). If the arrow of time cannot be defined by means of physical operations a mental definition is required using, for example, the content of memory in the following sense: of two perceived events, A and B, A is said to precede B if we can remember A when B happens but not B when A happens. Of course, the past is what we can remember but we cannot remember the future.

In order to constitute causality we must be able to identify patterns of events. If a number of events, say A, B, C, and D follow each other always at typical intervals independent of when the first one occurs (i.e. if the pattern is an invariant of translation in time), then we say that there must be a causal relationship between the events concerned. Otherwise the perceived regularity could not be explained. Causal relations, then, are defined as invariant patterns of time. This, however, requires more than just having a topology of events as provided by our memory. We also must be able to distinguish between shorter and longer intervals of time, i.e. we need a time metric defined by a mental metric-generator implemented physiologically somewhere in our brain. For example, that we say lightning is the cause of thunder but not the contrary, is based on the fact that the time between lightning and the next thunder is usually much shorter and varies less than the time between thunder and the following lightning. But the length of time intervals can be defined only by means of a time metric. *The mental time metric-generator is therefore responsible for the causal order established and for the prognostic capability derived from it.* This also applies to continuous sequences of events such as the development in time of a mechanical system described by the equation of motion, by means of which we can forecast the future development of the system.[3]

PHYSICAL EXTENSIONS OF PERCEPTIONS AND ALGORITHMIC EXTENSIONS OF MATHEMATICAL THINKING

Typical of most empirical sciences is the use of instruments and measurement devices by means of which we extend the range of natural perception in ways similar to those we use to extend our inborn physical capabilities by means of tools and engines. Here we have to distinguish between two important cases (Diettrich, 1994a).

We will speak of *quantitative extensions* if the perception operators and the measurement operators are commutable in the sense of operator algebra. In this case both operators will have the same spectrum of invariants. This means that the results of the measurement operations can be presented in the terms of invariants of the inborn cognitive operators, i.e. in terms of our classical world picture.

We will speak of *qualitative extensions* if perception operators and measurement operators are not commutable. Then the results can no longer be presented in a classical manner and require new, non-classical theories that accept classical theories as only approximations in certain borderline cases. As the set of possible measurement devices is, in principle, unlimited, what can never be excluded from considerations is that qualitative extensions of previously established operators will bring about modifications of the previously established world picture and of the theories associated with it. So there will never be a definitive world picture and there will never be a definitive 'theory of everything'. No objective laws of nature will ever be formulated. Those laws that we have, we have 'constructed' in a human-specific way in the course of human evolution; they will never converge towards a definitive set of laws except within the context of a limited set of operators, i.e. if we desist from further experimental research exceeding these limitations. What we actually do when we do science is to construct a world that we then believe we analyze by doing science. In other words: analytical in the sense of deepening our knowledge is characteristic of science only within quantitative extensions.

The notion of a theory of everything is equivalent to the notion of reality being characterized by objective laws of nature. According to what has been said in this paper, such a notion cannot be defined operationally. To require of reality structures that are independent of all human action, i.e. structures that are invariant under all possible operators, would deprive reality of just the specificity necessary for being a non-trivial notation. A non-trivial reality can result, thus, only through being invariant under particular operations. In other words, the operator that is to operationalise the notion of reality has to be commutable with every other possible operator. Unfortunately, only the trivial unity operator meets this requirement. So, what we can operationalise is, at best, a kind of current reality which refers to all the operations applied up to now (rather than to all possible operations), i.e. a reality which reflects all the perceptions and experiences man has ever had. This is just what we do when we speak in ontological terms about a reality which — according to our current knowledge — has this or that structure. From this it becomes evident that what we call reality cannot be brought about by adaptation to an independently extant or ontological reality. Under these circumstances we may well ask why, then, did cognitive evolution bring about the category of reality. A possible answer to this question is that we have to immunize our perceptions against doubts and distrust, particularly in situations where quick reactions are required. This is exactly what the notion of reality does. Within our day-to-day realism we consider our perceptions as representations of what is real rather than as the outcome of deliberate cognitive interpretation. In this way, time consuming (and, therefore, possibly dangerous) considerations as to whether these interpretations could be improved on do not arise.[4] Thinking in terms of reality can be regarded as a kind of "cognitive burden" incorporated during the course of cognitive evolution. I.e.: in whatever direction our cognitive evolution may proceed, reality remains an irreversible category — similar to the developmental constraints in organic evolution called "genetic burden".

As already mentioned, CEE requires that not only observational terms be defined operationally but also that mathematical and logical terms be so defined. If the cognitive operators generating our perceptions are phylogenetically related to those generating mathematical thinking, then their respective results, i.e. perceived structures and mathematical structures, must evidence certain similarities. This would explain why we can simulate the extrapolation of perceptions (i.e. predictions) by means of mathematical extrapolations (problem of induction, see Diettrich, 1991) and why the mathematics we use is so well suited to the descriptions of what we call nature. In this context Davies (1990) speaks of the 'algorithmical compressibility of the world' (i.e. that in so many cases the obvious complexity of the world can be described by means of rather simple mathematical expressions) and Wigner (1960) of 'the unreasonable effectiveness of mathematics in the natural sciences'.

If there is really a relationship between mathematics and perception as postulated here, then the phenomenon of qualitative extensions must occur also in mathematics (Diettrich, 1994b). Similar to the operators of sense perception which can be experimentally extended, also the operators constituting our elementary mathematical notions and concepts can be extended by higher and more complex mathematical calculi. This is what mathematics does as science. Here, as well, we have to make the same distinction.

We will speak of *quantitative extensions* if the truth value of the terms achieved can be derived from the axioms used.

We will speak of *qualitative extensions* if the truth value of the terms achieved cannot be derived from the axioms concerned, although the calculi used are completely based on these axioms. In this case the axioms themselves will have to be extended in order to make the truth value in question derivable. The incompleteness theorem of Gödel states that this can happen again and again.

Qualitative extensions are emergent phenomena which cannot be predicted since, by definition, they cannot be derived from previous knowledge. The blueprints of quantum mechanical devices are entirely classical in character and nothing provoked the idea that the results they may bring about could no longer be interpreted within classical theories. The same applies to mathematics. There is no general criterion telling us whether a given calculus will exceed its own axiomatic basis.

With this, the existence of non-classical theories in physics and the incompleteness theorem of Gödel are homologue cognitive phenomena. Neither is there a definitive set of physical theories (i.e. a theory of everything) explaining and describing all (also future) physical problems, nor is there a definitive set of mathematical axioms determining the truth value of all possible mathematical statements. In other words: there is neither a physical reality in the sense of objective laws of nature, (though laws may be universal in so far men will find them confirmed wherever they may be in the universe) nor is there a mathematical (or 'Platonic') reality in the sense of objective axioms. Both any set of physical theories and any set of mathematical axioms will be incomplete.

The demand that subjects of scientific interest be accepted only if they can be operationally defined has to be applied also to individuals and organisms. It is suggested to base the defining operator on the sum of interactions necessary to constitute the organism or individual. Multicellular organisms, for example, are defined by the interactions between and among their cells. Cells are defined by the interaction between and among their organelles, etc. (see Diettrich, 1989). Accordingly the interactions are what we have to consider as being the units of selection. From the Darwinian point of view the replicator rather is what represents the unit of selection (see Hull, 1988). Further to this, this constructivistic approach (by which the interactions are the units of selection) allows us to define a more general notion of units and individuals in cases where the notion of replicator hardly can be applied. Social groups, for example, are defined by the interaction between their members, biological species by genetic interaction and theories by the interrelation between their statements. We know little up to date about the technical realization of the cognitive operators for pattern recognition. But if it should turn out that they are constructed in a modular way as well, then, from the structuralist point of view, the interaction of elements which themselves are defined by interactions would be the generating principle by means of which any hierarchy in organic, cognitive, cultural and social areas could be described. That a modular concept would also allow a better understanding of the problems of homology has been discussed by Wagner (1995).

PERCEPTION AND ACTION AND THEIR COGNITIVE REPRESENTATION

The task of perception is to allow the formation of theories by means of which we can predict the effects of action. As the effects of action can be presented only in terms of perceptions we can say: by means of perceptions we will find out how perceptions may change under the influence of action. In other words, actions are operators which act upon perceptions.

Within classical realism this is described by means of what we call the laws of nature. The effects of our acting, we say, are determined by the laws of nature, and by means of our (natural and experimental) perceiving we acquire knowledge about just these laws. From this the scientific decoding of nature is seen as the crucial prerequisite for mastering nature. Only through knowledge of its laws can we master nature. This is the legitimation for all natural sciences in so far it aims at the exploration of natural laws.

Searching for the laws of nature succeeded whenever it was used for mastering nature. As there are, however, no objective laws (as we have seen), this way despite all its success is not legitimized heuristically. So there must be yet another link between perception and action that does not depend on the intermediate level of objective laws. If, nevertheless, we continue speaking about the laws of nature, we can expect that only those laws will help us predict perceptions and the perceived effects of action but not that they will provide us with objective statements about the world. According to classical understanding, action and perception are two entirely different categories: action refers to the individual's input to the world, whereas perception means the world's input to thc individual. With CEE, however, both refer to the same mechanism with the consequence that there is no essential difference between them. This sounds strange but is easy to illustrate. Visualize a hammer. It is an instrument designed primarily to alter certain objects. But a hammer, in its quality as an operator, also has invariants: objects and properties which would resist the hammer's strokes of a given strength. The hammer, then, can be used to measure mechanical properties such as the strength of materials. So, both perceiving and acting mean applying the same operator. The only difference is that, in the case of perceiving, we seek the invariants of the operator in question, i.e. we seek what remains unchanged under the application of the operators, whereas in the case of acting we seek what changes under the operator's influence. Our inborn world picture (i.e. the inborn interpretation of the sensorial input), therefore, depends on the phylogenetic 'decision' about which operators we use to construct the cognitive reference frame and which operators we use to modify what is described within the reference frame. (One of these phylogenetic decisions we have already mentioned: to use locomotion to give our perceptual space a third dimension rather than using it to modify the size of structures in a two-dimensional world.) The decision made about our actual cognitive reference frame, however, was not an evolutionary accident. It rather was made according to a relatively simple scheme as we will see.

In case of a physical measurement (which is an action, of course) the result, in physical parlance, is the invariant of the measuring process. In other words, we use the invariants of a process to describe the effect of just that process, i.e. we describe the covariants of an operator by means of its invariants. This can be generalized into the cognitive area. The actions by means of which we explore the world can be considered as measurements (i.e. as perceptions in the broadest sense). Results of measurements (or, as one could say, the results of our experiences) then are pictures of the world and theories representing what we call the unchangeable and, therefore, the objective world (i.e. what is invariant under all our doing and acting). If we look however for the covariants of our action, i.e. what changes under the influence of our actions, we have to refer to what we said about the relationship between the covariants and invariants of measuring processes: the effect of action can be described only in terms of the invariants of action, i.e. in the terms of our world picture. This is exactly the direct relationship between perception and action we looked for and which does not rely on the concept of an objective world.

If this is true, then the elementary categories of our perception must be the invariants of our most elementary action operators. But what are the most elementary action operators? They are not, as one might think, our hands and the tools guided by hands. They are, rather, our legs. By means of a few appropriate steps, we can change the environment of the room we are in into the environment of a blooming garden. Of course, we could achieve the same also by using our hands if we employ them do the necessary reconstruction work. But this is trouble-some and time consuming. So, one of the most important human-specific operators is locomotion. Our world picture, as a result, must be based on the invariants of this operator — and this is indeed the case. The most elementary descriptive category of our world view is the identity of extended objects and spatial structures defined as an invariant of locomotion.

This provokes the assumption that, from a phylogenetic point of view, the categories of description can be understood only through their capability to cope with the covariants of certain operators. From this it follows that the cognitive phenotype was fashioned by evolution not in order to explore the world but, instead, to extend the action possibilities of the organic phenotype.

Spatiality and the spatial metric, as we have seen, are categories that are necessarily defined by the process of motion. On the other hand, motion cannot be explained without the notion of space in which motion takes place. From this it follows that motion itself will have brought about the mental category of spatial structures necessary to deal with motion. Exactly this is what we maintain: what an operator is doing can be explained only in terms of its invariants.

We encountered a similar problem when we dealt with the operational definition of the arrow of time. An operational definition was impossible, because the notion of operators themselves would require a prior definition of the arrow of time. We therefore proposed referring to the memory and to events stored there. But, from the cognitive point of view, events themselves are already operators transferring the status before the event into the status after the event. So events, just as any operator, require the prior definition of the arrow of time. Without a definition of the arrow of time events and all we store in our memory in order to write history remain undefined.

By this, time turns out to be a mental modus which itself needs to have been brought about by operational means. In the same way as the spatial metric was generated by the process of motion (i.e. motion bringing about the category of space which is necessary for describing of motion), the category of time has to have been generated by operators (i.e. operators bringing about the category of time which is necessary for describing what operators are).[5] We can conclude: physical actions, and the cognitive operators we use to describe them, are brought about by the same organic operators (i.e. organic tools). Perceived patterns or regularities, and the instruments of mathematical thinking we use to describe them are brought about by the same cognitive operators. So mathematical patterns, perceived patterns and the results of our actions are literally homologue in so far as they have a common ontogenetic root, and this is the reason why they can 'cooperate' so well with one another — as well as the various physiological mechanisms having brought about each other.

CIRCULARITY OR THE REPRODUCTION OF THE COGNITIVE PHENOTYPE

The difficulty we have in accepting the notional character of our experiences as human-specific constructs differs with space and with time. As to the notion of space, undoubted (except perhaps by naive realists) is that the spatial patterns we perceive are not objective in the sense of their being considered pictures of real structures, i.e. the world is not necessarily as what it appears to be. Here, with space, we quite readily attribute to our world picture a reduced objectivity. Not so with time: The recorded time topology of events we consider to be real; the order of events is as we have perceived it, actual. The past is as it was and even God cannot change it a posteriori, we are used to say. Weizsäcker (1985) called this the 'facticity of the past'. Factually, however, events can only be defined as the results of cognitive or scientific interpretations, just as visual patterns can only be defined as invariants of cognitive operators. Events, as such, have as less clearly defined outlines than visual patterns have. A modification of the interpretations of events used (for example, in the presence of a novel theory) may well effect the past. An experiment may have been made in the beginning of this century documenting unambiguously a speed faster than that of light. After the appearance of the theory of relativity, the protocol of the same experiment would have had to be rewritten in using a speed less than that of light. A similar revision would have to be made if evolution would have changed our cognitive operators. But because this has not happened during historical times, the illusion arose of both the facticity of the past and the objectivity of the laws of nature. The allegation that the historicity of the world is a human specific artifact is the more problematic as it is based (through CEE) just on what is known about biological evolution, and this deals explicitly with the historical order of phylogenetic events. Said another way: on the one hand our world picture is the construct of our cognitive and experimental apparatus, on the other hand, just this world picture is what physics and biology refer to, particularly when describing the development of the human brain and the operators established there. What, then, is hen, and what is egg? Is it the real world we live in and which developed in the course of biotic evolution up to and including the brain's functions, or is it just these brain functions which bring about the picture of a real world as a tool for both articulating and solving our problems? Formulated differently: are perceptions brought about by nature, or is nature a category brought about by our cognitive apparatus? This dichotomy is the reason for the frequent accusations which say that the EE is circular in so far as not only the categories of space, time and causality are interpreted in phylogenetic terms but also the notion of reality and nature — the latter comprising phylogeny itself. So, phylogeny is interpreted by phylogeny, which is circular.

Actually, however, no real dichotomy exists as long as there is certainty that perceptions and nature condition one another through generating one another. This certainty is provided by the fact that our cognitive phenotype constructs a world picture which permits an understanding of the genesis of just this cognitive phenotype by means of evolution within the framework of just this world picture. Thus, not only organic ontogenesis but also cognitive evolution have to be understood as circular, autoreproductive processes in the subsequent sense:

In the biotic area the following holds: the epigenetic system of an organism is what determines how the genome's structure is to be interpreted and expressed into the phenotype. Identical reproduction is possible, however, only if the epigenetic system brings about a phenotype comprising the epigenetic system itself.

In the cognitive area the following holds: the cognitive apparatus (and all the science based on it) is what decides how the sensory input is to be interpreted and which world picture will be conveyed. The knowledge acquired in this manner is consistent and reproducible, however, only if the cognitive/scientific apparatus generates a world picture that includes containing the cognitive/scientific apparatus itself.

A genome on its own cannot determine the phenotype in the sense of providing a 'blueprint' — it rather represents one of several levels in the process of auto-reproduction — nor can the sensory input dictate its own interpretation, and, by this, the reactions it will effect. This limitation does not contradict the fact that, within the context of a given organic or cognitive phenotype having a given interpretative machinery, a genetic mutation as well as a new perception may lead to reproducible modifications of our physical constitution or of our theories. This means that, as long as the epigenetic system remains unmodified, a given genetic mutation will always produce the same phenotypic change; and as long as our cognitive apparatus and our scientific theories also remain unmodified, a given sensorial input will always lead to the same reading. What we have to avoid, however, is concluding that what mutations and perceptions initiate is also what they determine. Determinism is possible only within a given scheme of interpretations, i.e. outside qualitative extensions changing interpretations. The same limitations hold for adaptation. Adaptation makes sense only within a context of quantitatively extended acting. Acting will be modified positively in order to better meet the requirements given. The requirements themselves remain unchanged. Qualitatively extended acting, however, will affect the requirements and, by doing this, the relevant boundary conditions (i.e. the selective pressure). Qualitatively extended acting, thus, will become a source of emergent results. The world seen as the sum total of the boundary conditions of our acting is subject to a permanent actualization, as acting aims at changing just these conditions in order to make further and more ample acting feasible. This begins with the organic phenotype which defines the constraints for evolutionary 'acting', which in turn changes the constraints for further evolution (evolution meaning the evolution of its own boundary conditions). And it ends with the cognitive phenotype that defines, through our world picture, which kind of scientific acting is possible due to which the world picture itself may be affected — a paradigmatic shift in the sense of Kuhn, so to speak. The world as object of adaptation can be defined only for the time between two 'paradigmatic changes', i.e. between two qualitative extensions.

Circularity, a devastating objection for any theory within the context of classical realism, becomes (in the sense explained here) a necessary prerequisite for any complete constructivistic approach. The role of circularity constitutes the key difference between realism (of whatever kind) and constructivism as presented here. Realism requires of life mastering methods consistency with an independent outside world. A 'radically' constructivistic interpretation of the world, however, needs only to reconstruct itself. Hence, the so-called radical constructivism of Von Foerster, Glasersfeld, Maturana and Varela is not genuinely radical. In order to avoid conflict with the well established term 'radical constructivism' one should rather speak in terms of 'complete constructivism' when characterizing CEE.

The various epistemologies mentioned here can be explained more clearly when they are classified according to how they meet their functional requirements:

1. The most elementary position taken is that cognitive constructs (perceptions) have to delineate correctly the structures of the environment, since the strategies devised to meet the requirements of the environment are believed to be derivable from those structures. This is the basis for most kinds of *realism*.

2. In *constructivism, as well,* cognitive constructs have to contribute to meeting the requirements of the environment, but not necessarily by means of delineating environmental structures.

Both views, therefore, refer to the requirements of the external world — one in a structural way, the other functionally. The view that scientific theories are pictures of the world and that the laws of nature are rules to be followed by everyone legitimizes the program of empirical sciences searching for the world's structures and requirements.

3. *Complete constructivism* (CEE) proceeds on the assumption that not only the ways we see the world and the methods we use to meet the requirements of the world are constructed, but also the requirements of the world themselves, since these requirements — may be cognitive or organic — are not defined by an external world but rather by their previous organic, cognitive or even cultural evolution. For example, our phylogenetic predecessors explicitly designed the law of energy conservation when they "decided" upon which type of physical mechanism should emulate the mental clock defining the metric of time. So, the regularities men perceive and condense to what they call the laws of nature (or reality) represent nothing more than their own previous developmental history. We can use the constructive character of what we perceive to define the difference between actuality (Wirklichkeit) and reality. Actuality is what we can change or construct by means of acting in the literal sense. Reality is what we have constructed by means of evolutionary 'acting'. As the latter is irreversible, reality seems to be universal and therefore objective. In other words: 'home-made' is not only the way we see reality but also reality itself.

Darwinism and CEE have an interesting element in common. Darwinism tries to explain the actual forms of life as the result of evolution. Evolutionary models which fail at doing so will be victimized by selection. Darwinian models, so to speak, apply their own content to themselves. Something similar happens with CEE. It tries to simulate the evolution of stable forms of life. The main criterion of stable living forms is by definition their capacity to reproduce themselves. The same applies for CEE. CEE is stable or 'complete' only if it can reproduce itself in the sense just explained. So CEE as well applies its own content to itself. Selection and auto-reproduction can be considered the key concepts constituting the Darwinian and the constructivist approaches respectively. Both approaches represent meta-theories which, as such, do not make experimentally verifiable statements through which they could contradict one another. Neither excludes the Darwinian view that the selecting environment is a construct of the organisms concerned (see next section) nor excludes the CEE view that specific constructs can be selected out because they lack coherence with other constructs.

ASSIMILATION AND ACCOMMODATION VERSUS ACTION AND PERCEPTION

According to Piaget, assimilation means modifying or using external data in order to meet internal needs. Accommodation means modifying these internal needs in order to be met more easily. Let us apply the terms assimilation and accommodation to general evolution. According to the synthetic theory, successful evolution means constructing or modifying an organism so that it can satisfy external requirements. Evolution is thereby understood to precede primarily by means of accommodation, from its early commencement through to human technical and scientific achievements for managing life.

However, what evolves are not only internal needs for meeting external requirements but also the competence for acting, i.e. the capacity for modifying the environment, i.e. for modifying external requirements according to previously defined internal needs . Seen in this way, evolution is both accommodation and assimilation — with an increasing tendency towards assimilation: the more complex and "higher" organisms are, the more difficult it becomes for them to modify the phylogenetically acquired physiological and other basic strategies, and the more likely it is, therefore, that evolution tends toward assimilating strategies, i.e. toward improving the methods for modifying the environment. Warm-blooded animals, for example, do not react to climatic changes by altering their physiologically defined body temperature. Instead they conserve their internal climate by means of better isolation or through higher (internal or external) energy investments. Humans, after all, do not react any longer by means of evolutionary accommodation. Whenever a conflict arises between biologically defined human requirements and the environment, the conflict will be solved at the environment's expense.

One of the most popular methods for changing the environment of all animals is locomotion. Paramecia began early to use locomotion for escaping from adverse local conditions. That the relevant environment and its selection pressure is an artifact of the various species occupying it rather than an objective and external issue was seen already by Waddington (1959, p. 1636): "Animals are usually surrounded by a much wider range of environmental conditions than they are willing to inhabit. They live in a highly heterogeneous 'ambience', from which they themselves select the particular habitat in which their life will be passed. Thus the animal by its behavior contributes in a most important way to determining the nature and intensity of selective pressures which will be exerted on it. Natural selection is very far from being an external force, as the conventional picture might lead us to believe". Regarded from this aspect, life is a mode of world construction in the sense of

Goodmann (1984) rather than a process of exploring the world or of acquiring knowledge about the world as Lorenz (1983) said. In other words, evolution seems to aim at assimilation rather than at accommodation.

Actually, however, this picture is as biased as the pure-accommodation picture because the capacity for acting and reacting does not represent eo ipso a successful assimilation strategy. Strategies, as well as organic features, must accommodate themselves to given external conditions. This is what accommodation strategies aim at — not to explore the environment and modify one's constitution accordingly, but to improve the capability of changing one's environment in order to meet the requirements of one's constitution. These requirements, however, mean first of all making optimal use of existing assimilation techniques. So, accommodation must orient itself to the techniques available rather than to the environment concerned. The most elementary example is the evolutionary extension of homeostasis for adapting to a larger variety of external data rather than finding special solutions for each special case. Accommodation, therefore, will aim at extending the set of controllable data independent of what is actually required. Whether a species can profit from this strategy or not will depend on whether new conditions or a new environment can be found in which these new achievements will pay off. (A cultural example would be basic research that provides solutions for problems which do not yet exist and which will be successful only if appropriate applications can be identified.) So, the interplay characteristic of evolution is not only that between mutation and selection (supply strategy, defined by the supplies and constraints of the environment) but also that between extension of competence and applicability (demand strategy, defined by the requirements and the possibilities of the organism). If we, nevertheless, want to continue using the notion of environment as the instance which articulates the boundary conditions for physical and evolutionary acting, we have to consider it as a construct in a double sense: (a) by acting in the proper sense we can modify the relevant boundary conditions — for example, by means of external heating we can reduce the demands on internal temperature management; (b) by changing the internal requirements the same environment can come to represent different boundary conditions — for example, by the developing from anaerobic to aerobic respiration, a previously irrelevant content of oxygen in the atmosphere became the key survival factor. More generally we can say that what counts is the 'distance' between organism and environment and this can be reduced at both 'ends'.

Here we have an analogue in cognition. We said that the cognitive phenotype is entirely a construct of the organic phenotype, that was brought about to extend the functional possibilities of the phenotype rather than to 'recognize' the world or to explore what Vollmer called the cognitive niche. Nevertheless, here, as well, we can continue using the notion of environment if we consider it (as we did in the organic case) as being constructed in a double sense. (a) We can displace objects, change our position, practice agriculture or construct streets with traffic lights. By all this we change our environment and if we do this in the interest of our needs we practice assimilation. By this we construct what we call actuality (Wirklichkeit). (b) How we see the environment, which regularities we register and to what 'laws of nature' we will condense them is a matter of our cognitive phenotype. The development of the cognitive operators up to the development of our world picture (including the formation of scientific theories), is, therefore, an act of accommodation to the conditions of actuality brought about by assimilation. Our cognitive environment, thus, is a human-specific construct characterized by the laws of nature as comprised in the notion of reality. It is not nature which tells us how we have to see it (for example, as spatial objects moving in a world of three dimensions). It is rather the phylogenetically emerged mental operators and their invariants which define the regularities we perceive and the laws we derive from them. If we consider that the biological development of these operators has been completed, but not the development of their possible physical extensions by means of novel experimental devices with novel invariants which require novel theories, then we can say that we continue more than ever to construct the object of scientific research, i.e. reality. Reality, in so far it is articulated in terms of laws physicists formulate and try to explore in order to make predictions possible, is a human-specific artifact. What we see depends both on what we do and how we interpret the sensory reflexion of our doing.

In strict analogy to the organic case we can say that actuality and reality are the two ends which characterize the cognitive distance between individual and environment. Actuality is the result of our doing (including locomotion), i.e. of assimilation. Reality is the result of our phylogenetically acquired ways of interpreting of experiences with a view to making predictions. So reality is the result of accommodation. As we cannot describe actuality without a previously defined interpretation of our sensory input, we cannot speak of assimilation without accommodation. Just as in the organic context: the results of accommodation (i.e. the actual phenotype) define what certain assimilating activities will mean for the organism concerned. This applies, as well, in the opposite direction: accommodation is not possible without assimilation. We cannot speak of reality and its purpose for making predictions (accommodation) without reference to the object of those predictions, the actions (assimilation). The same argument holds in the organic case: action means the accommodation of assimilation strategies. Action and perception cannot be defined independently of one another. More particularly, it does not make sense to speak in terms of perception or recognition of an independent and objective outside world. In other words: the covariants (effect) of an operator cannot be defined separately from the operator's invariants describing this. This we referred to implicitly when we said that perceiving and acting means applying the same operator and that their only difference lies in whether we ask for the operator's invariants or the operator's covariants. Something very similar holds for accommodation and assimilation. A certain operation can be considered as an act of assimilation if we look for what it may effect in the environment, and as an act of accommodation if we look for how it may change our situation in the environment.

ORGANIC THEORIES

Assimilation strategies, as we have seen, do not aim primarily at controlling a particular situation but rather at expanding the spectrum of controllable situations so as to include, as well, a particular situation. The homeostatic temperature management of warm blooded animals is not designed to compensate for an actual temperature of, say, 5°C, but for all temperatures within certain limits. So, assimilation strategies are synonymous with extensions of acting competence.

The cognitive analog is the formation of theories. The purpose of a theory is not only to predict the result of certain actions for their repetition in the future, but also to extend the spectrum of predictable results of actions exceeding the set of those actually realized. Theories are 'true' for the set of actions leading to results they can predict. Prediction, i.e. the cognitive mastering of various actions (comprising also those not yet realized) is, therefore, an analogue of the control of different external conditions (comprising also those that have not yet occurred), i.e. prediction is the analogue of the extension of action competence. If an action mechanism can control more than just one outside condition, we might call it, in analogy, an 'organic theory'. Organic theories are 'true' for the set of outside conditions they can control. The question arises why the mechanisms actually implemented in organisms are 'true' just for the set of frequently occurring conditions rather than for any others. The answer might be as the following: in the cognitive area we have seen that the covariants of an operator (i.e. the effect of acting) can be described appropriately only in terms of the operator's invariants. The organic analog would be: an acting operator masters a set of outside conditions if and only if these conditions are invariants of the operator concerned — and, by definition, the actual conditions are precisely those. Whenever we speak about outside conditions we formulate them in such a way that they will be relevant for the organism and its activity — the invariants of organic perception, so to speak, in the same way that the objects of our sensory perceptions are defined as invariants of our physical activity.

A typical example is the set of possible outside temperatures which have to be mastered by the organic temperature-management devices. The energy invested in order to stabilize the required inside temperature under the condition of a given outside temperature can be regarded as a measure of the outside temperature, and the heat management device itself as an outside thermometer. In this way the notion of temperature is operationally defined by a certain physiological regulatory mechanism. That this mechanism is able to compensate for varying temperatures is therefore not a matter of phylogenetic adaptation. Rather it is a matter of definition: what we call temperature is what can describe the effect of this particular regulatory mechanism or, in the words used above: operators bring about not only certain effects but also the cognitive tools by means of which their effects can be described.

So, there is no need to wonder why the organism is so well adapted to the world in which it finds itself; the way the world is seen is just a reflex of the requirements and possibilities of that organism. In other words, what we call the world, which is said to guide our adaptive efforts, is rather a phylogenetically developed cognitive model by means of which we explain our reproductive success.

Earlier explained in this paper is that theories are invariants of our perceptional and experimental devices and that new devices which are not commutable with those already extant (qualitative extensions) may lead to novel theories that can no longer be described in terms of their predecessors. The question then arises as to whether or not something equivalent can be said about what we have called organic theories.

I propose to speak of a *quantitative extension* of an acting operator if that operator brings about a better or more effective mastering of the outside conditions without changing the goal envisaged. Example: a bird species, the members of which live by cracking nuts, may adapt their bills phylogenetically according to the nuts available.

I will speak of a *qualitative extension* if the extension will change implicitly also the goals previously envisaged, i.e. if the modification concerned leads to new applications in new areas and, therefore, to the transfer into a new niche with new adaptive pressures. If the bills of the (hypothetical) bird species, were to diminish phylogenetically towards an unsuitable size they would no longer be able to crack nuts, but they might well learn to pick the grain they found in their own or in a new biotope. From then on modifications would be evaluated and selected according to their grain-picking power. In other words, the original niche is no longer an invariant of the modified beak and therefore has to be transferred into a new niche. The cognitive analog is the effect of measurement devices on their objects as found in subatomic regions. This problem has been solved by using only those variables which are invariants of the measurement operators applied and which, therefore, will not be affected by the measurements made. The analog of the case discussed here would be a feeding habit that is independent of (and therefore an invariant of) the beak's shape, either due to selecting appropriate food stuffs or due to preparing the foodstuff by means of appropriate tools or techniques (e.g. by rolling hard-shell eggs on stony ground until they break; in this case the beak's shape is largely irrelevant). Another example comes from evolutionary biology: genetic mutations can be called quantitative if they affect only the phenotype and, therefore, are reversible. We will speak of a qualitative mutation if it affects also the epigenetic system that interprets and expresses genomic structures. So there is a kind of co-evolution between genomic structures and their interpretation by the epigenetic system which itself is subject to genomic modifications. This may lead to non-stable recursive processes which have been described as non-linear genetics (Diettrich, 1989, 1992). This means that long-term evolutionary processes can develop their own dynamic, a dynamic that does not need to depend on consecutive genomic mutations or environmental changes. A further example is provided by modular extensions such as multicellularity, i.e. combinations of equal or different elements or techniques evoking novel competencies and capabilities: the combination of manual and visual abilities, as well as the combination of electrostatics, magnetostatics and mechanics into electrodynamics with the novel ability to describe electromagnetic waves (such as light waves). Müller and Wagner (1995) have pointed out the possible importance of modular approaches in developmental biology.

CONCLUSIONS

The most promising effort of the constructivist approach proposed here (CEE) aims at a coherent description of organic, cognitive and scientific evolution. However, the price to be paid is high. We have to accept that the laws of nature are phylogenetically acquired human-specific artifacts; that there is no 'natural selection epistemology' (sensu Campbell); that there will be no complete set of physical theories ('theory of everything') and that there will be no meaningful context-free communication (such as that with extraterrestrial beings). On the other side, some explanations are offered that could hardly have been expected from outside CEE: the incompleteness of physical theories and the incompleteness of mathematical axioms discerned by Kurt Gödel have the same cognitive roots; the algorithmic compressibility of the world (which is equivalent to the success of induction) is due to the homology of cognitive mechanisms and the mechanisms of mathematical thinking. Of particular interest is the link between organic and cognitive tools: if cognitive tools are to describe the covariants of an operator (i.e. what the operator effects), they have to be designed in terms of the operator's invariants (a principle which has been reinvented by physicists within Hamilton-Jacoby and quantum mechanical formalism). Cognitive evolution (including the epistemology here applied), therefore, is brought about by organic evolution (and by the evolution of experimental tools) rather than by trial and error and selection. And, vice versa, organic (also experimental) evolution is guided by the possibilities provided by cognitive tools.

This relationship between action and perception (or organic and cognitive development) is the same as the relationship between assimilation (modifying the environment according to previously defined internal needs) and accommodation (modifying the internal constitution and the resulting needs according to externally defined requirements): assimilation aims at changing what cannot be approached by accommodation (i.e. at the invariants of accommodation) and accommodation aims at approaching what cannot be achieved by assimilation (i.e. at the invariants of assimilation). In terms of the action/perception dichotomy, this reads as follows: acting requires the view that what we see of the world has a real (i.e. unchangeable) background. And what we consider as real of the world (such as the laws of nature) requires the view that this cannot be changed by our acting. Assimilation, as long as it is in action, requires a fixed goal which is not modified by accommodation, and accommodation, as long as it is in action, requires a fixed goal which is not modified by assimilation. This is also so in scientific evolution: experimental efforts require a well defined theoretical basis, and theoretical efforts require given experimental facts to which they refer. It is like walking: at the same time we can precede only with one of our legs whereas the other one has to be at rest. We nevertheless can precede over unlimited distances if we change alternately the active and the passive role of our legs. Evolutionary 'walking' in this sense means the endless sequence of what we called qualitative extensions comprising the permanent co-evolution between accommodation and assimilation: tool-inventing and finding applications (in both organic and scientific evolution); perceiving and acting;

experiments and their interpretation; general research and the goal defining paradigms. A particular case is the co-evolution between (say organic) units and their interactions. The development of interactions between units may bring about new forms of units which then could develop higher forms of interactions defining higher units etc. as represented within the course of self-organization, and vice versa.

ACKNOWLEDGEMENTS

The author is grateful to Gertrudis Van de Vijver and Dr. Robert W. Kickert for many helpful comments on earlier drafts of the manuscript, and particularly to the Konrad Lorenz Institute for Evolution and Cognition (Austria) and its chairman, Prof. Dr. Rupert Riedl, where the author, then guest professor, could write this paper.

Research Unit "Complexity" and Evolution, University of Gent
Gent, Belgium

NOTES

[1] The question whether modifications of visual perceptions should be interpreted geometrically or physically was addressed in another case in physics: the orbits of planets could be considered as the effect of explicit gravitational forces (the physical solution) as well as geodetic lines within a 4-dimensionally curved 3-dimensional space (the geometrical solution within the theory of general relativity). As this is just a different interpretation of the same observations we cannot come to a decision on empirical grounds, nor was adaptation by selection relevant when the cognitive evolution of primates came to decide whether to see the visualized world in 2 or 3 dimensions. In other words: perceptional spaces and systems of categories are purely descriptional systems which may tell us something about how we see the world but which tell us nothing about the world itself. Hence, they cannot be the outcome of adaptation to the world.

[2] An interesting analogon can be constructed to the theory of special relativity. Visualize locally fixed plants that have eyes and can see, living in a world of no motion. For these plants there is no need to distinguish between objects that are small because they are really small, and those that are small due to their greater distance (as we would say). These plants do not know the phenomena of perspective and their world of visual perception is 2-dimensional. As soon, however, as they learn to communicate, they will find out that objects that are small for one observer might well be big for another one. They may, then, construct a theory of relativity of size, saying that size depends on the relative position of the observer — difficult to understand for someone who is used to live in a 2-dimensional world. The same happened to physicists when empirical evidence forced them to construct the theory of special relativity saying that time intervals depend on the relative motion of the observer — difficult to understand for someone who is used to live in a Newtonian world.

[3] The specificity of the metric generator has direct effects on the laws of conservation we record in physics (energy, momentum, etc.). Following Noether's theorem, these laws can be derived from the invariance properties of the equation of motion: invariance under a translation in time (i.e. physics is the same yesterday and today) implies the conservation of energy; invariance under translation in space (i.e. physics is the same in America and Europe) implies conservation of momentum; invariance under spatial rotations implies conservation of angular momentum. In other words: from the homogeneity of space

follows the conservation of momentum and from the homogeneity of time follows the conservation of energy. What 'homogeneous' means, however, is exclusively a matter of the mental metric-generator concerned. This applies also to the other conservation laws which, therefore, are human specifics rather than objective properties of nature. As will be seen below, the conservation laws constitute what one could call the cognitive reference-frame we use to describe actions and what those actions will bring about. Other conservation laws based on other cognitive operators would effect a different cognitive phenotype, but this would not mean that the methods and life strategies based on other operators would be less consistent or efficient. What is excluded, however, is communication between representatives of different cognitive phenotypes. This may apply to extraterrestrial beings. Though they may well survive with their particular cognitive phenotype, no communication is possible between them and us; and means invested by space research to this end is money thrown down the terrestrial drain. The organic analog is that representatives of different species cannot communicate genetically, i.e. they cannot interbreed.

[4] This is close to what H.A. Simon (1983) called 'restricted rationality' saying that higher animals and men in many cases concentrate their capabilities on solving actual problems while disregarding higher interests and long-term requirements.

[5] The program describing the covariants of operators by means of their invariants is well known and is often used in physics. Within the framework of Hamilton-Jacoby-formalism, the variables of a mechanical system are chosen so that conservation laws (invariants) will apply for them. With this prerequisite cared for, the transformations describing the system's development in time can be found easily and explicitly. On the other hand, the conservation laws themselves can be shown to be generated by the transformations considered. So the canonical total momentum (in this paper identified, in a more general way, as 'motion') brings about spatial translation, and the total energy (represented by the Hamiltonian) brings about translations in time. Something very similar applies for quantum mechanics. The system's development in time is generated by the Hamiltonian and the eigenvectors of the Hamiltonian constitute the reference frame by means of which this is described. If we replace the physical transformations and operators by their cognitive/mental equivalents we just get what we said: space and time have no independent or ontological quality but are rather products of human-specific operators.

REFERENCES

Campbell, D.T., 1973, Evolutionary epistemology, in *The Philosophy of Karl Popper*, P. Schilpp (ed.), Part I, Open Court, La Salle, pp. 413-463.

Davies, P.C.W., 1990, Why is the physical World so comprehensible?, in *Complexity, Entropy and the Physics of Information*, Santa Fe Institute studies in the Sciences of Complexity, W.H. Zurek (ed.), Vol VIII, Addison Wesley, pp. 61-70.

Diettrich, O., 1989, *Kognitive, organische und gesellschaftliche Evolution*, Berlin, Parey.

Diettrich, O., 1991, Induction and evolution of cognition and science, in *Teleology and Self-organization*, G. Van de Vijver (ed.), *Philosophica*, vol. 47, no. 2, pp. 81-111.

Diettrich, O., 1992, Darwin, Lamarck and the Evolution of Life and Culture, *Evolution and Cognition*, vol. 2, no. 3.

Diettrich, O., 1993, Cognitive and Communicative Development in Reality free Representation, *Cognitiva*, vol. 5, no. 2, pp. 219-243.

Diettrich, O., 1994a, Heisenberg and Gödel in the Light of Constructivist Evolutionary Epistemology, *Ludus Vitalis*, vol 2, no. 2, pp. 119-130.

Diettrich, O., 1994b, Is There a Theory of Everything?, *Bulletin of the Institute of Mathematics and its Applications*, vol. 80, pp. 166-170.

Hawking, S.W., 1979, *Is the end in sight for theoretical physics?*, Inaugural Lecture for the Lucasian, Chair. University of Cambridge.

Hull, D., 1988, *Science as a Process. An Evolutionary Account of the Social and the Conceptual Development of Science*, University of Chicago Press. Chicago.

Goodmann, N., 1984, *Weisen der Welterzeugung*, Frankfurt/M. Suhrkamp.

Lorenz, K., 1966, *Über tierisches und menschliches Verhalten. Gesammelte Abhandlungen*, München,

Piper.

Lorenz, K., 1983, Kants Lehre vom Apriorischen im Lichte gegenwärtiger Biologie, in *Die Evolution des Denkens*, K. Lorenz & F.M. Wuketits (Hrg.), München, Piper.

Müller, G.; Wagner, G., 1995, *Amer. Zool*, vol. 36, pp. 4-13.

Piaget, J., 1974, *Die Bildung des Zeitbegriffes beim Kinde*, Frankfurt/M. Suhrkamp.

Piaget, J., 1970, *Genetic Epistemology*, New York, Columbia University Press.

Riedl, R., 1980, *Biologie der Erkenntnis*, Berlin, Hamburg, Parey.

Vollmer, G., 1980, *Evolutionäre Erkenntnistheorie*, Stuttgart, S. Hirzel.

Üxküll, J. von, 1921, *Umwelt und Innenleben der Tiere*, Berlin, Springer.

Waddington, C.H., 1959, Evolutionary systems-animal and human, *Nature*, vol. 182, pp. 1634-1638.

Wagner, G., 1995, *Homologues, Natural Kinds and the Evolution of Modularities*, Internal Report YALEU/CCE/#23, Yale Institute for Biospheric Studies.

Weizsäcker, C.F. von, 1985, *Aufbau der Physik*, München, Hanser.

Wigner, E., 1960, The unreasonable effectiveness of mathematics in the natural sciences, *Comm. Pure Appl. Math.*, vol. 13, p. 1.

Wuketits, F., 1984, Evolutionary Epistemology, in *Concepts and Approaches in Evolutionary Epistemology*, F. Wuketits (ed.), Dordrecht, D. Reidel Publishing Company.

LUIS MATEUS ROCHA

SELECTED SELF-ORGANIZATION AND THE SEMIOTICS OF EVOLUTIONARY SYSTEMS

EIGENBEHAVIOR AND EMERGENT REPRESENTATION

Heinz von Foerster (1965, 1969, 1977) equated the ability of an organization to classify its environment with the notion of *eigenbehavior*. He postulated the existence of some stable structures (*eigenvalues*) which are maintained in the operations of an organization's dynamics. Following Piaget (von Foerster, 1977), he observed that any specific instance of observation of such an organization will still be the result of an indefinite succession of cognitive/sensory-motor operations. This reiterated the constructivist position that observables do not refer directly to real world objects, but are instead the result of an infinite cascade of cognitive and sensory-motor operations in some environment/subject coupling. Eigenvalues are self-defining, or self-referent, through the imbedding dynamics — implying a complementary relationship (circularity, closure) between eigenvalues and cognitive/sensory-motor operators: one implies, or defines, the other. "Eigenvalues *represent* the externally observable manifestations of the (introspectively accessible) cognitive (operations)" (*ibid.*, p. 278, italics added). Further, "Ontologically, Eigenvalues and objects, and likewise, ontogenetically, stable behavior and the manifestation of a subject's 'grasp' of an object cannot be distinguished" (*ibid.*, p. 280). *Eigenbehavior* is thus used to define the behavior of autonomous, cognitive systems, which through the closure (self-referential recursion) of the sensory-motor interactions in their nervous systems, give rise to perceptual regularities as objects (Varela, 1979, ch. 13).

"Eigenvalues are discrete (even if the domain of (their observables) is continuous)". In other words, even if the domain of an observable is continuous, its cognitive representation through cognitive/sensory-motor operators into eigenvalues must be discrete. This is a result of the stability of eigenvalues in the recursive chain of cognitive operators, if an eigenvalue changes its structure, thus ending the frame of stability, it will either revert to unstable structures (varying at each cognitive operation), in which case the eigenvalue representation is lost, or form another frame of stability with a new eigenvalue representation. In summary, eigenvalues are discrete representations of observables maintained by the successive cognitive operations of a cognitive agent. Notice that the representations and their stability are specific to the particular cognitive operations and how they recognize observables. That is, these discrete representations exist only in relation to the very same operators that define them. Any system, cognitive or biological, which is able to relate internally, self-organized, stable structures (eigenvalues) to constant aspects of its own interaction with an environment can be said to observe eigenbehavior. Such systems are defined as *organizationally closed* because their stable internal states can only be defined in terms of the overall

G. Van de Vijver et al. (eds.), Evolutionary Systems, 341–358.

dynamic structure that supports them. Organizationally closed systems are also *informationally open* (Pask, 1992), since they have the ability to classify their constructed environment in what might be referred to as *emergent representation*.[1]

Attractor behavior, self-organization, and constructivism

An eigenvalue of an organizationally closed system can be seen as an attractor of a self-organizing dynamical system. The global "cooperation" of the elements of a dynamical system which spontaneously emerges when an attractor state is reached is understood as self-organization (von Foerster, 1960; Haken, 1977; Prigogine, 1985; Forrest, 1991; Kauffman, 1993). The attractor behavior of any dynamical system is dependent on the structural operations of the latter, e.g. the set of Boolean functions in a Boolean network. Speaking of an attractor makes sense only in relation to its dynamical system, likewise, the attractor landscape defines its corresponding dynamical system. Further, attractor values can be used to refer to observables accessible to the dynamical system in its environment and therefore perform relevant classifications in such environment (e.g. neural networks). Naturally, and this is the crux of the constructivist position in the theory of organizationally closed systems, not all possible distinctions in some environment can be "grasped" by the autonomous system: it can only classify those aspects of its environment/sensory-motor/cognitive interaction which result in the maintenance of some internally stable state or attractor (eigenvalue). In other words, not everything "out there" is accessible; only those things that a particular physiology can construct with the stabilities of its own dynamics are. As with eigenvalues, attractors must be discrete even if used to refer to continuous observables, since attractors are defined by specific, typically small, volumes of the state space of a dynamical system.

Emergence and levels of description

There are three levels that need to be addressed when dealing with the notion of emergent representation. First, there is the material, dynamical, substrate, which will be the causal[2] basis for all other levels that we may further distinguish. Secondly, we have the attractor behavior of this dynamics. Finally, we have the utilization of the set of attractors (eigenvalues) as referents for some aspects of the interaction of the dynamical system itself with its environment, that is, as tokens for eigenbehavior. This indirect, constructed, "referring" results from the structural coupling (Maturana & Varela, 1987) of the dynamical system with the environment, and can be understood as a semantic relation.

The level of eigenvalues is emergent to the dynamics because it cannot be explained solely by a description of the latter. Stability of dynamical states is not expressed in the language of the interactions between the components of a dynamical system. At this lower level, there is no distinction between a stable and an unstable state. For instance, the transition rules of Conway's *Game of Life* cannot describe what "blinkers" and "gliders" are. Likewise, the level of eigenbehaviors, or the *function* of attractors as referring to some constructed reality, is emergent to the eigenvalues since the latter can only describe stabilities of the dynamics and not any "standing for" relation necessary for eigenbehavior (e.g. streams of gliders as information carriers in a universal computer built out of *Life* patterns (Poundstone, 1987)). No physical or formal description of the dynamical system and its attractors alone will completely explain this functional dimension (see Rocha, 1995b, 1996, 1997a; Rosen, 1995; Pattee, 1995). Hence, we need complementary descriptions of the several levels involved in such organizationally closed, emergent, systems (Pattee, 1978).

EMBODIMENT AND SELF-ORGANIZATION

Varela, Thompson, and Rosch (1991) have proposed an embodied, inclusive, approach to cognition which acknowledges the different levels of description necessary to effectively deal with emergent representation, or in von Foerster's terms, eigenbehavior. Cognitive science was traditionally concerned solely with those aspects of cognitive representation which can be described as symbolic. In other words, it was concerned with the semantic relation between cognitive categories and their environmental counterparts through some direct representational relation (intentionality), without taking into account any sort of material or internal organizational constraints: real-world categories directly represented by discrete symbols which could be freely manipulated. The connectionist, emergent, or self-organizing paradigm has changed this focus to the lower level of attractor behavior. That is, cognitive systems are defined as those systems capable of self-organizing their components into discrete basins of attraction used to discriminate the environment they are able to construct. Classifications become sub-symbolic and reside in some stable pattern of activation of the dynamic system's components, instead of based on some higher level symbols (emergent representation).

Selected self-organization: structural change and increasing variety

What is usually referred to as self-organization is the spontaneous formation of well organized structures, patterns, or behaviors, from random initial conditions. The systems used to study this phenomenon are referred to as dynamical systems: state-determined systems. They possess a large number of elements or variables, and thus very large state spaces. However, when started with some initial conditions they tend to converge to small areas of this space (attractor basins) which can be interpreted as a form of self-organization. Since such formal dynamical systems are usually used to model real dynamical systems such as the non-equilibrium thermodynamic behavior of chemical networks of reactions (Nicolis & Prigogine, 1977), the conclusion is that in nature, there is a tendency for spontaneous self-organization which is therefore universal (Kauffman, 1993).

This process of self-organization is also often interpreted as the evolution of order from a disordered start. Self-organizing approaches to life (biological or cognitive), in particular second-order cybernetics (see Pask, 1992), take chaotic attractors as the mechanism which will be able to increase the variety (physiological or conceptual) of organizationally closed systems. External random perturbations will lead to internal chaotic state changes; the richness of strange attractors is converted to a wide variety of discriminative power. Dynamic systems such as Boolean networks clearly have the ability to discriminate inputs. Generally, the attractors of their dynamics are used to represent events in their environments: depending on starting conditions, the network will converge to different attractors. However, for any classification to have survival value, it must relate its own constructed states (attractors) to relevant events in its environment, thus, similar events in the world should correspond to the same attractor basin. Chaotic systems clearly do not have this property due to their sensitivity to initial conditions. Ordered systems follow this trait.

Kauffman (1993, p. 232) further hypothesizes that "living systems exist in the (ordered) regime near the edge of chaos, and natural selection achieves and sustains such a poised state". This hypothesis is based on Packard's (1988) work showing that when natural selection algorithms are applied to dynamic systems, with the goal of achieving higher discriminative power, the parameters are changed generally to lead these systems into this transitional area between order and chaos. This idea is very intuitive, since chaotic dynamical systems are too sensitive to parameter changes, that is, a single mutation leads the system into another completely different behavior. By contrast, ordered systems are more resilient to damage, and a small parameter change will usually result in a small behavior change which is ideal for smooth adaptation (hill-climbing) in correlated fitness landscapes. However, even though very ordered systems can adapt by accumulation of useful successful variations (because damage does not propagate widely), they may not be able 'step out' of certain areas of their fitness landscapes. It is here that systems at the edge of chaos enter the scene; they are not as sensitive to damage as chaotic systems, but still they are more sensitive than fully ordered systems, and thus, some mutations will accumulate (by causing minor changes) and some others will cause major changes in the dynamics allowing more distant searches in fitness spaces. These characteristics of simultaneous mutation buffering (to small changes) and dramatic alteration of behavior (in response to larger changes) is ideal for evolvability (Conrad, 1983, 1990).

Chaotic classifications cannot grasp an ordered interaction with an environment, while point attractors and simple limit cycles may not allow enough behavior change for a good increase in variety. The edge of chaos regime seems to offer a good, intuitive, compromise. However, whatever the regime of a dynamic system, self-organization alone cannot escape its own attractor behavior. A given dynamic system is always bound to the complexity its attractor landscape allows. For a dynamic system to observe genuine emergence of new classifications (conceptual or of functionality) it must change its structure. Creativity, or open-ended variety can only be attained by structural perturbation of a dynamical system. One way or another, this structural change leading to efficient classification (not just random change), has only been achieved through some external influence on the self-organizing system. Artificial neural networks discriminate by changing the structure of their connections through an external learning procedure. Evolutionary strategies rely on internal random variation (mutation) which must ultimately be externally selected. In other words, the self-organizing system must be structurally coupled to some external system which acts on its structural changes and induces some form of explicit or implicit selection of its dynamic representations: *selected self-organization*.

Memory and selected self-organization

The dynamical approach of von Foerster (1965) to cognition emphasized the concept of memory without a record. By utilizing functionals to change the functions of state-determined systems, von Foerster formalized the idea that memory can be observed in systems which are able to change their own structure and its dynamics and attractor behavior. Today, we name this kind of memory *distributed*, and the kind of models of memory so attained connectionist. The categories a distributed memory system classifies are not stored in any particular location, they are nowhere to be found since they are distributed over the entire dynamics established by some network of processes (van Gelder, 1991). They exist however in the form of attractors (eigenstates) which are nonetheless discrete as earlier observed. Categories are not stored in any one particular location of the network, but are identified with particular dynamic attractors for which we need an emergent level description.

For a self-organizing system to be informationally open, that is, for it to be able to classify its own interaction with an environment, it must be able to change its structure, and subsequently its attractor basins, explicitly or implicitly. Explicit control of its structure would amount to a choice of a particular dynamics for a certain task (the functional would be under direct control of the self-organizing system) and can be referred to as *learning*. Under implicit control, the self-organizing system is subjected to some variation of its structure (including its distributed memory) which may or may not be good enough to perform our task. Those self-organizing systems which are able to perform the task are thus *externally selected* by the environment to which they are structurally coupled. If reproduction is added to the list of tasks these systems can produce based on their dynamic memories, then we have the ingredients for natural selection: heritable variation and selection.

This form of situated, embodied, self-organization can be referred to as *distributed memory-selected self-organization*. Its relying on some system-environment coupling of structure has been stressed most notably within second-order cybernetics and systems research. Maturana and Varela (1987) propose structural coupling as the general mechanism for variety increase, Pask (1976) refers to it as conversation in the cognitive realm. Both of these approaches owe a lot to von Foerster's eigenbehavior notions. More recently, in the realm of complex systems and evolutionary systems theory, Kauffman (1993) and others have relied on the notion of autocatalytic sets which are mutable, heritable, self-replicating, self-organizing systems evolvable through natural selection.

So far I have maintained that eigenvalues or attractors represent the building blocks of any system capable of discriminating its environment through some thus embodied construction. However, eigenbehavior (emergent representation) and its variety increase need a structural coupling of these eigenvalues with some externally selective environment. This kind of selected self-organization obliges us "to understand perception not just as an interactive dynamical structure, but as a process that arises from a more fundamental embodiment that makes it possible for evolution to create structures that are internally assigned interactive roles. This process carries with it an increase of complexity of the way the environment is perceived and acted upon" (Etxeberria, this volume). It also seems to offer a minimum requirement for evolution and cognitive categorization (Lakoff, 1987; Rocha, 1997, 1998).

Perhaps the most important characteristic of this distributed memory-selected self-organization is the fact that its specific embodiment both constructs the classification of the environment and ultimately dictates functionality and thus selection. The consequence of this for biological systems is that natural selection (acting on this form of self-organization) is not free to evolve any organism, but is constrained by the self-organizing properties of the materiality of the organisms it acts upon — evolution with both a self-organizing and selection component. The consequence for cognitive systems, is that what can be classified is also constrained by the particular materiality of the classifying system at stake — not everything "out there" can be grasped. In other words, the particular self-organizing dynamics of a particular classifying system constrains the universality of its classification. However, we should look into how this process can be made more efficient, and allow for genuine open-ended emergence of variety in classification.

VON NEUMANN: DESCRIPTION BASED SELECTED EVOLUTION

Von Neumann (1966) defended that a threshold of complexity exists, under which complexity degenerates, and above which complexity can increase in an open-ended fashion. He proposed a self-replicating automata scheme based on the notion of a memory-stored description $\Phi(A)$ that can be interpreted by a universal constructor automaton A to produce A itself. However, to avoid a logical paradox of self-reference, the description, which cannot describe itself, must be *both* copied (*uninterpreted* role) and translated (*interpreted* role) into the described automaton. This way, in addition to the universal constructor, an automaton B capable of copying any description, Φ, is included in the self-replication scheme. A third automaton C is also included to effect all the necessary manipulation of descriptions. To sum it up, the self-replicating system contains the set of automata $(A + B + C)$ and a description $\Phi(A + B + C)$. Automaton C feeds a description to the copier B which copies it, and another to the constructor A which produces another automaton $(A + B + C)$; the copy is then handled separately to the new automaton which together with this description is also able to self-reproduce.

Descriptions and open-ended evolution

As Von Neumann (1966) discussed, if the description of the self-reproducing automata is changed (mutated), in a way as to not affect the basic functioning of $(A + B + C)$ then, the new automaton $(A + B + C)^*$ will be slightly different from its parent. Von Neumann used a new automaton D to be included in the self-replicating organism, whose function does not disturb the basic performance of $(A + B + C)$; if there is a mutation in the D part of the description, say D^*, then the system $(A + B + C + D) + \Phi(A + B + C + D^*)$ will produce $(A + B + C + D^*) + \Phi(A + B + C + D^*)$. Von Neumann (1966, p. 86) further proposed that non-trivial self-reproduction should include this "ability to undergo inheritable mutations as well as the ability to make another organism like the original", to distinguish it from "naive" self-reproduction like growing crystals, or domino type reproduction (Kampis, 1991). Von Neumann's model clearly does not rely on a distributed but on a local kind of memory. Descriptions entail a symbol system on which construction commands are cast. These commands are not distributed over patterns of activation of the components of a dynamic system, but instead localized on "inert" structures which can be used at any time — a sort of random access memory. By "inert" I mean material structures with many dynamically equivalent states, in other words, the semantic relation, or what the structures are used to refer to, must possess a large degree of arbitrariness so that certain representations are not much more probable than others. In the genetic system, any sequence of nucleotides is possible, and its informational value is not dependent on the particular attractor behavior of DNA or RNA dynamics.

Why then is there an advantage of local memory over distributed memory self-replication? Von Neumann's argument maintains that if we do not have symbolic descriptions directing self-replication, then an organism must replicate through material self-inspection of its parts. In other words, the dynamics must be able to produce copies of itself by template identification of parts existing in its environment. The simplest way would be to have every part of the structure individually heritable. Clearly, as systems grow in complexity, self-inspection becomes more and more difficult (Pattee, 1995). The existence of a language, a symbol system, allows a much more sophisticated form of communication. Functional, dynamic structures do not need to replicate themselves, they are simply constructed from physically non-functional (dynamically inert) descriptions. For instance, for an enzyme to replicate itself, it would need to have this intrinsic property of self-replication "by default", or it would have to be able to assemble itself from a pool of existing parts, but for this, it would have to "unfold" so that its internal parts could be reconstituted for the copy to be produced (Pattee, 1995). With the genetic code, however, none of these complicated "gimmicks" are necessary:

functional molecules can be simply folded from inert messages. This method is by far more general since any functional molecule (with limitations to be discussed ahead) can be produced from a description, not merely those that either happen to be able to self-reproduce, or those that can unfold and fold at will to be reproduced from available parts. The evolution of distributed memory-based self-organizing systems is restricted to this type of trivial (in von Neumann's sense) or through self-inspection (self-description (Kampis, 1991)) reproduction.

The symbol system, with its utilization of inert structures, opens up a whole new universe of functionality which is not available for purely dynamical self-replication. In this sense, it can evolve functions in an open-ended fashion. The threshold of complexity proposed by Von Neumann is taken by some (e.g. Pattee, Cariani, Kampis in Rocha (1995a), as another category of self-organization which is capable of creative organization and selection from outside. Following the rationale above, we can call it *local memory-selected self-organization*, or *description-based selected self-organization*. In biology, this means that living systems can follow a largely open-ended evolutionary history (von Neumann's threshold of complexity). In the cognitive realm, the introduction of symbols also opened up a whole new world of communication possibilities as the aspects of an environment that can be communicated between individuals is not restricted to only those things we can "show" or otherwise somehow physically mimic: the displacement of local observations.

Semantic closure: open-endedness, materiality, and universality

The notion of description implies a self-referential linguistic mechanism. A description must be cast on some symbol system while it must also be implemented on some physical structure. Since many realizations of the same symbol system are possible, viewing descriptions only as physical systems explains nothing about their symbolic nature in the control of construction. When *A* interprets a description to construct some automaton, a *semantic* code is utilized to map instructions into physical actions to be performed. When *B* copies a description, only its *syntactic* aspects are replicated. Now, the language of this semantic code presupposes a set of material primitives (e.g. parts and processes) which the instructions are said to "stand for". In other words, descriptions are not universal, as they refer to some material constituents which cannot be changed without altering its significance. We can see that a self-reproducing organism following this scheme is an entanglement of symbolic controls and material constraints which is closed on its semantics. Pattee (1982, 1995) calls such a principle of self-organization *semantic closure*.

It is also important to understand that when we say that a description-based selected self-organizing system is endowed with open-ended evolutionary potential we do not believe it is universal, that is, that any physical system can be evolved. A given semantically closed system is based on some sort of coding mechanism between inert and functional structures. However, the code and the associated construction are built on some material substrate constraining the whole semantic closure: there is a finite number of functional structures which may be constructed with a given set of parts. The degree of open-endedness will be dependent on the representational potential of this code. In other words, the larger the number of possible equally dynamically inert structures, the larger the universe of functionality that can be represented in them. For instance, living systems cannot evolve any functional structure whatsoever, but still the number of possible functional combinations attainable with the DNA-protein code system is very large, far beyond computational limits. In this sense, the emergence of functionality is open-ended though not universal.

It is here that the emphasis on the symbolic level of open-ended evolutionary systems must be tamed. Strong Darwinism has emphasized the nature of the symbolic description of living systems, as much as strong cognitivism has emphasized the symbolic nature of cognition. However, semantic closure with its description-based selected self-organization is *not* reiterating this position. The symbolic component of open-ended evolutionary systems is stressed, but the material, dynamic, self-organizing characteristics of matter are equally stressed. It is the ultimate inclusive approach which is neither reductionist nor dualist (Pattee, 1995). While it is maintained that a purely physical description or dynamics will not explain symbolic function (as several material systems may implement the same function), it is also maintained that different material structures will not have identical domains of potentially evolvable functions. The important idea is that evolution relies *both* on self-organization and selection, and only those self-organizing systems able to harness their dynamics to obtain a symbolic dimension can have open-ended evolutionary potential.

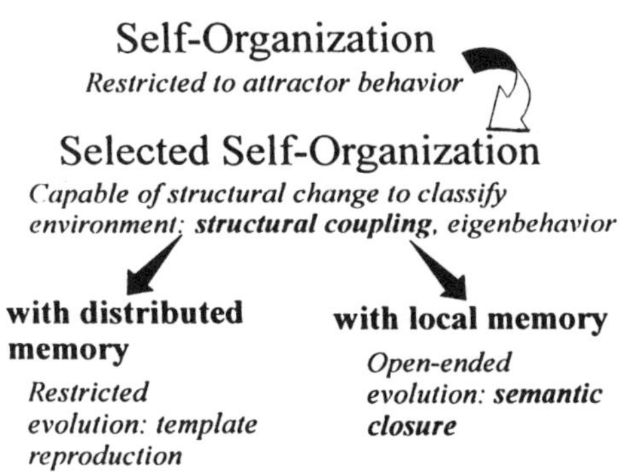

Figure 1. Memory-based self-organization

EVOLVING SEMIOTICS: A CONCEPTUAL FRAMEWORK FOR INCLUSIVE SELF-ORGANIZATION

Semiotics concerns the study of signs/symbols in three basic dimensions: syntactics (rule-based operations between signs within the sign system), semantics (relationship between signs and the world external to the sign system), and pragmatics (evaluation of sign system regarding the goals of their users) (Morris, 1946). The importance of this triadic relationship in any sign system has been repeatedly stressed by many in the context of biology and genetics (e.g. Waddington, 1972; Pattee, 1982, 1995; Salthe, 1995); in particular, Peter Cariani (1989, this volume) has presented an excellent discussion of the subject. We can understand the semiotics of the genetic system if we consider all processes taking place before translation (from transcription to RNA editing) as the set of syntactic operations; the relation between mRNA (signifier) and folded amino acid chains (signified), through the genetic code, as the implementation of a semantic relation; and finally, the selective pressures on the obtained proteins as the pragmatic evaluation of the genetic sign system.

Semiotics with two symbol types

Until now, the semiotics of DNA has been considered strictly unidirectional: DNA stands for proteins to be constructed. In other words, the symbolic DNA encodes (through the genetic code) actions to be performed on some environment. Naturally, through variation and natural selection (pragmatic evaluations) new semantic relations are created which are better adapted to a particular environment, however, real-time contextual measurements are not allowed by this unidirectional semiotics.

If in addition to symbols standing for actions to be performed, the genetic system is also allowed a second type of symbols standing for contextual, environmental, measurements, then a richer semiotics can be created which may have selective advantage in rapidly changing environments, or in complicated, context dependent, developmental processes. Figure 2 depicts such a sign system. The top plane contains two different types of symbols which are combined in different ways (symbolic operations). *Type 1* symbols stand for actions through a *code* ϕ (e.g. the genetic code) and *type 2* symbols stand for measurements through a different *code* γ which is being hypothesized here.

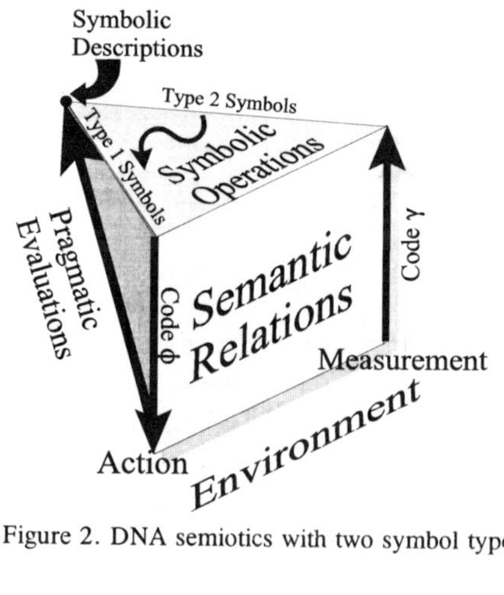

Figure 2. DNA semiotics with two symbol types

In Rocha (1995c) evidence was presented to show that RNA editing may be seen as a mechanism for this contextual input, at least for certain well known living organisms like the African trypanosomes (Benne, 1993), and as a potentially important mechanism in the morphogenesis of highly evolved animals (Lomeli et al., 1994). We can think of DNA as a set of symbolic descriptions based on two types of symbols: *type 1* symbols are expressed in mRNA molecules and stand for actions to be performed; *type 2* symbols are expressed in some sort of editing mechanisms (e.g. gRNA molecules in the genetic system of the African trypanosomes) which stand for contextual observables. RNA editing can be seen as a set of symbolic operations performed with symbols of both types, resulting in symbols of *type 1* to be translated into actions by the genetic code.

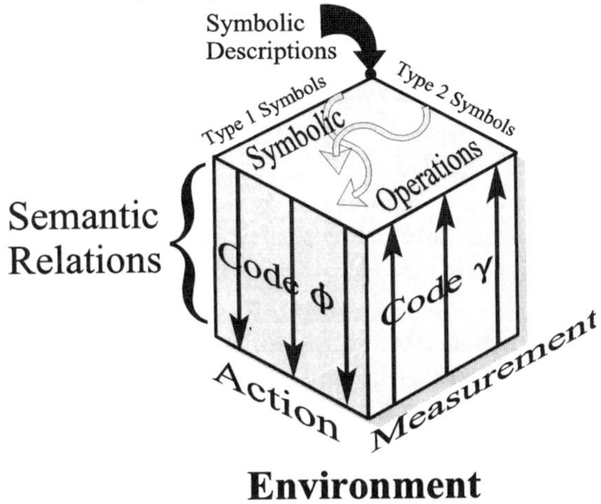

Figure 3. Semantic Closure with 2 symbol types

Notice that *code* γ is proposed here as an abstraction referring to the set of mechanisms which will link environmental measurements (context) into *type 2* symbols. It is *not* expected to function as a proper genetic code. Jon Umerez (1995) has stressed the importance of a code in any form of evolving semiotics. In simple terms, what I refer to as a code here is any mechanism able to relate "inert" material structures to other material structures with some functional dynamics "by virtue" of a larger organizational closure. In other words, the function of the first material structures is not dependent on its particular materiality, but on what they are used to refer to for the imbedding, material, self-referent semantic closure (Pattee, 1995). Again, a semantically closed system, endowed with this kind of symbol/matter code is able to establish open-ended evolution (Pattee, 1995; Umerez, 1995). Leaving pragmatic evaluations (selection) out of the picture momentarily, the semantic closure with two symbol types, which is able to act as well as perform measurements on its environment can be represented by the cube in Figure 3. The semiotic triadic relationship is only complete when individual semantic closures are coupled to an environment (measured and acted upon by each one of them) which ultimately selects (pragmatic evaluation) the most fit amongst these symbol-matter closures (e.g. in natural selection, those that reproduce the most, see Figure 4).

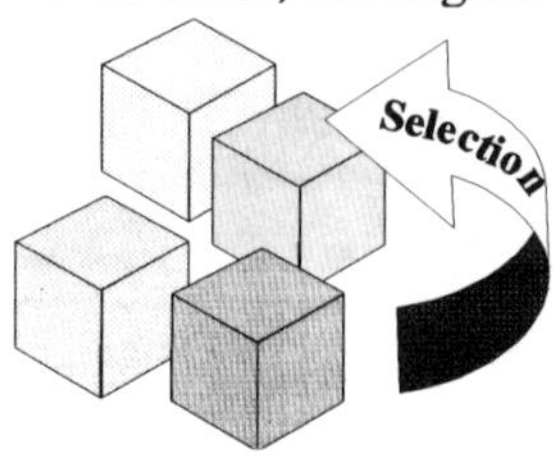

Figure 4. Selection on Semantic Closure

Materiality and implementation dependence: self-organization and selection come together

The issue of materiality is extremely important for two reasons: (i) all which can be represented in this evolutionary semiotics is restricted to what can be constructed by the specific, material, semantically closed system in the first place; and (ii) selection is ultimately performed on this specific material organization capable of performing a number of functions in an environment. The conceptual framework put forward by this material, evolutionary, semiotics forces self-organization and selection together as two indispensable dimensions of evolutionary systems. Pragmatic evaluations or selection takes place on particular dynamics, on the other hand, open-ended evolution is only possible through the existence of a symbolic dimension mediated through a code. Moreover, this code must be built out of some materiality which constrains its representation power and which also ultimately defines eigenbehavior, or an organism's ability to construct and discriminate its environment.

A semantically closed system is not implementation independent because matter constrains its eigenbehavior as well as its evolutionary potential. The second constraint is clear when we realize that two distinct closures which at some point establish the same eigenbehavior (the same representational function), if materially different, will evolve differently. The first constraint is not so clear since we hypothetically allow the idea that two different closures can have the same representational function. However, this equivalence can only be established between formal symbol systems which by definition are not materially constrained and are therefore universal, that is, the set of possible semantic relations is infinite (Figure 5).

Material symbol systems do not have this property. A coding relation must be formed out of certain available material parts in each domain (e.g. nucleotides and aminoacids in the genetic code), and no semantic relation can escape them. In the genetic system we can represent any protein, but we cannot represent and construct any other material structure which is not made out of aminoacid chains. Thus, our semiotics are necessarily constrained by matter, not just due to selection pressures, but on account of the parts available for the symbol system itself (Figure 6).

Material sign systems are not universal and cannot represent anything whatsoever, but this turns out to be their greatest advantage. The price to pay for the universality of formal symbol systems is complete specificity, that is, full description of its components and behavior. Conversely, material sign systems are built over certain building blocks which do not need a description. For instance, DNA does not need to encode anything but aminoacid chains, there is no need to include in genetic descriptions information regarding the chemical constituents of aminoacids nor instructions on how to fold an aminoacid chain — folding comes naturally from the dynamical self-organization of aminoacid chains. Notice how a logical simulation of these genetic mechanisms needs to include all this information that comes free when the self-organizing characteristics of matter are actually used rather than simulated (Moreno et al., 1994).

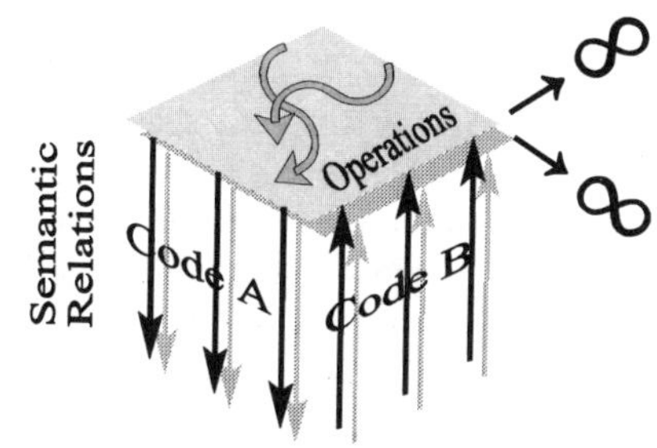

Figure 5. Formal Semantics

Figure 6. Material Semantics

WHAT DOES IT MEAN FOR APPLICATIONS?

Evolutionary strategies: selection alone

The underlying idea of computational evolutionary strategies (ES) is the separation of solutions for a particular problem (e.g. a machine) from *descriptions* of those solutions through a code. Genetic algorithms (GA's) work on these descriptions and not on the solutions themselves, that is, variation is applied to descriptions, while the respective solutions are evaluated, and the whole (description-solution) selected according to this evaluation. This separation follows von Neumann's self-reproducing scheme which is able to increase the complexity of the machines described. This leads to the conclusion that the form of organization attained by GA's is not self-organizing in the sense of a Boolean network or cellular automata. Even though the solutions are obtained from the interaction of a population of elements, and in this sense following the general rules usually observed by computationally emergent systems, they do not strictly *self*-organize since they rely on the selective pressures of some fitness function. The order so attained is not a result of the internal dynamics of a collection of interacting elements, but dictated by the *external* selection criteria. To say that the populations of descriptions of solutions self-organize *at all* in ES may stretch the concept of self-organization a bit too far. ES rely on different concepts: first, with the description-solution dichotomy the concept of local memory is introduced; second, the transition rules of ES are not state-determined — variation is stochastic; third, as already discussed, selection is external to the populations of descriptions. This way, we can hardly say that a population of memories is interacting with any sort of *"self-dynamics":* the solutions reached by a GA do not self-organize but are a result of external variation and selection. For all these reasons, it is therefore natural to think of ES as completely distinct from self-organization. It is perhaps useful to think of ES as modeling a very different aspect of biological systems that has to do with natural selection. Self-organizing systems model the abstract, internal, characteristics of matter, while ES model the existence of, external, selective pressures on populations of varying memory based descriptions of some system.

Artificial semantic relations: the origin problem

The coded relationship between descriptions and solutions for some task in ES is imposed at the onset by the users of such systems. Likewise, the database symbols of some artificial intelligence program are externally related to some categories its users are interested in. Both have to do with the issue of representation in computational domains. All formal systems must have their symbols related to some meaning by the external intervention of some user (Rocha, 1995b), in other words, a formal system cannot change the signifier/signified primitives imposed when it is started, and create new observables (Cariani, 1991). In the field of GA's some (Mitchell & Forrest, 1994) have been calling for more research to be done on schemes that may allow the evolution of the description/ solution relationship itself, that is, the evolution of a code. The same quest takes place in cognitive science for some way to ground the symbols of artificial intelligence models (Harnad, 1990).

Basically, everyone is one way or another dealing with the origin of symbols problem, or in other words, the matter/symbol problem. Some explain symbols away by searching for explanations in the dynamics of cognitive and biological systems (e.g. Churchland and Sejnowski, 1991) while others, usually in strong computationalist fields, will look solely at the purely symbolic aspects of complicated systems. Few have been calling for the inclusion of both aspects into complementary approaches (Pattee, 1978; Lakoff, 1987; Cariani, 1987; Etxeberria, this volume). This latter view calls for an embodiment of models of life and cognition in such a way as to be able to study the origin problems within an inclusive framework where material and computational aspects are intertwined.

In any case, however far we may be from solving any problems of origin, we may still recognize that both life and cognition rely on complementary dynamical and symbolic characteristics. Even if we do not yet know how these aspects ever came to be brought together, we should build artificial models using both of these aspects (or their simulations) to our advantage, since they have proved to be immensely powerful for natural organisms. For instance, in Rocha (1995c, 1997a), even though using fixed computational coding relations between descriptions and solutions in a GA, I proposed the establishment of, stochastic, contextual constraints on this coding relation following the basic mechanisms of RNA editing found in a variety of living organisms. These contextual GA's, though completely computational, are able to change the way they produce solutions from the same genetic description, according to changes in their environments. They are an instance of the two symbol type semiotic model discussed in section 3, and can be said to evolve an internal control of genetic expression which may be of use for organisms whose environment is subjected to cyclic changes.

Genetic algorithms and development: self-organization and selection in Artificial Life

Lately much attention has been posited on evolutionary strategies that bring together self-organizing systems and natural selection inspired algorithms. Particularly in the field of Artificial Life, Kitano (1994), and Dellart and Beer (1994), have proposed GA's which do not encode directly their solutions, but rather encode generic rules which develop into Boolean networks simulating given metabolic cycles. With these approaches, GA's no longer model exclusively selection, but also a self-organizing dimension standing for some materiality. The GA does not search the very large space possible solutions, but a space of basic rules which can be manipulated to build different self-organizing networks. These networks are then started (sometimes with some learning algorithm) and will converge to some attractor behavior standing for a solution of our simulation. Rather than directly encoding solutions, the GA harnesses a space of possible self-organizing networks which will themselves converge to a solution — *emergent morphology*.

The computational advantage of these systems lies in the tremendous reduction of the algorithm's search space since the solutions do not have to be encoded in all details, the emergent morphology "takes care" of details we do not need to encode. In particular, I have proposed a developmental scheme (Rocha, 1995c, 1997a) which uses the same search space (based on fuzzy rules) for whatever number of simulation primitives we desire, in other words, a generic GA which uses the same state space regardless of the simulation task by utilizing an emergent morphology scheme based on fuzzy logic. By simulating both selection and self-organization, the size of descriptions is dramatically reduced, and an avenue is opened for studying the simulation of both the symbolic and material aspects of evolutionary systems.

Categorization and constructivism: uncertainty and belief in artificial intelligence

Eleanor Rosch (1978) and George Lakoff (1987), among others, have stressed the importance of an embodiment of cognition to deal with its representation issues. In Rocha (1994, 1997a, 1998) I have introduced a set structure called evidence set based on fuzzy logic and the Dempster-Shafer (Shafer, 1976) theory of evidence. These structures allow the inclusion of all forms of uncertainty recognized in information theory (Klir, 1993) as well as a formalization of belief and contextual dependencies in a set format. Evidence sets do not specifically include an account of materiality, however, the formalization of belief and context allows larger imbedding models of cognitive categorization to base the belief and contextual strengths of concept membership on specific material constraints, or its simulation through, say, neural networks.

The contextual pointers of evidence sets are related to Pask's P-individuals (1976) in his conversation theory and are thus embedded in a constructivist framework which emphasizes the construction of a reality in terms of a cognitive system's specific materiality and environmental coupling (Rocha, 1997c). This is also a direct result of von Foerster's formulation of eigenbehavior and an aid to establishing another instance of the semiotic model of section 3 in the cognitive realm. It can be seen to offer a constructivist position of representation which stresses embodiment, but must also, on the other hand, concede that in an evolutionary context, the construction of categories must have a representational relation to aspects of the organism's environment, or its categorization would not have survival value in that particular organism/environment structural coupling (Medina-Martins & Rocha, 1992; Rocha, 1995d). In other words, embodiment does not eradicate the necessity to still explain some sort of representational relation between constructed categories and the cognitive system's context.

CONCLUSIONS: SELECTION MEETS SELF-ORGANIZATION

I have stressed that though self-organizing systems with distributed memory represent a minimum requirement for evolutionary systems, their evolutionary potential is much larger, possibly open-ended, if further endowed with dynamically "inert" structures to be used by their classification mechanisms which are effectively symbolic. It was stressed that this by no means entails a return to purely symbolic approaches to cognition, nor a belief in the absolute sovereignty of natural selection in evolution. Rather, it is a call for more inclusive, hybrid approaches to such evolutionary and learning systems. In Artificial Life this implies building models which bring together self-organizing mechanisms, such as cellular automata or Boolean networks, with genetic algorithms (with varying degrees of control of their genetic expression). In Artificial Intelligence it implies the establishment of models able to go beyond connectionist classification, by inclusion of higher level accounts of cognitive categorization. A semiotic model for selected self-organization with two symbol types was proposed as a conceptual framework for systems with open-ended evolutionary potential, given a particular material implementation.

Los Alamos National Laboratory
Los Alamos, USA

NOTES

[1] The implications to Second-Order Cybernetics of the current ideas are developed in Rocha (1996).

[2] Causality is here solely understood in the strict mechanist, not Aristotelian, sense. It is beyond the scope of the present work to discuss causality. Others in this volume Juarrero, Kampis, Minch, Riedl, Van de Vijver address the issue in detail.

REFERENCES

Benne, R. (ed.), 1993, *RNA Editing: The Alteration of Protein Coding Sequences of RNA*, Ellis Horwood.

Cariani, P., 1989, On the Design of Devices with Emergent Semantic Functions, Ph.D Thesis, SUNY, Binghamton.

Cariani, P., 1991, Some Epistemological Implications of Devices which Construct their own Sensors and Effectors, in *Towards a Practice of Autonomous Systems*, F. Varela & P. Bourgine (eds.), MIT press, pp. 484-493.

Churchland, P.S.; Sejnowski, T.J., 1992, *The Computational Brain*, Cambrige MA, MIT Press.

Conrad, M., 1983, *Adaptability*, New York, Plenum Press.

Conrad, M., 1990, The geometry of evolutions, *BioSystems*, vol. 24, pp. 61-81.

Dellaert, F.; Beer, R.D., 1994, Toward an evolvable model of development for autonomous agent synthesis, in *Artificial Life*, vol. 4, R. Brooks & P. Maes (Eds.), MIT Press.

Haken, H., 1977, *Synergetics*, Berlin, Springer-Verlag.

Harnad, S., 1990, The symbol grounding problem, in *Emergent Computation*, S. Forrest (ed.), MIT Press.

Kampis, G., 1991, *Self-Modifying Systems in Biology and Cognitive Science*, Oxford, New York, Pergamon Press.

Kauffman, S., 1993, *The Origins of Order: Self-Organization and Selection in Evolution*, Oxford UP.

Kitano, H., 1994, Evolution of Metabolism for Morphogenesis, in *Artificial Life*, vol. 4, R.A. Brooks & P. Maes (eds.), MIT Press. pp. 49-58.

Klir, G.J., 1991, *Facets of Systems Science*, Plenum Press.

Klir, G.J., 1993, Developments in uncertainty-based information, *Advances in Computers*, M. Yovits (ed.), vol. 36, pp. 255-332.

Lakoff, G., 1987, *Women, Fire, and Dangerous Things*, Chicago University Press.

Maturana, H.; Varela, F., 1987, *The Tree of Knowledge*, New Science Library.

Medina-Martins; Pedro R.; Rocha, L., 1992, The in and the out: an evolutionary approach, in *Cybernetics and Systems Research '92*, Robert Trappl (ed.), World Scientific Press. pp. 681-689.

Mitchell, M.; Forrest, S., 1994, Genetic algorithms and Artificial Life, in *Artificial Life*, vol. 1, pp. 267-

289.

Moreno, A.; Etxeberria, A.; Umerez, J., 1994, Universality without matter?, in *Artificial Life*, vol. 4, R. Brooks & P. Maes (eds.), Cambridge, MIT Press, pp. 406-410.

Morris, C.W., 1946, *Signs, Language, and Behavior*, New York, G. Braziller.

Nicolis, G.; Prigogine, I., 1977, *Self-Organization in Nonequilibrium Systems*, New York, Wiley & Sons.

Packard, N., 1988, Adaptation to the edge of chaos, in *Complexity in Biological Modelling*, S. Kelso & M. Schlesinger (eds.), vol. 21.

Pask, G., 1976, *Conversation Theory: Applications in Education and Epistemology*, Amsterdam, Elsevier.

Pask, G., 1992, Different kinds of Cybernetics, in *New Perspectives on Cybernetics*, G. Van de Vijver (ed.), Kluwer Academic Publishers, pp. 11-31.

Pattee, H.H., 1978, The complementarity principle in biological and social structures, *Journal of Social and Biological Structures*, vol. 1, pp. 191-200.

Pattee, H.H., 1982, Cell psychology: an evolutionary approach to the symbol-matter problem, *Cognition and Brain Theory*, vol. 5, no. 4, pp. 325-341.

Pattee, H.H., 1995, Evolving self-reference: matter, symbols, and semantic closure, *Communication and Cognition - AI*, vol. 12, nos. 1-2, pp. 9-27.

Prigogine, I., 1985, New perspectives on complexity, in *The Science and Praxis of Complexity*, United Nations Library, Reprinted in Klir 1991, pp. 483-492.

Rocha, L.M., 1994, Cognitive Categorization revisited: extending interval valued fuzzy sets as simulation tools for concept combination, in *Proc. 1994 Int. Conf. of NAFIPS/IFIS/NASA*, IEEE. pp. 400-404.

Rocha, L.M. (ed.), 1995a, *Special issue in Self-Reference in Biological and Cognitive Systems, Communication and Cognition - Artificial Intelligence*, vol. 12, nos. 1-2.

Rocha, L.M., 1995b, Artificial semantically closed objects, *Communication and Cognition - AI*, vol. 12, nos. 1-2, pp. 63-90.

Rocha, L.M., 1995c, Contextual Genetic Algorithms: Evolving Developmental Rules, *Advances in Artificial Life*, F. Moran, A. Moreno, J.J. Merelo & P. Chacon (eds.), Springer Verlag, pp. 368-382.

Rocha, L.M., 1996a (in press), Eigenbehavior and symbols, *Systems Research*, vol. 12, no. 3.

Rocha, L.M., 1997a, Evidence Sets and Contextual Genetic Algorithms: Exploring Uncertainty, Context, and Embodiment in Cognitive and Biological Systems, Ph.D Dissertation, SUNY Binghamton.

Rocha, L.M., 1997b (in press), Relative uncertainty and evidence sets: a constructivist framework, *International Journal of General Systems*.

Rocha, L.M., 1997c (submitted), Categorizing databases, Evidence Sets, and Evolutionary Constructivism, *International Journal of Man-Machine Studies*.

Rocha, L.M., 1998 (in press), Evidence Sets: Modeling Subjective Categories, *Int. J. of Gen. Syst.*

Rosch, E., 1978, Principles of Categorization, *Cognition and Categorization*, E. Rosch & Lloyd (eds.), Hillsdale.

Rosen, R., 1995, The mind-brain problem and the physics of reductionism, in Rocha 1995a, pp. 29-43.

Shafer, G., 1976, *A Mathematical Theory of Evidence*, Princeton Unversity Press.

Umerez, J., 1995, Semantic Closure: A guiding notion to ground Artificial Life, *Advances in Artificial Life*, F. Moran, A. Moreno, J.J. Merelo & P. Chacon (eds.), Springer Verlag, pp. 77-94.

Varela, F., 1979, *Principles of Biological Autonomy*, Amsterdam, Elsevier North Holland.

Varela, F.; Thompson, E.; Rosch, E., 1991, *The Embodied Mind*, Cambridge, MIT Press.

Von Foerster, H., 1960, On self-organizing systems and their environments, in *Self-Organizing Systems*, M.C. Yovits & S. Cameron (eds.), Pergamon Press, pp. 31-50.

Von Foerster, H., 1965, Memory Without Record, in *Anatomy of Memory*, D.P. Kimble (ed.), Science and Behavior Books, pp. 388-433.

Von Foerster, H., 1969, What is memory that it may have hindsight and foresight as well?, in *The Future of The Brain Sciences*, Samuel Bogoch (ed.), Plenum Press, pp. 19-65 and 89-95.

Von Foerster, H., 1977, Objects: tokens for (eigen-)behaviors, in *Hommage a Jean Piaget: Epistemologie Genetique et Equilibration*, B. Inhelder, R. Gracia & J. Voneche (eds.), Delachaux et Niestel; Neuchâtel.

Von Neumann, J., 1966, *The Theory of Self-Reproducing Automata*, University of Illinois Press.

Waddington, C.H., 1972, *Biology and the History of the Future*, Edinburgh University Press.

PETER CARIANI

TOWARDS AN EVOLUTIONARY SEMIOTICS: THE EMERGENCE OF NEW SIGN-FUNCTIONS IN ORGANISMS AND DEVICES

TOWARDS A THEORY OF SYMBOLS

Signs, symbols, and signals are basic to our existence on many organizational levels, from the biological to the psychological to the social. The 'semiosphere', the realm of symbolically-mediated processes, envelopes and incorporates us at every turn (see papers by Hoffmeyer, Umerez, Exteberria, and Joslyn in this volume; Hoffmeyer, 1997). Symbolic nucleotide sequences lie at the root of our biological organizations, neural pulse codes subserve the coherent functional organizations in our brains that permit us to think, while the symbol sequences of our languages afford the complex communications that make human society possible. Semiotic concepts, properly developed, are critical for a deep understanding of the organization of life, the functioning of the brain, and the functional organization of the observer.

This paper presents some of the basic semiotic concepts that are needed in order to understand how symbols function in signalling systems, scientific models, adaptive devices, and biological self-production systems. Using these concepts, we will outline how the symbol-functions in these respective kinds of systems can evolve over time.

The conceptual framework that will be used here incorporates ideas from theoretical biology, cybernetics, general systems theory, information theory, and semiotics. Its foundations come most directly from the work of theoretical biologists Howard Pattee and Robert Rosen on fundamental problems of living organization: what kinds of functional organizations define "life", what roles genetic "codes" play in such organizations, what constitutes biological "information", and what makes the emergence of qualitatively new functions possible. Theoretical biology and biological cybernetics have long concerned themselves with the means by which "modelling relations" might be embedded in material systems and how these relations might evolve over time (in other words, how an evolutionary epistemology grounded in the concrete organism or device is possible). Extensive parallels have been drawn between the operational structure of the scientific model, involving measurements, formal computations, and physical constructions, and that of informational processes in biological organisms (Pattee, 1982, 1985; Rosen, 1991, 1985; Kampis, 1991b, 1991a). In my own work in this field, I have strived 1) to analyze the modelling relations in terms of the semiotic framework of Morris (Morris, 1946) 2) to connect these abstract modelling relations with classes of physical adaptive devices that can be built, 3) to outline the inherent epistemic capabilities and limitations of these device classes, 4) to formulate a set of operational definitions sufficient to recognize modelling relations in natural systems, and 5) to apply these operation-

G. Van de Vijver et al. (eds.), Evolutionary Systems, 359–376.

al tests to an observer-based theory of functional emergence (Cariani, 1989, 1992a). The result has been to construct a biosemiotic, biocybernetic framework that is grounded in the material organism/device, and centered on the limited observer. The general perspective is not far from the biologically-grounded epistemology of von Uexküll (1926), and the operationalist, pragmatist philosophies of Mach, Bridgman, Dewey, and Bohr (Murdoch, 1987). It is hoped that this framework can provide additional linkages between theoretical biology and biological cybernetics on one hand and the field of biosemiotics on the other. Biosemiotics has evolved from the study of animal communication to more general considerations of biological codes (see Nöth, 1990, 1994; Anderson and Merrell, 1991; Sebeok and Umiker-Sebeok, 1991; Emmeche, 1994; Hoffmeyer, 1997; Uexküll et al., 1993, for entry-points into the field), such that the gaps between these fields appear to be narrowing as time goes on.

FUNCTIONAL RELATIONS: SYNTACTICS, SEMANTICS, AND PRAGMATICS

It is useful at the outset to define what a sign is and what kinds of relations that it can have with other signs and the world at large. Every sign conveys a distinction (a "difference"), and every sign system utilizes sets of distinctions in order to achieve some purpose. Any object, event, or state-of-affairs can serve as a sign as long as it can be "recognized if it occurs again" and distinguished from other signs. Thus signs are above all functional entities that are bound up in the functional organization of an informational system. In effect, they are the operational states of that system. For the sake of simplicity, we will discuss sign-systems with discrete, well-defined sign-distinctions, leaving iconic, analog processes for other discussions.[1] Such distinctions must incorporate at least two alternative distinguishable sign-states, and in order to be useful, a distinction has to have consequences beyond its mere recognition; in Bateson's terms, it must be a "difference that makes a difference". While a sign-distinction is physically described in terms of a switch (Pattee, 1973), the description of switching by itself does not capture the sign's functional role — the manifold effects it has on the world and the purposes it serves.

Signs can engage in several basic kinds of functional, informational transactions, both within the sign-system itself and between the sign system and the external world (Figure 1). A sign can have rule-governed relations to other signs in the system (syntactics), it can have linkages to the world outside the system (semantics), and through its use advance or retard the purposes of those using it (pragmatics).

Charles Morris first introduced this tripartite set of sign-relations: "(...) pragmatics is that portion of the semiotic which deals with the origin, uses, and effects of signs within the behavior in which they occur; semantics deals with the signification of signs in all modes of signifying; syntactics deals with combinations of signs without regard to their specific significations or their relation to the behavior in which they occur. When so conceived, pragmatics, semantics, and syntactics, are all interpretable within a behaviorally oriented semiotic, syntactics studying the ways in which signs are combined, semantics studying the signification of signs, and so the interpretant behavior without which there is no signification, pragmatics studying the origin, uses, and effects of signs within the total behavior of the interpretants of signs. The difference does not lie in the presence or absence of behavior but in the sector of behavior under consideration. The full account of signs will involve all three considerations" (Morris, 1946, p. 219; see also Nöth, 1990).

Syntax can be conceived as the set of operationally-deterministic rules which govern the manipulation of signs. A syntactic operation is one in which only the sign's operational state — its *type* — determines the outcome of the procedure (i.e. the manifold material properties of the sign are irrelevant except as they contribute to the recognition of type). Since type is an undifferentiated, unitary property, and the operation's outcome depends only on this property, there can be but one successor state for every predecessor state. All truly syntactic operations thus consist of "operationally determinate" transitions between operational states that depend solely upon the type-identities of other operational states. In effect all physically-realizable syntactic systems are describable in terms of chains of these transitions, i.e. as deterministic finite-state automata. All formal procedures and computer programs are purely syntactic operations, wherein the application of the rules governing the manipulation of the signs does not depend upon their "meaning".

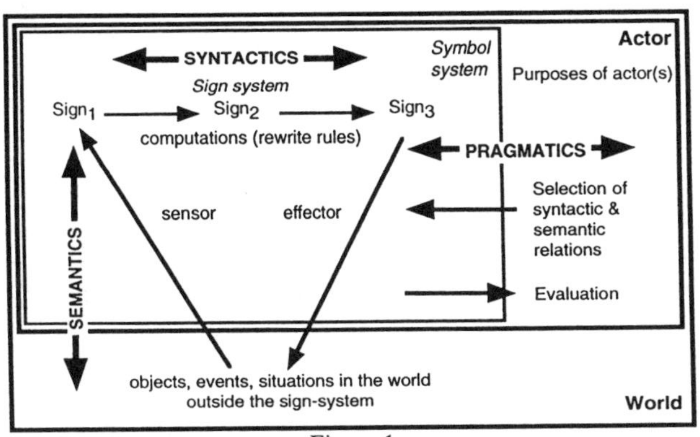

Figure 1

Semantics concerns those aspects of signs associated with "meaning", here taken as the linkage between signs on one hand and objects or states of affairs in the world on the other (i.e. "external semantics"). In its broadest sense semantics is the relation of signs to the world beyond the sign-system, encompassing the manifold consequences which flow from their use, both from the world to the sign (measurement, sensing) and from the sign to the world (action). In contrast to the determinate operational structure of syntactic rules, measurement (sensing) has a contingent structure in which an initial ("null", "reset" or "reference") state transits to one or more successor states. Multiple possibilities are collapsed to one outcome, thereby reducing the uncertainty of the system's subsequent trajectory. The distinction between computation and measurement is therefore that of the (logically) necessary vs. the (empirically) contingent.

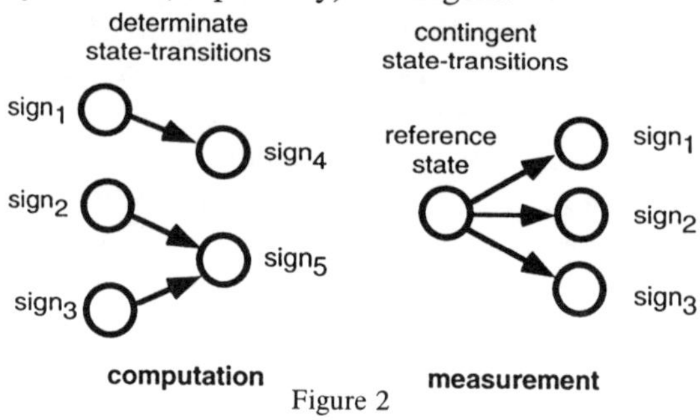

Figure 2

Sometimes the semantics of a sign can be distinguished according to the directionality of the sign-world relation. If interaction with the world determines the sign-state, then the relation is one of "measurement". Similarly, if the directionality flows from the sign to an effect in the world, the relation is one of "action". Syntactic processes in percept-action systems allow particular signs involved in either perception or action to be separated so as to enable their flexible coordination (all possible percept-action mappings). In effect, this decoupling allows for a sign to have a unidirectional semantics, with some signs devoted to measurement (receptor states), others those devoted to action (effector commands), and still others devoted to the coordination of these two semantic sign-types.

Pragmatics is concerned with questions of why particular meanings and syntactic conventions are useful, how they fit into the purposes of symbol-using agents. For human communication this involves what goals or desires an utterance facilitates, why it is useful in a given context. Pragmatics and semantics have often been conflated, but they should not be if purposes are to be kept distinct from percepts and actions. The linkage of the word "snow" to some particular state of water or weather condition does not tell us why it is useful to have such a linkage (e.g. to communicate whether skiing is possible or flying is advisable), nor does the utility of having a sign for snow determine what that particular sign-world relation will be. While syntactics deals with determinate, rule-governed relations between signs, and semantics deals with contingent, interaction-dependent relations of signs to the world, pragmatics deals with the relations of signs to their intended purposes. Syntactic, semantic, and pragmatic thus are complementary kinds of semiotic relations that a given sign may or may not possess. Uninterpreted formal systems have only syntactics, but they can acquire semantic and pragmatic linkages if their human users provide particular interpretations for their signs, and use them for a particular purpose.

The complementarity of syntactic, semantic, and pragmatic relations means that each provides a different mode of explanation for why things behave as they do. These semiotic relations can be related to Aristotelian modes of explanation, the "four causes" (Graham, 1987; cf. Van de Vijver, this volume; Minch, this volume).

Behaviors explained in terms of syntactic operations are formal causes, those explained in terms of semantic operations are material causes (because the contingent action of a sensor is explainable only in terms of its specific material interactions), and those explained in terms of pragmatic operations are final causes (teleologies). Those explained by the action of a semiotic agent on some other agent or object are efficient causes. Thus constructed, syntactic, semantic, and pragmatic operations involve independent aspects of a sign-system that demand different modes of explanation.

Given the triad of semiotic relations, an "evolutionary semiotics" asks how a "sign" becomes a "symbol", i.e. how new semiotic relations come into being ("semiogenesis"). This involves both how existing relations are adaptively altered by experience ("learning"), and how new sign-distinctions themselves might arise de novo ("functional emergence").

THE SEMIOTICS OF SIGNALLING

A simple schematic of a signalling system illustrates the semiotic triad (see Miller, 1951). A signal-code is shown in Figure 2, while the mappings of the signs and their states to situations in the world, their external semantics, are given in the table. The decoder of such a signal is assumed to be capable of distinguishing inter-pulse intervals, and pulse-bursts are identified as a succession of similar intervals. Intraburst intervals encode the categories of alternatives while the number of pulse-intervals in a burst encodes the specific alternatives.

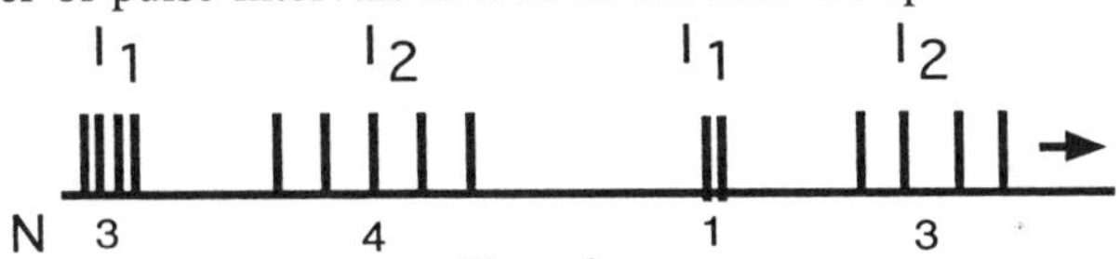

Figure 3

The syntax of this system is described by the rules that bursts must alternate between those containing long and short intervals and that each burst of long intervals is preceded by a burst of shorter intervals with which it is associated. (i.e. a "non-grammatical" message would be a string of bursts with only short intervals).

The semantics of the system are that bursts of short intervals (Il) refer to cities, while bursts of longer intervals (I2) refer to weather conditions. The pragmatics of this signalling system involve its function to inform travelers of weather conditions in the cities involved. The particular message shown [(3,4)(1,3)] translates as "snow in New York and rain in London". The syntactic structure requires a means of parsing the signal and associating the signs properly. Without syntax, the message would be ambiguous, and it would be necessary to make a distinguishable sign for each of the 12 global message possibilities.

Semantic linkages require reporters (humans or instruments) making observations about the weather in each place. Without semantics, the message would have no empirical connection to what is going on in the world. Here the pragmatics relate to how the use of the system furthers the purposes of those who use it (both senders and receivers of signals). While the syntactics (e.g. signal structure) and semantics (e.g. which cities, which weather conditions) are related to the purposes of the users, by virtue of design or adaptation, they should not be conflated with those purposes (analytic conventions ≠ empirical information ≠ desires).

semantic category	**sign**	**sign status**
	type of intervals in burst	# intervals in burst
city	burst with short intervals (I_1)	1: London 2: Tokyo 3: New York
weather	burst with long intervals (I_2)	1: clear 2: cloudy 3: snow

Traditionally, information theory did not deal with the creation of new signalling channels, but an evolutionary account of this signalling-system could do so. This would involve describing how the particular sign-distinctions (pulses, bursts, intervals) arose (pulse-coded transmitters and receivers were most available), how the particular semantic linkages came to be established (reporters with instruments were stationed at each city), and how the system relates to the purposes of its users (so that travelers could pack appropriate clothing). One could then also explain how new sign-distinctions might arise (perhaps, pulses of different amplitude), how new semantic linkages might be formed (reporters could add thermometers to their repertoire and link their temperature readings to the new distinction of pulse-height), and how the pragmatics might change (the user suddenly wants to predict crop failures).

THE SEMIOTICS OF SCIENTIFIC MODELS

The semiotic structure of scientific models sheds light on the functional organization of observers and their epistemic relations to the world. Helmholtz, Hertz, and Mach a century ago made explicit the relationships between observations and predictions that constitute the functional correspondences of the symbols in scientific models with events in the world (Cassirer, 1955). In

Hertz's scheme, the "commutation diagram", a model must have two parts: 1) a set of measuring devices which, through their interactions with the world, generate a set of signs ("pointer readings") which constitute the "initial conditions" and 2) a mathematical algorithm which takes the initial conditions and generates a prediction. The measuring devices form the bridge between the mathematical part and the undifferentiated world outside the sign-system. Here the model functions not as a mirror of the world, but as a transformation whose detailed structure may not even be completely understood, since the sign-distinctions in the model need not share similarities or even identifiable correspondences with the world (von Helmholtz).

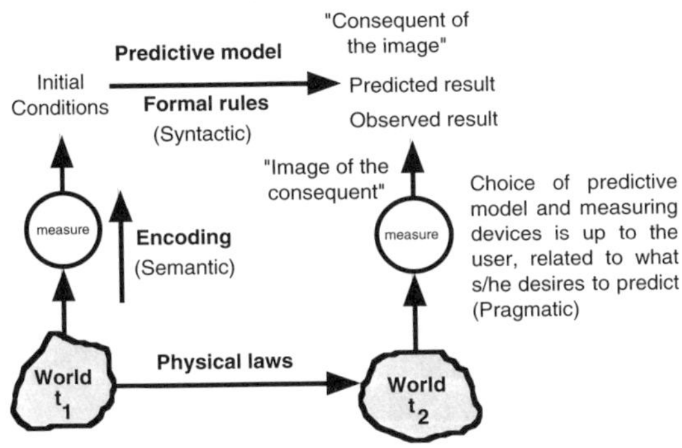

Figure 4

Once the modelling relation is described in operational terms, the semiotic relations inherent in its functional organization become immediately apparent. The formal, mathematical part is described completely in terms of rule-governed syntactic operations on signs, while the measurement part is described in terms of semantic relations between the signs and the physical world. The measuring devices realize the "observables" and the "initial conditions" for the formal model. The model is evaluated by carrying out measurements under two sets of distinguishable situations or experimental arrangements (i.e. "prepared states"), often differing only in time or in space. One of the measurement outcomes is used the initial state of the formal procedure which produces a sign as its output (prediction). The sign generated by the formal part is then compared with the sign generated by the other measurement. If the two always agree, then the model effectively predicts the outcome of the other measurement. To the extent that predictions and measurements diverge, the model becomes less predictive, less reliable in anticipating the outcome of the second measurement, hence less useful to the user. Thus pragmatic relations are described in terms of what predictive goals are achieved with the model.

Scientific models evolve when predictions do not match observations. When the two signs do not correspond, the user has several choices: 1) change the formal part of the model (Δ syntactics), 2) change the observables by altering the measuring devices (Δ semantics), or 3) redefine one's desires/goals (Δ pragmatics). In practice we readily adopt the first solution, less readily the second, and seldom the third (i.e. decide to change what we are trying to predict).

Altering model algorithms and/or observables is an adaptive process, in Spencer's terms "the adjustment of inner to outer relations," i.e. a change in the internal structure that leads to better performance vis-à-vis the external environment. In general, adaptivity requires plasticity of structure coupled with some means of evaluating the performance of the given structure and some means of altering the structure contingent upon that evaluation. When our models don't work properly, we choose to alter algorithms first because it is relatively easy and cheap to perform a different computation, and the set of alternative computations is known to us. Changing the observables, on the other hand, means physically altering the sensors, which can be harder to do and more expensive. Worse still, the set of possible measurements is ill-defined in a way that the set of possible (operationally realizable) computations is not — we may not understand what physical sensor parameters or measurement conditions need to be altered. In the worst case alteration of model algorithms and observables can be carried out using blind variation, and those configurations that yield better predictions can be selected. This process of structure change, evaluation, and selection is a pragmatic operation, because it is driven by the goals implicit in the evaluative process. Thus the structure of the model evolves through the action of both the observer's mutational-constructional capabilities and the selective pressures brought to bear by the observer's goals.

From this example and through experience, it is obvious that these operations (measurement, computation, evaluation) are independent and complementary. One cannot make measurements and gain empirical information solely by carrying out syntactic operations on tokens (computations); minimally to use computation to predict the state of the world one needs initial conditions, and these must be measured. Similarly one cannot carry reliably out computations using procedures which themselves are contingent upon events in the environment (e.g. if the output of one's computer depended on the amount of car traffic on the streets nearby). By grouping together unreliable elements, one can make more reliable assemblages of elements (à la von Neumann), but then the functional states of the assemblage are no longer contingent on environmental fluctuations. Finally, one cannot create new predictive goals by simply performing computations or by simply making measurements; while an evaluation is a contingent process like a measurement, unlike a measurement, evaluations alter the structure of the modelling relation itself ("feedback to structure") rather than simply changing a particular predictive outcome.

THE SEMIOTICS OF ARTIFICIAL DEVICES

The semiotics of the modelling relation in science can be extended to incorporate organisms and devices that sense the world and act on it contingent upon what they sense. While scientific models are externalizations of the structure of the individual observer, modelling relations are also embedded in the internal structure of individual organisms and devices. Sensory organs and sensors correspond to measuring devices, while coordinative parts of organisms and devices (nervous systems, computational parts) correspond to formal, predictive algorithms (Figure 5). While modelling relations include only measurements and computations, organisms and devices also act directly on the world through effector organs. Effector organs convert signs into action on the material world (labelled "control" in the figures). Thus the basic informational operations of signs (semiotic functionalities) present in organisms and devices can be described in terms of measurement (sensing), computation (coordination), and effecting (action) (Uexküll, 1926; Cariani, 1989; Nöth, 1990). We will touch on several kinds of devices here: computers, robots, trainable machines, and those which construct their own sensors (for fuller discussions see Cariani, 1989, 1992a, 1992b, 1993). The semiotic organization of organisms is more difficult to assess and discuss, primarily because the nature of the informational operations performed by nervous systems are presently very poorly understood and may simultaneously involve more than one kind of operation (cf. Etxeberria, this volume). We are still at the stage where we must make analogies to artifacts whose functioning we understand.

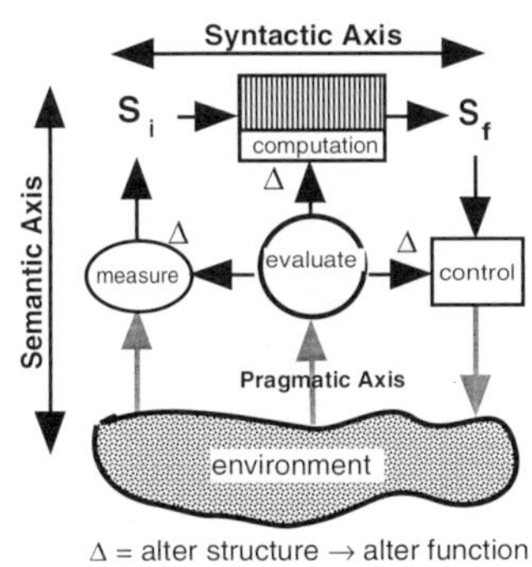

The semiotic of the digital computer, absent its human user and any sensors or actuators, consists solely of syntactic operations (Figure 6). Physically-realized computational devices operate only on signs, and are describable in terms of finite state automata. Thus each total machine state of the device leads to a unique successor state — what the user does is to load into the machine a program and initial conditions ("data"), and let the machine run from state to state until a terminal state (or sequence) is reached. In order for a computer or formal system to solve a real world problem, the situation must first be encoded into symbolic form by a programmer who has independent access to the real world. Those problems which cannot be effectively encoded into a symbolic notation cannot be solved by computer (Dreyfus, 1979).

Computing devices cannot implement modelling relations because they cannot make measurements. Neither can they act on the world by performing computations; some sorts of effectors are needed. Because they themselves they have no inherent linkages to the world at large, human beings mediate between the world and the computer, choosing the encodings of inputs, measuring their values, interpreting outputs, and taking action. Since they are finite-state devices everything which can occur in the machine is bounded by the finite set of machine states and state-trajectories. Computations within the set of machine states do not create new states, nor can they create new linkages to the world outside the machine. Thus no new syntactic or semantic primitives are created (Carello et al., 1984; Cariani, 1989), and as a result their behavior can be described in a closed notational system ("closed world assumptions"). The best that can be done is to combine pre-existing syntactic primitives into logical combinations, and to search the space of combinations for "interesting" patterns. "Formal times formal is formal." (Kampis, 1991b).

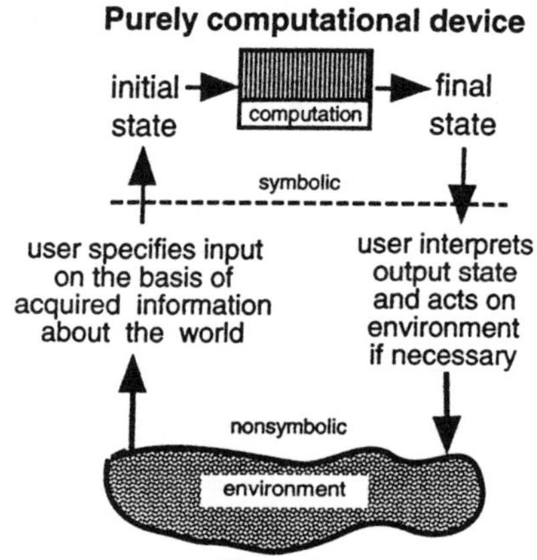

Figure 6

Despite these limitations, pure computation is nevertheless an exceedingly powerful tool. Whatever problems can be effectively encoded, can potentially be solved by purely computational devices, within the physical limits of the device (speed, memory, reliability), the computational demands of the encoded problem, and the foresight of the designer in choosing the appropriate encodings needed.

Robotic devices are computational devices with sensors and/or effectors which connect them directly to the world, giving them inherent external semantics (Figure 7). Robotic devices thus do not need human interpreters to provide meaning for their internal states, or to act directly on the world. By virtue of their sensors and effectors, robotic devices can solve problems and produce behaviors that are not already encoded in symbolic form, (e.g. recognizing sights and sounds, assembling cars, walking). Whatever, problems can be solved by combinations of sensing, computing, and effecting elements can be solved by the appropriate robotic device. Still, the designer is left the responsibility of finding which sensors, computations, and effectors are suitable for solving a given real world problem. In these nonadaptive devices there is no way of altering the parts, and as a consequence, these devices are limited to the foresight of their designers. If the designer failed to incorporate the sensors needed for a particular task, then the device will not have access to needed sensory information. If the designer failed to foresee the optimal percept-action mappings, then the device will act sub-optimally in some situations. The only solution to this "problem of specification" is to make the device adaptive, to allow it to change its structure contingent upon its performance. The structures and functionalities involved can be either in the syntactic or the semantic realm or both.

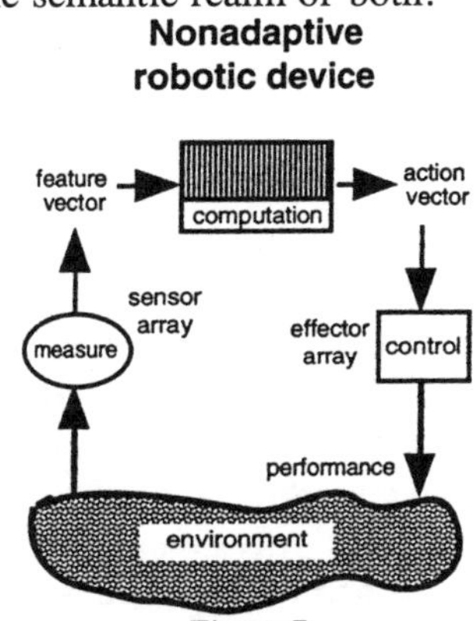

Figure 7

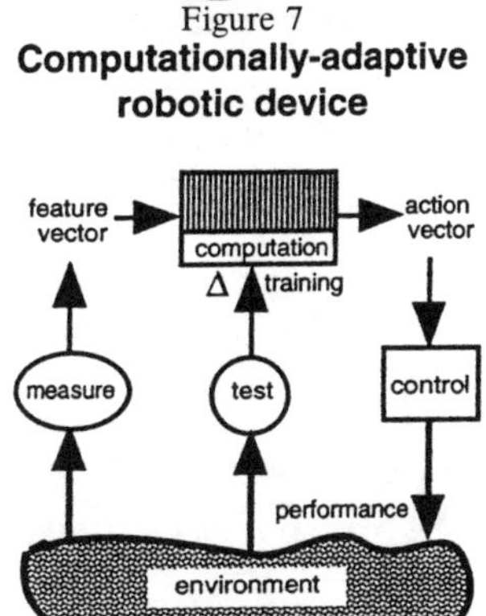

Figure 8

Trainable machines are devices which have adaptation computational parts (Figure 8). Altering the computation contingent upon experience is a process of "syntactic-adaptation". Contemporary trainable machines include neural networks, genetic algorithms, adaptive classifiers, Boltzmann machines, and many others. Like the purely computational counterparts, problems must first be symbolically encoded by the designer, but unlike computers, such machines receive (contingent) feedback from their outputs, which then directs the adjustment of their decision function (i.e. the input-output mapping, the computation performed). For example, the designer of a trainable classifier must decide which aspects of the world must be encoded ("feature primitives") such that the machine can find a successful classification rule. The semantics of the states of the adaptive classifier, are therefore supplied by the designer. If the classifier is embedded in a robot, these semantics are determined by the sensors and effectors of the robot. The machine uses feedback from its performance ("supervised learning") in order to steer the search for better classification rules. This feedback is an evaluative process ("test" in the figure) that reflects the goals of the device's designer (i.e. what classification is desired). The consequences of this process are that the internal functional structure of the device changes (i.e. Δ in the input-output mapping, Δ syntactics).

The trainability of the machine means that it can adapt to situations unforeseen by the designer. The designer is freed from having to directly anticipate and specify all appropriate percept-action contingencies. Without the means of altering the sensors and effectors (Δ semantics), trainable machines are ultimately limited by their fixed percepts (feature primitives) and actions. Since these determine the categories in which it operates, the trainable machine can only be as good as its percept and action primitives allow. Since computations alone cannot create new empirical information nor can they create new actions on the world (new semantic linkages), these machines are semantically bounded.

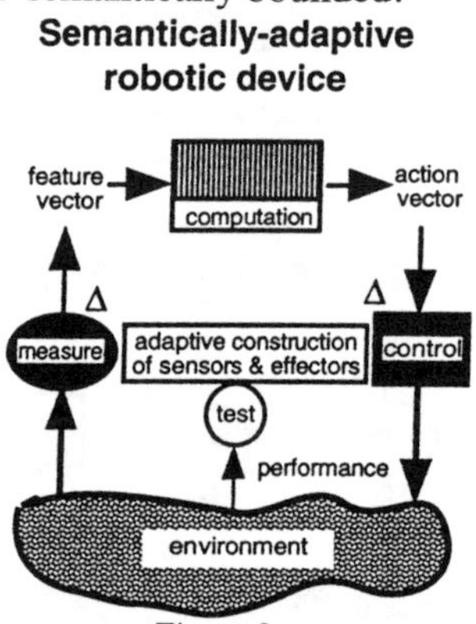

Figure 9

Sensors and effectors are the crucial points at which a real world situation is encoded into a symbolic representation and at which an action decision is transformed from symbolic representation into physical action. It is only by virtue of actual connection to the world via sensors and effectors that symbolic representations become semantically grounded. Adaptive devices that change their sensors and effectors contingent upon experience are also possible (Figure 9). Semantic categories would then be adaptively selected by altering these parts ("semantic adaptation"). Such devices would in effect automatically implement what a scientist does when s/he builds a new measuring device or what happens when entirely new sensory and effector organs arise through biological evolution. Within individual organisms, the immune system is a semantically-adaptive process wherein molecular sensors are constructed, evaluated, and selected.

The various device types outlined above manifest different kinds of emergent behaviors and functions. Semantic adaptation is related to the emergence of new sensory distinctions (e.g. evolving color vision) and new action-alternatives (e.g. evolving capability for flight). Syntactic adaptation is related to emergence of new percept-action mappings ("learning"). In terms of behavior, emergence can be rigorously defined in terms of the deviation of a material system from a model of it (Rosen, 1985); emergent behaviors are those which defy a given model of the system. A given class of devices with emergent functions can then be specified in terms of what an observer has to do to "track" (predict) the device's behavior as it changes its structure and its relations to the world. If the observer must change syntactic state-transitions in his/her model to track the device's behavior, then the device is "syntactically-emergent"; if new observables are needed, then the device is "semantically emergent". Whenever the device adds a new independent sensory distinction, a new observable is required of the observer, and the dimensionality of the apparent behavior of the device increases (Chen & Conrad, 1994; Conrad, this volume). Trainable machines such as neural nets are thus syntactically emergent, while those which construct their own sensors are semantically-emergent (Cariani, 1989, 1992a, 1993).

When an organism or device has the capacity to determine its own syntactic and semantic relations we have a situation of semantic closure (Pattee, 1982, 1985; Rocha, 1996: Umerez, this volume), of organizational closure (Maturana, 1981; Pask, 1981; Rosen, 1985; Minch, this volume) wherein a self-modifying system (Kampis, 1991b) attains a degree of epistemic autonomy (Cariani, 1989, 1992a) or semiotic freedom (Hoffmeyer, this volume): it chooses its semantic categories as well as its computational states within those categories (semantic and syntactic autonomy). If the device is capable of changing its pragmatic relations by redefining its evaluative criteria (goals, desires), then it attains a degree motivational or pragmatic autonomy (freedom to change its own desires).

EVOLUTIONARY SEMIOTICS AND THE SEMIOTICS OF EVOLUTION: SYMBOLICALLY STEERED CONSTRUCTION

Critical to the adaptive process is the modification of the "hardware" that subserves syntactic and semantic functions. In order to alter semantic functions of sensing and effecting, sensors and effectors must be physically altered; in order to augment the mapping capabilities of the computational part, more sign-distinctions must be enabled (e.g. adding RAM to a computer), and this must occur by physically constructing more accessible states (one cannot not create new RAM by running computations, or our computers could expand by continuously "growing themselves"). Optimally the means by which these physical substrates are altered should result in stable alternatives (changes persist, making learning possible) and inheritable structures (new generations can build on old experience). Arguably, symbolic control of the physical construction process is the most efficient means of attaining these ends (Pattee, 1982, 1985), and all biological organisms now in existence have by one means or another exploited this mode of functional organization.

The cycle (Figure 10) is one of symbol string → construction process → formation of physical parts of the device → action of the device → performance in the environment → differential survival → selection of symbol strings (plans). Within the self-construction process are mutational mechanisms for generating variability in the symbol strings, which permits an expanding the portion of the space of symbol strings to be tested.

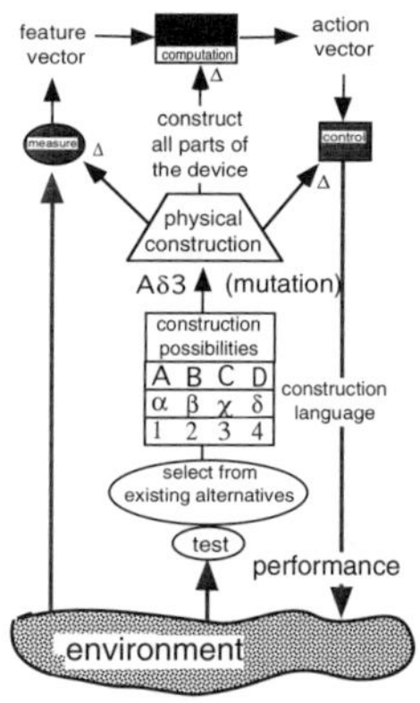

Figure 10

Thus, the most striking aspect of symbols in biological organisms is their central role in self-production, in reproduction and in evolution. Virtually all physical systems that we intuitively recognize as living organisms engage in DNA-directed construction processes. We *cannot understand the organization of living things without understanding the role symbols play in biological organization.*[2] The converse is also true, that virtually all symbols are associated with biological organisms, whether for communication, coordination of action, or construction, and whether at a cellular, organismic or social level. We *cannot understand symbols fully until we understand their role in the organization of life*.

One can outline the semiotic aspects of biological construction languages (Figure 11). The sign-distinctions are DNA sequences, and their most obvious syntactic relations involve transcription-translation rules, the mapping of a DNA nucleotide sequence into an amino acid sequence of a protein.

The semantics of a DNA sequence involve the genotype-phenotype relation, i.e. the relation of the nucleotide sequence to the three dimensional protein structure it codes for and all of the manifold consequences of this folded protein on the rest of the cell, the organism, and the world beyond. While the transcription process can be explicitly described in terms of a syntax of discrete nucleotide and amino acid sequences, the semantics of these sequences are bound up the ill-defined analog dynamics of protein folding, enzymatic action, and still more distal consequences. Analog and digital aspects of proteins thus coexist and complement each other (Pattee, 1979, 1982) in what has been termed a "code- duality" (Hoffmeyer, this volume).

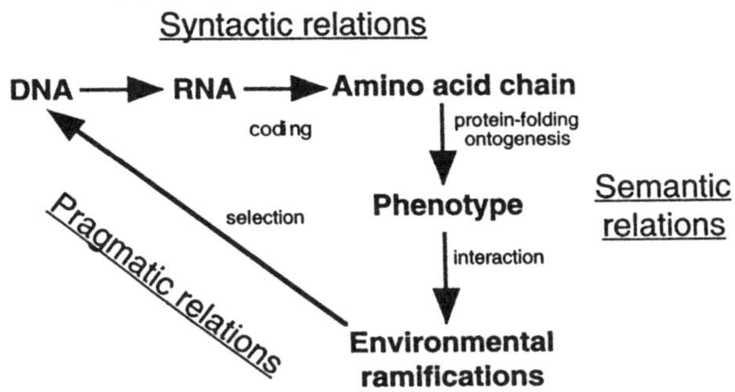

Figure 11

The pragmatics of genetic sequences involve the manifold consequences of gene products on the well-being, survival, and reproduction (i.e. the evolution-sculpted "goals" of the organism). These three categories, syntactic, semantic, pragmatic, are what biologists usually call the genetic, the phenotypic, and the selective aspects of life. All are essential requirements for evolution. In effect, these categories define the organism at the molecular genetic level as a semiotic entity.

One might ask why living organisms need be semiotic organizations at all. After all, there are self-organizing metabolic systems that are capable of a large range of life-like behaviors (cf. Minch, Joslyn, this volume). Are genetic codes critical for the stability and evolution of complex biological organizations? While we lack definitive answer to this question, some general functional roles for symbols in cells can be postulated (Cariani, 1989). Some possible advantages of utilizing discrete codes over purely analog dynamics involve the enhancement of internal stability (organizational anchors), greater resistance to perturbation and greater amenability to error repair, provision of memory through gene-switching, more reliable inheritance of acquired information, and the reliable construction of specific, complex amino acid sequences. Discrete genetic coding permits nearly identical copies to be reliably made and inherited over a large number of successive generations, an accomplishment that would be very difficult in an analog system.[3] DNA coding permits an inordinately rich variety of structure by permitting arbitrary (and improbable) amino acid sequences to be reliably manufactured and replicated.

Do symbols enhance evolvability? Perhaps the strongest argument for such a role is that symbols permit universal construction and generalized encoding

(Cariani, 1989; Rocha, 1996). This allows for any genetic sequence to code for a product, regardless of whether it is already involved in an autocatalytic loop. Once an apparatus is in place for replicating genetic sequences and the replication apparatus itself, all other sequences can "piggyback" on those which code for the self-reproduction apparatus. This means that a protein need not be connected into its own special, autocatalytic network in order to be replicated in the next generation. Von Neumann recognized this ability of a self-reproducing apparatus to produce other, non-self-replicating "byproducts" (Figure 12): "Let X be A+B+C+D, where D is any automaton [and A+B+C is a self-replicating combination of automata]. Then (A+B+C) + F(A+B+C+D) produces (A+B+C+D) + F(A+B+C+D) [the object itself is in parentheses, e.g. (X), and its construction description as a function, e.g. F(X)]. In other words, our constructing automaton is now of such a nature that in its normal operation it produces an object D as well as making a copy of itself. This is a normal function of an auto-reproductive organism: it creates byproducts in addition to reproducing itself" (Neumann, 1987, p. 489). In terms of the genetic code, A+B+C constitute all of the apparatus needed for gene expression, while F(A), F(B), and F(C) are the genetic "plans" read by the expression apparatus. Once such a generalized "universal constructor" is in place, new proteins need not be self-reproducing in order to persist long enough to have useful functions for the organism. Their DNA is translated and transcribed along with the generalized machinery for transcription and translation. The ability to enhance stability and survival of the organism and the ability to self-replicate are thus separated for genes and their associated proteins. Coupled with the indefinitely rich possibilities of arbitrary peptide polymers and their combinations, this separation allows DNA sequences to become a common "informational currency" of the organism.

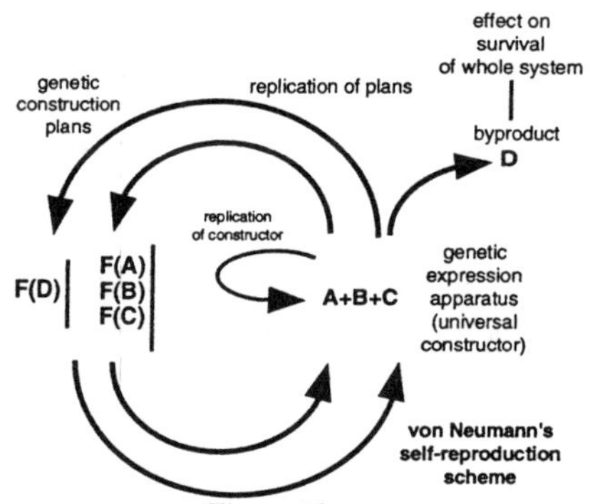

Figure 12

Without generalized encoding, evolution proceeds much more slowly, since every component must both catalyze its own formation and provide some other survival benefit. The alternative to generalized encoding is an autocatalytic ("autopoietic") network in which each component is involved in the production of some other network component, such that all components are produced. Once the requirement that each protein be linked into an autocatalysis network is removed, the number of proteins with useful functions which are propagated to subsequent generations increases enormously. The enormous stabilizing effects and evolutionary advantages conferred by symbols may explain why there are very few natural, complex self-production networks which are not tied in some way to genetic construction. Semiosis, through symbolically-directed construction, is arguably the functional organization that permits complex self-producing biological entities to persist. These complex functional organizations then permit networks of self-producing signals that form the organizational substrates for self-directing epistemic agents such as ourselves.

CONCLUSIONS

We have discussed basic syntactic, semantic, and pragmatic relations of symbols in signalling, systems (communication), modelling relations (epistemology), adaptive systems (self-adjustment), and constructive systems (self-production). The semiotic relation between symbol and matter forms the basis of the functional organization for both life and information-use. We have outlined basic mechanisms for how new syntactic, semantic, and pragmatic relations can come to be incorporated into organisms and devices, how an "evolutionary semiotics" that spans both the natural and the artificial might be possible.

ACKNOWLEDGMENTS

I would like to acknowledge a profound intellectual debt to Howard Pattee, who brought me into the realm of symbols and their functions.

Eaton Peabody Laboratory, Massachusetts Eye and Ear Infirmary
Boston, USA

NOTES

[1] This is not to say that analog, iconic, or continuous signals are unimportant, only that the definition of what constitutes a distinction in such systems is a much more involved process. Most of the neural codes that have been investigated in the brain have a graded, analog character to them, such that the brain is probably best seen in terms of a mixed digital-analog device (with heavy emphasis on the analog part). To the extent that discharges of primary sensory neurons follow the time structure of their respective stimuli, there exists an iconic representation of the stimulus in the time intervals between spikes (Cariani, 1995).

[2] To quote Howard Pattee (1979, p. 218): "The question is whether information is to be treated in biology as just another physical variable, or as the characteristic and exclusive aspect of living systems and their artifacts that distinguish living systems from all other physical systems", and "Life depends upon records" (Pattee, 1972).

[3] One difficulty with analog processes is that they tend to accumulate small errors. Digital systems have sets of attractor basins that constantly reduce the effects of small perturbations.

REFERENCES

Anderson, M.; Floyd Merrell, 1991, *On Semiotic Modelling*, Berlin, Mouton de Gruyter.

Carello, C.; Turvey, M.T.; Kugler, P.N.; Shaw, R.E., 1984, Inadequacies of the computational metaphor, in *Handbook of Cognitive Neuroscience*, M. Gazzaniga (ed.), New York, Plenum Press.

Cariani, P., 1992a, Emergence and artificial life, in *Artificial Life II*, C.G. Langton, C. Taylor, J.D. Farmer & S. Rasmussen (eds.), Redwood City, CA, Addison-Wesley, pp. 775-798.

Cariani, P., 1992b, Some epistemological implications of devices which construct their own sensors and effectors, in *Towards a Practice of Autonomous Systems*, F. Varela & P. Bourgine (eds.), Cambridge, MA, MIT Press, pp. 484-493.

Cariani, P., 1993, To evolve an ear: Epistemological implications of Gordon Pask's electrochemical devices, *Systems Research*, vol. 10, no. 3, pp. 19-33.

Cariani, P., 1989, *On the Design of Devices with Emergent Semantic Functions*, Ph.D., State University of New York at Binghamton.

Cassirer, E., 1955, *The Philosophy of Symbolic Forms*, vols. 1-3, R. Manheim (transl.), New Haven, Yale University Press.

Chen, J.-C.; Conrad, M., 1994, A multi-level neuromolecular architecture that uses the extra-dimensional bypass principle to facilitate evolutionary learning, *Physica D*, vol. 75, pp. 417-437.

Dreyfus, H.L., 1979, *What Computers Can't Do*, New York, Harper & Row.

Emmeche, C., 1994, The Garden in the Machine, Princeton, Princeton University.

Hoffmeyer, J., 1997, *Signs of Meaning in the Universe*, Bloomington, Indiana University Press.

Kampis, G., 1991a, Emergent computations, life, and cognition, *World Futures*, vol. 32, pp. 95-110.

Kampis, G., 1991b, *Self-Modifying Systems in Biology and Cognitive Science*, Oxford, Pergamon Press.

Maturana, H.R., 1981, Autopoiesis, in *Autopoiesis: A Theory of the Living*, M. Zeleny (ed.), New York, North Holland.

Miller, G.A., 1951, *Language and Communication*, New York, McGraw-Hill.

Morris, C., 1946, *Signs, Language, and Behavior*, New York, George Braziller.

Murdoch, D., 1987, *Niels Bohr's Philosophy of Physics*, Cambridge, Cambridge University Press.

Neumann, J. von, 1987, Re-evaluation of the problems of complicated automata — problems of hierarchy and evolution, in *Papers of John von Neumann on Computing and Computer Theory*, W. Aspray & A. Burks (eds.), Cambridge, MIT Press, pp. 477-490.

Nöth, W., 1990, *Handbook of Semiotics*, Indianapolis, Indiana University Press.

Nöth, W., 1994, *Origins of Semiosis: Sign Evolution in Nature and Culture*, Berlin, M. de Gruyter.

Pask, G., 1981, Organizational closure of potentially conscious systems, in *Autopoiesis: A Theory of Living Organization*, M. Zeleny (ed.), New York, North Holland, pp. 265-308.

Pattee, H.H., 1972, Physical problems of decision-making constraints, *International Journal of Neuroscience*, vol. 3, pp. 99-106.

Pattee, H.H., 1973, Discrete and continuous processes in computers and brains, in *The Physics and Mathematics of the Nervous System*, W. Guttinger & M. Conrad (eds.), NY, Springer-Verlag.

Pattee, H.H., 1979, The complementarity principle and the origin of macromolecular information, *Biosystems*, vol. 11, pp. 217-226.

Pattee, H.H., 1982, Cell psychology: An evolutionary view of the symbol-matter problem, *Cognition and Brain Theory*, vol. 5, pp. 325-341.

Pattee, H.H., 1985, Universal principles of measurement and language functions in evolving systems, in *Complexity, Language, and Life: Mathematical Approaches*, J.L. Casti & A. Karlqvist (eds.), Berlin, Springer-Verlag, pp. 268-281.

Rocha, L., 1996, Eigenbehavior and symbols, *Systems Research*, vol. 12, no. 3, pp. 371-384.

Rosen, R., 1985, *Anticipatory Systems*, Oxford, Pergamon Press.

Rosen, R., 1991, *Life Itself*, New York, Columbia University Press.

Sebeok, T.A.; Umiker-Sebeok, J. (eds.), 1991, *Biosemiotics: The Semiotic Web*, Berlin, M. de Gruyter.

Üxküll, J. von, 1926, *Theoretical Biology*, New York, Harcourt, Brace & Co.

Üxküll, T. von; Geigges, W.; Hermann, J.M., 1993, Endosemiosis, *Semiotica*, vol. 96, nos. 1-2, pp. 5-51.

JON UMEREZ

THE EVOLUTION OF THE SYMBOLIC DOMAIN IN LIVING SYSTEMS AND ARTIFICIAL LIFE

> "Life makes sense only in some physical milieu from which it originated and over which it gains some control. The intriguing question about life is why it is so different from non-living physics."
>
> - Pattee, unpublished handout, 1995

INTRODUCTION

The prospective answer I intend to explore in this paper can be introduced through another quote from H. Pattee: "I have argued for many years that life is peculiar, fundamentally, because it separates itself from non-living matter by incorporating, within itself, autonomous epistemic cuts (Pattee, 1969, 1972, 1982, 1993, 1995a). Metaphorically, life is matter with meaning. Less metaphorically, organisms are material structures with memory by virtue of which they construct, control and adapt to their environment. Evolution entails semantic information (Eigen, 1992) and open-ended evolution requires an epistemic cut between the genotype and the phenotype, i.e. between description and construction" (Pattee, 1995b, p. 24).

The central hypothesis at work here is that the elementary emergence of symbols from matter happened along with the origin of life constituting what can be called the first and most basic instance of an *epistemic cut* (Pattee, 1995b), which is autonomously embodied within an organism by virtue of the *semantic closure* (Pattee, 1982), between the description and the construction of the living system. Better than seeing it happening at once and for ever at any precise moment, we may say that it began with life and then underwent an evolutionary process of "becoming" together with the whole system. This "continuum" can be viewed as punctuated first with the origin of the genetic code and the genotype/phenotype distinction and with several later turning points as, for example, the origin of language. Hence, the reason why life is "so different" is because it autonomously embodies, in different stages of accomplishment, symbolic and inherently meaningful structures.

These allegations do not imply that we know how this happened nor even that we fully understand the particulars of the process. Their main value is as radical hypotheses about the nature of life and evolution, offered 1) as a better way to research their intricacies, 2) as an evolutionary path to pursue a naturalist approach to the origin of meaning, and 3) as the means to envisage what might be the best course towards creating Artificial Life (AL).

In this paper, I confine myself to the biological domain, though some of the questions may transcend in some respect this deliberate restraint. I am aware that this is not a totally neutral option for more "unitarian" perspectives inasmuch as it

G. Van de Vijver et al. (eds.), Evolutionary Systems, 377–396.

implies highlighting the living phenomenon which I certainly regard as a very distinctive and specific event emerging from the inorganic world. However, the restriction to the biological is also motivated by methodological purposes in that I will address some problems and point out some considerations which are born around the reflection on *what is life*. This also implies that I will assume Salthe's distinction between development and individuation and, in talking of *evolution*, I have in mind some kind of "irreversible accumulation of historical information" instead of a more general (developmental) sense as "predictable irreversible change" (Salthe, 1993, pp. 27-33, 147 ff.).[1] Moreover, I intend to contrast questions about evolution in the natural domain with its consideration in the artificial domain and more specifically in the field of AL. The point is that this displacement can contribute to clarify, discern, and frame some of the open and controversial issues with respect evolutionary systems in general and biological evolution in particular.

OPEN DEBATES REGARDING EVOLUTIONARY SYSTEMS AND ARTIFICIAL LIFE

The field of research dealing with evolutionary systems is nowadays productively engaged in a lot of discussion and polemics, both in a biological and in a more general sense. There certainly are numerous open questions and some very deep conceptual issues at stake. To mention just one: the currently traditional (even if not much more than one century old) conceptual distinction between evolution and development (Bowler, 1989; Lewontin, 1982, 1983) is being challenged again from very different perspectives and with very diverse goals. For instance, Salthe's recent proposal aims to encompass self-organization and evolution under a more general perspective of development (Salthe, 1993); other approaches strive for something similar by conflating (again) both senses and keeping the term of evolution (Richards, 1992).

In what has been called general evolution (theories) we have global proposals covering from cosmogony and the evolution of the universe(s) to the evolution of societies and cultural traditions.[2] In some of these proposals we find an extension of the abstract features of biological evolution in the neo-Darwinian synthetic tradition whereas in others a total reconsideration of the very idea of evolution is intended under very diverse general principles.

In the biological realm we assist to lively disputes around more specific issues about the structure, causality, agency and nature of the Darwinian theory. As Burian wrote in an assessing review of some of the literature allegedly challenging the orthodox view (if there is one), we could note: "(...) evolutionary issues raised by the complexity of organisms, the internal dynamics of the genome and of ontogeny, cladistics, rational morphological theories of form, the alleged incorporation of environmental information into the genome, epigenetic analysis implicating the involvement of environmental information in development, and the potential contributions of near-equilibrium thermodynamics to evolutionary theory" (Burian,

1988, p. 150).[3]

Most of them share a common concern regarding the more global understanding of the living phenomena as complex self-organizing, adaptive systems. In this sense, the issues addressed are similar to the ones AL has to deal with. In both cases we are dealing with the consideration of the very fundamentals of such systems without restricting ourselves to the current standard approaches in biological sub-disciplines. It is not casual that some people claim that AL should become theoretical biology[4], nor that discussions of evolutionary aspects refer to the core and center of stage in the more conceptual realms of biology (as in Dobzhansky's (1973) motto that "nothing in biology makes sense except in the light of evolution" which is sustained in most general accounts of biology).

I do not wish to imply that this is so because AL might or might not be a promising avenue of research. It just obliges us to face some features of the problems which, in a purely biological approach, may remain ambiguous, hide fundamental assumptions, or appear as opposed alternatives or compatible requirements when they are neither one or the other. I am not implying either that AL's claim of aiming "towards a theoretical biology" is more (nor less) sound than anyone else's. It is simply that AL can provide us with some useful technical and conceptual tools to understand the fundamentals of life and evolution and can, therefore, contribute to the building of an enhanced theoretical biology.

On the other hand it is equally important to realize that AL by itself is confronting some foundational issues. I have elsewhere stressed the relevance of two of them (Umerez, 1995). One is the lack of agreement on an intuitive notion of life as a working hypothesis. The other has to do with the absence of efficient criteria to "ground" the interpretation of its artificial achievements in the computational or robotical domains. Just as Artificial Intelligence has to deal with what is called the Symbol Grounding problem (Harnad, 1990), AL has to give account of the intrinsically meaningful autonomy of natural life.

In addressing the first issue — regarding the notion of life — we need to accommodate two conflicting approaches: on the one hand we have to appeal to Biology or "life-as-we-know-it" in order to get some clues, whereas, on the other hand, we are prepared to transcend it in search of more general principles and definitions of "life-as-it-could-be". Therefore we want to look to the more general features of actual life and we try to discern which could be the more basic conditions for any kind of life. Self-organization (which might be too general) and evolution (which might not be general enough) are the main sources of inspiration.

In dealing with the second one — i.e. the intrinsically meaningful autonomy of life — researchers are turning towards evolutionary theory in search of useful hints and understanding. The underlying assumption is that the process of evolution on Earth has been able to bring forth the actual living systems with all their complexity and "meaningfulness", i.e. materially and dynamically grounded symbolic structures at all levels from the genetic to the cognitive. Here, as well, we find ourselves dwelling in conflict inasmuch as current evolutionary theory can be said to be only certain about the more specific, contingent and detailed aspects of the process of evolution whereas it remains silent or utterly divided and inconclusive about any more general principles.

It turns out that both of these problems are closely linked to the debates on evolutionary systems as we realize when we consider the methods and objectives that groups in both fields share. In particular we find several approaches looking for more general and lawful explanations, and many discussions are centered around divergent understandings of both self-organization and selection and, most especially, of their mutual relationship. Let us introduce the first problem, which will lead us to the second.

LOOKING FOR A CONCEPT OF LIFE

One of the first things that had to be done to substantiate Artificial Life as a full-fledged field of research was to define its object of study. That was not an easy task (it never is): 'artificial' was introduced as 'human made' instead of made by nature, and 'life' was left as relying in our intuitive ideas, with the intention, however, of "transcending" the one case study of biological life (*life-as-we-know-it*) into an abstract exploration of any kind of life (*life-as-it-could-be*) (Langton, 1989b). In addition to this we were given a kind of methodological characterization: a synthetic and formal approach was explicitly preferred to the (allegedly) analytic and material one of biology and this vision was "(...) strongly based on themes such as *bottom-up* rather than *top-down* modeling, *local* rather than *global* control, *simple* rather than *complex* specifications, *emergent* rather than *pre-specified* behavior, *population* rather than *individual* simulation, and so forth" (Langton, 1989a, p. xvi).

This "methodological" essence of AL brings forward a clear research program which "(...) involves the *realization* of lifelike behavior on the part of man-made systems consisting of *populations* of semi-autonomous entities whose *local interactions* with one another are governed by a set of *simple rules*" (Langton, 1989a, p. xxii).

This is certainly a good working project but it does not contain any substantive statement about its object of research. Moreover it already entails a functionalist and behaviorist kind of evaluation of the results. Of course, the 'lifelike' reference still invites one to take a look into biology, at least as a point of reference, but biology as a whole doesn't have a definition much clearer or more homogeneous. What we can find in first instance is that, depending on the particular discipline of Biology, we come across several candidates to a definition. Emmeche has listed at least five which are used in one or another branch of biology: physiological, metabolic, biochemical, genetic and thermodynamic definitions of life (Emmeche, 1994, pp. 33-37). Therefore, it seems that we should try to put them together and find whether there is a common denominator. Unfortunately this does not converge so nicely into a more or less defined single concept, as several attempts show.

Let us select two of these attempts in order to compare their results. They simultaneously constitute coincident and opposed points of view. None of them find it appropriate to ask for a precise definition and hence they choose to offer some list of common features. In both, it is chosen to offer only a list of common features. Yet one seeks to point out the particular distinguishing traits of life (-as-we-know-it) whereas the other aspires to indicate the general properties associated with life (-as-it-could-be).

Coming from a broad biological perspective, Mayr, in his comprehensive survey of "biological thought" (1982), estimates that life cannot have a proper definition (because it is no substance) and opts for characterizing living systems through the attributes they possess and "by which living organisms differ from inanimate matter". Since he cannot find an exhaustive list in the literature, he proposes a tentative one (Mayr, 1982, pp. 53-59) including: complexity and organization, chemical uniqueness, qualitative nature, uniqueness and variability, possession of a genetic program, historical nature, natural selection, and indeterminacy (derived from factors such as hierarchical novelty, randomness or emergence).

Farmer and Belin (1992), in one of the early assessments of Artificial Life as a whole, confront the same problem of needing "a general definition of what it means to be alive" in order to introduce AL, and not having any available. Therefore they propose a "list of properties that we associate with life" (Farmer & Belin, 1992, pp. 817-818) which contains: life as pattern in space-time, self-reproduction, information storage of a self-representation, metabolism, functional interactions with the environment, interdependence of parts, stability under perturbations, ability to evolve, and growth.

Emmeche has worked out these lists in different occasions (e.g. 1991, 1992, 1993, 1994) and has added a central and fundamental property for the artificial case: *autonomy*.

This last addition can give us a clue to begin contrasting both listings. We could not properly say that 'autonomy' is lacking in Mayr's list due to the clear assumption that we are dealing with actual, natural living systems which are identifiably autonomous (at least to a certain undefined extent), but in some deeper sense it *is* lacking (Varela, 1979). It could be retorted that this underlying sense is somehow accounted for by several of the characteristics, in particular the first one referring to complexity and organization, which incorporates metabolism and integration, or the connotation of identity carried upon by the fourth one on individuality, among others. But, then, why should we think it to be a necessary addition to Farmer and Belin's list if theirs also contains these properties? The reason is not, of course, that the idea of autonomy is better covered in the first list, but it is that we know categorically that the second one is purported to indicate a potentially artificial living system whose autonomy from its programmers or constructors needs to be explicitly stated in order to be accepted. We could then ask about the nature of this autonomy in the biological realm, where it is taken for granted, or about the nature of the Genetic Program — both questions being the same.

Besides, if we examine all the points on both lists, we can notice that several factors do not entail absolute differences with respect, at least, to some properties of inanimate matter in non-linear and non-equilibrium physics. The difference implied by the first four properties as well as the last one including its several factors in Mayr's list is just a matter of degree and does not even have any clear threshold when compared, for instance, with abiotic dissipative systems. Yet the other three — genetic program, historicity and natural selection — are different stuff. The three of them are linked around the genotype-phenotype duality and they *do* constitute an absolute difference regarding the inorganic world. In Farmer and Belin's list, on the other hand, we do not have any absolutely differential property because even the ones which are the closest to be so — information storage of a self-representation and ability to evolve — are formulated in a way ambiguous and general enough as to leave many more possibilities open.

The question is that we can roughly distinguish among the properties featuring in those lists two distinct sets: those properties shared with other dissipative systems even if manifested at a greater degree of complexity and precision and those which are not found at all among the more complex non-living dissipative systems. The "genetic program" (Mayr) represents the keystone of these distinctive properties and is the one which, therefore, needs to be related to the physico-chemical realm. This compels us to confront the second question, the "grounding" problem, if we want to advance in our understanding of what can life be.

DICHOTOMIES IN BIOLOGY AND THE PROBLEM OF INFORMATION GROUNDING

In relation to the previous classification of properties into two sets, there is a fairly recent diagnose of the state of the art in Biology in general which can be said to thoroughly apply to the situation of both Artificial Life and Evolutionary Systems as more or less defined fields of research dealing with complex and self-organizing adaptive systems. In his exposition of the "problems of biology", Maynard Smith states from the very beginning that: "(...) two distinct pictures of living organisms have emerged. One is a population of entities which, because they possess a hereditary mechanism, will evolve adaptations for survival. The other is of a complex structure which is maintained by the energy flow through it. (...) Perhaps the hardest part of biology is to see how the two pictures fit together" (Maynard Smith, 1986, p. 7).

Of course, we see that in current research actually made in the different areas of biology (genetics, systematics, paleontology, embryology, ecology, etc.), this is not an extant problem as far as most biologists have to work under the frame of one of the two pictures and can easily forget the other for practical purposes. This is actually a perspective consciously embraced by some thinkers with a more encompassing view of biology as a whole as, for instance, Mayr himself who identifies a similar (yet not identical) duality but is more concerned with explaining their "interplay" (Mayr, 1982, p. 52) than with connecting them in a unified view. Based on his classical distinction between *proximate* and *ultimate* causes (Mayr, 1961), he holds that the fact that "organisms, in contrast to inanimate objects, have two different sets of causes" involves having two corresponding and equally legitimate biologies, functional and evolutionary, dealing with each of them (Mayr, 1982, pp. 67-73). Accordingly, "(...) no biological problem is fully solved until both the proximate and the evolutionary causation has been elucidated. Furthermore the study of evolutionary causes is as legitimate a part of biology as is the study of the usually physico-chemical proximate causes" (Mayr, 1982, p. 73).

However, this easygoing perspective is not so widespread as it might seem when we inquire into other equally theoretical provinces. In particular the claim of "autonomy" for evolutionary biology is not welcomed either by the most orthodox and reductionist views in biology nor by the alternatives looking for different unifying principles. Yet the problem is that this genetic program has to be accounted for somehow; it has to have an origin and a relationship with those aspects over which it is supposed to act. In this sense, Maynard Smith feels that he can convey an optimistic conclusion to his survey of the problems of biology by pointing out the way in which "the two pictures fit together": "It is appropriate to have ended this book with a discussion of the code. I started by suggesting that there are two ways of viewing living things: as dissipative structures, and as entities capable of transmitting hereditary information. (...) Those two aspects of life are brought together by the genetic code" (Maynard Smith, 1986, p. 124).

Or, in other terms, "the transition point between physics and history is the origin and development of the code" (Smith & Morowitz, 1982, p. 265). I wholeheartedly endorse this conclusion which I find accurate. It is certainly the genetic code which fulfills the connecting role and which "grounds" the hereditary information into physical structures and processes.

Unfortunately, I think that we have at least three problems regarding this "solution": (i) the first one is our ignorance about the code, specially its origin; (ii) the second is that most of the time this connecting nature of the code is simply stated and then forgotten for any further purpose, preferring general laws which cover one or the other aspect mentioned by Maynard Smith and leaving the remaining one as a mere detail; (iii) the third one is that we need a radical review of our understanding of the code, a review which puts its relational feature at the center of any consideration. I will address these in the remainder of the paper.

GENETIC CODE

Before reaching the previously quoted conclusion, Maynard Smith addresses the problem of the origin of life and explains the different experimental strategies that have been employed to attack the problem[5]: "Yet each of these experiments fails, in crucial ways, to be a complete solution. They all leave unsolved the central problem of the origin of a hereditary mechanism based on nucleic acids, able to code for all the enzymes needed for their own replication" (Maynard Smith, 1986, p. 111).

This formulation expresses the essence of the logic of the circular relation which Pattee (1982, 1986, 1995a) calls Semantic Closure. Important research efforts are devoted to the investigation of this problem but, despite the confident impression that Maynard Smith (pp. 122-124) conveys before reaching the previously quoted conclusion, others think that "the transition between such prebiotic forms and the earliest genetically coherent cells represents a large gap even in our ability to write scenarios. (...) This transition point between physics and history is identified with the introduction of the genetic mechanism and it has been long recognized that the origin of the code is central to our understanding of this transition. (...) It is, therefore, this important transitional region between physics and history, *the region most uniquely biological*, about which we can make *the fewest definitive statements*" (Smith & Morowitz, 1982, p. 275, italics added).

Our ignorance about the origin and workings of the genetic code is remarkable. We do not know, of course, how it could can come to happen neither if it is totally casual (historical) or to what extent might it be causal (physical). It is certain that it is a degenerate code and therefore it is reasonable to consider it arbitrary. But, then, we cannot view it as an evolutionary or self-organizing phenomenon of prebiotic material structures? Are we condemned to consider it just a frozen accident (at least if we do not want to leave Earth and postulate an unknown origin somewhere else followed by a shedding to other places)?

But we do not even know the principles of most of the workings of the genetic code. We understand the transcription (replication) operation fairly well and we even know how the translation is attained. However we have no clue to solve the core issue of the translation process which is not the codon-anticodon coupling but the way in which the aminoacyl-tRNA synthetases are able to connect an amino acid to the tRNA carrying the "right" anticodon. Some researchers are beginning to talk about a second code — *paracodon* — in order to account for this connection, a code which, curiously, could be less degenerate and more causal than the traditional one (de Duve, 1988). Likewise, new mechanisms of "editing" of the sequences (Benne, 1993) are found every day which render the picture yet more complex. There are also unsolved issues about how the intracellular coding (protein specification) relates to the epigenetic coding (epigenesis description).

Even without going so deeply into the physical structure we have a much more ominous problem in guessing how the meaning of the genetic code is expressed beyond this plain codon-anticodon connection, into real organisms. This connection between gene and organism, between genotype and phenotype (which is the real *code* as we will insist later), is the bigger issue at the very center of our understanding of living systems. As Lewontin said long ago (1974) and repeatedly insisted thereafter (1982, 1983, 1992): "(...) such an assessment will depend on an understanding of the relation between gene and organism that far transcends any present knowledge of development, physiology, and behavior" (Lewontin, 1974, p. x).

Unfortunately the situation is mostly the same now: the relation between the genotype and the phenotype is still a double (two-directional) many-to-many mapping too complex to be grasped in any simple explicative or predictive law: "Because of the Mendelian rupture between the processes of heredity and the processes of development, the explanation or prediction of the hereditary passage of *phenotype* requires a three stage inferential and deductive process involving both genotypic and phenotypic spaces of description. Beginning with the phenotypic description of the parents, the first step is to infer the correct genotypic description. The second step is the deduction of the genotypes of the offspring, given the genotypes of the parents and the Mendelian laws of genetic inheritance. Finally, the genotypic description of the offspring must be mapped back into the space of phenotypic description by the forward application of the epigenetic rules. *The only unproblematic part of this inferential process is the second step of carrying parental genotypes into offspring genotypes*" (Lewontin, 1992, pp. 138-139, italics added).

It would seem, therefore, that "the hardest part" is not solved at all with the mastering of the laws of inheritance, neither with the discovery of the double helix structure or the elucidation of the chart of conversion (the dictionary) of nucleotide triplets into amino acids.

UNIVERSALITY AND GENERAL PRINCIPLES

The second problem is that this very relational feature is normally thrown away as soon as it is recognized (if it is) and one of the pictures is emphasized letting the other one linger as a subservient or subsumed aspect of the purportedly more general one. Modern biology has been able to make a very odd and apparently successful combination of those pictures by means of the alliance between molecular biology and population genetics, reinforced by an undeniable empirical success and a very loose usage of informational concepts. But, in the last term, this situation conceals a total relegation of the dissipative and self-organizing aspects behind the hereditary and informational ones. Furthermore, this view does not give a real account of what the evolutionary process is. Due to the difficulties of dealing with the genotype/phenotype relation that we have seen in the previous point, we end up with a situation which Hull (in contrasting Dawkins' approach) has described as follows: "With the rarest of exceptions, current expositions of population genetics are set out entirely in terms of the transmissions of genes — pious allusions to the role of the phenotype notwithstanding (...) Biologists would like to account for the role of the phenotype in their theories, but the task is too difficult" (Hull, 1982, p. 276).

The lack of the organism as a real system is then substituted with data from molecular biology under the undue assumptions of genetic determinism and almighty adaptationism. This is the way to achieve the reductionist and supposedly unifying picture which was intended. However, the end result of this failed connection is a version in disguise of the "two biologies" picture defended by

Mayr, but both of them being reductionist.

There are, of course, other approaches to this problem which seek a unification and explicitly do not care to let one of the two aspects aside. This is the case of most of the "ultra-Darwinian" views (Eldredge, 1995) which embrace a totally informational and functionalist view and do not really care about the phenotypic domain, giving birth to a peculiar form of reductionism which postulates the gene as a virtual basic atomic entity in exclusivity. This view only requires a dual language which in its more chemically acquainted versions reduces evolution to molecular evolution through the replicative abilities of nucleic acids (RNA in particular due to its catalytic potential). It is a peculiar form of reductionism because it takes the informational realm for granted and as a primitive of the explanation.

In a logical swing, we encounter also a whole range of alternatives which focus on the dissipative and self-organizing aspect considering the genetic realm as, depending on the perspective, a logical consequence, a mere refinement or a secondary aspect in the explicative scheme. Here also we can distinguish the application of physical principles of self-organization in different contexts and formulations from the more informational ones (informational entropy, cellular automata, random Boolean networks, *baupläne*, catalytic RNA, ...) to the more physical ones (physical entropy, autocatalytic sets, protocells, ...). Curiously enough the organism as a genotype/phenotype compound is also absent in most of these approaches. Here the organism is substituted by elements belonging to some modern version of natural kinds as primitive structures, by condensations of simple phenomena under the regulation of a macroscopic law or by complex *informed* dissipative systems.

In general, unifying theories insist on one or the other of the aspects: self-reproducing informational structures (generalized through computational theories) or self-organizing dissipative structures (generalized through physical theories).[6] Without going into detail, we can advance the dangers for the first theories of not being able to give account of the information they take for granted and for the second ones of not being specific enough to account for the peculiarity of life (Umerez & Moreno, 1995).

Universality and formalism

The diversity of ways to handle this dichotomy that seems to remain at the very core of our understanding of life describes well the situation in biology. This circumstance can be phrased in terms of AL in a way which allows us to discern clearly the powerful influence of two goals not always explicit in those discussions: *lawlike universality* and *implementation independent formalism*. Let us quote at length one of the first justifications of AL: "Biology is the scientific study of life — in principle anyway. In practice, biology is the scientific study of life based on carbon-chain chemistry. There is nothing in its charter that restricts biology to the study of carbon-based life; it is simply that this is the only kind of life that has been available for study. Thus, theoretical biology has long faced the fundamental obstacle that it is difficult, if not impossible, to derive general theories from single examples. Certainly life, as a dynamic physical process, could "haunt" other physical material: the material just needs to be organized in the right way. Just as certainly, the dynamic processes that constitute life — in whatever material bases they might occur — must share certain universal features — features that will allow us to recognize life by its dynamic *form* alone, without reference to its *matter*. This *general* phenomenon of life — life writ large across all possible material substrates — is the true subject matter of biology. Without other examples, however, it is extremely difficult to distinguish essential properties of life — properties that must be shared by any living system *in principle* — from properties that are incidental to life, but which happen to be universal to life on Earth due *solely* to a combination of local historical accident and common genetic descent. Since it is quite unlikely that organisms based on different physical chemistries will present themselves to us for study in the foreseeable future, our only alternative is to try to synthesize alternative life-forms ourselves — *Artificial Life*: life made by man rather than by nature" (Langton, 1989b, p. 2).

In the first instance this can resonate as a declaration of principles of most structuralist and self-organizing approaches (in its several kinds). There are some obvious coincidences in claims and goals with different evolutionary theories more or less alternative to the synthetic view. Among the main ones we can stress the objective to formulate universal principles in a lawlike way. That is, not only descriptive ('narrative') principles able to account for contingent and historical phenomena (organisms and lineages) (Gould, 1986; Nitecki & Nitecki, 1992), but real "general theories" that should hold for any conceivable living system and determine the very range of what is possible and what transformations are allowed.

Nevertheless, there is another coincidence not so evident but very important regarding the formalist perspective. Without trying to sound paradoxical, we have to say that most of these structuralist approaches are thoroughly *functional* in a clear sense. Let us explain this point a little further. There are, at least, two senses of 'functionalist' that concern us here.[7]

One is the idea of functionalism as referring to the adaptation to external forces typical of the "panglossian" (Gould & Lewontin, 1976) versions of neo-Darwinism against which all kinds of structuralist approaches react, as exemplified by Goodwin's remark: "It is now necessary to recognize that biological process conforms not only to extrinsic functional stability criteria of the type expressed in the concept of fitness, but also to intrinsic principles of order and organization (...)" (Goodwin, 1984, p. 113)

The other is the idea of functionalism arising from arguments in (philosophy of) psychology and Artificial Intelligence, which refers to implementation independence.[8] It is in this sense that many structuralist versions conform fully to the formalist view expressed by the AL manifesto quoted before. Moreover this is a common ground in which the more orthodox expressions of the synthetic view could feel certainly at ease: leaving the actual materiality and historicity of organism's aside, it is much more rewarding — and neat — to dwell in formal abstractions that are easier to handle and amenable to satisfying generalizations.

Unity of science

The unity of science is certainly an appealing goal in our striving for understanding the world we inhabit. A powerful means to pursue this end has always been through theories unifying different classes of phenomena. This attempt has classically been linked to a reductionist approach aiming to find the most basic and general laws governing the more elemental and simple entities. This connection does not yet seem to be a necessary one since we can nowadays identify several perspectives which consistently enough propose the same strategy of theoretical unification without endorsing the reductionist corollary.

This is something that applies very specially in the case of our understanding of living systems where the formulation of several theories of general principles of self-organization are helping to overcome the shortcuts of the standard view. These approaches are good examples of unifying theories more respectful with the phenomenological complexity exhibited by very diverse structures and systems. We can always find common elements or properties looking to physics and expressing its laws in mathematical formal terms. What happened until recently was that these elements and properties could only be too basic and simple and were, therefore, mostly mute regarding aspects of organization at other levels. Nowadays, thanks to the discovery of new aspects of non-linear physics which are as common (if not as basic) as the linear ones we can also say something about more complex systems. In this sense, it is gratifying to have a "more emergent" and "less mechanistic" kind of physics but this does not make the perspective of an engulfing of biology any less unwelcome or any less "reductionist" (Campbell, 1988; Berry, 1995). In this sense, we cannot pretend — it would be as simplistic and of the same caliber of reductionism — that "there is nothing else".

However, there is still another way to pursue the very same objective — unity of science — which differs from the previous ones in that it does not ask for even more encompassing theories. In this sense, the unity of science should not be sought as much through commonalities applying everywhere but through the implementation of plausible and reliable mechanisms to connect the theories in different realms of knowledge. This view is congenial with autonomist claims (à la Mayr) but it is also methodologically compatible with some self-conscious criticisms of this position, like the ones exemplified, for instance, by Depew and Weber: "Yet adaptationist-inspired defenses of the autonomy of evolutionary biology (...) seem to conflict with the expectation that as science progresses their links with one another will grow increasingly close — even if it is well conceded that these links do not move in the reductionist direction once anticipated by the positivist *conception* of the unity of science" (Depew & Weber, 1988, p. 322).

Let us stress the idea of links among theories instead of the goal of unifying theories, following in this Depew and Weber who explicitly refer to the conception of "networks of theories bridged by interfield theories" formulated by Darden & Maull (1977).

In this sense, the versatility of very general dynamical or informational views should not make us forget the importance of the specificities or the links between commonalties and specificities. In the case of our concern with the living phenomenon this amounts to the equal consideration of the common traits with other dynamical and informational systems and of the different ones. This is, to a certain extent, what complementarity means: to give importance to the common and identify the specific. The comprehension of evolving systems depends on this conjunction and on the ability to find the right connecting perspective. This is precisely what a proper understanding of the relational nature of the genetic code embodied in actual organisms might offer.

BEYOND FUNCTIONALISM *VERSUS* STRUCTURALISM

The third problem is that a conceptual review of the meaning of the genetic code is necessary. Self-organizing and structuralist approaches seem to be orthogonal with functionalist and informational ones in the biological domain, but both of them may equally lack the symbol/matter connection. Moreover, from the point of view of AL, both are eventually similarly devoid of actual materiality and can perfectly match a merely computational universe. This might be due to a common aim to achieve universal and unified explanations.

However, the issue is not just one of reconciling functionalism and structuralism, asserting, as for instance Rieppel (1990) does, that they account for different Aristotelian causes, reserving the formal and final ones for the functional side and allotting the material and the efficient ones to the structural side. It is not only that this ecumenic solution is debatable.[9] It is also that the issue is not one of distribution of tasks, but one of asking how these two aspects are, and came to be, related. We need a reconsideration stressing the relational nature of the genetic code and the genotype/phenotype connection beyond the straightforward coupling between nucleotide and amino acid sequences, a perspective focusing on its sense as a code.

A code is a semiotic relationship which implies syntactic, semantic and pragmatic aspects which cannot be severed from one another (Cariani, this volume). A code only makes material sense in a context of rules of transformation, reference relations and performance evaluations. At the risk of enunciating a platitude, let us recall that a coded record has to be made of something, has to stand for something else and has to be interpreted by something, i.e. the "standing for" relation has to be incorporated in some system and has to be meaningful (operationally useful) for this system. I have insisted somewhere else (Umerez, 1995) on this general point. Now, I would like to center it on our concern with the "genetic program" (Mayr) or the "hereditary information" (Maynard Smith). A 'program' is a more or less precise concept but needs a programmer and has too heavily loaded anthropocentric connotations, while information, on the other hand, is a hopelessly ambiguous concept.[10] So, if we do not want to postulate some abstract entity in the void, we should tie it to some operative consequence. In this sense, to have heredity we need an hereditary record, and this has to be some kind of description amenable to construction or able to instruct some dynamical process in the context of an interpretative system — *an organism in a broad sense, from cell to multicellular or even superorganism.* To make this formulation more precise and explicit and not just analogical or metaphorical, and in order to demarcate it from other more general uses[11], let us stress, following Pattee (e.g. 1986, 1987), that we are talking about a symbolic relation for which an arbitrary mapping from pattern to action and, therefore, a separated and localized (not distributed) kind of memory is required. This implies that a record be rate independent or dynamically incoherent regarding the underlying process.

This view of living systems based on a really semiotic relation between description and construction which, being autonomous and self-referent, realizes a semantic closure (Pattee, 1982, 1986, 1995a) may better reinstate "the organism as the subject and object of evolution" (Lewontin, 1983) in its most consummate sense. In this approach the organism is still the object of internal and external forces[12] driving its evolution, but it is also the subject of this process. Nevertheless, it is not only the subject in the old transformational sense of development as unfolding or via the determination of the environment (Lewontin's sense), but through its autonomous implementation of this "code-duality" (Hoffmeyer & Emmeche, 1990), by being the *subject* of an epistemic cut that entails semantic information.

In this sense it is outrageously restrictive and narrow-minded to talk about the genetic code as something purely informational and with no dynamical dimension ("genocentric view" as Goodwin (1994) puts it). But, eventually, a purely dynamical view (even considering some kind of "informational dynamics" as in the computational domain) could similarly jeopardize a proper understanding of the code. In the last term it is as dynamical as it is informational because it is none of them exclusively, but is a connection between these two domains of phenomena and explanation (Moreno, Etxeberria & Umerez, 1993).

THE EVOLUTION OF THE SYMBOLIC DOMAIN

"Traditional philosophy sees this relation as the problem of reference, or how symbols come to stand for material structures (...). I have always found the complementary question of how material structures ever came to be symbolic much more fundamental" (Pattee, 1995a, p. 11).

It seems that in a kind of basic semiosis we can find a comprehensive sphere of discourse to talk about living systems in a way able to give full account of their nature as complex self-organized evolving entities. There is however a problem with this "solution". To be fair, the identification of the appropriate epistemological "stance" constitutes more the statement of the problem than its solution. This is so because we still need to find which may be the links available to connect this "epistemic" (Pattee, 1995a) realm with other scientific domains toward a network-like unity of science of the kind advocated above (Umerez, 1994).

We can take a *synchronic* perspective and then recognize that actual living systems respond simultaneously to two different sets of principles: dynamical universal laws and informational particular rules (Pattee, 1993). These two universes of explanation (canonically exemplified by physics and computation) are mutually exclusive. Their only connection, in our scientific models, is made by the observer through the act of measurement in which a dynamical pattern is converted into an informational record and the act of control in which an instruction is converted into an action (Pattee, 1989, 1993). This connection is provided by an external modeler in all our cognitive constructs, but it is immanently implemented by the organism as a whole. But, even if we can give an account of the basic epistemic cut performed by a cell in a measurement-like way, this does not constitute a good bridge theory since the very nature of measurement is severely problematic in physics itself (Wheeler & Zurek, 1983). This can only sustain a complementary and pragmatic solution of a kind in which we allow two universes of discourse linked through an epistemic concept which we do not fully grasp.

Notwithstanding, we can also try to build up a *diachronic* perspective willing to study the "long and gradual biological evolution of measurement and control processes" (Pattee, 1993, p. 12). This involves the study of the evolutionary emergence and elaboration of a symbolic universe from a physical one. We should have to begin with the first epistemic cut implemented by the primitive cell at the origin of life and this is, indeed, a problem of the first magnitude. But, once we get there, we could advance the kind of principle by which the process went on: *progressive disentanglement of symbol from matter*. That is, we begin with a very simple system in which matter and symbol are totally entangled and where we almost cannot separate dynamical interactions from informational ones nor both of them from connecting ones as measurement (recognition) or control (action). Then we engage in an irreversible process of complexification and progressive implementation independence of symbols from matter towards an increasing form of disembodiment (Etxeberria, 1995, this volume).

This is an approach which, despite all its difficulties, is worth pursuing and which could eventually constitute a powerful way to build links between different theories at very diverse levels of organization and complexity. By its very nature, it is an evolutionary view, and could even help to found a non-anthropomorphic and non-cognitive centered naturalist account of meaning which, without this recognition of a primitive epistemic cut at the very origin of life, could not easily overcome the "antinomy of naturalism" (Van de Vijver, 1991, 1994).

CONCLUSION

Darwin's account of evolution by natural selection could have been "wrong"[13] from our current point of view in being too organism centered, gradualist, adaptationist, etc., but he was, at least, dealing with real material historical entities and just assuming an undefined mechanism of heredity and a broad view of selection (Lewontin, 1974, pp. 171-172; 1974a). To a certain extent, the same cannot be said about the synthetic view, at least in its "hardened" versions (Gould, 1983). Unfortunately, with the advent and triumph of the gene, the materiality and even the historicity is lost in the informational-functionalist formalism and the exclusively statistical and optimizing way of measuring fitness. We cannot comfort ourselves thinking that materiality is kept through the phenotype (its study and evaluation) due to the way in which fitness is measured (more in abstract over genotypes) and the difficulties of the phenotype-genotype-phenotype mapping mentioned above. In practice, when the phenotype is related to the genotype a false genetic determinism or a pragmatic one-to-one or statistical mapping is used because only among genotypes (which are almost virtual and inferred entities) do we have some kind of lawfulness. Conversely, when the phenotype is really considered, only a "naturalist" kind of science seems possible, with no clear laws or general principles beyond those of physics or of idealist abstractions.

The proposal is that an avenue of research worthy to explore entails facing the radical peculiarity of a phenomenon — life — which instantiates from the very beginning operations of an epistemic nature which we normally only address in their more complex and highly evolved configurations. This certainly has both conceptual and operational consequences regarding the research strategies to be adopted in the study of evolutionary systems in general and in Artificial Life in particular. A common one is the semiotic nature of the phenomena studied. More specific outcomes could be, for AL, the idea that self-organization, by itself, does not guarantee the semantic grounding of interpretation any more than information could by itself. For the field of evolutionary systems a similarly specific one is that the actual materiality of the systems is something that we can not dispense of without a price.

ACKNOWLEDGMENTS

I wish to thank G. Van de Vijver and S. Salthe for their kind invitation to participate in the ISES and all the participants for the width of the issues and the depth of the discussions. At the time of writing this paper, I was beneficiary of a postdoctoral grant awarded by the MEC (Spain). This work has also been supported by funding for the research projects PB92-0456 from the DGCYT (MEC) and UPV 003.230-HA160/94 from the UPV/EHU.

Department of Logic and Philosophy of Science, University of the Basque Country
San Sebastian, Spain

NOTES

[1] I cannot pretend here to answer to Salthe's multiverse and book length challenge to *evolutionary* thinking in those terms, neither to the more specific one about the impossibility of making a substantive discourse based on evolution so defined.
[2] See Csányi (1989) and Laszlo (1987) as most significant and complete examples.
[3] Complementary listing of debates and open questions can be found in equally detailed renditions of more or less overlapping samples of literature by Gayon (1990) and Grene (1990), being all fairly coincident in their denial of a wholesale overturn of the neo-Darwinian approach, at least in its potentially broader and "expanded" versions. Besides their more or less specific claims about biological evolution, some of these issues, in the last term, relate to some broader picture belonging to the realm of theories of general evolution. Others relate with more "internal" disputes in a neo-Darwinian perspective and with the ways open to expand it. An exceedingly long list of references should be needed in order to cover all these aspects which include, besides the items in Burian's list, discussions on the rate of evolution (punctuated equilibria), macro-evolution and speciation, levels of selection (hierarchical and pluralistic views), etc., as well as others with a further influence in the public domain, such as sociobiology or evolutionary psychology.
[4] See Langton (1989) for original claims, Taylor & Jefferson (1994) for a favorable opinion and Miller (1995) for a critical but supporting assessment.
[5] Strategies which, as Fox (1988) calls them, proceed either backwards, from known living systems towards more and more simple systems, or forward, from inorganic structures into prebiotic systems, and which leave a gap in between. See also Cairns-Smith (1985), Dyson (1985), Shapiro (1986) as general reviews for the state of the art about life origins.
[6] Mathematical equality of thermodynamical and informational entropies allows a sort of unification of informational and physical concepts in some approaches (i.e., despite their differences, Brooks & Wiley, 1988; Salthe, 1993; Wicken, 1987) but always under the unifying and encompassing principle of a physical theory (even if macroscopic and statistical).
[7] There is also a third one which is closely related to these issues but is not going to be discussed here: the sense more directly linked to the idea of function within an integrative system (Moreno, Umerez & Fernández, 1994; Umerez, Etxeberria & Moreno, 1993).
[8] See Sober (1992) for a recent presentation focussing on AL and Umerez (1995) for a critical discussion of implementation independence in terms of degrees of multiple realizability. See Moreno, Etxeberria & Umerez (1994) for a plea on materiality in AL.
[9] I cannot discuss it here but I could mention, for instance, the dubious case for structuralism which is only apparently recovering the material cause, or the danger of converting the final cause into something just formal (see Minch, this volume, for a discussion on final causes).
[10] And usually no less anthropocentric. See, for instance, Stuart (1985a, 1985b) for a critique of what he calls "the bio-informational equivalence" and its potentially harmful influence when accepted as an innocent "metaphorical way of talking".
[11] For instance, Salthe's formulation of the specification hierarchy which allows him to talk about "modelling dust devils", "learning drainage systems" or, in general, "dynamic systems as systems of interpretance" (Salthe, 1993).
[12] Mutation (unlimited or constrained) and selection (creative or eliminative) in the functionalist version, or different overlapping factors in the structuralist versions.
[13] I think, notwithstanding, that considering his view as a specific case of a more general theory — the modern and increasingly plural one — (Sober, 1985) is a more fair assessment of his actual achievements.

REFERENCES

Benne, R. (ed.), 1993, *RNA Editing: The Alteration of Protein Coding Sequences of RNA*, New York, Ellis Horwood.

Berry, S., 1995, Entropy, Irreversibility and Evolution, *J. of Theor. Biology*, vol. 175, pp. 197-202.

Bowler, P.J., 1989 (2nd ed.), *Evolution: The history of an Idea*, Berkeley, CA, Un. of California Press.

Brooks, D.R.; Wiley, E.O., 1988 (2nd. ed.), *Evolution as Entropy. Toward a Unified Theory of Biology*, Chicago, ILL, University of Chicago Press.

Burian, R.M., 1988, Challenges to the Evolutionary Synthesis, in *Evolutionary Biology,* M.K. Hecht & B. Wallace (eds.), vol. 23, New York, Plenum, pp. 247-269.

Cairns-Smith, A.G., 1985, *Seven Clues to the Origin of Life*, Cambridge, Cambrige University Press.

Campbell, J.H., 1988, Evolution as Nonequilibrium Thermodynamics, Halfway There? in *Entropy, Information, and Evolution. New Perspectives on Physical and Biological Evolution*, B.H. Weber, D.J. Depew & J.D. Smith (eds.), Cambridge, MA, MIT Press, pp. 275-284.

Csányi, V., 1989, *Evolutionary Systems and Society: A General Theory*, Durham, NJ, Duke Un. Press.

Darden, L.; Maull, N., 1977, Interfield Theories, *Philosophy of Science*, vol. 44, no. 1, pp. 43-64.

Depew, D.J.; Weber, B.H., 1988, Consequences of Nonequilibrium Thermodynamics for the Darwinian Tradition, in *Entropy, Information, and Evolution. New Perspectives on Physical and Biological Evolution*, B.H. Weber, D.J. Depew & J.D. Smith (eds.), Cambridge, MA, MIT Press, pp. 317-354.

Dobzhansky, T., 1973, Nothing in Biology Makes Sense Except in the Light of Evolution, *American Biology Teacher*, vol. 35, pp. 125-129.

Duve, C. de, 1988, The Second Genetic Code (Transfer RNA's), *Nature*, vol. 333, pp. 117-8.

Dyson, F., 1985, *Origins of Life*, Cambridge, Cambridge University Press.

Eigen, M., 1992, *Steps Toward Life*, Oxford, Oxford University Press.

Eldredge, N., 1995, *Reinventing Darwin*, New York, John Wiley.

Emmeche, C., 1991, A Semiotical Reflection on Biology, Living Systems and Artificial Life, *Biology and Philosophy*, vol. 6, pp. 325-340.

Emmeche, C., 1992, Modeling life: A note on the semiotics of emergence and computation in artificial and natural living systems, in *Biosemiotics*, T.A. Sebeok & J. Umiker-Sebeok (eds.), Berlin, M. de Gruyter, pp. 77-99.

Emmeche, C., 1993, Is life as a multiverse phenomenon?, in *Artificial Life III*, C.G. Langton (ed.), Reading, MA, Addison-Wesley, pp. 553-68.

Emmeche, C., 1994, *The Garden in the Machine. The Emerging Science of Artificial Life*, Princeton, NJ, Princeton University Press.

Etxeberria, A., 1995, Representation and Embodiment, *Cognitive Systems*, vol. 4, no. 2, pp. 177-196.

Farmer, J.D.; Belin, A.d'A., 1992, Artificial Life: The coming evolution, in *Artificial Life II*, C.G. Langton et al. (eds.), Reading, MA, Addison-Wesley, pp. 815-40.

Fox, S.W., 1988, *The Emergence of Life. Darwinian Evolution from Inside*, New York, Basic Books.

Gayon, J., 1990, Critics and Criticisms of the Modern Synthesis. The Viewpoint of a Philosopher, in *Evolutionary Biology,* M.K. Hecht, B. Wallace & R.J. Macintyre (eds.), vol. 24, New York, Plenum, pp. 1-49.

Goodwin, B.C., 1984, Changing from an Evolutionary to a Generative Paradigm in Biology, in *Evolutionary Theory: Paths into the Future*, J.W. Pollard (ed.), Chichester, J. Wiley, pp. 99-120.

Goodwin, B.C., 1994, *How the Leopard Changed its Spots. The Evolution of Complexity*, New York, Charles Scribner's Sons.

Gould, S.J., 1983, The Hardening of the Modern Synthesis, in *Dimensions of Darwinism: Themes and Counterthemes in Twentieth Century Evolutionary Theory*, M. Grene (ed.), Cambridge, Cambridge Uuniversity Press, pp. 71-93.

Gould, S.J., 1986, Evolution and the Triumph of Homology, or Why History Matters, *American Scientist*, vol. 74, pp. 60-69.

Gould, S.J.; Lewontin, R.C., 1979, The Spandrels of San Marco and the Panglossian Paradigm: A critique of the Adaptationist Program, *Proc. of the Royal Soc. of London B*, vol. 205, pp. 581-598.

Grene, M., 1990, Is Evolution at a Crossroads?, in *Evolutionary Biology*, M.K. Hecht, B. Wallace & R.J. Macintyre (eds.), vol. 24, New York, Plenum, pp. 51-81.

Harnad, S., 1990, The symbol grounding problem, *Physica D*, vol. 42, pp. 335-346.

Hoffmeyer, J.; Emmeche, C., 1990, Code-Duality and the Semiotics of Nature, in *On Semiotic Modelling*, M. Anderson & F. Merrell (eds.), Berlin, M. de Gruyter, pp. 117-66.

Hull, D.L., 1982, The Naked Meme, in *Learning, Development, and Culture*, H.C. Plotkin (ed.), New York, John Wiley & Sons, pp. 273-327.

Langton, C.G., 1989a, Preface, in *Artificial Life*, Langton (ed.), pp. xv-xxvi.

Langton, C.G., 1989b, Artificial Life, in *Artificial Life*, Langton (ed.), pp. 1-47.

Langton, C.G., (ed.) 1989, *Artificial Life*, Reading, MA, Addison-Wesley.

Laszlo, E., 1987, *Evolution. The Grand Synthesis*, Boston, MA, Shambhala.

Lewontin, R.C., 1974, *The Genetic Basis of Evolutionary Change*, New York, Columbia Un. Press.

Lewontin, R.C., 1974a, Darwin and Mendel — The Materialist Revolution, in *The Heritage of Copernicus*, J. Neyman (ed.), Cambridge, MA, MIT Press, pp. 166-183.

Lewontin, R.C., 1982, Organism and Environment, in *Learning, Development, and Culture*, H.C. Plotkin (ed.), New York, John Wiley & Sons, pp. 151-170.

Lewontin, R.C., 1983, The Organism as the Subject and the Object of Evolution, *Scientia*, vol. 118, pp. 65-82.

Lewontin, R.C., 1992, Genotype & Phenotype, in *Keywords in Evolutionary Biology*, E.F. Keller & E.A. Lloyd (eds.), Cambridge, MA, Harvard University Press, pp. 137-44.

Maynard Smith, J., 1986, *The Problems of Biology*, Oxford, Oxford University Press.

Mayr, E., 1961, Cause and Effect in Biology, *Science*, vol. 134, pp. 1501-1506.

Mayr, E., 1982, *The Growth of the Biological Thought*, Cambridge, MA, Harvard University Press.

Miller, G.F., 1995, Artificial Life as Theoretical Biology, *Cognitive Science Research Paper*, no. 378, School of Cognitive and Computing Sciences, University of Sussex.

Moreno, A.; Etxeberria, A.; Umerez, J., 1993, Semiotics and Interlevel Causality in Biology, *Rivista di Biologia - Biology Forum*, vol. 86, no. 2, pp. 197-209.

Moreno, A.; Etxeberria, A.; Umerez, J., 1994, Universality without Matter?, in *Artificial Life IV*, R. Brooks & P. Maes (eds.) Cambridge, MA, MIT Press, pp. 406-410.

Moreno, A.; Umerez, J.; Fernández, J., 1994, Definition of Life and Research Program in Artificial Life, *Ludus Vitalis. Journal of Life Sciences*, vol. II, no. 3, pp. 15-33.

Nitecki, M.H.; Nitecki, D.V. (eds.), 1992, *History and Evolution*, Albany, New York, SUNY Press.

Pattee, H.H., 1969, How does a molecule become a message?, *Developmental Biology Supplement*, vol. 3, pp. 1-16.

Pattee, H.H., 1972, Laws and constraints, symbols and languages, in *Towards a Theoretical Biology 4. Essays*, C.H. Waddington (ed.), Edinburgh, Edinburgh University Press, pp 248-58.

Pattee, H.H., 1982, Cell Psychology: An evolutionary approach to the symbol-matter problem, *Cognition and Brain Theory*, vol. 5, no. 4, pp. 325-341.

Pattee, H.H., 1986, Universal principles of measurement and language functions in evolving systems, in *Complexity, Language, and Life*, J.L. Casti & A. Karlqvist (eds.), Berlin, Springer-Verlag, pp. 268-81.

Pattee, H.H., 1987, Instabilities and information in biological self-organization, in *Self-Organizing Systems. The Emergence of Order*, F.E. Yates (ed.), New York, Plenum, pp. 325-38.

Pattee, H.H., 1989, The measurement problem in artificial world models, *BioSystems*, vol. 23, pp. 281-90.

Pattee, H.H., 1993, The limitations of formal models of measurement, control, and cognition, *Applied Mathematics and Computation*, no. 56, pp. 111-30.

Pattee, H.H., 1995a, Evolving self-reference: Matter, symbols, and semantic closure, *Communication and Cognition - Artificial Intelligence*, vol. 12, nos. 1-2, pp. 9-27.

Pattee, H.H., 1995b, Artificial Life Needs a Real Epistemology, in *Advances in Artificial Life*, F. Morán, A. Moreno, J.J. Merelo & P. Chacón (eds.) Berlin, Springer, pp. 23-38.

Richards, R.J., 1992, *The Meaning of Evolution*, Chicago, ILL, University of Chicago Press.

Rieppel, O., 1990, Structuralism, Functionalism, and the Four Aristotelian Causes, *Journal of the History of Biology*, vol. 23, pp. 291-320.

Salthe, S.N., 1993, *Development and Evolution. Complexity and Change in Biology*, Cambridge, MA, MIT Press.

Shapiro, R., 1986, *Origins*, New York, Summit Books.

Smith, T.F.; Morowitz, H.J., 1982, Between History and Physics, *Journal of Molecular Evolution*, vol.

18, pp. 265-82.

Sober, E., 1985, Darwin on Natural Selection. A Philosophical Perspective, in *The Darwinian Heritage*, D. Kohn (ed.), Princeton, NJ, Princeton University Press, pp. 867-899.

Sober, E., 1992, Learning from functionalism - Prospects for strong Artificial Life, in *Artificial Life II*, C.G. Langton et al. (eds.), Reading, MA, Addison-Wesley, pp. 749-65.

Stuart, C.I.J.M., 1985a, Physical Models of Biological Information and Adaptation, *Journal of Theoretical Biology*, vol. 113, no. 3, pp. 441-54.

Stuart, C.I.J.M., 1985b, Bio-informational Equivalence, *Journal of Theoretical Biology*, vol. 113, no. 4, pp. 611-36.

Taylor, C.; Jefferson, D., 1994, Artificial Life as a tool for Biological Inquiry, *Artificial Life*, vol. 1, no. 1, pp. 1-13.

Umerez, J., 1994, *Jerarquías Autónomas. - Un estudio sobre el origen y la naturaleza de los procesos de control y de formación de niveles en sistemas naturales complejos*, Ph.D. Dissertation, University of the Basque Country (Spain).

Umerez, J., 1995, *Semantic Closure*: A guiding notion to ground Artificial Life, in *Advances in Artificial Life*, F. Morán, A. Moreno, J.J. Merelo & P. Chacón (eds.), Berlin, Springer, pp. 77-94.

Umerez, J.; Etxeberria, A.; Moreno, A. 1993, Emergence and Functionality, in *Proceedings of the XIII International Congress on Cybernetics*, J. Ramaekers (I.A.C.) (ed.), Namur, Belgium, International Association of Cybernetics, pp. 519-523.

Umerez J.; Moreno, A., 1995, Origin of Life as the First MST - Control Hierarchies and Interlevel Relation, *World Futures. The Journal of General Evolution*, vol. 45, pp. 139-154.

Van de Vijver, G., 1991, The Emergence of Meaning and the Antinomy of Naturalism, *Uroboros*, vol. 1, no. 2, pp. 153-175.

Van de Vijver, G., 1994, Emergencia del sentido morfológico y semántico, *Ludus Vitalis*, vol. 2, no. 3, pp. 151-171.

Varela, F.J., 1979, *Principles of Biological Autonomy*, New York, Elsevier North Holland.

Wheeler, J.A.; Zurek, W.H. (eds.), 1983, *Quantum Theory and Measurement*, Princeton, NJ, Princeton University Press.

Wicken, J.S., 1987, *Evolution, Thermodynamics, and Information. Extending the Darwinian Program*, New York & Oxford, Oxford University Press.

ARANTZA ETXEBERRIA

EMBODIMENT OF NATURAL AND ARTIFICIAL AGENTS

INTRODUCTION

The term embodiment suggests a return to the body (or to a physical or perceivable realm) of something that was (but should not be) previously separated from it. This phenomenon can be found in a wide range of contexts; for example, abstract entities, such as computer programmes, may acquire dynamics when executed in material devices; theoretical ideas can become operative when put in relation to practical or contingent situations; or, similarly, when considered as properties of bodies (including brains), mental capacities recover a physical nature. The return we refer to has an explanatory character: it is motivated by an assumption that embodiment may throw light upon areas where disembodied explanations are unsatisfactory. Many scientific and philosophical traditions have postulated privileged realms (e.g. Platonic worlds) deprived of materiality, dynamics, interactions or praxis for explanation, but they priorise the *know that* in front of the *know how* (Ryle, 1949) and may thus side-step the more complex problems. This is the reason why it is important to explore a differently motivated epistemology, one able to approach phenomena in their original embodied situations. Then, a claim for embodiment would not be a demand for a restitution, but an urge to start from the beginning, from the *things themselves*.

A proper treatment of embodiment stems from a consideration of the role played by the organization and the physical structure of the body in cognitive processes. Therefore, the task of building artificial systems does not have a practical goal only, it is also a way to evaluate the understanding of the phenomena our models reflect. A motivation underlying all the sciences of the artificial is to test our models by reproducing the complex phenomena they refer to. In the case of natural agents, embodied explanations require a consideration of different stages or scales. Organisms are self-organizing physical things, they may reproduce, they are homeostatic, they have experiences (for example, pain and pleasure help them protect their integrity), some of them can move to find food, escape from predators or look for mates, some can guess what the behavior of others will be and devise strategies according to it, etc. Many of these capacities are essential to understand life, but it is a matter of inquiry which of them, and to what point, are important to explain the origin of minds or of cognition. In the case of artificial systems, a proper notion of embodiment poses additional problems related to how to handle complexity. So far, there is no procedure to build living beings and it is doubtful that such a specification will ever be obtained. Our ideas about how possible artificial living creatures would look like or behave are fuzzy. While embodiment is relevant to understand the meaningful interactions of natural systems, it is crucial when we try to build artificial systems that present the same capacities.

G. Van de Vijver et al. (eds.), Evolutionary Systems, 397–412.

An underlying question of this paper will be whether embodiment facilitates the appearance of cognition at different levels, including the higher ones, or it is just a set of constraints that shape the behavior of real material systems, but could be neglected in formal models. Our argumentation starts by considering the nature of cognition in respect to life and the next three sections will examine three different questions basical to characterize the approach of embodied cognition in an evolutionary frame. The last two sections identify two different ways to understand embodiment, an externalist and an internalist one, and discuss briefly the role of representations in cognitive processes from an evolutionary point of view.

LIFE AND COGNITION

Many attitudes to understand cognition can be classified in two loose groups: the one that takes a global organismal perspective and the one that focuses on specifically functional organs. In the first case all cognitive capacities are more or less sophisticated ways to structurally change the organism or the environment so as to articulate the system-environment transaction and possibly improve the viability of the system or the quality of experience. In turn, an approach that focuses on organs — like sensors or the nervous system — will try to identify the components of the relation that act as vehicles of function or meaning: the symbols or representations. The difference between both attitudes might imply the need of accepting complementary ways to explain a process as complex as the embodiment of cognition. We say more about this in the last section.

In both cases, a naturalist notion of what cognition is depends on judgements of when cognitive phenomena start in the evolution of life on earth or in an incremental scale of complexity in the case of artificial systems. From certain viewpoints, cognition already exists in single-celled organisms; others think that it only appears with the development of the nervous system; and, still others, require the origin of language in hominisation. The definition of what counts as a cognitive process changes in each case.

In recent times there has been some controversy on whether biologically inspired epistemologies can distinguish life and cognition in a consistent way. Defences that they are basically the same process arise from at least two different motivations. One of them, evolutionary epistemology, contends that evolution is the only source of novelty for organisms, therefore no cognitive gain takes place in ontogenetic time (for an example, see Heschl, 1990). If the origin of minds in living systems has an adaptive value, then mental capacities, like the other traits, should be a result of evolution by natural selection. A similar position is held by evolutionary psychology. Adaptationist approaches to this question hold externalist views of cognition in which natural selection is the only force considered relevant to guarantee the cognitive adaptedness of the agent.

The enactive approach to cognitive phenomena (Varela et al., 1991; Stewart, 1992) maintains a similar position in what respects life and cognition starting from a very different background and with a radically different motivation. From this perspective, life is self-production (autopoiesis) in the form of a dynamical organization of component production that defines an identity by separating itself from the environment. This organization is already cognitive (like others, such as the immune system (Varela et al., 1987). If it is the autonomy of the global behavior of the system what makes it unlike inanimate things — for example, if its behavior cannot be characterized by straightforward input-output mappings —, then this autonomy implies the emergence of cognitive-like phenomena; for example, attribution of meaning to some of the *perturbations* detected in the environment. Therefore, even if different kinds of cognitive phenomena can be defined, they all start and ground in the same autonomy.

From another perspective, however, life and cognition seem to be distinguishable according to the properties of different processes that take place at different time scales within the same global system. This point of view maintains that the biological organization pre-exists any cognitive capacity and its existence is a pre-condition for its appearance (Moreno & Etxeberria, 1992). Cognition then deserves a special treatment: the biological basis of the organism adapts to its environment via evolutionary processes (variation and natural selection acting on genotypes), unlike the cognitive capacity, which depends on ontogenetic processes of maturation and learning at the level of the nervous system.

Following a classification by Godfrey-Smith (1994), the first two positions vindicate a strong continuity between life and cognition, saying either that life is a condition of mind (evolutionary epistemology) or that to live implies capacities that are intrinsically cognitive (enactive approach), while the second only sees a weak continuity, life is a condition for mind but mind implies more than just life. Both of them embrace a methodological continuity, because generic properties of life are considered important to explain cognition in all these cases.

In the following sections three questions important to characterize an embodied approach to cognition are examined. The three of them legitimize somehow a conceptual perspective that plays down or suppresses the need for embodiment in explanations of the nature of cognitive systems. The issues to analyze are:

1. The nature of the physical systems able to embody cognitive processes. The relation between causality and computation (between the material and the formal or the ontology and the description) is, of course, the central issue. A claim of isomorphism of causality and computations often appears in the literature, usually discussed in terms of whether computations can emulate physical processes. Here it will be posed in a slightly different way, by inquiring whether, and to which extend, causal processes embody computations. The notion of constraint appears to be a keyword.

2. Understanding complex properties of living systems as adaptations by natural selection. Explanations of the complex design of living beings as generating from a passive adaptation to the environment hide the question of embodiment, because most of the properties of organisms (bodies as well as minds) are studied from the point of view of external forces that shape them.

3. The idea that a cognitive process consists mainly in an informational transaction with the environment, so that other considerations of this transaction, generically energetic, may be left aside. A consideration of the nature of sensors arises some alternatives that may be interesting.

All three problems might help clarify epistemological assumptions underlying the adoption of disembodied epistemologies, which are obstacles to develop truly embodied ones. The problem of the origin and role of symbols or representations in cognition appears at the end of the paper.

FIRST PROBLEM: CAUSAL SYSTEMS COMPUTE

An interesting locus to start the analysis of possible equivalences between formal and causal processes, and to examine artificial embodiments of living systems, is the Church-Turing thesis. It was based upon independent analyses of the general notion of an effective procedure proposed by A. Turing and A. Church in the 1930's. The thesis is a statement of the theoretical limits of computation, but, in some of its interpretations, it constitutes also a claim about embodiment. Originally, the thesis stated that there are no number theoretic functions which cannot be computed by a Turing machine (TM), but can be computed by some other effective procedure. Since then, other interpretations take it to represent an ultimate limit to all possible computations. Hence, it has been construed as applying to functions in general, as opposed to just the number theoretic functions, but, more strikingly, it has also been maintained that its domain of application extends to the production of physical and/or mental phenomena. Turing himself argued that, because mental processes are procedural, anything that can be produced by a mental process can also be produced by a TM procedure. Finally, the thesis has also been applied to physical systems in general, therefore, the formulation with a maximum extension would be that any physical system, including the brain-mind, is an embodiment of some TM procedure. Here the original sense, that the TM can compute any function or that computation can emulate any process, is inverted. Now it states that nothing exists which is not TM computable or that any physical process computes.

Recently, the validity of the thesis for the extreme case of physical systems has been ruled out on the basis of the difference between formal and causal processes (Cleland, 1993). Causal processes, it is argued, are more powerful than the TM, therefore no formal process can perfectly simulate or duplicate a physical system, which are more powerful: "we have good reason to suppose that the computational capacities of causal processes exceed those of TMs and we have no idea as to the ultimate capacities of causal processes to compute functions" (Cleland, p. 287). In order to defend this point, Cleland proposes to consider well-defined procedures that cannot be realized by a TM: *mundane procedures*, those whose steps involve causal processes. If we compare formal procedures with everyday mundane procedures such as recipes, directions, etc., each step of the latter, and not of the former, initiates a causal process and its effectiveness depends on it. For example, when following a procedure that indicates how to prepare a Hollandaise sauce, the success of the performance does not depend only on the actions taken (which are not necessarily sequential), but also on the fact that each of them initiates and sustains some causal process. Therefore, causal processes differ from TM actions in that the former, at least as they are ordinarily construed, are neither temporally nor causally discontinuous "(...) their constituent parts (sub-processes) are not, as in the case of TM actions, thought to be temporally separate (...) Causal processes just do not seem to be procedure-like, (they are) 'continuously evolving' (...) 'self-generating' and 'self-determining'" (*ibid.*, p. 296).

I basically agree with this point, as far as it implies that formal processes may be less powerful than causal processes. Yet, the whole discussion seems to avoid an important problem, for it remains unclear who performs the computation and, in general, the sense in which the term computation is used, i.e. whether it is a human mental operation or an objective real process of natural systems. For example, it is said: "if, as many believe, there exist genuinely continuous causal processes, then the capacities of actual physical systems to compute (i.e. mirror the structure of) functions exceeds that of TMs (...) we have no idea as to the ultimate limits of causal processes to mirror precisely the structure of functions" (*ibid.*, p. 309). For Turing (1936-37) the action of computing was performed by a person realizing a calculus. When computing machines appeared, and as their operation emulated the calculus specified by formal procedures (such as those of the TM), the machine itself could be interpreted as physically performing the computation. But, how far can physical systems in general be interpreted as computations? No analysis of how causal processes can manifest by themselves as computations is given in Cleland's paper. The problem is not trivial though, some authors have considered it to be the core of epistemology: the origin of the modelling relation in which a formalism results from an abstraction of certain causal processes as action steps (see, for example, Rosen, 1985).

For the operation of a physical machine to be equivalent to the calculus (to the formal machine), the causal processes have to be fully constrained, to the extent that each step of the causal process corresponds to a step in the procedure. In effect, the causal processes of physically computing machines are completely constrained, so that all their operations are isomorphic with the steps required by the formal procedure. No knowledge of the causality of the physical system is required to calculate the next state. If it is a special-purpose machine, it may only emulate one specific formal process; if it can potentially emulate any procedure, then it is said to be capable of universal computation. The latter requires the physical system to be *structurally programmable* (Conrad, 1987), so that it can be modified to adjust the causal processes of the physical device to any formalism. In this case, these constraints are such that they can be specified as a procedure, so that any causal change of state of the physical device always corresponds to a logical step. As a consequence, the physical and the formal aspects of a computing device never interact, they are isolated, except when errors take place. And this constitutes a very special relation indeed. Conrad (1987, 1989) enunciates a trade-off principle between the programmability and the evolvability of a system. Structurally programmable systems are not robust in front of errors, whereas evolvable systems must be capable of integrating slight variations and perhaps taking advantage of them.

At this point, when the initial distinction between a TM procedure and a mundane procedure is revisited, it is readily manifest that embodiments of TMs share the causal characteristics associated with mundane procedures. Nothing can happen in the world that is not causal, therefore the difference between the two, as embodiments, cannot be placed in the fact that one involves causal processes while the other does not. Yet, we still find a difference between the effectiveness of computer programmes to calculate the identity function, for example, and a recipe to prepare Hollandaise sauce. The difference may have to do with the degree of constraint we would require to artificially emulate a process and the characteristics of the material substrate that is being constrained. Even if we eliminate the human component of the mundane procedure and we consider the operation of a fully automated machine or a robot, capable of performing all the actions specified by the recipe in a highly skilful way, the whole procedure seems to rely in causal processes to a point where the computer does not. The causal structure of a computing device performing a computation can be controlled by technology to an extent where the process of mixing egg yolks with butter to do a Hollandaise sauce might not. In what respects the fact of being a causal process it is only a difference of degree, not of principle. Both of them are causal processes, and this remains an important difference that should preclude any affirmation that physical systems "compute" in any way. Their behavior may be more or less accurately formalized, even emulated by a computational simulation, but we should not speak of computation in the case of physical systems unless we can specify how its behavior is constrained by a formal procedure.

SECOND PROBLEM: ORGANISMS ARE ADAPTATIONS

To understand how the origin of constraints of living systems is understood in evolutionary theory we should refer to discussions about externalism and internalism. Both positions, that organisms are shaped by the environment (externalist) and that adaptivity results from processes within the organism (internalist) can be found in the history of biology (Godfrey-Smith, 1994). Evolutionary materialists, such as Lamarck, emphasize the capacity of living matter to complexify and, thus, to adapt to environmental changes. For Lamarck, increases of complexity are inevitable in nature, its progression is only disturbed by the effects of the environment, only the adaptability of living systems to different environments can give reason of departures from the orderly progression. However, others, like for example Spencer (who was otherwise a follower of Lamarck), are externalists. Spencer proposed a general law of evolution that applies to the evolution of solar systems, planets, species, individuals, cultural artefacts, and human social organizations. According to it, there is a universal course of change from a state of indefinite, incoherent homogeneity towards a state of definite, coherent heterogeneity. This ideas were influenced by theories of development and thermodynamics. From this perspective, all increases in complexity are produced by changes in the environment: nothing would happen if the environment remained constant and simple. Natural systems change from a state of little differentiation of parts and little concentration of matter, towards a state in which there is a variety of clearly distinguishable parts, where the individual parts differ from each other and are densely structured as an adaptation to the conditions of the environment (*ibid.*, p. 80).

A naturalist view of the origin of cognitive processes regards evolution as the source or process where constraints regulating the causal processes of living beings generate. Externalism is criticized by those who consider that knowledge and cognition need to be explained starting from life, to overcome dualistic accounts of relations between mind and world and artificial separations between them (Godfrey-Smith, 1994). For example, Dewey and other "progressive Darwinians", regarded the organism plus the environment as constituting a single system. Organic activities tend to preserve the pattern of activities between organism and environment, rather than just the organism itself. Cybernetics can be counted among those approaches that focus in this self-maintaining perspective for life: first order Cybernetics maintained an externalist approach similar to that of Spencer; but second order Cybernetics was worried about the identity of the artificial systems and preferred internalist ideas, an example being the theory of autopoiesis (Maturana & Varela, 1992).

Evolution by natural selection explains that variations among organisms give some of them advantages to survive or have offspring in certain environments. Selected traits are adaptations, evolved because they provided advantages to their bearers under their environmental conditions (Burian, 1982). The simplified understanding of the evolutionary process as driven solely by natural selection, exemplified by optimality models, is usually called adaptationism (Gould & Lewontin, 1979). Assumptions underlying adaptationism are that the properties of organisms can be explained as a sum of atomic features, each of them an optimization resulting from natural selection acting on alternative configurations.

Understanding organisms as a sum of atomic features is problematic in at least two senses. One is that if the evolution of each one is not independent of the others, then the constraints imposed by the interactions should be taken into account. This view demands to expand the theory of evolution to take into account evolutionary forces other than natural selection and historical events. Two types of constraints, constraints on adaptation and constraints of the appearance of form (Amundson, 1994) should be added to the standard models of population genetics to have a richer image of evolutionary processes. The view of evolution as a process of natural selection of each relevant trait fails to take into account evolutionary forces different from natural selection, such as genetic drift or pleiotropy, which are constraints on adaptation. On the other hand, the influence of developmental processes in evolution requires models of the capacity of matter to self-organize. Natural selection acting on genetic variation is not creative enough to account for the origin of form and organization of living systems (see Kauffmann, 1993; Depew & Weber, 1995). Thus, a consideration of both the historical and contingent processes and the intrinsic self-organizing capacities of living systems can expand the explanatory resources of evolutionary biology. Not every force in evolution is natural selection and not everything on which selection acts is random variation.

THIRD PROBLEM: COGNITION IS AN INFORMATIONAL PROCESS

In the recent work in Cognitive Science, there have appeared some criticisms against the functional decomposition of the cognitive systems in three subsystems: the perceptual, the processing, and the motor. Most theories of cognition tend to focus mainly on the structure of the nervous system and ignore other aspects of the body. Often it is said that the biological system (the body at large) is *energetically* coupled to its environment, whereas the cognitive system (the nervous system) is *informationally*. A different way of organizing the internal structure of the organism according to principles inspired in the incremental organization of evolved systems was proposed by Brooks (1990). Instead of dividing the work into functional modules he proposes a behavioral decomposition in which modelling starts from layers of simple complete behaviors on the top of which new more complex behaviors can be added. Thus, behavioral decomposition is a step forward in what respects the organization of the structure of a cognitive system, but it is still unsatisfactory to account for the autonomy of the physical system. Living systems are autonomous because they behave in an articulated way without the interference of any creator, they separate themselves from the outside and internally define ways of transaction with the environment. The relation between energetic and informational (biological and cognitive) capacities is closely entangled. In the organization of perceptual information in living systems the internal energetic state (hunger, fear, etc.) plays a role in the way sensory stimuli from the environment become relevant (or informational) for the organism. Also, many transactions with the environment involve mainly forces acting on the agent and the environment, rather than information (Smithers, 1994).

There are different degrees of the relative importance of energy and information in the sensory projection of organisms. This gradation depends on the specificity of the physical element with which the sensor binds or couples. Some types of interactions are made possible by sensors that bind or couple with a given domain detected only by energetic means. Others require more indirect relations in which the outcome is transformed into active information within the organism. Both kinds of couplings may improve the adaptedness of organisms or enrich its experience, but, in the first case, the sensor is specialized to detect specific substances (required for the organism or that it has to avoid), while in the second case the detection is only indirect and the detected domain may not be linked in a direct way with the needs of the organism.

Enzymes, which bind with very specific substrates, are an example of the first case. Many regulating factors affect them in subtle ways and determine the metabolic paths in which components enter. Nevertheless, the basic reaction is very specific, the recognition of a specific pattern. The sensors that help bacteria swim in favor (or against) chemical gradients belong to this very class, protein receptors of the bacterial surface bind with the attractive (or repellent) substance and stimulate the locomotor system. Smell and taste sensors of mammals belong to the same category, because, even if the mechanisms involved are more complex, the pattern recognition is highly specific, achieved through comparison and association with previously acquired patterns. For example, rats can eat very different food and need a variety of different substances to maintain their metabolism, but they do not have a general procedure to help them distinguish what is convenient before hand. Their method for recognition is a trial and error learning that involves the whole organism, not only a single perceptual mode. The procedure acquires specificity because rats try any new food in tiny amounts and, if they feel sick after they have eaten some particular substance, this will be always rejected in the future (Richter, referenced by Bonner, 1980).

Indirect detection is a different case because there we can find a deferred relation between the kind of binding the sensor realizes and the domain with which the organism interacts in a meaningful way. The organism detects spheres of significance after an internal elaboration of data and an interpretation that makes possible the interaction. For example, in many organisms vision does not enable a relation with light, but with something else. This case can be made more clear by using a borderline example: the primitive acquisition of an eye by a paramecium (Wachterhauser, 1984). Some single-celled organisms can use light as nourishment, via photosynthesis, in which light is absorbed directly as energy. They present forms of phototropism: they can sense light and move towards the light. This detection is not usually considered to be vision, it is a case of the mode considered above, of direct detection. Other organisms cannot nourish out of light directly, but by eating plants they indirectly obtain substances they cannot produce themselves, including vitamins, which enable them to be sensitive to light and to develop eyes. Thus, in evolutionary terms, the search for light starts as a search for nourishment, not of information. Wachterhauser reports the case of a paramecium, a single-celled creature, which feeds on green algae and who uses one of them (a chlorella) both as food and literally as eyes. The paramecium sticks the chlorella on itself and maintains a symbiotic relationship with it: as an eye, the chlorella is used to steer the movement of the organism and, as a stomach, it gives back part of its light nourishment to its host. The chlorella couples energetically with the light, but, for the paramecium, the same chlorella is a visual sensor, an eye, used as a means to detect light and thus co-ordinate its locomotion.

This example can show the difference between coupling only with specific substances or specific patterns, and elaborating further the physical, causal effect that certain processes trigger on an organism. The highly evolved perceptual systems of animals can be confusing in what respects what it is to be a sensor or a perceptual system. As a consequence, often perceptual operation is supposed to be completely separated from the energetic/metabolic structure of the organism. This is not the case for those sensors responsible for direct detection.

In fact, not all the highly complex perceptual systems share common signal-processing strategies, and this difference is related to the specificity of the information acquired by the organism. A comparison of the visual and olfactory systems of insects shows many differences between them, different neural processing strategies and different neural architectures (Osorio et al., 1994). Visual perception is influenced by the structure and statistical properties of optical signals, which are complex but highly constrained. Olfactory signals, on the other hand, carry less information, they are less constrained and predictable. Arbitrary patterns of excitation on the olfactory receptors have to be parsed and learnt, and then recognized against a complex background of smells. Vision is hard-wired, neural circuits tailored to specific behaviors, while olfaction lacks the ordered and highly differentiated neural circuits used for vision and may use an associative network at an early stage to recognize patterns. This difference suggests a further difference in the functions both systems accomplish. Probably the olfactory system has to recognize specific substances, while the main task of visual elaboration is not to recognize *objects* but to elaborate different types of behaviors, for example locomotion. Vision has evolved to allow many different behaviors, some related to locomotion, others to the recognition of a variety of features (relevant for the organism but not always specifically directed to recognition of objects) through an organism-specific elaboration of the properties of one single physical phenomenon: light.

As a consequence, perception cannot be reduced to simple informational tasks, many perceptual couplings are directly involved with the energetic maintenance of the organism. Both kinds of perception are important to understand how organisms involve in meaningful interactions with the environment. Vision provides detailed and coherent ways to interact with the surrounding world through locomotion, it is closely related to motion in the environment, and of the agent itself. This specialization might have provided structures to develop anticipatory behavior; motion detection requires high sensitivity to time constraints and perception of the relevant dynamical features of space. Consequently, it can be expected that neural structures have specialized to detect motion in relation to the movements that the agent itself must make to succeed in a moving world. This implies a great amount of embodiment of perception, so that the capacity of anticipation is a consequence of movement, not represented by the agent (in any kind of Cartesian map), but put in relation with the movement of the agent itself. If perception is not mainly a matter of object recognition, anticipatory behavior could be explained as a dynamical coupling to tune the movement of the body to the moving world. This coupling or tuning does not involve to represent the world as it will be in the future, but to perceive the movement of the world in relation to one's own and act accordingly.

INTERNALIST AND EXTERNALIST EMBODIMENT

Now the moment has arrived to somehow summarize the consequences of the three problematic aspects considered so far, to develop what I would call an epistemology of embodiment for cognition. The three of them legitimize conceptual perspectives that avoid or suppress the need for embodiment in explanations of the nature of cognitive systems even in the field of Artificial Life (ALife). ALife is a discipline that studies life by emulating living phenomena in artificial media, it is a "distinctive attempt to explain evolutionary and adaptive systems, including (ultimately) the phenomena that we group together with labels such as intelligence, mind, and cognition" (Wheeler, 1995, p. 65). Some of the features that make it distinctive are its synthetic flavor (opposed to the analytic approach of Artificial Intelligence), the use of evolutionary methods to construct individual cognitive agents or complete artificial worlds, and the attention paid to emergence. Many Alife related theories of mind and cognitive phenomena are motivated by anti-Cartesian ideas and they try to develop embodied models. Nevertheless, many problems, some of them conceptual and related to the aspects rehearsed before, make this task very difficult.

The discussion of these points brings us back to the question of the two ways of facing embodiment, as a restitution or as a start from the beginning. ALife models of cognitive processes have understood embodiment more as a restitution, developing a notion of embodiment that does not require to consider living properties to understand cognition. Externalism and internalism are polysemic conceptual dichotomies used in several different contexts (one classification appears in Oyama, 1992). So far, the forms of embodiment tried in ALife can be considered externalist in many of these senses.

In respect to the dichotomy causality-computation, a consideration of the complexity of living systems may be a good starting point. Complexity, it is well known, is difficult to measure. However, a notion that seems to be useful for our purposes is Bennett's concept of *logical depth* defined as "the length of the logical chain connecting a phenomenon with a plausible hypothesis (e.g. the minimum description) explaining it" (Bennett, 1988). From this perspective, complexity would depend on the time it necessarily takes to construct a phenomenon. Complex entities should obey a "slow growth law" (this eliminates many trivial complexities such as smashed glass) and even if it takes a long time to form a given substance in the earth, if it could be made more quickly using shortcuts (as in a laboratory) it would not be complex (see Salthe, 1993, p. 4). Similarly, a minimum description will be complex in this sense if it takes a long time to boil it down from a larger description and it would describe something complex if it took a long time to synthesize it using the minimum description as instructions. Can this notion be related to our discussion on embodiment? The machine metaphor has been an inspiration to explain living beings. Polanyi (1968) described the design of machines in this way: "the machine as a whole works under the control of two distinct principles. The higher one is the machine's design, and this harnesses the lower one, which consists in the physical-chemical processes on which the machine relies (...) the constructor of a machine restricts nature in order to harness its workings" (p. 1308). Machine constraints are built with clear design, and, in general, the restriction they exert in motion or change (in the causal processes) is well-known (at least, while the system works as expected). The complexity of the organization or design of organisms could now be measured according to the constraints limiting or directing underlying causal processes: are these easy to simulate or emulate? If living systems could be viewed as very complex cases of *mundane procedures*, it would imply that there exist ways in which causal processes can be artificially orchestrated to give rise to living phenomena. Yet, the problem rests in the notion of procedure itself: causal processes of living systems do not follow fixed functions, they vary according to many regulating processes which take place at different levels. These may not be amenable of short description, also, it may not be possible to impose them externally in the causal system. An artificial origin of life requires, then, a synthetic method to discover experimentally how to instantiate different causal processes (to discover a mundane procedure) and also a capacity or criterion to acknowledge or recognize the living aspect of the resulting system (Stengers, 1995). As a consequence, to say that living processes embody computations is misguiding. Further, computer simulations can produce evolution faster than real evolution, but the complexity of evolution should not be trivialized to the point in which important stages and factors are taken for granted.

Externalism appears in ALife also as adaptationism. Anti-adaptationists claim that the theory of evolution should expand its explanatory scope to consider other evolutionary forces such as pleiotropy, alometry or historical constraints. Thus, the dispute between adaptationists and anti-adaptationists involves a methodological aspect: to decide which factors must be included in evolutionary models (Sober, 1993). If factors omitted by adaptationists (history, self-organization) are important, then adaptive forces will not be sufficient to explain evolution. This is an important question when properties of life are to be emulated in artificial media like in ALife. Until now, most of the artificial models developed to study the evolution of behavior in ALife have an adaptationist character, their structure mimics the evolutionary explanations of adaptation by natural selection. Reasons for this are diverse: on the one hand, ALife models are often based on research described in textbooks of behavioral ecology which are clearly adaptationist; on the other, the simplest way to model artificial evolution implies fitness evaluations dependent on objective functions and thus, experiments cannot include phenomena independent of the externally imposed selection.

In what respects the third discussion, because it is considered important that the behavior of artificial agents arises from perception-action cycles, and is not imposed by the programmer, agents are situated in real or simulated environments, so that their actions depend on the perceived state of the environment. Still, they are not autonomous in the strong sense. Simulated agents or robots are built as special purpose devices, able to measure the relevant parameters to control certain degrees of freedom of the system from the very situation in which the system finds itself (and not from the point of view of an external observer). This way the results of sensors causally influence the motor devices. This strategy can produce situatedness but, can it generate autonomous cognition? In these ALife cases, unlike in organisms, there is no relation with the physical functions responsible for the capacity the creature has to "stay alive". If, on the contrary, an internalist approach grounded in a notion of biological organization were attempted, situatedness would emerge as a function of the need to energetically maintain the organism, from those processes that endow the agent with identity (i.e. the self-organizing processes that maintain it as a physical system in a physical domain). Thus, behavior would be a manifestation of the self-maintaining structure and meaningful interactions would emerge from the whole pattern of system-environment relations. In sum, for an externalist perspective, a coupling of perception and action might be enough, but radical embodiment demands a broader consideration of the body and its evolutionary history. This type of embodiment is a property of causal evolvable systems and cannot be achieved by means of restituting certain capacities to physical devices that are *per se* devoid of them, embodiment has to be investigated from the very start in the material properties of evolvable systems.

REPRESENTATIONS IN EVOLUTION

There is still a point about the embodiment of cognitive processes that is considered to be central for this discussion, and has received a considerable attention in the literature. We refer to the problem of the vehicles of meaning, of the reasons to ascribe it to specific carriers (symbols or representations) and the relation those maintain with the organism as a whole. Even when meaning is considered as an emergent phenomenon, emergence can either be studied at the level of the whole activity of the organism, or at the level of the perceptual system itself. The latter requires to study cognition as an activity that involves representations.

In Cognitive Science, the role of internal representations in cognitive processes has raised long debates. Sometimes research has urged to ground the meaning of representations or to find different explanations of their nature and function, others to preclude their use for, at least, some cognitive tasks. A position that disclaims relevance to internal representations is research in adaptive behavior, where roboticists claim that robots do not have to be provided of representations to be able to perform simple tasks such as navigation. Another comes from criticisms of the Cartesian mind-body duality and its need of a subject-object distinction to explain cognition (for example, Varela et al., 1991). Both of them allege that representationalism fails to adequately explain phenomena that should be embodied.

Many of the activities that humans and some animals perform are the production of representations of the world with a great variety of purposes: to show the location of things, to alert others, to escape from predators, etc. These are, of course, external representations. We are able to produce representations in a meaningful relation with the world and our activities require them. Yet, meaning itself arises from many other bodily functions, like, for example, reflections and analogies that can be drawn from the fact that different individuals share the same body structure (Sheets-Johnstone, 1990).

There have been attempts to clarify the nature of internal representations, for example, Clark and Toribio (1994) have tried to give away many pre-conceptions of what representation is, namely, that representations are explicit, that they can be manipulated literally as a text and that they bear familiar intuitive concepts. Thus, they suggest that "it may be fruitful to stop thinking in terms of a dichotomy (representation/no-representation) and instead imagine a continuum of cases" (p. 403). In this continuum, they are ready to concede that no representation is needed as "insofar as the bulk of our cognitive activity is (...) defined in terms of 'perceiving' and 'acting'" (p. 426), while the need of representations starts when the system must dilate and compress (select and manipulate) the input space and is definite when this job is realized by a "systematically related body of intermediate representations" (p. 427). These authors are probably thinking on a continuum of cases within a single organism (probably of the animal kingdom), but, of course, it is possible to analyze the problem starting with all biological beings and maybe even before.

Some of the stages for the evolution of cognitive systems in the Earth might have been be the following. The first organisms were protocells formed by self-organizing processes, which cannot reproduce reliably. The origin of life would imply the slow stabilization of a type of code for self-reproduction and a semantic closure between dynamic and symbolic components to define an identity (Pattee, 1982). The origin of sex would bring about an exchange of genetic material and, in some species, the appearance of dimorphism and reproduction from two parent-cells, instead of cell division. The origin of pluricellular organisms and the development of the germ/soma duality would be a further step after which, with the development of the nervous system, animals capable to produce and interpret external signs would appear. In hominids, natural language evolves as a system of signs shaped by bodily thinking (Sheets-Johnstone, 1990). Finally, formal symbol systems as human productions which, as far as can be implemented in different materialities and give rise to different embodiments, constitute the most sophisticated natural example of arbitrary symbols. The evolution of complexity in evolution can be thus seen as a process of externalization or disembodiment of symbols at different levels, in the form of a transformation of causal relationships into relations mediated by symbols. This process has been called epistemic cut (Pattee, 1995). This perspective can explain how symbols appear in natural systems, it constitutes an evolutionary way to overcome or dissolve the world-mind (or matter-symbol) duality. The evolutionary continuous process extended in time helps to understand levels or scales of complexity without drawing strict discontinuities between different types of systems. This way, complexity can be given a materialistic explanation, for dualism is often a consequence of conceiving matter as inert and signs as "informing" entities.

Organisms are not passive objects of descriptions in terms of symbols, they are active producers of them (Emmeche, 1994).

For example, in Cariani's (1992) study of the semiotic properties of different subsystems of a cognitive system, sensors transform continuous and non-symbolic physical properties into discrete tokens, due to causal processes taking place in the device. Sensors are naturally occurring measuring devices that produce this transformation from causal processes to signs or symbols (Pattee, 1973). However, sensors of living systems are not fixed, they can produce a variety of transformations and participate in a number of system-environment interactions. The open-ended capacity of natural sensors should be an evolutionary property. Those who consider that only the behavior of the whole system can be seen as emergent defend an organismal view, because, it is alleged, it is not possible to understand emergence in terms of particular functions. These would be too arbitrary to provide relevance criteria, isolated sensors lack relevance criteria, unless they are considered as prostheses (like it is the case of Cariani). Probably both types of emergence have to be considered to understand meaning. Models and theories of hierarchical organization might allow research to flip from functional or semiotic approaches to global operational views of embodied processes, and back.

ACKNOWLEDGEMENTS

I am very grateful to G. Van de Vijver for her kind invitation to attend the ISES Seminar. Comments by R. Lemmen, S. Salthe and G. Van de Vijver helped to clarify some dark passages of older versions. This research was supported by a Contract associated to a DYGCIT Project Number PB92-0456 from the MEC (Ministerio de Educación y Ciencia, Spain).

Logika eta Zientziaren Filosofia Saila, Euskal Herriko Unibertsitatea
Donostia, Spain

REFERENCES

Amundson, R., 1994, Two concepts of constraint: Adaptationism and the challenge from developmental biology, *Philosophy of Science*, vol. 61, pp 556-578.

Bennett, C.H., 1988, Dissipation, Information, computational complexity and the definition of organization, in *Emerging Syntheses in Science*, D. Pines (ed.), Addison-Wesley, pp. 215-233.

Bonner, J.T., 1980, *The evolution of culture in animals*, Princeton, University Press.

Brooks, R., 1990, Elephants don't play chess, *Robotics and Autonomous Systems*, vol. 6, pp. 3-15.

Burian, R., 1982, Adaptation, in *Dimensions of Darwinism*, M. Grene (ed.), Cambridge University Press, Cambridge, pp 287-327.

Cariani, P., 1992, Some epistemological implications of devices which construct their own sensors and effectors, in *Toward a Practice of Autonomous Systems*, F. Varela & P. Bourgine (eds.), Cambridge, MA, MIT Press, pp 484-493.

Clark, A.; Toribio, J., 1994, Doing without representing?, *Synthese*, vol. 101, pp. 401-431.

Cleland, C.E., 1993, Is the Church-Turing thesis false?, *Minds and Machines*, vol. 3, pp. 283-312.

Conrad, M., 1987, Rapprochement Of Artificial Intelligence And Dynamics, *European Journal Of Operational Research*, vol. 30, pp. 280-290.

Conrad, M., 1989, The Brain-Machine Disanalogy, *Biosystems*, vol. 22, pp. 197-213.

Depew, D.; Weber, B., 1995, *Darwinism evolving. Systems Dynamics and the genealogy of natural selection*, MIT Press, Cambridge, MA.

Emmeche, C., 1994, The computational notion of life, *Theoria*, vol. 9, no. 21, pp. 1-30.

Godfrey-Smith, P., 1994, Spencer and Dewey on Life and Mind, in *Artificial Life IV*, R. Brooks & P. Maes (eds.), MIT Press, Cambridge, MA, pp. 80-89.

Gould, S.J.; Lewontin, R., 1979, The spandrels of San Marco and the Panglossian Paradigm: A critique of the Adaptationist Program, *Proceedings of the Royal Society of London*, CCV, pp. 581-98.

Heschl, A., 1990, L=C. A simple Equation with astonishing Consequences, *J. Theor. Biol.*, vol. 185, pp. 13-40.

Kauffman, S.A., 1993, *The Origins of Order: Self-organization and Selection in Evolution*, Oxford, Oxford University Press.

Maturana, H.; Varela, F., 1992, *The tree of knowledge*, Shambala, Boston, MA.

Osorio, D.; Getz, W.M.; Rybak, J., 1994, Insect vision and olfaction, Different neural architectures for different kinds of sensory signals?, in *From animals to animats 3*, D. Cliff, P. Husbands, J.A. Meyer & J.S. Wilson (eds.), Cambridge MA, MIT Press, pp. 73-81.

Oyama, S., 1992, Ontogeny and Philogeny: A case of Metarecapitualtion?, in *Trees of Life. Essays in Philosophy of Biology*, P. Griffiths (ed.), Dordrecht, Kluwer Academic Publishers, pp. 211-239.

Pattee, H., 1982, Cell Psychology: An Evolutionary Aproach to the Symbol/Matter Problem, *Cognition and Brain Theory*, vol. 5, no. 4, pp. 325-341.

Polanyi, M., 1968, Life's irreducible structure, *Science*, vol. 160, pp. 1308-1312.

Rosen, R., 1985, *Anticipatory Systems*, Pergamon Press.

Ryle, G., 1949, *The concept of mind*, Penguin, London.

Salthe, S., 1993, *Development and Evolution, Complexity and Change in Biology*, Cambridge MA, MIT Press.

Sheets-Johnstone, M., 1990, *The roots of thinking*, Philadelphia, Temple University Press.

Smithers, T., 1994, What the Dynamics of Adaptive Behaviour and Cognition Might Look Like in Agent/Environment Interaction, Notes of the III Int. Workshop on Artificial Life and Artificial Intelligence, *On the Role of Dynamics and Representation in Adaptive Behaviour and Cognition*, 9 and 10 December 1994, San Sebastian, Basque Country.

Sober, E., 1993, *Philosophy of Biology*, Oxford, Oxford University Press.

Stengers, I., 1995, *God's heart and the stuff of life*. Paper read at ECAL-95, European Conference on Artificial Life.

Stewart, J., 1992, Life=Cognition. The Epistemological and Ontological Significance of Artificial Intelligence, in *Toward a practice of autonomous systems*, F. Varela & P. Bourgine (eds.), Cambridge MA, MIT Press.

Turing, A.M., 1936/7, On computable numbers, with an application to the Entscheidungsproblem, *Proceedings of the London Mathematical Society*, vol. 42, no. 2, pp. 230-265; correction, *ibid.*, vol. 43 (1937), pp. 544-546.

Varela, F.; Coutinho, A.; Dupire, B.; Vaz, N., 1987, Cognitive Networks: Immune, Neural, and Otherwise, in *Theoretical Immunology*, A. Perelson (ed.), Redwood City CA, Addison Wesley, vol. II, pp. 359-376.

Varela, F.; Thompson, E.; Rosch, E., 1991, *The embodied mind*, Cambridge MA, MIT Press.

Wachterhauser, G., 1984, Light and Life: On the nutritional origins of sensory perception, in *Evolutionary Epistemology, Theory of Rationality, and the Sociology of Knowledge*, G. Radnitzky & W.W. Bartley (eds.), La Salle (IL), Open Court, pp. 121-138.

Wheeler, M., 1995, Escaping from the Cartesian Mind-Set: Heidegger and Artificial Life, in *Advances in Artificial Life*, F. Moran, A. Moreno, J.J. Merelo & P. Chacon (eds.), Springer, Berlin-Heidelberg, pp. 65-76.

Williams, G.C., 1992, *Natural selection: Domains, Levels and Challenges*, New York, Oxford University Press.

CLIFF JOSLYN

ARE LIFE AND MEANING COEXTENSIVE?

INTRODUCTION

The purpose of this paper is to lay out the structure of an argument about what our understanding of the proper relation between the categories of living and semiotic (semantic) systems is or should be. I will attempt to identify important conceptual distinctions which could be made, decision points in the argument, and what's at stake at each step.

My fundamental perspective is that of a systems scientist, attempting to develop theories and components of theories (concepts, explanatory principles, modeling methods) which span system types and disciplinary boundaries. Thus my attitude is necessarily interdisciplinary and conceptual, and concerned with *general* evolutionary theory.

From within that research context, in order to test our ideas about the nature of complex, evolved systems, we pursue a number of different directions simultaneously: we generalize to the broadest possible interpretations; simplify as much as possible in terms of the real systems being examined; and consider the discrete evolutionary steps from which these qualitatively new forms of organization arose. For example, when looking at the nature of human cognition, we are drawn to consider non-human cognition, then non-human cognition in very simple animals like worms, and finally the origin of neural systems in general.

Continuing in this manner leads quickly to a knot of the questions which are central to systems science: what is the nature and origin of information? of meaning? of control? ultimately, of life? Furthermore, what are the relations among these ideas?

METHODOLOGICAL PRELIMINARIES

I would like to begin with a bit of "meta-argument", examining the methodology which is brought to bear in this kind of discussion.

Necessity, sufficiency, and contingency

In this paper we will consider whether there is any necessary relation between the classes of living and semiotic systems. In so doing, I will distinguish between two different senses of necessity. What I will call "extensional necessity" is contrasted with "sufficiency". Saying that a class or property A is extensionally necessary for a class or property B is simply saying that the class B is included in the class A. So asking about the extensional necessity between the classes A and B leads to questions as to whether they overlap, whether one is included within the

G. Van de Vijver et al. (eds.), Evolutionary Systems, 413–422.

other, or whether they are identical classes.

This is an expression of material entailment as a matter of *fact*, that all existing B's *are* A's, for whatever reason. Perhaps it is because of an accident of history; perhaps it is because of some other causal factor or force. For example, there are no New World elephants, so living in the Old World is extensionally necessary for being an elephant. It remains to be seen whether there is something in the Americas which is deadly to elephants, or whether they could be introduced and thrive.

Extensional necessity contrasts with what I will call "ontological necessity", which is itself contrasted with "contingency". Saying that a class or property A is ontologically necessary for a class or property B is saying that B requires A because of the *nature* of A and B, because of some causal relation or property of natural law. Otherwise the relation would be ontologically contingent. This is clearly a stronger expression of material entailment as a matter of *truth*, that all B's *must* also be A's, or otherwise they simply *could not* be B's.

For example, breathing air is ontologically necessary for me, but living in the United States is (ontologically) contingent. Asking about ontological necessity between the classes A and B leads to questions as to whether it *could have been otherwise* that there is a given relation between A and B.

On definitions

In this discourse there is a great deal of discussion about a wide range of very complicated and difficult ideas, and it is not uncommon to have raging arguments about the meanings of terms. So perhaps in this community more than any other, it is absolutely vital to pay close attention to definitions, and to try to forge some kind of consensual meaning. When I expound a theory of "life", or "information", what am I actually talking about? Whatever it is, is it the same as what you mean? If not, then we will simply be arguing past each other.

It should also be clear just what is at stake in such semantic argument, that is, in argument about the meanings of terms. Definition-making is an action of people taken for the specific purpose of parsimoniously capturing appropriate and meaningful distinctions, and identifying them with particular linguistic markers (terms). Further, the purpose of propounding certain definitions within a linguistic community is to bring others to make those same distinctions, and for those terms to be shared among them.

It should be clear that the purpose of working with definitions is not to *discover* the "true" meaning of a term. Nor should the purpose be (only) to win an argument. Thus there is a clear distinction between "mere" semantic "quibbling" over the usage of terms and a reasoned, deliberate, valuable, and fundamental argument about which semantic distinctions are significant, and what their appropriate linguistic labels should be.

Little is at stake in the choice of specific words for specific concepts, except, of course, for the *rhetorical* value gained in the battle of the *politics* of ideas. Perhaps the most cogent example from within this community is Shannon's mathematical measure of the spread of a probability distribution. Whether we call this quantity a "dispersion", an "information", or an "entropy", the formalism remains unaffected. But the result is decades of confusion among generations of scientists.

Nor should semantic argument necessarily go on prior to or be cleanly separated from the rest of an argument. That is, any hope that we would all sit down together, decide on the usage of terms, and only then go on to engage argument using those terms, is vain. Rather, argument should proceed at both levels complementary and simultaneously, with a vigorous interplay between argument *within* and *about* the linguistic frame.

The point is that there are vastly more *concepts* that we wish to discuss than there are specific *terms* to use. Therefore the key is to clearly distinguish *senses* of terms from each other, and then appropriately and consistently identify them with specific qualified terms or phrases. So, for example, if two scientists A and B are arguing about the proper use of the term "complexity", they should simply identify two senses complexity A and complexity B. It may turn out through their discussion that one sense is a case of another, or that a different term (say "organization", or "information") would be more appropriate for one sense or the other. The goal is to reduce the overall set of required terms, where possible, and where not, to achieve the "null consensus" of simply agreeing to disagree. In this way, a linguistic community can move towards a consensual basis for usage and meaning.

SEMANTICS AND LIFE

In surveying the broad spectrum of scientific problems and issues, it is clear that there are a small number of key developments in general evolutionary history which our attention is drawn to. One of those is the nature and origin of living systems. Another is the nature and origin of what I will call "semiotic systems". We can a priori consider these classes of systems to be independent of each other, and then go on to consider what their proper relation is. In particular, we are interested in deciding what are the relations of necessity, sufficiency, and contingency among them, if any.

The category of living systems has a common sense meaning (whether or not it is a good one), and in the spirit of the discussion about definitions from the previous section, this is what will be adopted uncritically here. But the latter category of semiotic systems requires some elaboration.

Living systems

When it comes to the class of living systems, we have a general idea of what we mean. There is, of course, significant debate over an appropriate definition of life, and to what extent pursuing such a definition is feasible or worthwhile. For example, is it useful to consider applying either the crisp "living" or "nonliving" to an organism, or is it more valuable to introduce degrees of livingness (for example, for a virus)?

It is useful here to identify at least the following schools of thought about definitions of life:

1. From the perspective of the relational biology school of Rashevsky and Rosen, Rosen defines life as "closure to efficient causation" (Rosen, 1991).

2. In the biosemiotics school, including Sebeok (1992) and Hoffmeyer (1991), the presence of life in a system can be understood as equivalent to the presence of meaning in that system. Whether the simultaneous origins of life and meaning are contingent or necessary is partly the question of this paper.

3. Similarly, in the control theory school of William Powers, the presence of life in a system can be understood as equivalent to the presence of control in that system (Powers, 1973, 1989, 1995). There is some further question as to what extent the presence of control is equivalent to the presence of meaning.

4. Finally, Morowitz (1992) favors a gradual view of a *movement towards* life, rather than an *origin of* life. His perspective emphasizes energetic and metabolic processes over information, meaning, and reproduction. This will be discussed below.

Rules, codes, and semiotic systems

I think of a semiotic system as a real entity (typically, although perhaps not necessarily, an organism) which manifests, embodies, or in other ways "makes actual", semantic relations (Joslyn, 1995). In turn, a semantic relation is a particular regularity or stability in a system which cannot be explained by natural law alone, but rather results from some form of "rule following" (to use Pattee's (1991) terminology), or "meaningfulness".

Consider a system S wherein two sets of phenomena $A = \{a\}$ and $B = \{b\}$ are related according to such a regularity or stability. This situation can be represented by denoting $S = \langle A; B; f \rangle$, where f is a functional relation or entailment $f: A \rightarrow B$, where for each a in A, there is a corresponding $b = f(a)$ in B. While in general there are many such possible functions f, at any particular time only one particular f is present.

When f is a law then it is ontologically necessary in all ways (in this universe, at least): nature supplies us, or rather we discover in nature, the particular regularity f which holds *within* S, and from that for a given a can predict the associated $b = f(a)$. But when f is a rule, then it is extensionally necessary at one explanatory level, but contingent at another. As with the law, once f as a rule is established within S, then f must necessarily be followed *by* S: $f(a) = b$. But *which* rule f among a class of rules is or has been selected or manifested by S is *contingent* at least on that system's evolution and history, and perhaps on its design by a human agent.

This combination of fixedness and arbitrariness is familiar to us from semiotics as being a hallmark property of symbols, and codes (Deely, 1990; Saussure,

1959). It is clear that a acts to "stand for" b in virtue of the coding f; a is the signifier or sign-vehicle, b the signified or object, and f the sign function; a is what is interpreted, b the interpretation, and f the interpreter; a is the token, b the referent, and f the code.

But further, in a symbol (for Peirce) the relation between the signified and the signifier, that is the nature of the code or sign-function, is necessarily an *arbitrary* one, without "motivation". At the individual level, I am free to construct any coding I choose, to have anything stand for anything else I wish. At the group level, each linguistic community evolves in a unique historical direction, with a unique lexicon. Codes act as rules, not laws.

Thus all semiotic systems, as so characterized, are dependent on the semiotic frame of reference established by the coding f. The manifestation of the code is the *action* of interpretation, of the system taking a's for b's in virtue of the coding f(a) = b. The semantic boundary imposed by f also marks the limit of the semiotic system's subjectivity: it is that entity which establishes the code which is the system's interpreter. Pattee (1982) refers to this property as "semantic closure", which can be briefly stated as the requirement that the interpreter of a semiotic system is also a referent (an a) of that system. It is through the manifestation of a code, that is, through the action of interpretation by an interpreter, that meaningfulness is introduced into the world: within S, b means a to the subject, that is, to the interpreter.

Coding in living systems

The perspective of "biosemiotics" (Deely, 1992) examines the role of semiotics in biology, for example in intra-organismal communications systems (hormones, neuroreceptors, genetic bases, and biomolecular recognition and memory in general); and similarly the role of biology in semiotics (to what extent are sign-functions determined by an organism's biology and evolution).

Thus the core question for systems theory is a question of biosemiotics: to what extent are the classes of biological and semiotic systems identical or distinct? But living systems are complex, with many aspects and components. So in answering this question, it is first necessary to consider *where* in an organism coding is present.

Obviously, genetics is the first thing which comes to mind. The genetic code is itself a prime example of a semantic relation, mapping codons to amino acids in an ontologically contingent manner. Since its discovery, this analogy has fueled generations of scientific speculation about the informational basis of life, the role of meaning and interpretation, and biological semiotics in general.

At a completely separate level of analysis is the active, ongoing control relations manifested between an organism and its environment. Unlike genetics, it is not at first clear why this is a semantic relation. As an example, consider a simple organism O which lives near an oceanic thermocline with warm water above and cold water below. O has an internal variable of temperature which can be either too hot, too cold, or just right, and it can take the actions of going up, going down, or doing nothing.

There are many possible relations between the internal temperature and the action performed, many different ways that O can react to a change in temperature. But only some are consistent with life. In particular, if O goes up when hot, and down when cold, then positive feedback will lead either to a floating, baked O or a sunken, frozen O; other relations result in negative feedback, and a comfortable swimming O.

O will perform behaviors consistent with life because it possesses certain biological structures, typical neural, which provide it with certain instincts. But what is crucial to realize is that the nature of these biological structures, the functions of these instincts, are not ontologically necessary. There is no a priori physical reason for O not to rise when hot, no ontological necessity or fundamental natural law for it to sustain itself near the thermocline: O *could have been* wired differently. But if so, such organisms would die rather quickly, and certainly not survive to reproduce. Thus the effect of the reproductive process is that O's behavior be established within the O as a contingent rule in virtue of the evolution of the O's. Thus the relation between the perception and the action of the organism is a semantic relation, and in fact a control relation (Joslyn, 1995).

It should be clear that this semantic relation is completely distinct from that of the genetic level. Although no doubt O's instinctive behavior is genetically coded, presumably other genes could result in the pathological behavioral pattern.

Furthermore, it should be recognized that a consequence of this position is to say that for O "too hot" actually *means* "go down", and "too cold" actually means "go up". While this can be argued, it is a very interesting view.

THE RELATION OF LIFE TO MEANING

So finally we are ready to consider the proper relation between livingness and meaningfulness in systems. The question in general is whether, and how to determine whether, coding is either extensionally or ontologically necessary for life, and vice versa. More specifically, we would wish to ask this question of the various components of life: metabolism, replication, evolution as variation with selective retention, and the control relations between organisms and their environment. Such a detailed task is perhaps beyond the scope of this paper, but we can take a shot at the general question.

A working null hypothesis is that there is an ontologically necessary extensional equivalence between the class of living systems and the class of semiotic systems; in other words, that it is not contingent that living systems are both extensionally necessary *and sufficient* for semiotic systems. Or, stated yet another way, that (extensionally) all living systems are semiotic, and vice versa; but further that this is not an accident of history, but rather an ontological necessity. We will now go on to consider this hypothesis in detail, and see what is at stake at each stage.

Extensional necessity

First, we consider the question "is life necessary for meaning?" in the sense of extensional necessity. For this to be true, then all semiotic systems must be biotic. As we review the cases of semiotic systems, at first this appears to be the case: semantic relations appear only in the living world.

But then we consider tools and machines, and more generally our semiotic productions like languages and texts. These are unquestionably semiotic systems, in that there is a rule-like, semantic relation between a tool and its use. But also clearly a hammer is not alive.

However, it can be observed that extant rule-following systems are either organisms or *require the presence and action of* organisms. That is, nonliving semiotic systems exist only as *used by* living systems. So while a linguistic utterance is a semiotic production of a person, so is the written form or the tool produced by that person. They are thus logically considered as *extensions* of the semantic space of the organism. For Pattee (1982), all of these entities are part of the *semantic closure* of the organism.

Now given the extensional necessity of life for meaning, what about the ontological necessity? To say that it is ontologically necessary that life is extensionally necessary for meaning is to say that only living systems are *capable* of manifesting codes. And this is indeed a strong statement. What is it about the *ontological* nature of living systems, as opposed to their historical development, which results in this? This is indeed a mystery to me at this point.

But on the other hand, if life's necessity for meaning is contingent, then it must be possible for there to be semiotic systems which are not alive, and do not include living systems in their semantic closures. If so, then why have they not been realized on the Earth? Is it possible for them to be so realized in the future, either on the Earth or elsewhere? Is it possible for them to be constructed or grown by us, but *outside* or our semantic closures? Indeed, what would that even mean, if we attempted this move? What is there that we affect which is *not* within our semantic closure ?

Extensional sufficiency

The converse of the previous case is the question "is life sufficient for meaning", or equivalently "is *meaning* necessary for *life*", again in the sense of extensional necessity. For this to be true, then all living systems are semantic. Again, on the surface this appears to be true. As argued above, any organism using the genetic code manifests semantic relations in virtue of that coding. The universality of basic biochemistry and the metabolic chart would then seem to clench the argument.

So here, the real question is about the ontological necessity or contingency of this result. Consider that it would be ontologically necessary for living systems to be semantic, that there would be some law of nature which required life in the presence of semantics. The results are quite striking: given that the universe is so constructed that life requires semantics, and furthermore (from above) that conversely all semantic systems are biological (whether this is ontologically necessary or not), then it is natural to advance semantic relations as the *key explanatory principle of life*.

Indeed, there is little to separate this view from the desire to advance semantics as the much sought-after *definition* of life, the "holy grail" of theoretical biology. Of course, there is a vitalistic danger in this view, reducing one phenomenon (life) whose origins and nature are difficult to explain to another (semantic relations). But on the other hand, if semantics does in fact emerge as the necessary *and sufficient* condition for life, and furthermore that this is an ontological necessity, then there would be little to stand in our way from making this move.

But on the other hand, there is no *a priori* reason to accept the ontological necessity of the extensional sufficiency of life for meaning. Rather, we could entertain the view that coding is "merely" a catalyst for the processes of living systems, simply a vast expedient. And indeed, there are good arguments for this view, and possibly good counterexamples to the converse.

This view is, in fact, that of Morowitz (1992) mentioned above. He has long challenged the traditional "RNA world" view of the origins of life by theorizing about the relatively late appearance of nitrogen, and therefore nucleic acids and genetic coding, in pre-biotic evolutionary history. Morowitz's program is a bold effort to approach the origins of life problem the "orthodox" way: through chemistry. Thus his view of Artificial Life (AL) would be as an attempt to produce *real* AL by forming objects resembling organisms (protocells) in the laboratory setting: "Biogenesis must be pursued as an experimental science" (*ibid.*, p. 97). A brief appendix suggests that he and his colleagues have made some progress towards this goal, in particular, by producing non-enzymatic catalysis in a non-biotic flask-type experiment. Indeed, Morowitz has recently said that they have been hesitant to publish their recent experimental results in order to avoid a "cold fusion effect" (personal communication, November 1994.)

What would a counterexample of a non-semantic, but nevertheless biotic, system look like? Morowitz has offered some tantalizing suggestions. Morowitz's basic premise is to turn the traditional Miller-Urrey (1959) approach to biogenesis on its head. Rather than a "primordial soup" of pre-biotic organic monomers and macro-molecules, within which is established a complex set of reactions eventually leading to vesicle formation, Morowitz proposes the formation of closed ampiphilic (lipid) bilayer membrane vesicles very early. Echoing the autopoietic theory of Varela and Maturana (1974), these protocellular vesicles provide the key property of systems formation: a thermodynamic boundary dividing system from environment, and establishing a three-way phase separation across which energy and molecules are asymmetrically transported. Just by itself, this could result in protocellular growth, division, and replication.

A number of conditions are required, including:

1. Catalysis in the absence of nitrogen, and thus of enzymes;
2. Early proto-photosynthesis, as an early source of energy input to the protocells, in the form of photonic energy transport mediated by a variety of non-chlorophyl pigments (chromophores and retinals); and
3. Electron and proton transport, mediated by a variety of non-protein molecules absorbed within the protocellular walls, to facilitate redox and acid-base reactions respectively.

Thus Morowitz is able to articulate the startling image of proto-biotic cells within which the emergence of non-genetic molecules provides increased fitness through the reflexive action of autocatalysis, increased energy transduction, or increased transport of its own chemical constituents. But through the differential production and faithful reproduction of various forms of protocells with different species and concentrations of molecules, and therefore different chemical networks and cycles, all of these cases would result in a form of non-genetic, but nevertheless Darwinian, variation, selection, inheritance and evolution.

Note that Morowitz claims that the early stages of cellular evolution went on for millions of years without any genetic coding. This is accomplished strictly through carbon-based reactions, independent of amino acids and peptides, not to mention the far more complex machinery of enzymatic catalysis, RNA transcription, and protein synthesis. For him, these necessarily semiotic components of extant living systems are simply *not relevant* for biogenesis.

CONCLUSIONS

We have considered what the proper relation is between the classes of semiotic and living systems, both extensionally and ontologically, in the context of a circumspection about the making of definitions. Morowitz provides a major challenge to the null hypothesis that the constellation of properties including meaning, control, information, semantics, and semiotics is roughly equivalent to that of life, and furthermore that this is ontologically required. The essence of that challenge is the idea that proto-life evolved for millions of years in a non-genetic context, and that this implies a non-coded, non-semantic, non-semiotic context. The question then becomes, if coding and semantics are not *necessary* for life, what *is* the role of coding in biological evolution?

Of course genetic (coded) variation and selection has been the central mechanism for the vast preponderance of evolutionary history. Further, von Neumann (1966) has made the strong argument that open-ended evolution *requires* symbolic representations and coding. But it is well known that complex self-organization exists in purely physical systems, and Morowitz also strongly argues that heritable variation and selection are possible without coding. So presumably there is some kind of qualitative difference between Morowitzian evolution and genetic evolution, and this remains to be explicated. But the hypothesis that there is a large *quantitative* advantage to genetic evolution — a quantity which, as they say, has a quality all its own — can also not be rejected.

Given Morowitz' claimed empirical results, this question may shortly be decidable. If it turns out that this is a valid criticism, then the danger of identifying meaning and information as the vitalistic principle of life may be able to be avoided.

Computer Research Group, Data Mining Team, Los Alamos National Laboratory Los Alamos, USA

REFERENCES

Deely, J., 1990, *Basics of Semiotics*, Bloomington IN, Indiana UP.

Deely, J., 1992, Semiotics and Biosemiotics, in *Biosemiotics: The Semiotic Web 1991*, T.A. Sebeok; J. Umiker-Sebeok (eds.), Berlin/NY, Mouton de Gruyter, pp. 46-75.

Hoffmeyer, J.; Emmeche, C., 1991, Code-Duality and the Semiotics of Nature, in *On Semiotic Modeling*, M. Anderson & F. Merrell (eds.), New York, Mouton de Gruyter, pp. 117-166.

Joslyn, C., 1995, Semantic Control Systems, *World Futures*, vol. 45, pp. 87-123.

Miller, S.L.; Urey, H.C, 1959, Organic Compound Synthesis on the Primitive Earth, *Science*, vol. 130, pp. 245-251.

Morowitz, H.J, 1992, *Beginnings of Cellular Life*, New Haven, Yale University Press.

Pattee, H., 1982, Cell Psychology: An Evolutionary View of the Symbol-Matter Problem, *Cognition and Brain Theory*, vol. 5, pp. 325-341.

Pattee, H., 1991, Limitation of Formal Models of Measurement, Control and Cognition, *Applied Mathematics and Computation*, pp. 111-130.

Powers, W.T., 1973, *Behavior, the Control of Perception*, Chicago, Aldine.

Powers, W.T. (ed.), 1989, *Living Control Systems*, CSG Press.

Powers, W.T., 1995, Origins of Purpose: The First Metasystem Transition, *World Futures*, vol. 45, nos. 1-4, pp. 125-138.

Rosen, R., 1991, *Life Itself*, New York, Columbia University Press.

Saussure, F., 1959, *Course in General Linguistics*, New York, McGraw-Hill.

Sebeok, T.; Umiker-Sebeok, J. (eds.), 1992, *Biosemiotics: The Semiotic Web 1991*, Berlin/NY, Mouton de Gruyter.

Varela, F.G.; Maturana, H.R., 1974, Autopoiesis: The Organization of Living Systems, Its Characterization, and a Model, *Biosystems*, vol. 5, pp. 187-196.

INDEX